AIR POLLUTION

THIRD EDITION

VOLUME V

Air Quality Management

ENVIRONMENTAL SCIENCES

An Interdisciplinary Monograph Series

Editors: Douglas H. K. Lee, E. Wendell Hewson, and Daniel Okun

A complete list of titles in this series appears at the end of this volume.

AIR POLLUTION

THIRD EDITION

VOLUME V

Air Quality Management

Edited by

Arthur C. Stern
Department of Environmental Sciences and Engineering
School of Public Health
University of North Carolina at Chapel Hill
Chapel Hill, North Carolina

ACADEMIC PRESS New York San Francisco London 1977
A Subsidiary of Harcourt Brace Jovanovich, Publishers

ACADEMIC PRESS, INC.
111 Fifth Avenue, New York, New York 10003

United Kingdom Edition published by
ACADEMIC PRESS, INC. (LONDON) LTD.
24/28 Oval Road, London NW1

Library of Congress Cataloging in Publication Data

Stern, Arthur Cecil.
Air quality management.

(His Air pollution, third edition ; v. 5)
(Environmental sciences ;)
Includes bibliographical references and index.
1. Air quality management. I. Title.
II. Series: Environmental sciences ;
TD883.S83 1976 vol. 5 363.6 77-24094
ISBN 0-12-666605-9

PRINTED IN THE UNITED STATES OF AMERICA

To Katherine

and

Rachel, Larry, Sunshine, and Jacob
Hazel, Stephen, and Naomi
Lois, Hal, and Sarah Rebecca

Contents

3. Episode Control Planning

Darryl D. Tyler

4. Organization and Operation of Air Pollution Control Agencies

Jean J. Schueneman

5. Public Information and Educational Activities

Edward W. McHugh

6. Air Pollution Surveys

August T. Rossano

Part B National and Worldwide

7. Energy and Resources Planning—National and Worldwide

William A. Vogely

12. Emission Standards for Mobile Sources

John A. Maga

13. Emission Standards for Stationary Sources

Arthur C. Stern

List of Contributors

Numbers in parentheses indicate the pages on which the authors' contributions begin.

William A. Campbell (355), Institute of Government, University of North Carolina, Chapel Hill, North Carolina

Noel H. de Nevers (3), Department of Chemical Engineering, University of Utah, Salt Lake City, Utah

Milton S. Heath, Jr. (355), Institute of Government, University of North Carolina, Chapel Hill, North Carolina

Edward W. McHugh (209), Bay Area Air Pollution Control District, San Francisco, California

John A. Maga (505), Sacramento, California

Robert E. Neligan (3), Environmental Protection Agency, Monitoring and Data Analysis Division, Research Triangle Park, North Carolina

Vaun A. Newill (445), Medical Department, Exxon Corporation, Linden, New Jersey

Goran Persson (381), National Swedish Environment Protection Board, Solna, Sweden

P. W. Purdom (415), Environmental Studies Institute, Drexel University, Philadelphia, Pennsylvania

August T. Rossano (247), Department of Civil Engineering, University of Washington, Seattle, Washington

Jean J. Schueneman (109), Control Programs Development Division, U. S. Environmental Protection Agency, Research Triangle Park, North Carolina

Herschel H. Slater (3), Environmental Protection Agency, Monitoring and Data Analysis Division, Research Triangle Park, North Carolina

Arthur C. Stern (571), Department of Environmental Sciences and Engineering, School of Public Health, University of North Carolina, Chapel Hill, North Carolina

Darryl D. Tyler (67), Office of Air Quality Planning and Standards, United States Environmental Protection Agency, Research Triangle Park, North Carolina

Ronald A. Venezia (41), Cary, North Carolina

William A. Vogely (293), Department of Mineral Economics, The Pennsylvania State University, University Park, Pennsylvania

Preface

This third edition is addressed to the same audience as the previous ones: engineers, chemists, physicists, physicians, meteorologists, lawyers, economists, sociologists, agronomists, and toxicologists. It is concerned, as were the first two editions, with the cause, effect, transport, measurement, and control of air pollution.

So much new material has become available since the completion of the three-volume second edition that it has been necessary to use five volumes for this one. Volumes I through V were prepared simultaneously, and the total work was divided into five volumes to make it easier for the reader to use. Individual volumes can be used independently of the other volumes as a text or reference on the aspects of the subject covered therein.

Volume I covers two major areas: the nature of air pollution and the mechanism of its dispersal by meteorological factors and from stacks. Volume II covers the effect of air pollution on plants, animals, humans, materials, and the atmosphere. Volume III covers the sampling, analysis, measurement, and monitoring of air pollution. Volume IV covers two major areas: the emissions to the atmosphere from the principal air pollution sources and the control techniques and equipment used to minimize these emissions. Volume V covers the applicable laws, regulations, and standards; the administrative and organizational strategies and procedures used to administer them; and the energy and economic ramifications of air pollution control. The concluding chapter of Volume II discusses air pollution literature sources and gives guidance in locating information not to be found in these volumes.

To improve subject area coverage, the number of chapters was increased from 54 of the second edition (and 42 of the first edition) to 71. The scope of some of the chapters, whose subject areas were carried over from the second edition, has been changed. Every contributor to the second edition was offered the opportunity to prepare for this edition either a revision of his chapter in the second edition or a new chapter if the scope of his work had changed. Since 8 authors declined this offer and one was deceased, this edition includes 53 of the contributors to the second edition and 46 new ones.

The new chapters in this edition are concerned chiefly with aspects of air quality management such as data handling, emission inventory, mathematical modeling, and control strategy analysis; global pollution and its monitoring; and more detailed attention to pollution from automobiles and incinerators. The second edition chapter on Air Pollution Standards has been split into separate chapters on Air Quality Standards, Emission Standards for Stationary Sources, and Emission Standards for Mobile Sources. Even with the inclusion in this edition of the air pollution problems of additional industrial processes, many are still not covered in detail. It is hoped that the general principles discussed in Volume IV will help the reader faced with problems in industries not specifically covered.

Because I planned and edited these volumes, the gap areas and instances of repetition are my responsibility and not the authors'. As in the two previous editions, the contributors were asked to write for a scientifically advanced reader, and all were given the opportunity of last minute updating of their material.

As editor of this multiauthor treatise, I thank each author for both his contribution and his patience, and each author's family, including my own, for their forbearance and help. Special thanks are due my secretary, Susan Bigham, and her predecessors, who carried ninety-nine times the burden of the other authors' secretaries combined, and Eleanor G. Rollins for preparing the Subject Index for this volume. I should also like to thank the University of North Carolina for permitting my participation.

Arthur C. Stern

Contents of Other Volumes

VOLUME 1 AIR POLLUTANTS, THEIR TRANSFORMATION AND TRANSPORT

VOLUME II THE EFFECTS OF AIR POLLUTION

Part A Effects on Physical and Economic Systems

Part B Effects on Biological Systems

Part C Air Pollution Literature Resources

VOLUME III MEASURING, MONITORING, AND SURVEILLANCE OF AIR POLLUTION

Part A Sampling and Analysis

Part B Control Devices

Part C Process Emissions and Their Control

Part A

LOCAL, STATE, REGIONAL

1

Air Quality Management, Pollution Control Strategies, Modeling, and Evaluation

Noel H. de Nevers, Robert E. Neligan, and Herschel H. Slater

I. Air Pollution Strategies and Tactics

An air pollution strategy is a master plan that provides a solution to a municipal, state, provincial, national, or international air pollution problem. Air pollution tactics are the detailed procedures for carrying out this master plan. Tactics may be of a broad scale, involving the control of air pollution in a city, or of a more limited nature, e.g., determining the type of control equipment to install on a specific emission source. An air quality strategist would ask, "How shall we develop future master plans for the control of air pollution?" while an air pollution tactician would ask, "What shall we do this week, this month, and this year to insure the best possible implementation of the strategy that has been developed?"

This distinction between air pollution strategy and tactics, and the separation of the strategic and tactical functions is of recent origin. Before the 1960s, no distinction was made between strategy and tactics, or between strategists and tacticians. For example, the air pollution groups which pioneered smoke abatement efforts in St. Louis, Missouri, in the 1940s (*1*) had to simultaneously decide what the objectives were to be, how best to obtain them, what legislation was needed, and what technical assistance was to be rendered to emitters.

Those who pioneered in the field of air pollution probably seldom thought of air pollution strategy, because they dealt with visible, local sources of pollution. Their strategy was control of the most obvious problems first. This *is* a strategy; and at the time probably the best strategy that could have been developed. It may still be the best strategy for the less industrialized countries. However, in industrially advanced countries, the results which can be accomplished by that original strategy have generally now been accomplished, and further progress depends on controlling the many numerous but smaller emission sources.

In some countries, local agencies have total responsibility for both strategy and tactics. However, in other countries, e.g., the United States and the United Kingdom, the strategy is written into national legislation, so that the legislative body became the strategist.

As with other master plans for politics, warfare, or social betterment, there are competing schools of thought as to which is the best overall strategy for air pollution control. There are competing ideas on tactics as well. Four basically different strategies of air pollution control will be discussed in an attempt to make apparent their strengths and weaknesses. Although each strategy is considered separately as a distinct entity, it would be incorrect to assume that combinations are not usual. Rather, many have combined two or more of the strategies; many combinations are possible. Some of the strategies may be applied in pure form, as indicated here, or may be used as tactics subservient to a more primary overall strategy. For clarity, examples of the various strategies will be included.

One of these strategies, air quality management, forms the basis of air pollution control in the United States. Therefore, a concerted and detailed effort has been made in the United States to implement this comprehensive air pollution control strategy (*2*). Thus, more is known about the practical problems of trying to implement this strategy than is known about the others discussed in this chapter, and more detail will be given about it. However, our objective is to make clear to the reader that there are fundamentally different master plans and that other master plans and strategies are worthy of consideration.

There is no obvious reason why one must or should use the same strategies or the same tactics for all air pollutants. In the United States the air quality management strategy is being used for the control of pollutants called "criteria pollutants,"* Sections 109 and 110 of the Clean Air Act (*3*); and the emission standard strategy for another list of pollutants called "hazardous pollutants,"† Section 112. Various tactical approaches are being used for the different criteria pollutants covered by the air quality management strategy.

Because air quality management is the dominant strategy in the United States, there has been a tendency to use the expressions "air quality management" and "air pollution control strategy" interchangeably. This practice will not be followed here. It will be shown that air quality

* Criteria pollutants are defined as those pollutants for which National Ambient Air Quality Standards have been established; i.e., as of 1977, total suspended particulate matter, sulfur oxides, photochemical oxidants, carbon monoxide, and nitrogen dioxide.

† Hazardous pollutants are defined as pollutants that have been formally promulgated by the administrator of the United States Environmental Protection Agency as being able to cause or contribute to an increase in mortality or an increase in serious irreversible or incapacitating reversible illness; i.e., as of 1977, beryllium, mercury, and asbestos.

management, as the most commonly used in the United States, is the description of *one* of several possible strategies, but not a good description of some of the others. By making this distinction, it is possible to clarify the range of possibilities and alleviate possible confusion.

Strategies and tactics are interrelated. One cannot make sense of strategies, nor the tactics used to carry them out without some knowledge of individual pollution sources (Volume IV, Chapters 10, 12–21), the geographical distribution of pollution sources (Volume III, Chapter 17), the control devices and options available for various sources (Volume IV, Chapters 1–6, 8, 9), the chemistry of pollutants once released to the atmosphere (Volume I, Chapter 6), the atmospheric dispersion of pollutants (Volume I, Chapters 8 and 9), the measurement of atmospheric concentrations of pollutants (Volume III, Chapter 9), the effects of various pollutants on various kinds of receptors (Volume II, Chapters 1, 2, 4–7), and the setting and enforcement of pollution control regulations (Volume V, Chapters 4 and 8). But to bring all that knowledge to bear on the question, "What shall we do?" one needs a strategy.

A. Limits of a Pollution Control Strategy

It is the desire of all to have a totally pollution-free environment at no cost. That is only possible if one can repeal the laws of nature. Since the laws of nature cannot be repealed, a rational goal must be to try to limit air pollution to an appropriate level, by an appropriate expenditure of pollution-control resources, with an appropriate distribution of those expenditures. The controversies over public pollution control policies are usually centered on the definition of the "appropriate" values for these variables.

The ultimate goal of any air pollution strategy must be a reduction or reapportionment of pollutant emissions. Thus, a strategy must be empowered with tactics which result in a specific set of directives designed to realize strategic goals by regulation of various emission sources. As shown in Section I,C, there are several ways of arriving at these directives, which are essentially different strategies.

B. Qualities of a Pollution Control Strategy

The perfect strategy and the resulting tactics for carrying it out would have the following qualities:

a. It would be cost-effective and fair, gaining the maximum benefit for the resources (financial and social) expended and distributing costs and benefits equitably.

b. It would be simple and understandable to all concerned.

c. It would be easily enforceable.

d. It would be flexible, allowing some reasonable method for dealing with special situations, problems, and hardship cases without recourse to extensive litigation.

e. It would be evolutionary, allowing new data on pollution effects and new advances in control technology to be incorporated rapidly.

No air pollution control strategy proposed thus far possesses all of these virtues. The strategies discussed below will be compared with this ideal list in Section VII.

C. Available Alternative Strategies

The four air pollution control strategies which have received the greatest public attention to date are air quality management, emission standards, emission taxes, and cost–benefit strategies.

1. Air Quality Management Strategies

The air quality management approach specifies a set of ambient air quality standards (Volume V, Chapter 11). These are the heart of the strategy; once they have been developed and adopted, the quality of air is *managed to meet these standards.* This management takes place through regulation of the amount, location, and time of pollutant emissions. Air quality management is described in some of the literature of the late 1960s as "air resource management"; the two are practically identical.

2. Emission Standards

An emission standard strategy establishes permitted emission levels for specific groups of emitters and requires that all members of these groups emit no more than these permitted emission levels. These standards can be applicable to any selected group of emitters and can be local, regional, or national in application. They can be based on some air quality standards (thus making them a tactical subsidiary to an air quality management strategy) or can be entirely independent of any such air quality standards, serving as an entirely separate and different type of strategy from air quality management. When used independently of air quality management, this strategy may be called the "best practicable means" approach (Volume V, Chapters 12 and 13).

Air quality management and emission standard strategies are reasonably well known and have been studied and discussed at length. Emission tax and cost benefit strategies are less well known and understood and, as such, generally represent future, rather than current, possibilities.

3. Emission Taxes

An emission tax strategy would tax each emitter of major pollutants according to some published scale related to its emission rate, e.g., x¢ per pound of a pollutant for all emitters. The tax rate is set so that most major polluters would find it economical to install pollution control equipment rather than pay the taxes. In its "pure form" there would be no legal or moral sanction against an emitter who elected to pay the tax and not control his emissions. Czechoslovakia has a form of emission taxes in which an emission norm is established for state-owned industries, and there are fines for exceeding the norm. Thus, in the pure form, it is clearly a strategy quite different from the air quality management or emission standard strategies.

Emission taxes have also been proposed in combination with air quality management strategies; in this combination, emission taxes act as an incentive to reduce emissions to lower levels than those required to meet air quality standards (*4*). In this case, the two strategies work in parallel. In alternative versions, the tax rate steadily increases until some predetermined ambient air quality level is reached. In this version emission taxes are not a strategy, but a tactic being used in the air quality management strategy.

Emission taxes can be considered as one member of a larger class of strategies called "economic incentives." The other members of this class are tax rebates and public loans for the installation of air pollution control equipment and direct public subsidies for pollution control. These latter seem in most cases not to have been proposed as separate and complete strategies (i.e., they have no "pure form"), but rather have mostly been proposed and applied as ways of distributing the costs of implementing air quality management or emission standard strategies.

4. Cost–Benefit Strategies

Cost–benefit strategies attempt to quantify the damages from various pollutants and the cost of controlling those pollutants, and then select those pollution-control alternatives which lead to a minimum in the sum of pollution damage and pollution control costs. This leads to much more stringent air quality requirements for urban air (where large numbers of people are exposed to damage) than for rural air (where the number of persons exposed is small).

One could construe this as a variant of the air quality management approach with differing air quality standards for various areas, as contrasted with having one set of standards uniformly enforced. Thus, the cost–benefit strategy appears to be a different strategy from the air quality management approach as previously described.

D. Local, Regional, National, or World Control

The history of air pollution (*5*) indicates that the area considered for regulation under one authority has steadily grown. The earliest regulations were for a city, later there were regional regulations (e.g., County of Los Angeles, California, in the 1960s), and in 1970, the United States adopted regulations requiring national minimum standards, but allowing states and regions to adopt more stringent ones if they wished.

Sweden (*6*) has indicated that the majority of the acid rain falling in southern Sweden is due to sulfur emissions in other countries; from this, the Swedes argue that international control is needed. The arguments in favor of local regulation are that it allows the greatest flexibility, and the greatest possibility for each region to adopt pollution control methods best suited to local requirements.

The arguments for national or international regulations are that pollutants do not respect political boundaries and that in densely populated areas like the northeastern United States and western Europe, pollutants emitted in one jurisdiction may have their most significant impact in another (*7*). The other major argument is that localities and nations are often in economic competition, which has led to relaxed regulations in an attempt to attract and/or maintain industry. Thus, the lax regulations are allowed to drive the more stringent ones out of the marketplace. To counter this tendency, some minimum standards, applicable to all, seem appropriate (*8*).

For the United States, this argument was settled by the Clean Air Act of 1970,* whereby minimum standards not only for air quality but also for new industrial plants were set. The states and localities were free to adopt more but not less stringent standards. Of the fifty states, twenty-three have adopted air quality standards which are significantly more stringent than the federal minima. The arguments for and against international air quality standards are summarized by Yanagisawa (*9*).

* The 1970 Act was extensively amended in 1977. These amendments did not change the concept of minimum standards set forth in the 1970 Act.

II. Air Quality Management

A. Definition

Of the various definitions now used for the air quality management strategy, the following seems broad enough to encompass the ideas of most. Air quality management is the regulation of the amount, location, and time of pollutant emissions to achieve some clearly defined set of ambient air quality standards or goals. It includes the evaluation of various sets of emission control schedules to determine consequences to air quality and the formulation of alternative emission control schedules to meet air quality goals subject to some other constraint, e.g., technological feasibility or minimum cost.

B. Information Inputs

From the definition, it follows that to practice air quality management in a city, state, or region the air quality manager requires the following:

a. A set of air quality standards or goals to be achieved. These can be locally, nationally or world-determined, and may relate to long- or short-term concentrations of various pollutants, but must be directly measurable if one is to manage air quality. Thus, a standard for total suspended particulate matter of "not more than 75 $\mu g/m^3$ annual average, as determined by high volume sampling" is a perfectly clear and manageable standard. A standard like "the incidence of asthma shall not exceed 20 per 100 population" or "petunias shall not suffer more than 10% leaf damage" are air quality standards, but would be very difficult to manage as they require a subjective quality evaluation (Volume V, Chapter 11).

b. An inventory of the emissions from various sources in the region, including man-made (anthropogenic) emissions as well as emissions from natural sources (biopogenic). Ideally, the emission inventory would indicate not only the location of the source (grid coordinates and stack height), but also the emission schedule (emission rate and plume-rise parameters versus hour of day and day of year) for each source. This latter requirement is seldom met and estimates are regularly used. Historically, a greater effort has been made to obtain precise information for large sources than for small ones, because the large sources are more easily surveyed and are large-percentage contributors to the overall problem. In the future, as the larger sources come under tight control, more effort will inevitably be directed toward the smaller sources, which are larger in number and cumulatively classed as "area sources" (Volume III, Chapter 17).

c. Predictive methodology to relate air quality to emissions. Normally, this will be some type of diffusion model (Volume I, Chapter 10), but other possibilities exist, e.g., rollback. Diffusion models usually include local meteorology (Volume I, Chapters 8 and 9). The ideal model is one which would accurately predict the air quality at any location at any time for any set of emissions and meteorological conditions. It would provide both short- and long-term average predictions and show the impact of each emission source on air quality at each location. The real diffusion models now available fall far short of that goal. It seems appropriate to emphasize at this point that all air quality management procedures are based on some type of a predictive model. In the past, some air pollution control workers have implied that this was not true, but when their position has been examined, it was always found that they were relying on an intuitive model, though frequently of lesser sophistication and accuracy than the more formal models regularly used in air quality management. Air quality management, therefore, is inherently dependent upon air quality models.

d. Monitoring data to determine the status of ambient air quality. Adequate storage, retrieval, and analytical procedures for these data are necessary to be sure that they are correct, accurate, and properly interpreted (Volume III, Chapters 9 and 12). Since the ultimate goal of air quality management is to regulate emissions to meet a clearly defined air quality standard, the air quality manager must test his performance by measuring ambient air quality. Some day in the distant future, enough may be known about emission inventories, meteorology, and predictive models to allow simple prediction of air quality without measuring it; but this appears to be a goal for the next century, not this one.

e. Plausible sets of emission control tactics which can be evaluated to see if they will allow the standards to be met, subject to constraints of technical and economic feasibility, enforceability, etc. The first workers in air pollution and the group attempting to control some newly identified pollutant never before subject to control must undertake this *de novo,* but others can begin with the compendia of existing sets of tactics (Volume V, Chapters 12 and 13).

f. Data on the cost and effectiveness of various control devices and options (Volume IV, Chapters 1–6, 8, 9). In evaluating the sets of emissions tactics (item e above), the air quality manager must estimate the cost of each set of tactics, including both capital expenditure for control equipment and operating cost. To seek the most cost-effective way to meet the ambient air quality standards, he must also inquire whether the control cost falls unreasonably on some group(s) of potential emitters.

g. An enforcement procedure which allows the air quality manager to implement his planned emission control program. In the United States, this involves regulation setting, public hearings, compliance scheduling, etc. (Volume V, Chapter 4).

A graphic presentation of air quality management requirements is given in Figure 1.

C. *Tactics*

Once the decision has been made to use the air quality management strategy and the set of ambient air quality standards has been determined, the air quality manager has numerous tactical options for achieving these standards. The best known sets of tactics are discussed below:

1. Tactics Based on Rudimentary Models

As previously discussed, there can be no air quality management unless there are definable standards or goals and objective methods of determining the emission reductions necessary to meet the standards. In effect, there must be a model. Strategies used worldwide before 1950 were not air quality management strategies. These early efforts were based on the rudimentary "model" which says, "If you reduce emissions, the air will become cleaner." This leads to emission standard types of strategy, discussed in Section III below.

The simplest mathematical model which has been used to attempt quantitative air quality management is the rollback, or proportional, model (*10*). It assumes that the pollutant concentration at the point of interest is linearly related to emissions in the surrounding area. If the measured concentration is above the ambient air standards, this model leads to a direct computation of the percentage reduction in emissions for the area necessary to meet these standards. The formula normally given is

$$R = 100\% \frac{gc_{\max} - s}{gc_{\max} - b} \tag{1}$$

where R = percent reduction in emission per emitter; g = growth factor, i.e., the anticipated number of emitters at the time the standard is to be met divided by number of emitters at the time air quality was measured; $c_{\max}$ = highest measured value of ambient air pollutant concentration in area of interest; s = applicable ambient air quality standard; and b =

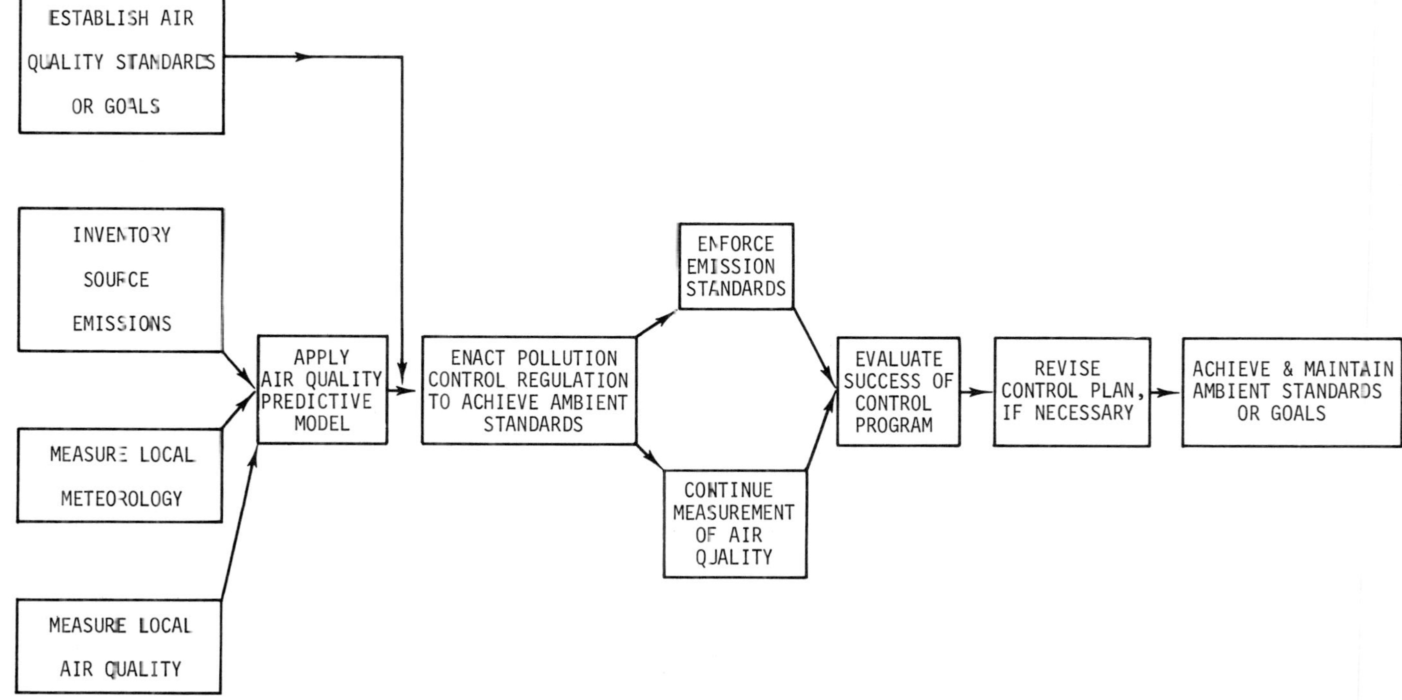

Figure 1. Air quality management requirements.

background concentration in air entering area of interest. As an example of the application of this formula, Barth (*11*) reported that the maximum 1967 value of 8 hour CO concentration in Chicago, Illinois, was $c_{\max} = 51$ mg/m^3, that a suitable standard for 8 hour CO concentration was $s = 10$ mg/m^3 (= 9 ppm), that the anticipated growth in number of vehicles between 1967 and 1990 was $g = 2.18$, and that the background concentration anticipated was 1 mg/m^3. Substituting these in Equation (1), he found

$$R = 100\% \frac{2.18 \cdot 51 - 10}{2.18 \cdot 51 - 1} = 92\% \tag{2}$$

Thus, according to this computation, if the proposed standard is to be met in Chicago in 1990, the average emission of CO per emission source would have to be reduced by 92% below that in Chicago in 1967.

This approach has been widely used because it is simple, understandable, and requires very little input data (Fig. 2). It is the simplest model with which one can attempt air quality management. Because it is so widely used, it seems appropriate to spell out here its limitations in some detail:

a. It is a purely theoretical model for which no experimental verification has ever been attempted in a metropolitan area and which can probably never be subjected to experimental verification in a metropolitan area. The reason that the experimental verification for a metropolitan area has not been attempted, and probably never will be, is that the relation between concentration and emission rate assumes that all other factors remain unchanged, including the spatial distribution of emissions. Thus, to test the equation, one would have to reduce the emission of each and every emission source in the area by the same percentage. For practical reasons, this does not seem possible in a metropolitan area.

This is not as severe a shortcoming as it might appear, because the theoretical basis of Equation (1) is quite plausible. However, it could err in several ways. If emissions influence climate (e.g., by changing turbidity of the atmosphere), then the linear assumption leading to Equation (1) would probably prove false. If pollutant disappearance (e.g., agglomeration or photochemical reaction) is not a linear function of pollutant concentration (it probably is not) then a nonlinear relationship would be expected between emissions and concentration. There are certainly other causes which could lead to nonlinearity in this relation as well.

b. The application of the rollback equation requires that the air quality manager know the value of $c_{\max}$, the highest concentration of pollutant in the area. Normally, he substitutes the highest observed con-

Figure 2. Proportional, or rollback, model with no growth.

centration $c_{max\ obs}$ for c_{max}. These two will only be the same if one of the air quality measuring stations is located at the point of maximum concentration. This assumption is nonconservative, i.e., leads to less stringent regulations than would be called for if the true value of c_{max} were known.

c. The growth factor (g) as used in Equation (1) assumes that all emission rates will grow proportionately with the increase in the number of emission sources, but does not take into account changes in other significant parameters (e.g., distribution of emissions, city size, stack heights). If the value used in Equation (1) is the projected increase, for example, in vehicle miles per day per square miles of the downtown part of the same city, and there is reason to believe that the percentage increase in vehicle miles per day per square mile will be the same for each square mile of the area of interest, then this is a satisfactory way to use the growth factor. If, however, the value used is simply the projected increase in vehicle population or vehicle miles per day in the entire metropolitan area, and there is no reason for believing that growth will be uniformly distributed, then there are reasonable grounds for assuming that the model will give misleading results.

If the air quality standard to be met is a short-term standard, which is most severely tested by meteorological situations which mix the pollutants thoroughly in a finite volume of air (for example, under inversion conditions in a completely enclosed valley), then the growth factor as shown in the simple model is probably satisfactory. An additional provision is that the boundaries of the area considered are the same as the boundaries of the area whose emissions are trapped in this body of air. If, on the other hand, the standard to be used is an annual average standard or some other standard which does not represent thorough mixing of all emissions in a fixed, finite volume of air, then the growth factor used in simple rollback would be very conservative. This would lead to much more restrictive standards than would be needed for a model which took into account growth in emissions per unit area in the areas of greatest interest, rather than total emission in some arbitrarily defined metropolitan area.

The growth factor as used in Equation (1) is simply the ratio of the population of emitters (residences, cars, factories, etc.) at the time when the standard is to be met, divided by the population of emitters at the time $c_{\max}$ was measured. There has been some discussion over whether this future population should be obtained by linear or logarithmic extrapolation of existing population trends. This is really a question outside of the basic rollback model. The model requires that a value be specified for the population of emitters on the appropriate date. It is the responsibility of demographers, planners, etc., to determine the most reliable way of estimating that value and to determine the retirement rate of existing sources (Fig. 3).

d. Simple rollback is applicable for short-lived pollutants (those whose half-life in the atmosphere is comparable to their travel time across a metropolitan area at the wind velocity of interest) only if the spatial distribution of emissions is unchanged. This means that any growth by expanding into the suburbs cannot be modeled for a short-lived pollutant by simple rollback. Short-lived pollutants can be modeled by simple rollback for the complete mixing within a fixed finite volume

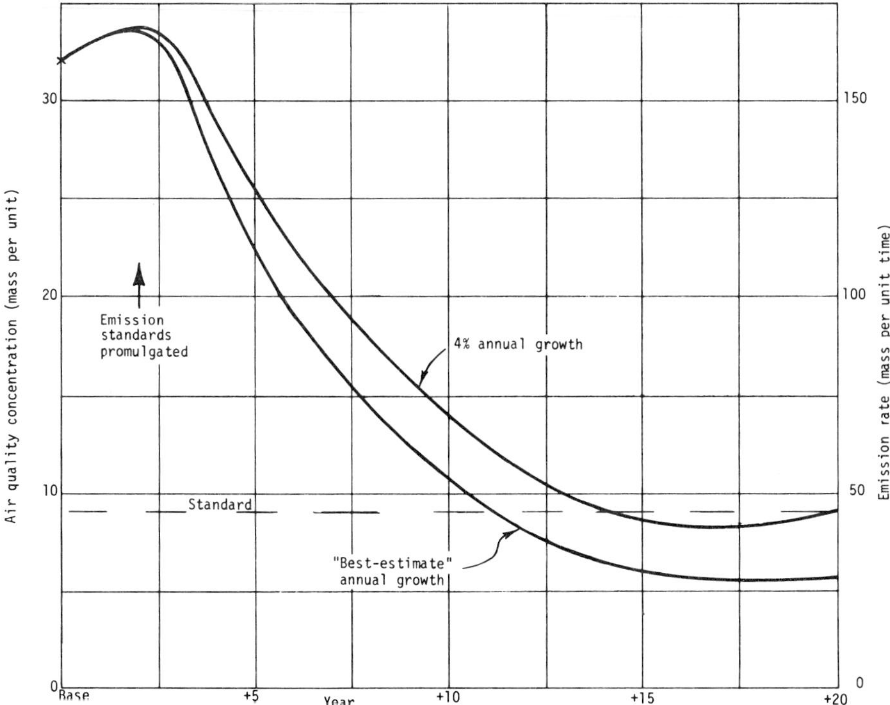

Figure 3. Effect of assumed growth rate of automobile vehicle miles traveled on emissions and future air quality, assuming imposition of emission controls on base year cars (based on rollback model).

of air situation only if there is instantaneous horizontal mixing of pollutants over the entire area being modeled.

e. Simple rollback is applicable to the problem of determining the effect upon air quality due to a change in emission rate, only if one of the following three conditions exists: (1) The class of sources being considered is by far the largest contributor of the pollutant in question. If this condition prevails, the effects of other classes of emitters can be ignored. (2) The class of sources being considered has the same temporal, spatial, and vertical distribution of emissions as the average of all other emissions in the area. If this condition prevails, a change in the emission rate of the class has the same effect on the emission distribution in time and space as a properly scaled reduction in the emission rates of all classes of sources. (3) The standard for the pollutant being considered is a short-term standard which is most severely tested in periods of excellent mixing within a limited air volume. If the area under consideration experiences excellent atmospheric mixing, yet is located in a completely enclosed valley above which an inversion aloft prevents exchange of air with the surrounding area, then the difference between this excellent mixing in a limited volume described here and the perfect mixing required by rollback can be ignored.

If none of these conditions can be shown, or reasonably assumed to exist, then the application of simple rollback or proportional modeling to the question of the impact of changes in the emission of one class of emitters on ambient air quality is totally without theoretical or experimental foundation.

From the viewpoint of an air quality manager, this is its worst drawback. Only very rarely will he have as an available control tactic the possibility of reducing the emission rate of all sources by the same proportion. This is possible, for example, for carbon monoxide in areas where motor vehicles are the only significant source, and where we can assume that the vehicle mix is uniform over the whole area. In that case, reduction in carbon monoxide emissions per vehicle-mile traveled (caused by changes in vehicle design or by control devices) will cause an apparently uniform percentage reduction of emissions without changing the spatial or temporal pattern of these emissions. In contrast, because pollutants such as total suspended particulate matter or sulfur dioxide have some large point-source emitters subject to add-on control devices as well as many small area sources not amenable to such devices, it seems impossible that the air quality manager will have as a viable option any emission-reduction program which will make this uniform percentage reduction on all emitters, and

f. Simple rollback assumes that the meteorological conditions which existed when c_{max} was measured and the emissions calculated are those which will exist on the date when the standards are to be met. Climate does change without human intervention, and growth of cities and growth of energy release does influence climate, so this is not necessarily a sound assumption.

A series of modified rollback models has been proposed (*10*) which are more complex than simple rollback, but much less complex than the diffusion models described in the following section. These models have had some utility in problems where this intermediate level of complexity seemed appropriate.

2. Tactics Based on Diffusion Models

Because simple rollback is very limited as an air quality management tool, more advanced tactics based on diffusion models of pollutant behavior have been developed. This section considers in some detail one such model, the implementation planning program (IPP) (*12*). This is a large-scale computer system designed to assist the air quality manager in managing emissions of two pollutants, total suspended particulate matter (TSP), and sulfur dioxide (SO_2). This was clearly a first try at this idea; future efforts in this area should and will be better. However, what was done is instructive, and it is believed that future air quality management programs will probably use the same basic management approach and same basic organization as IPP. As the various modules in the program are replaced with more accurate and better-validated advanced versions, the program will become a better air quality management tool.

Figure 4 is a schematic of the IPP program. The figure is divided into three categories—inputs on the left, internals in the middle, and outputs on the right. The inputs correspond to five of the seven kinds of information listed in Section II,B. Air quality standards and enforcement procedures are omitted. The program begins (1 in the figure) by using the local values of the meteorological variables, the measured local air quality data, and the local emission inventory in a diffusion model to compute a receptor concentration table (2), a list of computed annual average concentrations of TSP and SO_2 at selected locations, assuming current levels of emission control. The diffusion model also presents a source-receptor table (3), a listing of the ambient air concentration contribution of each source at each of the specified receptor locations, assuming current levels of emission control.

The diffusion model used here is very similar to the air quality display

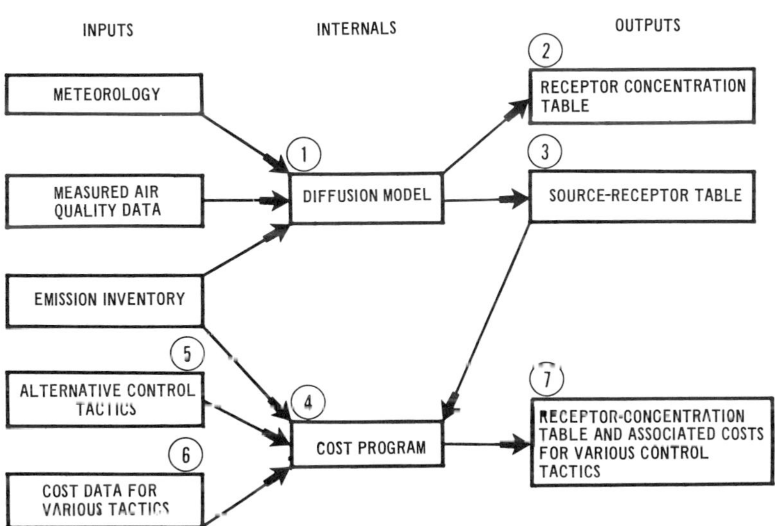

Figure 4. Schematic of the IPP program.

model (AQDM) described in some detail in Volume I, Chapter 10. The model is an annual average model; it uses the method suggested by Larsen (*13*) to estimate short-term concentrations. These are reported for the air quality manager's information, but not for further use in the program.

Next, the cost program (4) applies each of the alternative sets of control tactics to the various sources using the cost data and the source–receptor table to produce a new receptor-concentration table for each of these sets of tactics proposed, as well as the cost of each set. The source–receptor table (3) makes it unnecessary to rerun the diffusion model for each assumed set of control tactics, because the concentrations at each point are assumed by this model to be the sums of the individual source contributions, and each contribution is a linear function of the source emission rate.

The alternative control tactics (5) consist of lists of emission standards or fuel switching requirements for various kinds of emitters in each political jurisdiction of a control region. The built-in control tactics in the program are based on weight-rate and fuel sulfur regulations in common use at the time the program was written. The air quality manager has the option of either using these or entering different tactics as he chooses. The cost data (6) section consists of formulas for estimating installed and operating costs of various kinds of control devices as well as of fuel changes. These are built in but modifiable by the manager; the locally estimated fuel prices must be supplied by the manager.

The output of the program consists of the two tables mentioned previously (2 and 3), plus the listing (7) of the receptor concentration tables (2) and costs associated with those sets of tactics selected for evaluation by the manager.

The comparison of the predicted concentrations with the applicable air quality standard is made by the manager. (Although the program has the option of accepting the standard as an input and flagging the receptors in violation of the standard, this is not a particularly significant function, since it involves simply comparing a table to a fixed value.)

An example of the utility of IPP is shown in Figure 5, which shows isopleths of predicted annual average concentrations of total suspended particulate matter for the New York, New York, area that would result from the application of one proposed set of pollution control tactics. (The isopleths were drawn on the map after the predicted point concentrations from the receptor-concentration table had been entered in their proper places on the map). The standard to be met was 75 $\mu g/m^3$, so it is clear from this figure that under this set of tactics an area encompassing a large part of Manhattan would fail to meet the standards. Using this

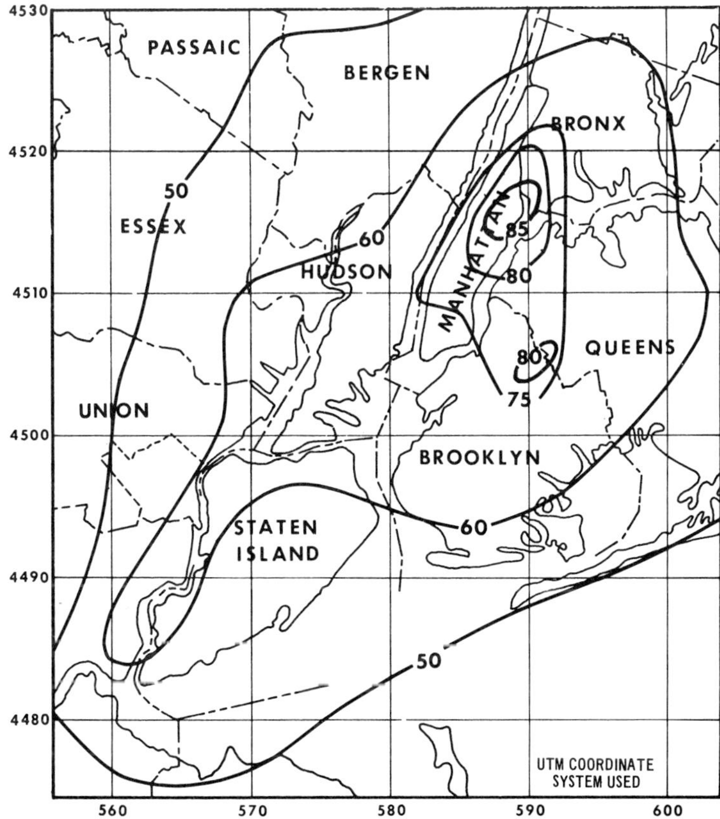

Figure 5. Typical computed suspended particulate matter isopleth output from IPP ($\mu g/m^3$).

plot (and others like it for various sets of proposed emission control tactics), it was possible to suggest alternative sets of tactics for which one could predict that the standards would be met and estimate the cost of doing so. Once these alternative sets of tactics were known, the political process of getting them adopted and enforced could begin. The principal drawbacks of the initial IPP model are:

a. It requires data which are difficult to obtain and are seldom available in computer-compatible form. As more accurate emission inventories, meteorology, and air quality data become available in computer-retrievable form, it will then be possible to run IPP or programs like it at a reasonable cost. Until then, the costs of collecting and processing the data into the required form will be high.

b. The diffusion model used is only a fair predictive tool. As discussed in Volume I, Chapter 10, its shortcomings are well known.

c. The control-cost module was written at a time when there seemed to be adequate supplies of any kind of fuel. By 1976, this was not the case, and the control-cost module must be rewritten to take fuel availability problems into account.

d. The area-source treatment in both the emission inventory and the diffusion model is of questionable accuracy.

e. The programmed method of predicting possible violations of short-term (rather than annual average) standards is not as reliable as the best short-term models now available (Volume I, Chapter 10).

f. An extension of the program to pollutants other than SO_2 and total suspended particulate matter is not currently available, nor is it likely that an extension to photochemical oxidants can be made without major overhaul of its fundamental structure.

An interesting variant on IPP was developed by the Consad Corporation (*14*). In it, they replaced the alternative control tactics module (5 of Figure 4) with an optimization program. This program, instead of applying uniform regulations to all emitters, applied controls selectively to the most important emitters, attempting to find a set of selective controls which would allow the standards to be met at lower costs than would be needed to meet them with uniform controls. Using the most efficient control equipment on the sources which cause greater degradation of air quality and omitting control equipment on the sources which have least impact can lead to attainment of air quality goals at a lower cost than a scheme which applies a uniform standard to a wider class of sources.

3. Land Use and Transportation Control Tactics

It is clear that land use and transportation systems influence air quality. Land use controls and transportation planning controls are beginning to become tactics available to the air quality manager. This subject is covered in Volume V, Chapter 2.

4. Intermittent Control Tactics

Intermittent control tactics recognize that there are certain times of the year (or times of the day) in which emissions are more likely to come to ground in high concentrations than other times of year or day. These tactics attempt to reduce emissions at those times, allowing emissions to return to normal rates at other, less-critical times. In most cases, the short-term emission reduction is brought about by a plant shutdown, fuel switching, or production curtailment during the period of control. There are two basic kinds of intermittent control tactics—predictive and observational.

a. PREDICTIVE INTERMITTENT CONTROL TACTICS. Predictive tactics are based on the knowledge that the atmospheric conditions likely to call for an emission reduction are normally ones which recur regularly and can be predicted with (some) accuracy. Frequently, the damaging situation is caused by "morning inversion breakup fumigation" (Volume I, Chapter 8) in which the pollutants that cause the standard violation are emitted several hours before the violation occurs. For emission reduction to be effective in this case, it is necessary to curtail emissions several hours before the predicted violation.

The most famous, and apparently first documented, predictive intermittent control tactic was instituted in 1941 at the lead–zinc smelter at Trail, British Columbia, Canada. This smelter is located in a narrow portion of the Columbia River valley, 7 miles (11 km) north of the Canada–United States border. During the night, down-valley flow of its emissions passed over agricultural lands on the United States' side of the border; during morning inversion breakup, the emissions were brought to ground level, resulting in crop damage. This led to the formation of an international arbitration tribunal to award damages; the tribunal instituted a study which ultimately led to the adoption of an intermittent control tactic to minimize the damage (*15*).

The control scheme used considers the following variables: growing and nongrowing season, wind direction, turbulence intensity, and time of day. The regulations require most stringent controls for the period between 3 A.M. to 3 hours after sunrise during the growing season, as this is the time when turbulence is low and winds cause down-river flow.

One may argue whether or not this is air quality management, but if one accepts as an air quality standard "short-term SO_2 concentrations during the growing season shall not be high enough to damage crops," then it is a classic, historical example of air quality management.

An entirely analogous predictive control scheme is that of the Paradise, Kentucky, steam power complex of the Tennessee Valley Authority (*16*). When the meteorological prediction is that the dispersion of the plant's plume will be severely restricted, threatening a violation of an ambient air quality standard, the plant's power output (and thus fuel consumption) is reduced several hours before the standard violation is predicted to occur. Other examples are the smelter operations of the ASARCO at El Paso, Texas, and Tacoma, Washington (*17*).

An additional method, widely used, which can be classified as an intermittent control procedure, are regulations that allow open burning only on days when the predicted meteorology leads to a minimum impact on the quality of the air (*18*). Two other forms of predictive intermittent control tactics are the proposals to allocate long-distance natural gas for interruptible users based on predicted daily air quality problems (*19*) and to allocate electric generating capacity on a similar basis (*20*).

An additional type of predictive, intermittent control tactic (which could possibly be classified as an entirely different type called a "time of emission" tactic) is one suggested for Los Angeles, California (*21*). It proposes that if the peak auto traffic, corresponding to commuters going to work, were delayed one hour in the summer, the resulting dispersion of the auto-related pollutants into an atmosphere with a greater mixing depth would cause the observed daily maximum ground-level oxidant concentrations to be reduced. Since the morning and evening traffic peaks are intermittent, it seems reasonable to classify this as an intermittent control tactic.

b. Observational Intermittent Control Tactics. An observational intermittent control tactic is a control scheme whereby the emissions are promptly controlled when an air quality sensor or a network of sensors indicate that air quality is deteriorating unacceptably. This tactic has limited application in many situations, particularly in circumstances where the emissions do not effect the sensor until some time, even hours, after they occur. A great many sensors are required if all areas in the vicinity are to be protected. The observational tactic is most useful when it supplements a predictive scheme. It can then serve as a fail-safe system for the predictive tactic.

The Trail, British Columbia, intermittent control procedure, in addition to the predictive part described above, also provides for continuous monitoring of SO_2 at an agricultural location in the United States and curtailment of emissions whenever this monitor shows a continued high value (*15*). Similarly, the ASARCO scheme at El Paso, Texas, contains both predictive and observational components (*17*). Apparently, the same kind of control, on a less formal basis, was practiced in the smelter industry as early as the 1920s (*22*). Figure 6 is a schematic of an integrated intermittent control system possessing features of both the predictive and observational tactics.

An entirely different kind of observational control tactic is the emergency episode control tactic. In this tactic, when observations indicate that there exists a very high ambient concentration of one or more pollutants, the emitting sources which release the pollutant are curtailed or

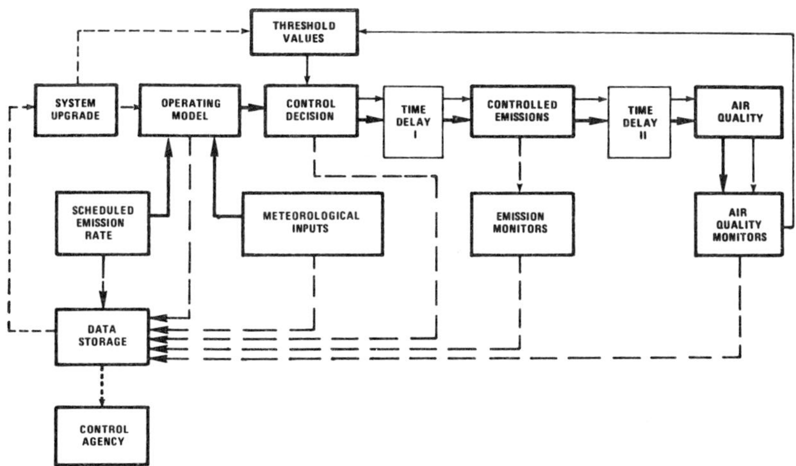

Figure 6. Schematic of an integrated predictive and observational intermittent control tactic.

shut down. Normally, the ambient concentration at which this tactic goes into effect is significantly higher than the applicable national ambient air standard, so that one can argue whether or not this is air quality management in the conventional sense. If ambient air quality standards are defined to include alert, warning, and emergency levels, this then becomes a form of air quality management. This subject is treated in Volume V, Chapter 3.

5. Tall-Stack Tactics

Tall-stack tactics are based on the fact that under most meteorological conditions, raising the height of emission will reduce the maximum ground level concentration (Volume I, Chapter 9). For an isolated point source (e.g., a smelter or power plant located in a rural area), one can clearly show that under most meteorological conditions the maximum ground-level concentration resulting from the emissions will be reduced by a tall stack (*23, 24*). Thus, one may consider a tall stack as an air quality management tool. The use of the tall stack is an air quality management tactic, rather than an emission standard tactic, because its goal is not to reduce the amount emitted, but rather to reduce the maximum ground level concentration to meet an ambient air quality standard.

One of the first such applications of a tall stack to solve an air pollution problem was at the Murray smelter near Salt Lake City, Utah, in 1918 (*22*). There, the problem was similar to the one with the Trail,

British Columbia, smelter discussed above, and the applicable ambient air quality standard was "SO_2 concentrations shall not be high enough to damage crops." Because of the meteorology of the site, a 455 foot (137 m) stack reduced the threat to crop damage.

Numerous states in the United States (*25*) and other countries in the world at one time had regulations requiring tall stacks for certain types of sources. This tactic makes sense for a single source in a remote area, but has been strongly challenged as a tactic for densely populated areas. In densely populated areas, the pollutants dispersed by a tall stack come to ground far away in areas which have air pollution problems of their own and generally are not in the same pollution control district (or even in the same country) as the emission source (*6*). For long-lived pollutants or precursors of long-lived pollutants, the arguments against tall-stack tactics for densely populated areas seem strong and are self-evident (*26*).

III. Emission Standard Strategies

Historically, emission standard strategies predate air quality management strategies. The first major air pollution control efforts were directed at the smoke from inefficient combustion of coal (Volume I, Chapter 1). The method most often used was to limit "visible emissions." This was selected as the criterion (in preference, for example, to weight rate of emission) because of its ease of detection and enforcement (*27*). Because visible emissions of smoke occur when the emission rate of particulates is large, a visible emission standard is relatable to a particulate emission rate standard. The effect of such a standard is to limit the amount of particulate matter emitted to the atmosphere.

The next logical development in emission standard strategies was the "good practice" or "best practical means" approach. This required that all members of a given class of emitters attain a level of control which corresponded to "good practice" or which utilizes the best practical means of control. There has been considerable debate over the meaning of "good practice." The differences between "common practice," "good practice," and "best feasible practice" will be found in Volume V, Chapter 13. The terms "adequately demonstrated control technology," "available control technology," and "best available control technology" have similar meanings. Regardless of which of these terms is used, the idea is the same, namely, that there is some level of emission control which all members of a class of emitters should be required to attain.

The philosophical basis of such standards is that emissions to the atmosphere should be held as low as possible (using "good practice") and that control measures demonstrated successfully on one plant are appropriate for others. Those who assert that this strategy is superior to air quality management claim that in air quality management we are reliant on air quality standards which will always be based on arguable data. Once the standards are set and the required emission reduction calculations made, the best that can be done in trying to meet the ambient standards is to get each emitter to reduce his emissions to meet some emission standard, which will ultimately be based on someone's judgment of what is "good practice." Thus, they argue, why not shortcut the development of ambient air standards, air monitoring, diffusion modeling, etc. and simply require all emitters to go directly to a "good practice" emission standard (*28*).

Emission standards can be applied as a pure strategy or as a tactic in enforcing another strategy. For example, the United States Clean Air Act of 1977 has three emission standard sections, two of which are pure strategies and one which is a tactic for implementing the overall air quality management strategy of the act. This latter is the mobile source regulation (Section 202), which limits the allowed emission rates of carbon monoxide, hydrocarbons, and nitrogen oxides from all motor vehicles sold in the United States (*3*). The emission limits were determined by an air quality management rollback model which computed the allowable emission rate for all motor vehicles in order that the national ambient air quality standards for these pollutants be met everywhere in the United States. Another example is where the emission standard is based on an area classification which in turn is determined by the ratio of existing air quality to the air quality standard. In the United States, under the Clean Air Act of 1977, the states were required to classify areas within the state based on existing air quality (*29*). Variable emission standards were then developed that were sufficient to achieve the ambient air quality standard (Fig. 7). Similar tactics have been developed in other countries such as East Germany.

The two pure emission standard strategies in the United States Clean Air Act of 1977 are the Hazardous Pollutant (Section 112) and New Source Performance Standard sections (Section 111) (*3*). The administrator of the United States Environmental Protection Agency is directed to issue what amounts to "best technology" emission standards for all sources of hazardous pollutants. The philosophy is clearly one of controlling emission sources to the best degree technologically possible, without economic consideration. The New Source Performance Standard section of the Clean Air Act of 1977 (Section 111) directs the administrator of the United States Environmental Protection Agency to establish "standards of performance" that are equivalent to emission standards for

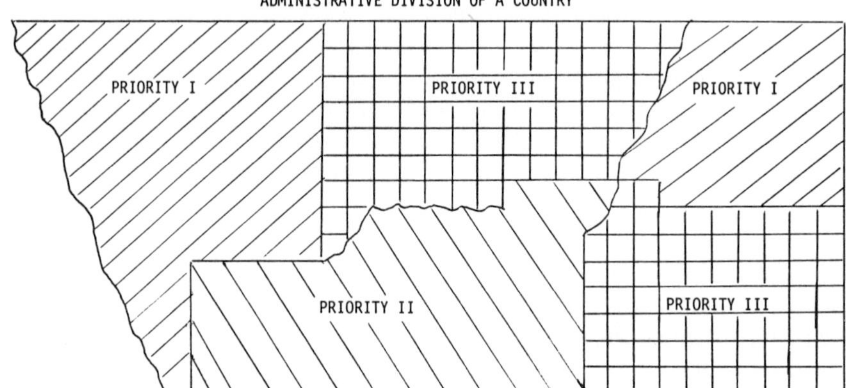

Figure 7. Variable emission standard based on area classification.

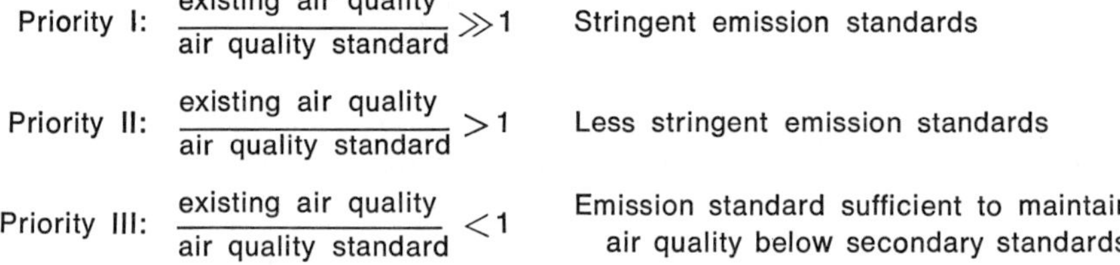

Priority I: $\dfrac{\text{existing air quality}}{\text{air quality standard}} \gg 1$ Stringent emission standards

Priority II: $\dfrac{\text{existing air quality}}{\text{air quality standard}} > 1$ Less stringent emission standards

Priority III: $\dfrac{\text{existing air quality}}{\text{air quality standard}} < 1$ Emission standard sufficient to maintain air quality below secondary standards

various classes of potential emitters. These are specifically stated to be based on "best technology." The philosophy here is also "as clean as possible"; however, costs of control must be considered. Although not strictly a pure emission standard strategy, a "no significant deterioration" or "nondegradation" strategy is related to the "as clean as possible" concept (see Section VI,A).

Emission standards generally fall into one of four categories—numerical rate, fuel specification, equipment, and prohibitive. The following examples will illustrate that each of these is an emission standard.

A typical numerical rate standard in the United States for coal-fired stationary steam generators requires that the emission rate of particulates shall not exceed 0.1 pound per 10^6 BTU (0.180 kg per 10^6 kcal) of fuel burned (*30*). This is easily recognizable as a numerical emission standard.

Fuel sulfur regulations are, in effect, emission standards because in most fuel-burning equipment, practically all of the sulfur in the fuel appears as SO_2, with a minor portion being some other oxidized compound of sulfur in the exhaust gas or ash. Thus, limiting the sulfur content of the fuel is equivalent to limiting the sulfur oxide emissions per unit of fuel burned. This is particularly clear in the aforementioned standards for coal-fired power plants, which limit the emissions of sulfur oxides to 1.2 pounds per 10^6 BTU (2.16 kg per 10^6 kcal) of fuel burned. This standard can be met by using low-sulfur coal (typically about 0.7% sulfur by weight), in which case it is a fuel standard; or by using some form of stack-gas cleaning or combustion modification to remove the sulfur oxides from the stack gas, in which case it appears as a classic numerical emission standard.

The Los Angeles County, California, Air Pollution Control District has long had regulations limiting the fraction of olefinic components in motor fuels (*31*) sold in that area. In effect, this is also an emission standard, because there is evidence that the hydrocarbon emissions from motor vehicles using high olefinic fuels are more photochemically reactive than those from vehicles using nonolefinic fuels. Thus, this is an emission standard on reactive hydrocarbons.

A typical "equipment standard" is Los Angeles County Air Pollution Control District Rule 56, which stipulates requirements for the storage of petroleum products (*30*). Rule 56 limits the storage of volatile petroleum distillates to containers of less than 40,000 gallon capacity unless the container is pressurized or equipped with a floating roof, a vapor recovery system, or any other approved vapor limiting device of equal efficiency. This is an emission standard of the "best technology" variety, because it requires all operations of such equipment to utilize control devices that limit the emission rate to no more than normally encountered in a floating-roof tank.

Examples of prohibitive standards are those on the burning of coal in residential furnaces and bans on open burning of garbage in dumps. Both of these types of burning produce copious particulate matter emissions. Their prohibition is a standard requiring the reduction of particulate matter emissions below the levels which are achieved in residential oil or gas furnace or landfill.

IV. Emission-Tax Strategies

This concept has been widely discussed in academic journals but has not been widely applied in practice except in Czechoslovakia (see Section I,C,3). Those who have supported this approach (*32–36*) favor it because it makes pollution control an internal matter for the polluter. In all the other types of strategy, the polluter stands to gain if he can avoid installing pollution control technology; by so doing he passes the damage cost on to someone else and gains a cost advantage over any competitor who does install pollution control equipment. In this circumstance, he has no economic incentive to reduce emissions and will only do so when law or public opinion forces him to. Under an emission-tax strategy, he could lower his tax bill by lowering his emissions, so that emission control becomes a profit-making operation and he will, at least in theory, devote more effort to it than under other strategies.

The most commonly proposed version of this strategy (*4, 37*) would consist of a flat rate tax on pollutants emitted (e.g., x¢/pound of SO_2), charged against all emitters in a metropolitan area or in the nation. At regular intervals, if an appropriate set of air quality goals have not been met, the tax rate would be raised until the goals were met. Presumably, as the tax rate was raised, more and more emitters would find it more economical to control than to emit, so that emissions would decrease.

A great benefit claimed for this approach is that it leads to market allocation of the dispersive capacity of the atmosphere. Those who can use this capacity most efficiently will bid more for it than those who can make less efficient use of it, so resources will be allocated in the best possible way. For example, if we wish to reduce SO_2 emissions in a whole region, the lowest cost per pound of SO_2 removed will occur in large installations. Thus, the small emitters have their break-even point at a higher tax per pound of emissions than the larger emitters, and will willingly pay more. Hence, we obtain optimal use of resources, as seen by economists.

Many other versions of emission taxes are clearly possible. Sales taxes, tax remissions, property taxes, subsidies, import restraints, and domestic production restraints could all function as emission taxes (Table I). These remain interesting theoretical concepts but are not close to application. This is an area of investigation in which much more work is required in order to determine feasibility (*38*).

V. Cost–Benefit Strategies

Like emission-tax strategies, cost–benefit strategies have been discussed in academic journals (*39–42*) but seldom applied. A cost–benefit strategy attempts to assess the damage functions for all levels of exposure to all pollutants and then weigh each emission in terms of the cost to abate it and the benefits, reduced damages, due to abatement at various degrees of control. Taking into account the interaction of such controls applied simultaneously to more than one source, which will be necessary if the damage functions are not linear, it attempts to set the levels of pollution control. Society may then choose between every pollution control expense, accepting those which will produce benefits greater than or equal to the costs and rejecting those which will produce benefits less than the control costs.

Table I Financial Incentives to Supplement or Replace Regulation[a]

Taxes	
Sales taxes	On fuel, fuel additives, ingredients in fuel, pollution-producing equipment
Ultimate disposal taxes	On automobiles or other objects requiring ultimate disposal
Land use taxes	For pollution-producing activities
Tax remission	
Corporate income tax	For investment in or operation of pollution control equipment; accelerated write-off of pollution-control equipment
Property taxes	For pollution control equipment
Fines, effluent charges, and fees	
Fines	For violation of regulations
Effluent charges	For permission to emit excessive quantities of pollution; paid after emission
Fees	For permission to emit excessive quantities of pollution; paid before emission
Subsidies	
Direct	Governmental production (e.g., nuclear fuel)
Grants-in-aid	For pollution-control installations
Indirect (low-interest bonds and loans)	For pollution-control process or equipment development
Import restraints	
Duties and quotas	On materials, fuels, and pollution-control and pollution-producing apparatus
Domestic production restraints	
Quotas, land and off-shore use restraints	On materials and fuels

[a] From reference (*38*).

VI. Other Strategies

A. Nondegradation

At least until the mid-1960s, the prevailing philosophical view in American air pollution was one of the air being a natural resource. This view holds that the atmosphere has natural processes for removing pollutants, principally rainfall, which collects particulates and gases, by the same mechanism as does a wet scrubber. This cleansing power of the atmosphere was viewed as a resource. This parallel was felt so strongly that the concept of an "airshed" comparable to the watershed which feeds a river was introduced into air pollution management. According to this view, the logical goal of air pollution control was to allocate this cleansing capacity of the atmosphere so that man-made pollutants were dispersed away from receptors that might be harmed, thus allowing this natural cleansing capacity to reduce and remove the emitted pollutants. The logical outgrowth of this philosophy was the location of pollution sources in remote areas where their emissions would affect the fewest receptors.

Between the mid-1960s and early 1970s, a different view emerged. This view held that because the world is finite, it should not be overloaded with pollutants, and hence, rather than relying on the cleansing (or absorbing) power of the environment to deal with pollutants, they should be controlled at their source. Similarly, this view held that the low level of pollution (i.e., very clean air) in remote areas was a valuable natural resource. Furthermore, it was believed that the use of this resource to help solve the air pollution problems of cities was a mistake. The conflict between these two views led to legal action which proceeded as far as the Supreme Court of the United States (*43*). It is far from clear what the final resolution of this conflict will be, and it will probably take years for a complete resolution of the variant views.

The Standards of Performance for New Stationary Sources section (Section 111) of the Clean Air Act of 1977 is in part a nondegradation strategy. Before its enactment, many industrial sources had installed less effective and less expensive pollution control devices on remote installations than on urban ones. Many nonindustrial parts of the country had tried to encourage industry to locate there by promising insignificant or nonexisting air pollution control regulations. By applying uniform national emission standards for new industrial installations, the United States has impeded these practices.

B. Emission-Density Strategies

Much of urban zoning is ultimately related to air pollution emission density. It has normally called for the location of air-polluting industries away from residential and commercial areas. Usually this has been done without explicit air pollution regulations, but rather for implicit air pollution emission-density reasons. Considerable effort has been made in recent years to formalize and upgrade this procedure. This effort is summarized in Volume V, Chapter 2 on Land Use and Transportation Planning.

VII. The Strategies Compared

This section compares the four kinds of strategies listed in Section I,C to the desirable characteristics of a strategy listed in Section I,B (Table II).

Table II Comparison of Strategies

Strategy	*Cost effective-ness*	*Simplicity*	*Enforce-ability*	*Flexibility*	*Evolu-tionary ability*
Air quality management	Good	Poor	Fair	Fair	Fair
Emission standard	Terrible	Excellent	Excellent	Poor	Fair
Emission taxes	Fair	Excellent	Excellent	Unnecessary	Good
Cost–benefit analysis	Excellent	Terrible	Unknown	Unknown	Good

A. Air Quality Management

More is known about the advantages and disadvantages of air quality management than about any of the other three strategies that have been described, as it has been the most widely applied in the world, and because a major concerted effort has been made in the past few years in the United States to implement it.

Its cost effectiveness is good but not excellent. It has the virtue of concentrating pollution-control expenditures in the areas with the worst pollution problems, and allowing higher emission rates (and lower pollution control expenditures) in areas with less serious problems. Its fault, concerning cost effectiveness, is that once a set of nationwide air quality standards have been set, they must be met everywhere, even in areas seldom or never now visited by people. Thus, it will require some control expenditures for which the damage reduction benefits will be small. Furthermore, its cost effectiveness is crucially dependent on the standard-setting process. If the standards are set in a manner that provides for the adequate protection of the public's health and welfare, they should be reasonably cost effective. But if they are set too stringently, the control costs will exceed the pollution reduction benefits. If they are not set stringently enough, the resulting air pollution levels could cause damages larger than the control costs of preventing those damages.

Air quality management is not simple. It attempts to control air quality by regulating emissions from millions of sources, which have varying operating characteristics, location, and use patterns. No simple way has been found to do this; the United States' best efforts to do it in a simple, straightforward way have resulted in a very complex set of control regulations.

The enforceability of air quality management is fair. There have been legal challenges, which are likely to continue. The procedure is indirect but has been upheld in court.

Its flexibility is fair. Because of the multiple ways ambient air quality standards can be met, those administering it have some flexibility, and each local agency can choose the tactics it considers best, within limits. Special cases and emergencies can be handled locally.

Its evolutionary ability is fair. As new data require, air quality standards can be changed. But new standards will probably require completely new emission regulations, which will be expensive and time-consuming. Improvements in control technology will increase the options open to the air quality manager.

B. Emission Standards

The cost effectiveness of a uniform emission-standards strategy is poor; it is clearly the least cost-effective way that one can proceed. The emission standards must be set in some way, probably on a "best-technology" basis. Once they are set they must be met by all emitters. This means that if they are set in a lax way, high pollution levels can be experienced in major urban areas and hence high pollution damage. This could be avoided, in a cost-effective way, by setting the standards more stringently. If the standards are set stringently, on the other hand, vast sums will be expended for controlling remote plants whose air pollution damage impact is slight.

The emission-standard strategy is the simplest strategy from an administrative viewpoint. The standards are promulgated and the test procedures are defined; then the regulation-setting part is accomplished. The enforceability of this strategy is excellent. Once standards are set and test methods defined, one knows whom to monitor and for what, and violation criteria can easily be written and penalty schedules formulated.

The flexibility of this strategy is poor. For major sources such as power plants, the standards must be announced several years before they take effect and probably cannot be changed more often than once every five or ten years. The control official will have difficulty dealing with the emitter who, because of technical or economic problems, cannot meet the standard. The control official would have to determine the severity of the area's pollution problem. If it is severe, he must then either allow the situation to become worse or shut down the emitter. Either decision is fraught with problems.

The evolutionary ability of this strategy is fair. If a new technology makes it possible to set a lower standard, it can be done for all sources built after a certain date. This works well for mobile sources, whose lifetime in the economy is short, but poorly for industrial plants, whose lifetime is thirty to fifty years. Mandating a lower emission standard for plants built after a certain date will help the air quality in areas undergoing growth after that date, but not those which do not have such growth.

C. Emission Taxes

Czechoslovakia has adopted a form of these strategies through the establishment of emission fees, and many versions of emission taxes have been proposed and discussed. The cost effectiveness should be fair, because it would allow each emitter the choice of controlling emissions or paying the taxes (or controlling to some economic level and paying for the rest). Making the decision whether or not to control and the level of control desired a matter of the internal economics of major emitters would probably result in a better overall level of cost-effectiveness than possible with uniform emission standards.

Most emission tax schemes proposed so far only envisage taxes on larger sources. For such schemes, the simplicity would be excellent. For larger sources, the tax rates and emission test methods comprise the whole of the regulations. If an attempt was made to extend the tax to all emitters (e.g., area and mobile sources or fugitive emission sources) the problem would become more complex (*44*). For sulfur oxides, for example, one could tax motor vehicle and home-heating fuels based on sulfur content at a rate comparable to the rate for sulfur emissions from larger industrial sources. This would be simple. But no comparably simple scheme for particulate emissions from home heating area sources, mobile sources, etc., is yet apparent.

If the tax schemes are limited to larger sources, enforceability should be excellent. The emission-testing industry would have to be expanded, and certification of testers (like the current certification of weights and measures) would have to be instituted, but once a certified body of independent emission testers was available, their test values would be readily accepted as the basis for tax payments.

Flexibility would be unnecessary. Flexibility is needed in the other strategies to deal with the problem of an emitter who cannot economically meet an area-wide standard, or who cannot meet it by a statutory deadline. In an emission-tax system the emitting source simply pays the tax.

Evolutionary ability should be good, because the tax rate could be changed as the need arose. This would have to be done with some caution, because industry has complained about their difficulties with changing standards. (They speak of the difficulty of "shooting at a moving target.") However, raising a tax rate for existing plants causes much less economic disruption than lowering an emission standard. In the case of a tax rate increase, an existing plant would probably elect to pay the higher tax, while with a lowered emission standard it would probably have to replace its existing pollution control equipment with more efficient equipment.

D. Cost–Benefit Analysis

The cost effectiveness of the cost–benefit analysis strategy should be excellent. However, it would be very complicated. For it to be successfully implemented, the true dose–damage curve for all pollutants and all classes of receptors for all exposure times must be known. Then dollar values must be assigned for all kinds of damage, including premature death, various kinds of health impairment, loss of visibility, and property damage to irreplaceable historic and scenic landmarks. To date this has seemed so far beyond current capabilities, with the possible exception of damage to farm crops, that this approach has not received much serious attention. Most who have discussed cost–benefit approaches (*45–51*) have really considered only the minimum-cost routes to meet various air quality goals which were established by some method other than trying to quantify the benefits of reaching them.

Its enforceability and flexibility are unknown, because so little is known about how it would operate. Its evolutionary ability should be good. As new control methods or new effect data become available, they could be entered into the standard-setting process easily.

E. Conclusion

The previous discussion assumes that one of these strategies will be used to the exclusion of the others. It appears possible to combine two strategies to obtain the benefits of both. For example, the Clean Air Act of 1970 is most heavily oriented toward air quality management, but it contains some emission standard provisions. Many other combinations are possible.

Table II shows that air quality management scores fair or good in all areas except simplicity. Emission taxes score fair to excellent in all areas. However, if as much were known about applying the other strategies as is known about air quality management, they might not be rated as highly.

VIII. Integration of Air Pollution Control Strategies into Overall Environmental Management

In the past, society has often caused new environmental problems while solving old ones. Examples are the air pollution problems with the incineration of municipal waste or the incineration of sewage treatment plant sludge, and the water pollution problems caused by discharge of effluents from wet scrubbers used for air pollution control.

A more subtle class of problems is those involving pollutants which pass through the air but subsequently cause water pollution. Examples are the acidification of streams and lakes caused by acid rain due to the burning of sulfur-bearing fuels and lead contamination of surface waters due to lead aerosol from motor vehicle exhausts. If these types of pollution are to be controlled, they must be controlled through some kind of air pollution emission regulation, since they can hardly be controlled by water pollution regulations, which apply to the direct discharge of contaminants into surface waters.

A different class of problem is posed by nuclear power plants. These clearly emit less atmospheric pollutants than fossil-fuel power plants, and this is frequently cited by their proponents as a major advantage. However, they produce radioactive wastes, which are very hazardous and must be isolated from the environment for millenia (*52*). Thus, with them, society trades one kind of problem for another. The calculus of this kind of trade-off is as yet uncertain.

Through the United States National Environmental Policy Act (*53*), an effort has begun to look at these problems in an integrated way. This has not yet resulted in the integration of air pollution management into a truly comprehensive overall environmental management.

IX. Future of Air Quality Management, Pollution Control Strategies, Modeling, and Evaluation

A. Improved Analytical Tools

The goal of the air quality manager is to be able to predict the detailed air quality consequences of any change in emissions. Using present predictive models, he can do that only within a very wide error band. Thus, the principal outstanding problem in air quality management is that of developing certain and accurate prediction techniques. A cost–benefit strategy would require even better analytical techniques than for air quality management. Emission taxes or emission standard strategies would not demand such predictive techniques.

To develop those techniques, air quality managers need a complete and closed information cycle. The information would include emission data, modeling results, and air quality monitoring. If the air quality manager has any real doubts as to the accuracy and validity of any of these three factors, he cannot be certain that he can predict the results of changes in emission rates. At present, there is considerable uncertainty about all three factors. Each must be improved, and the results of all three correlated in order to finally develop a predictive scheme in which he can have real confidence.

B. Air Quality Modeling as a Decision-Making Tool

When the improved analytical tools described above are widely available, it is reasonable to assume that each major development (new factory, new highway, new suburb) will be required to submit to the air quality managers a prediction of the air quality consequences of its operation. These predictions will then be weighed against the other costs and benefits of the project, and they will strongly influence the decisions on which projects to undertake.

C. Merger of Air Quality Management and Cost–Benefit Analysis

The air quality standards upon which most air quality management strategies ultimately rest are based on limited experimental and epidemiological data on the effects of air pollution. As more data become available, it will probably appear that some or all of the major pollutants do not have clearly defined threshold values below which there is no measurable effect. In this case, society will either have to decide to set standards at some level of effect, or replace air quality standards with some other criteria.

It is predicted that eventually the cost–benefit analysis approach will be the most satisfactory method to determine what an appropriate emission rate or air quality level will be. It is recognized that this will be very difficult to carry out, but all the other alternatives will be worse.

REFERENCES

1. R. R. Tucker, "Accomplishments in Air Pollution Control by the Control Agencies," Proc. Nat. Conf. Air Pollut. PHS No. 654. U.S. Govt. Printing Office, Washington, D.C., 1959.
2. N. de Nevers, *Sci. Amer.* **228,** 14–21 (1973).

3. "The Clean Air Act," 42 U.S.C. 1857, *et seq.* United States Environmental Protection Agency, Washington, D.C. 1970 and P.L. 95-95 (1977).
4. R. M. Nixon, "The President's 1972 Environmental Program," Weekly Compilation of Presidential Documents, Vol. No. 8, 7, p. 220. U.S. Govt. Printing Office, Washington, D.C., 1972.
5. E. C. Halliday, World Health Organ., *Monogr. Ser.* **46,** Geneva, Switzerland (1961).
6. H. Rhode, *Tellus* **24,** 128–138 (1972).
7. "Air Conservation," Publ. No. 80, Chapter 3. Amer. Ass. Advan. Sci., Washington, D.C., 1965.
8. L. Reed, *New Sci.* **54,** 756–757 (1972).
9. S. Yanagisawa, *J. Air Pollut. Contr. Ass.* **23,** 945-948 (1973).
10. N. de Nevers and R. Morris, *J. Air Pollut. Contr. Ass.* **25,** 943–947 (1975).
11. D. S. Barth, *J. Air Pollut. Contr. Ass.* **20,** 519–523 (1970).
12. "Air Quality Implementation Planning Program," TRW Syst. Group, Washington Operations (prepared for EPA under contract PH22-68-60). Superintendent of Documents, U.S. Govt. Printing Office, Washington, D.C., 1970.
13. R. I. Larsen, *J. Air Pollut. Contr. Ass.* **19,** 24–30 (1969).
14. "Investigation of Direct and Indirect Alternate Implementation Strategies." Consad Research Corporation, Pittsburgh, Pennsylvania (prepared for the United States Environmental Protection Agency under Contract No. 86-01-0509). Superintendent of Documents, U.S. Govt. Printing Office, Washington, D.C., 1971.
15. E. W. Hewson, *Quart. J. Roy. Meteorol. Soc.* **71,** 266–283 (1945).
16. J. M. Leavitt, S. B. Carpenter, J. P. Blackwell, and T. L. Montgomery, *J. Air Pollut. Contr. Ass.* **21,** 400–405 (1971).
17. K. W. Nelson, M. A. Yeager, and C. K. Guptil, "Closed-Loop Control System for Sulfur Dioxide Emissions from Non-Ferrous Smelters." American Smelting and Refining Company, Salt Lake City, Utah, 1973.
18. H. C. Johnson and H. A. James, *J. Air Pollut. Contr. Ass.* **20,** 530-533 (1970).
19. N. de Nevers and A. H. Wehe, *J. Air Pollut. Contr. Ass.* **24,** 211–215 (1974).
20. D. S. Shepard, *J. Air Pollut. Contr. Ass.* **20,** 756–761 (1970).
21. E. K. Kauper and C. J. Hopper, *J. Air Pollut. Contr. Ass.* **15,** 210–3 (1965).
22. G. R. Hill, M. D. Thomas, and J. N. Abersold, *Mining Congr. J.* 31, 21–34 (1945).
23. "Tall Stacks, Various Atmospheric Phenomena and Related Aspects," Publ. APTD 69-12. Nat. Air Pollut. Contr. Admin., U.S. Dept. of Health, Education, and Welfare, Arlington, Virginia, 1969.
24. F. F. Ross, A. J. Clarke, and D. H. Lucas, *in* "Proceedings of the Second International Clean Air Congress" (H. M. Englund and W. T. Berry, eds.), p. 1041. Academic Press, New York, 1971.
25. V. H. Sussman, *J. Air Pollut. Contr. Ass.* **19,** 73–76 (1969).
26. J. H. Ludwig, "Hearings Before the Committee on Air and Water Pollution of the Committee on Public Works of the U.S. Senate, May 15, 10, 17, 1967," Air Pollution-1967, Part 4. Superintendent of Documents, U.S. Govt. Printing Office, Washington, D.C., 1967.
27. L. R. Burdick, *U.S., Bur. Mines, Inform. Circ.* **IC 7718** (1955) (a revision of **6888** by R. Kudlich).
28. V. H. Sussman, *J. Air Pollut. Contr. Ass.* **21,** 201–203 (1971).
29. *Fed. Regist.* **36,** 15486 (1971).
30. *Fed. Regist.* **36,** 24876 (1971).

31. "Rules and Regulations." Air Pollution Control District, County of Los Angeles, Los Angeles, California, 1971.
32. R. Witherspoon, J. Hoicka, T. Trumbull, and D. Infeld, "Governmental Approaches to Air Pollution Control—A Compendium and Annotated Bibliography." Institute of Public Administration, Washington, D.C., 1971 (National Technical Information Service, Springfield, Virginia).
33. G. Fishelson, *J. Air Pollut. Contr. Ass.* **24,** 44–47 (1974).
34. R. Wilson, *Science* **178,** 182–183 (1972); see also **181,** 1202 (1973).
35. L. C. Talsky, *Environ. Sci.* **14,** 23–29 (1971).
36. "Air Conservation," Publ. No. 80, Chapter 6. Amer. Ass. Advan. Sci., Washington, D.C., 1965.
37. A. M. Schneider, *J. Air Pollut. Contr. Ass.* **23,** 486–489 (1973).
38. A. C. Stern, M. S. Heath, and M. M. Hufschmidt, *in* "Proceedings of the Third International Clean Air Conference," p. 1310–12. Int. Union Air Prev. Ass., Düsseldorf, Federal Republic of Germany, 1973.
39. R. D. Wilson and D. W. Minnottee, *J. Air Pollut. Contr. Ass.* **19,** 303–308 (1969).
40. L. R. Babcock, Jr., and N. L. Nagda, *J. Air Pollut. Contr. Ass.* **23,** 173–179 (1973).
41. P. H. Ableson, *Science* **170,** 1159 (1970).
42. L. B. Lave and E. P. Seskin, *Science* **169,** 723–733 (1970).
43. *Fed. Regist.* **38,** 18986 (1973).
44. L. Lees, M. Braly, M. Easterling, R. Fisher, K. Heitner, J. Henry, P. J. Horne, B. Klein, J. Krier, W. D. Montgomery, G. Pauker, G. Rubenstein, J. Trijonis, "Smog—A Report to the People," EQL Rept. No. 4, Sect. II/5. Environmental Quality Laboratory, California Institute of Technology, Pasadena, California, 1972.
45. R. G. Ridker, "Economic Costs of Air Pollution." Praeger, New York, 1967.
46. L. B. Barrett and R. E. Waddell, "Cost of Air Pollution Damage: A Status Report," Publ. No. AP-85. United States Environmental Protection Agency, Research Triangle Park, North Carolina, 1973.
47. J. Horowitz, *J. Air Pollut. Contr. Ass.* **23,** 273–276 (1973).
48. J. Horowitz, *J. Air Pollut. Contr. Ass.* **23,** 395–397 and 418 (1973).
49. F. C. Hamburg and F. L. Cross, Jr., *J. Air Pollut. Contr. Ass.* **21,** 66–70 (1971).
50. J. Trijonis, Jr., "An Economic Air Pollution Control Model-Application: Photochemical Smog in Los Angeles County in 1975." Environmental Quality Laboratory, Jet Propulsion Laboratory, California Institute of Technology, Pasadena, California, 1972.
51. "Cumulative Regulatory Effects on the Cost of Automotive Transportation (RECAT)," Final Report of the Ad Hoc Committee. Office of Science and Technology, Washington, D.C., 1972.
52. A. Weinberg, *Science* **177,** 27–34 (1972).
53. "National Environmental Policy Act," 42 U.S.C. 4321. D.C., 1970.

2

Land Use and Transportation Planning

Ronald A. Venezia

I. Introduction

Many issues are now being raised with regard to how we perceive land use and transportation. Previous positions on national, regional, and urban growth and development are being reevaluated. In the past, environmental problems have been approached independently, either to be solved or to be justified on their own merits. Instead of recognizing environmental problems as the inseparable part of the entire land use–

transportation complex that they are, they have been considered as unrelated to anything else. The pollution problem is integral with land use and transportation, including concerns such as community planning, highway design, traffic control, mass transportation, demography, topography, economy, and social concerns. Population growth and concentration significantly affect planning considerations to minimize environmental insult. During the decade ending in 1970, most metropolitan areas experienced an increase in population (Table I) *(1, 2)*. Generally, larger populations have severe pollution problems related to poor land use and transportation practices.

In free-enterprise economies, consideration of land as a commodity to be exploited for the highest price must be reappraised to consider land as a resource to be used for the benefit of society as a whole. The systems approach to land utilization is necessary if we are to maintain the environmental quality mandated by laws, regulations, and common sense. Some areas are seeking to limit growth and development as constrained by environmental capacity. This requires attitudinal change by the public. Implications of limited growth for environmental concerns turns planning the full circle. Traditionally, community planning was meant to encourage growth of industry and population while minimizing adverse impacts. Growth no matter what the cost is no longer an acceptable doctrine. There is a need for institutional reform. Governments must decide on how their land will be used within given environmental constraints. Failure to perceive environmental consequences associated with planning decisions has created many of today's problems. Land-use problems associated with urban growth, congestion, sprawl, and unnecessary decay of the central cities have evolved over many years and are in conflict with environmental goals.

Table I Population Increase for Urban Areas—Decade Ending 1970

Urban area	*Percent increase*
Buenos Aires, Argentina	19.3
Chicago, Illinois	17.1
Los Angeles, California	8.4
Mexico City, Mexico	56.8
Moscow, U.S.S.R.	37.6
Paris, France	25.6
São Paulo, Brazil	63.9
Tokyo, Japan	18.3
Toronto, Ontario	26.9
Washington, D.C.	58.2

These are problems that can not be corrected in a short time. There are many vital issues that planners must consider within a framework for decision making. Air pollution control is but one of them. To legislate the achievement of air quality standards in a very short time imposes severe hardships. Long-range programs will be required to solve the basic cause of the problem. Long-range solutions will allow time to lessen or eliminate previous programs that were socially or economically too severe or restrictive. Legacies of past land-use and transportation decisions must be considered when developing new plans and policies. Old laws and concepts relative to land use and transportation must be changed to remain viable as new laws and concepts relative to the environment are accepted. Personal preference and life style must also change, as some of the most difficult problems are due to individual choice, such as the desire to live a long distance from the place of one's work and a preference for a personal automobile over mass transit.

Land-use and transportation plans are subject to change and in no way bind jurisdictions to their long-range goals, which are often subverted for short-term gains. Successful maintenance of standards requires that land-use, transportation, and environmental planning be closely interrelated with common goals clearly defined and vigorously pursued. Attainment and maintenance of air quality standards depends largely on public attitudes toward changes in life style, additional legal requirements, and social and economic impacts.

The expanding metropolitan area, environmental inadequacies of suburban design, increasing use of the automobile, and the difficulty in correcting the problem through local and national action specifically directed at stationary sources and the automobile lead to the conclusion that air pollution must be approached with other associated problems on a regional basis.

II. Transportation Planning for Air Pollution Control

A. Transportation Controls

Short-term land-use planning generally precludes controls other than those directly related to transportation. Transportation controls have a high impact on the general public, and can be divided into two classifications—direct, where the action yields quantifiable reduction in air pollutants, and indirect, where the action does not relate directly to a quantifiable reduction (Table II). Direct controls are the most disruptive to and disliked by the general public. However, their emission reductions can

Table II Transportation Controls

I. Direct transportation controls
  A. Motor vehicle restraints
    1. Auto-free zones
    2. Selective vehicle entry permits
    3. Motorcycle limitation
    4. Heavy duty vehicle ban
    5. Restricted delivery times
  B. Gasoline limitations
  C. Motor vehicle emission reduction
    1. Inspection and maintenance
    2. Retrofit installation
    3. Gaseous fuel conversion
    4. Idling limitations
    5. Gasoline vapor controls
II. Indirect transportation controls
  A. Reduction of vehicle miles traveled
    1. Existing mass transit system improvement and expansion
      a. Bus lanes on city streets and freeways
      b. One way streets for buses only
      c. Park ride
      d. Cost reductions
      e. Service improvement
      f. Additional buses
    2. Regulation
      a. Parking policy
      b. Road user tax
      c. Parking surcharge
      d. Gasoline tax increase
      e. Car pool lanes
      f. Four day, forty hour week
    3. Voluntary approaches
      a. Car pool matching system
      b. Car pool incentives
      c. Staggered work hours
      d. Bicycle lanes
      e. Mass transit incentives
      f. Partial vehicle restriction
  B. Traffic flow improvements
    1. Traffic operations program to increase capacity and safety
    2. One way street operations
    3. Traffic responsive control
    4. Loading regulations
    5. Pedestrian control
    6. Reversible lanes and streets
    7. Parking restrictions
    8. Driver advisories
    9. Ramp control on freeways
    10. Signal progression
    11. Street and highway design improvement

be accurately predicted and technically verified. Attendant problems of direct controls, such as equity, regressiveness, and social and economic disruption must be balanced against the uncertainties of the indirect approach to produce a viable plan.

Indirect controls are trial and error, based on past experience. Many jurisdictions have already implemented transportation controls for other purposes, such as for reduction of congestion, through traffic flow improvements or other noncapital alternatives (Table III) (*3*). Four measures were utilized in more than 10% of all 296 United States urban areas reporting—staggered work hours, increase in central business district (CBD) daytime parking rates, unrestricted entry of taxis, and restrictions on curbside truck loading and unloading in congested areas. Use varies greatly with urban area size. Generally, smaller urban areas use fewer alternatives than larger areas. Major difficulties with indirect controls arise because results observed in one geographical area are not necessarily relatable to results in another area. Each metropolitan area has specific qualities that make data transition difficult. Therefore, indirect controls based upon human experience must be constantly monitored and adjusted to obtain maximum effectiveness.

Transportation controls have been studied extensively to determine their costs and effectiveness (*4–8*). The most significant observations are

Table III Noncapital Transportation Control Alternatives Presently Practiced in United States Urban Areas

	Percent of urban areas with program		
Alternative	*296 areas greater than 50,000*	*14 areas over 2,000,000*	*117 areas 50,000–100,000*
Staggered work hours	25.7	35.7	18.8
Measures to encourage car pools	6.8	14.3	6.8
Banning private autos from central business district	0.7	0	0
Increase in central business district daytime parking rates	19.6	28.6	15.4
Lower transit fares during off-peak	3.7	42.9	0
Unrestricted entry of taxis	21.6	50.0	21.1
Unrestricted entry of jitneys	9.8	7.1	9.4
Reserved bus lanes	7.8	50.0	1.7
Restrictions on curbside truck loading and unloading in congested areas	51.0	85.7	39.3
Evening truck deliveries in central business district	9.1	21.4	7.7

that voluntary controls require disincentives to make plans work and that the human response to indirect controls cannot be adequately predicted. This reemphasizes the fact that a trial-and-error approach is necessary to develop a workable plan that minimizes adverse impacts. A detailed investigation of each metropolitan area where transportation controls are required is necessary to develop a good initial plan. Examples are the transportation control strategy documents prepared for fourteen United States cities (*9*) in which air quality and emissions were determined and projected to target years; growth projections and land use were considered. Candidate control measures were evaluated as they apply to these fourteen urban areas; obstacles to implementation were evaluated for each proposed strategy; the percent reduction required and that attainable from each measure were determined using a suitable model, and a strategy yielding the total required reduction formulated.

When a transportation control strategy is selected there are many factors to consider. The type and severity of the problem is perhaps the most basic. Many controls are pollutant specific. For example, an auto-free zone for a central business district may work well for a carbon monoxide hot spot, but would have little effect on an area-wide oxidant problem. The amount of reduction required will dictate the strategy or strategies selected. For example, an area requiring only slight reductions to attain air quality standards might apply one or two control measures of only slight impact, while a highly polluted area might require many control measures including those with severe social and economic impacts, such as gasoline limitations. The availability of control measures must be considered within the time frame required for compliance. Some measures can be adopted and implemented in a short time, such as gasoline limitations, while others require longer times and possible extensions for compliance. Development, distribution, and installation of retrofitted devices on automobiles is an example of a control measure that takes a longer time to implement.

Local control activities must be evaluated to determine if they are flexible enough to adjust to the transportation control strategy. It may be necessary to shift priorities or acquire new staff and additional funds. Local agencies may not have adequate legal authority to carry out the provisions of the plan, and time may have to be allowed to obtain it. State, provincial, and local desires and preferences are important if a plan is to succeed. Many measures are based upon voluntary compliance and public cooperation. Therefore, public comments should be solicited either through the hearing process or invited public comment and incorporated in plans where appropriate. Full recognition should be given to socially and economically disruptive impacts. Adverse impacts must be minimized for efficient implementation of the plan. This can be accomplished by attaching highest priority to those measures having least impact.

In promulgating transportation control plans for the several states (Table IV), the United States Environmental Protection Agency used the following priority schedule to minimize adverse impacts: First, stationary sources were reexamined to determine if additional reductions were technologically feasible. In many cases, additional hydrocarbon controls were possible, while it was determined that additional carbon monoxide controls were not. Second, it was found that a 3–10% reduction in vehicle miles traveled was possible in most regions by 1975. This moderate reduction was combined in some instances with a limited inspection and maintenance program. The third priority was to increase vehicle miles traveled reductions to from 10 to 20% with a 2 year extension until 1977.

Table IV Transportation Controls Approved or Promulgated by United States Environmental Protection Agency, December 1973

State plan	*Inspection–maintenance*	*Catalytic retrofit*	*Other retrofit*	*Pricing policies*	*Gaseous fuel*	*Traffic flow improvements*	*Parking restrictions*	*Mass transit improvements*	*Motor vehicle exclusion*	*Bus–car pool locator*	*Gasoline limitations*	*Motorcycle emission regulations*	*Idling limitations*	*Bicycle lanes*	*Bus and car pool lanes*	*Mass transit incentives for employees*	*Miscellaneous measures*
Alaska	×	×	×			×	×		×				×				
Arizona	×	×	×		×		×			×							×
California	×	×	×	×		×	×			×	×	×			×	×	
Colorado	×		×			×	×	×			×				×		
District of Columbia	×	×	×				×	×		×				×	×		
Illinois	×					×	×										
Indiana	×		×				×			×					×	×	
Maryland	×	×	×				×	×	×	×	×			×	×	×	
Massachusetts	×	×	×	×		×	×			×							
Minnesota						×	×	×									
New Jersey	×	×	×				×		×	×	×				×	×	×
New York	×	×	×						×								
Ohio	×																
Oregon	×					×	×	×									
Pennsylvania	×						×			×				×	×		
Texas	×		×			×	×	×		×	×				×	×	×
Utah	×		×				×		×					×			
Virginia	×	×	×				×	×		×				×	×		
Washington	×		×				×	×		×				×			

It was determined that vehicle miles traveled reductions greater than 20% would be unreasonable in most metropolitan areas. Catalytic converter retrofits, because of their inherent operating problems and high cost were the fourth priority. The last priority item was gasoline supply limitation. In areas where other combinations of control measures would not achieve the standards, gasoline supply limitation would be instituted in 1977 to make up the difference. Control measures were selected within these priorities to achieve a balanced incentive–disincentive transportation control plan that would achieve the air quality standards within the short term. The 1975 and 1977 controls were not instituted as promulgated by the United States Environmental Protection Agency. Local preferences strongly influenced the changes.

The degree to which transportation controls will be necessary to achieve and maintain air quality standards will be a function of several factors, such as the extent to which the automobile industry produces and markets a clean car, the time the car will actually remain clean, the accuracy of vehicle usage projections and future operating characteristics in urban areas, and the accuracy of growth projections.

B. Transportation Policy

National actions can provide impetus to state, provincial, and local programs by making it politically easier for them to act. National policies can affect land-use and transportation planning directly; e.g., indirect source regulations *(10)* will affect major facilities, such as highways, parking lots, and shopping centers. In this case, each state or province must determine if the proposed facility will generate sufficient automobile traffic to exceed the national ambient air quality standards. This type of regulation is necessary because, in addition to transportation directly consuming a great deal of land, it also induces development that consumes additional land and generates additional traffic. Long-term planning for the transportation system and the area influenced by it will prevent significant deterioration of air quality and provide for maintenance of standards.

The United States Department of Transportation has approached the problem by requiring regional transportation plans. The implementation of these plans plays a significant role in determining the shape and character of the metropolitan area. Initially, these plans were highway dominated, which led to formation of the suburbs, increased vehicle miles traveled and excessive air pollution. Later, the plans began to become more truly comprehensive by giving realistic assessments of mass transit systems.

Development of the United States interstate highway system provides an example of inadequately conceived national goals having a profound effect on the environment. The transportation planning concepts of the 1950s are incompatible with goals of the 1970s. Routing of superhighways through metropolitan areas has produced just the opposite of clean air goals. Circumferential by-pass routes designed to divert traffic, and its attendent congestion, from the central city stimulated development by opening up new areas. Lack of mass transit planning compounded the problem. Trip demand still focused toward the central city. Programs were instituted to revitalize the central city. Revitalization programs were intended to benefit by shortening the work trip and reducing vehicle miles traveled. However, most revitalization development was new commercial space and civic centers, which attracted more commerce and tourism to the central city, with the displacement of still more city dwellers. The trend to live in the suburbs and work in the city continued unabated.

C. Mass Transit

Mass transit is consistently offered as a potential solution for urban air pollution problems. Selection of a mass transit system to serve an area will ultimately change land-use patterns. Different modal splits between highway use and bus or rail service will produce different land uses. For example, additional free space may be available under or near elevated or surface mass transit systems that could be used for recreational purposes. This "new" land could also be used for special urban needs. A definite bonus in aesthetics could also be achieved by making some of this land a park. Improved environmental conditions and reduced air and noise pollution may improve neighborhoods through improved housing and new commerce and industry attracted to the proximity of the transit line, which in turn may attract additional vehicular traffic that offsets any reduction in automobile traffic due to transit system use. The accessibility of areas served by the mass transit system will open them for further development.

Modal mix is important in the overall evaluation of a transit system, because environmental factors are directly related to it. As the mix varies, air pollution levels from transportation and related sources will change. There is an optimum mix that will minimize air pollution. However, this mix may not be acceptable when nonenvironmental factors are considered. Using air quality standards determined through evaluation of health, vegetation, and material damage and aesthetic factors, the planner can work toward the optimum modal mix. Assuming that total passenger volume will remain constant for a given time period, the mix can be varied, and it can be determined if air quality standards are met. It would be desirable, from an economic standpoint, not to require pollutant levels to drop below the standard in any part of the region, because strict attainment would represent minimum control costs. However, because emission rates vary for different pollutants and sources, air quality levels can not exactly equal the standard in all parts of the region. Some areas will have to be better than the standard. Trade-offs between convenient, unmaintained, highly polluting vehicles and high-cost or inconvenient mass transit may result in the decision being in favor of the highly polluting, convenient vehicle. The planner must evaluate the feeling regarding these trade-offs of the potential population of transit riders and the relation of demographic changes in the area near the transit system and changes on auto use and transit ridership before reaching and/or implementing a decision.

Mass transit must be subsidized if it is to survive as a vital community service. In most countries, it has not benefited from a subsidy to the same extent as has the private automobile. If mass transit were offered on a completely free basis, the annual subsidy per passenger for the peak-hour transit user would be far below that being given the automobile user *(11)*. In many jurisdictions, this basically unfair competition has prevented the development of a rational fare schedule for rapid transit systems. There are certain gains to be derived from having free service, such as savings in collection costs, improved service, and reduction of highway congestion. Operating revenues have to be obtained from other sources, such as special regional taxation. This would make it possible to spread the costs over the entire regional population and would provide maximum utility for the entire region because of the advantageous externalities, such as reduction in noise levels, air pollution, and congestion.

Solutions that force the majority of commuters to switch to mass transit may be too costly for the financially disadvantaged and may well so deplete their income as to prevent their operation of a car, which may be their sole practicable means of transportation. From current mass transportation studies, such as in the Washington, D.C. metropolitan area, it is evident that proposals for transit lines usually are not primarily to accomodate the disadvantaged. It is a paradox that mass transit, the most economical means of transportation for the poor, has through its inefficient distribution system made the poor more reliant on automobiles for transportation. Once having established this reliance, it now seems necessary to impose prohibitive restrictions on automobile use to reduce disadvantageous externalities, which may restrict the mobility of the poor still further. The collapse of effective mass transportation systems also has a crippling effect on the mobility of children and the elderly. This places an additional burden on the family automobile with its inherent problems.

An argument for subsidization of mass transit is that it requires a lesser subsidy than is at present provided for highways. The most often heard objection to subsidizing mass transit is that it would lead to inefficiency and would tempt labor to intensify demands for higher salaries. Labor costs already account for a sizeable portion of transit revenues. This is a cost that can be effectively reduced. However, it is anticipated that cost reduction through labor reduction might be difficult to attain.

A self-sufficient transportation system must meet competitive standards. Mass transit cannot pay for itself out of the fare box. Subsidies may be applied through government ownership of the transportation system or of some of its assets, with operating deficits made up from the general tax fund, or by allowing these assets to remain in private hands with government reimbursing the firm for losses. There are many gains in costs of transportation that are not readily stated in pecuniary terms, because by their very nature, transportation systems confer substantial benefits to nonusers. It follows that it is most desirable to recognize and consider the social benefits and costs of urban transportation systems in addition to their strictly pecuniary factors. Because the problem is regional by nature and in many instances crosses state or provincial lines, the national government should play an active role in working toward a solution. Coordinated transportation planning for metropolitan areas requires a regional political agency or authority having jurisdiction over the transportation planning and the administration of such plans irrespective of municipal or other political boundaries. Even if a transportation system is revenue producing, there is merit in a regional plan to provide the operating agency with a continuing source of public funds to assist in financing improvements to produce maximum community benefits.

Mass transit systems are operated in most large metropolitan areas. Unfortunately, most are inefficient and/or inadequate and do not meet the needs of the commuting public. Many attempts have been made to develop new and better rapid transit systems. In the United States, one of the most advanced systems for rapid mass transit is the San Francisco, California, Bay Area Rapid Transit District. Rapid transit planning evolves over long periods of time. The San Francisco Bay Area Rapid Transit Commission was created by the California legislature in 1951. All aspects of the mass transit problem were investigated, including in-depth studies of other public transit agencies, transit planning in other metropolitan areas, general mass transit trends, finances, engineering, and organization. These were then related to developments in the Bay area. It then took over twenty years for parts of the system to become operational. Clearly, innovative and efficient rapid mass transit systems are long-term projects. Therefore, they will maximize utility in maintaining air quality standards in the long term when population growth becomes more significant.

D. Highway Congestion Costs

In general, the basic difficulty with urban transportation is the economy of scale. It is a general rule that transportation is a decreasing cost industry, i.e., the more vehicles that use a highway, the lower the cost per person mile or per ton mile of use will be. A freeway that remains largely unused represents a major subsidization of a few by the many. On the other hand, this general rule may not hold true within highly urbanized areas. The automobile has been called the greatest generator of economic externalities the world has ever known. Some are positive, such as jobs and mobility, but the negative ones, such as air pollution, accidents resulting in death and injury, property damage, noise, loss of time, and aggravation, to mention a few, lead to the conclusion that transportation is not a decreasing cost industry in urbanized areas.

Highway congestion costs are another externality. Lost time and congestion costs are difficult to determine. Highways are designed for an optimum number of vehicles per road mile per lane per hour. In most metropolitan areas, these design figures are exceeded. It can be assumed that motorists impose congestion costs on each other and therefore they balance out. This assumption is not entirely valid, because some individuals value their time more than others. While one additional motorist on an already congested highway may add an additional amount of trip time to all users, two additional motorists may increase trip time exponentially. If travel costs for work trips do not include congestion costs, a true evaluation of competing transportation modes, such as mass transportation, is not obtained. It would appear that for the public to fully account for all the alternatives of competing transportation modes and to put them on an equitable basis, either the cause of congestion, the automobile, must be taxed in some way to pay for the decrease in utility the region suffers, or competing mass transportation must be subsidized to credit it with the decrease in congestion cost it provides.

There have been many ideas suggested on how to equitably charge a motorist for his use of the urban streets in times of peak congestion. In most countries, taxes are collected that pay part of the construction and maintenance costs of the highway system, but there is no tax for externalities. Some suggested ideas are prepaid permits to allow entry into the urban area during the rush hour, traffic counters or scanning devices to tie each automobile into a computer for monthly billing, high taxes on parking lots, auto-free zones, and mandatory car pools. All these suggestions are practical. However, the cost of their implementation may be as great as the revenue they would collect. Congestion tolls, if sufficiently expensive, may reduce automobile use in favor of less costly systems with fewer disadvantageous externalities. However, this should not be considered an effluent fee, allowing the payer the right to emit air pollution or generate noise.

Most metropolitan areas have had the idea that the best approach to reduced congestion is more and wider freeways to accommodate the ever increasing load of peak hour vehicles. It has been estimated that registration of motor vehicles in the United States will approach 150 million by 1985, and some estimates have indicated 200 million vehicles will be on the road by the year 2000 (*12*). Left to itself, the peak hour traffic problem can get no better. The question arises as to what capacity peak hour load should be provided for. There has been a suggestion that roads be built to have a capacity sufficient to handle the traffic level of the thirtieth busiest hour during the year.

In an analysis by Vickrey (*13*)—based on construction costs for grade-separated expressways per lane mile ranging from about $300,000 in favorable rural locations with few grade separations or building demolitions to from $1 to 5 million or higher in high-density cores of metropolitan areas—peak hour costs per land mile amounted to $100. With an average headway of 2 seconds, 1800 cars per lane per hour would cost 5.5 cents per car mile. User taxes of all kinds contribute only 1 cent per car mile. The full costs should not be charged to the peak hour user, even though the system would operate more efficiently without him. Certain beneficial externalities result to the off-peak user, such as greater safety due to extra lanes and the ability to get downtown much faster.

Highway users are subsidized in another, less obvious way. Accident costs on United States highways were estimated for 1967 at from $8 to 12 billion, depending, in part, on the extent to which allowance is made for pain and suffering as distinct from strictly monetary loss (*14*). Many parts of the community besides the highway user bear this cost through items such as medical insurance, loss of wages, and payments for social security and welfare. These costs may result in an additional subsidy to highway users of from $1 to 3 billion per year (*14*).

E. United States Federal Legislation

In the United States, transportation planning was first stressed as part of the 1962 Federal-Aid Highway Act, which called for the establishmest of a continuing, cooperative, comprehensive transportation planning process in every urban area with a population greater than 50,000. Commonly referred to as the "3-C process," it is a prerequisite for federal highway construction grants. The Highway Act requires the development and promulgation of guidelines to assure that highway construction is consistent with air quality implementation plans. It also establishes the means to increase the efficiency of highways in transporting people rather than cars through funding projects, such as exclusive bus lanes. The United States Clean Air Amendments of 1977 direct that grants for developing and maintaining vehicle inspection systems shall not be made to states unless the Department of Transportation certifies to the Environmental Protection Agency that such an inspection program is consistent with their highway safety program. This transportation environmental interaction at the federal level should stimulate the cooperation of transportation and air quality agencies on the state and local levels, as well as promote research on transportation–air quality relationships.

Section 204 of the Demonstration Cities and Metropolitan Development Act of 1966 makes comprehensive planning agency review a requisite for a wide variety of public facilities including highways. Its outgrowth, Part I of the Intergovernmental Cooperation Act of 1968, mandates the coordination of single-objective programs and policies to reduce overlap and conflict. Bureau of the Budget Circular Number A-95 (July 24, 1969, and subsequent revisions) established what is commonly referred to as the "A-95 review process" as a means of implementing the Intergovernmental Cooperation Act of 1968. Rules and regulations governing the formulation, evaluation, and review of federal programs and projects having a significant impact on area and community development were promulgated. The main thrust of the program was to establish state, regional (nonmetropolitan), and metropolitan clearing houses to give advice and assistance in comprehensive planning as it relates to potential federally funded projects. Section 701 of the Housing and Urban Development Act of 1954 established the comprehensive planning assistance program that still serves as the basis for most regional planning programs. The Department of Transportation has a program of intermodal planning, and related actions, including in addition to mass transportation and highways, airport development as provided for in the Airport and Airways Development Act of 1970.

With the passage of the United States Environmental Policy Act of

1969, the Congress set forth for the first time a broad national environmental policy calling for the alignment of all major proposed federal legislation, plans, and programs with a national commitment to improved environmental quality. Part of the act sets forth the rationale for this policy:

> The Congress, recognizing the profound impact of man's activity on the interrelations of all components of the natural environment, particularly the profound influences of population growth, high-density urbanization, industrial expansion, resource exploitation, and new and expanding technological advances and recognizing further the critical importance of restoring and maintaining environmental quality to the overall welfare and development of man, declares that it is the continuing policy of the Federal Government, in cooperation with State and local governments, and other concerned public and private organizations, to use all practical means and measures, including financial and technical assistance, in a manner calculated to foster and promote the general welfare, to create and maintain conditions under which man and nature can exist in productive harmony, and fulfill the social, economic, and other requirements of present and future generations of Americans.

Federal agencies are required to perform specific actions to implement this policy, for example, "include in every recommendation or report on proposals for legislation and other major Federal actions significantly affecting the quality of the human environment, a detailed statement by the responsible official on: (1) the environmental impact of the proposed action; (2) any adverse environmental effects which cannot be avoided should the proposal be implemented; (3) alternatives to the proposed action; (4) the relationship between local short-term uses of man's environment and the maintenance and enhancement of long-term productivity; and (5) any irreversible and irretrievable commitments of resources which would be involved if the proposed action should be implemented." By introducing transportation and environmental considerations into the planning process and forcing interagency coordination and solicitation of public comment it is anticipated that environmental crises can be avoided.

As originally passed, the Urban Mass Transportation Act of 1964 limited federal assistance for mass transportation to unified urban transportation systems that were part of a comprehensive plan for desirable urban development. In keeping with the policy of the National Environmental Policy Act, the Urban Mass Transportation Act of 1964 was amended in 1970 to include environmental protection. Section 14 sets the policy:

> It is hereby declared to be the national policy that special effort shall be made to preserve the natural beauty of the countryside, public park and recreation lands, wildlife and waterfowl refuges, and important historical

> and cultural assets, in the planning, designing, and construction of urban mass transportation projects for which Federal assistance is provided pursuant to section 3 of this Act. In implementing this policy the Secretary shall cooperate and consult with the Secretaries of Agriculture, Health, Education, and Welfare, Housing and Urban Development, and Interior, and with the Council on Environmental Quality with regard to each project that may have a substantial impact on the environment.

In addition, the Act requires that transportation planning be coordinated through a regional planning agency to consider all aspects of long-range transportation planning in order to determine what alternatives will be best for their region. Knowledge of the air pollution impact of these alternatives helps the planners reach their decisions. The United States Clean Air Act (1977) impacts directly on the design and operation of transportation systems through several provisions, including revised emission standards for motor vehicles; state air quality implementation plans that may include land-use and transportation controls to achieve national standards; fuel additive regulation; inspection, maintenance, and retrofit programs for in-use motor vehicles; and emission standards for aircraft.

The Clean Air Act also directs each state to develop an implementation plan that will demonstrate how air quality standards will be achieved. Transportation controls are required in many States because stationary source controls and the Federal motor vehicle pollution control program alone can not attain the national ambient air quality standards. Some of these controls (Table IV), such as parking restrictions and surcharges were a preemption of local land-use decisions. Many of these actions were challenged, and plans were changed to be more responsive to local desires. Once established, the legislative framework provides for implementation plan modification to keep pace with technical innovations and changing public attitudes.

The Federal-Aid Highway Act of 1973 for the first time allowed a portion of the highway trust fund to be used for nonhighway purposes. As much as $200 million could be used for bus-related improvements during 1975, with additional funds available for rail transit improvements starting in 1976. Also provided by the Act was an additional $3 billion for the existing Urban Mass Transit Capital Grant Program. The urban roadway system was expanded to include arterial and collector routes. This increased the percentage of total urban roadway mileage eligible for federal aid. Without adequate planning, the increased federal aid for highways could create additional air quality problems. To compensate for this, one-half of one percent of apportioned Federal-aid system funds was made available to regional planning agencies. The act, however, continued to approach the environmental problems caused by high density with encouragement for construction of new highways. Recognizing the need for additional nonhighway approaches, $40 million was made available annually for bicycle path construction in conjunction with highway projects under a bicycle transportation and pedestrian walking program. Through the act, the first step in providing a legislative basis for expanding mass transportation systems in the United States was made. Operating expenses were not included.

Congress has sought to increase the federal share of cost for comprehensive planning and to allow program funds to be used for management activities. The trend is to require comprehensive planning activities to include housing and transportation elements and to give grants for modernization and revitalization of state and local governments, including increased capital expenditures and operating expenses for mass transit systems. Consolidation of mass transportation and highway funding into one unified program would promote rational transportation systems in urban areas.

The National Environmental Policy Act, the Clean Air Amendments, the Urban Mass Transportation Act, and portions of the Federal-Aid Highway Act represent significant points in the evolution of legislation aimed at protecting the environment of the United States from adverse impacts of transportation. Together, these acts establish a framework for the proper inclusion of environmental objectives in the transportation planning process. With such a complex task, no single piece of legislation can be viewed as static. The necessity for readapting an approach that may be taken is fully contemplated in the Clean Air Amendments, which require the revision of state implementation plans and air quality standards when appropriate.

III. Land-Use Planning for Air Pollution Control

A. Zoning Codes and Performance Standards

In the past, common law nuisance doctrine in effect regulated land use to protect individuals from impacts originating from another's property. Initially, environmental problems were approached with this doctrine. Exclusionary practices for entire activities resulted. Protection of the individual, his property, and his economic opportunity required innovative approaches in land controls, because economic growth and development were synonymous with environmental problems. Land-use zoning was the initial attempt to accomplish this. However, because exclusionary practices developed for a better living environment were restricted to highest and best use zones, a proliferation of environmentally unsound practices were allowed to develop in less desirable areas. A shift away from prohibitions in zoning codes to permitted uses took place to better control the intrusion of unexpected undesirable land use. The ubiquitous nature of air pollutants required more than physical separation to protect the individual. In the early 1950s, performance standards were introduced into some zoning codes in an attempt to establish quantitative criteria for identifying potential problems. However, early standards were largely based on aesthetics or qualitative decisions.

B. Land-Use Controls

The limitations of the assimilative capacity of the environment must be identified for long-term planning. Once this is established, programs and techniques, such as land-use controls, can be developed that will prorate the capacity equitably among desired societal activities. Land-use controls do not have to be regulations. A government may influence the location of polluters and receptors through advice, economic incentives or disincentives, and acquisition and development of public land and facilities. Examples of land use controls include permit systems for review of new stationary sources, zoning regulations that segregate industrial and residential land uses, restriction by easement or purchase of land, and receptor location control. Land use controls also include emission-density zoning and emission allocation regulations.

Emission density for a given land area can be directly related to emission controls. Emission controls alone may be insufficient to attain and/or maintain desired air quality as new sources enter an area already congested with individual sources. New source performance standards and maximum control technology alone can not assure maintenance of desired air quality levels unless the emission standard is zero. Therefore, the emission density of a given area cannot be allowed to increase indefinitely. An emission density limitation must be established that will reinforce existing performance and emission standards. Possible alternatives might be prohibition of any new emission sources through a construction ban or reliance on progressively restrictive source controls. Both these alternatives would be expected to meet with considerable resistance. Emission limitations can be used in place of the source list in many zoning codes that prohibits certain activities by category from specifically zoned areas. Emission-density zoning can be defined as a regulatory system in which the maximum legal rate of emissions of air pollutants from any given area is limited by the size of the area and its zoning classification. Emission-density limitations, when actively enforced, are similar to zoning ordinances. There is opposition to emission-density zoning, because it places air quality as a significant constraint on land development. Significant questions arise; e.g., What is the basis for area selection? How are limitations equitably enforced? When should limitations be instituted vis-à-vis increased point source controls? Emission limitations may be a taking of personal property without just compensation.

Emission-density zoning can be utilized to maintain desired air quality in the long-term through limitation of additional emissions from individual point sources in planning areas. It can deal with both localized air quality problems and large-scale regional problems. Future growth and development can then be directed in a manner to avoid exceeding air quality goals through the establishment of maximum, rather than minimum, acceptable pollutant levels. As stationary sources congregate in an area, the interrelationship between land-use planning and air quality maintenance becomes more critical. Emission-density zoning can strengthen this interrelationship through its use as a land use constraint for environmental control. For example, new facilities could be examined within the context of presently existing facilities, and benefits could be derived from locational controls and open space planning to provide for pollutant dispersion prior to reaching receptors. Emission-density zoning regulations were included in the Cook County, Illinois, zoning ordinance. They were never effectively applied and were subsequently repealed.

Emission allocation procedures can be defined as a regulatory system that assigns a maximum legal emission rate to any given political jurisdiction or other area with provisions for sanctions, such as a construction ban to ensure that the areas' allocation is not exceeded. Emission allocation procedures are generally applicable to all pollutants from both stationary and mobile sources. In an allocation procedure, the distribution of pollutants within the area is not of critical concern. Therefore, jurisdictions are not constrained as how to distribute their allocation. Local governments tend to control their growth as they define their best interest. Emission allocations will be more equitably distributed by local governments in concert with their growth decisions. Emission allocation procedures are independent of predefined land use and will predefine air quality at any desired future date. Sources can then be allocated emissions within that time frame to assure maintenance of goals. Emissions can be allocated by area size and pollutant type. This allows for the development of a long-term maintenance plan that is responsive to the needs of any given region.

Emission allocation procedures have found limited application. They may enable existing source control regulations to be effective for a longer time and serve as an additional control tool for long-term maintenance of air quality goals. The question arises however, as to equitable distribution of the allocation. If new sources are allowed to move into an area that is at the emission lid, or has used its allocation, existing sources would be required to further reduce their emissions to maintain the area's allocation. It is reasonable to assume that existing controlled sources would resist reducing emissions further to allow competitors to move in. The potential for "hot spot" control of stationary sources can make this an effective tool for air quality maintenance.

C. Housing and Land Development

Some of the most severe air pollution problems today in the United States are the result of past federal policies. This is especially true of housing policies. Beginning in the 1930s and lasting until the 1950s, the federal government sought to make home ownership available to many who could otherwise not afford it. Many new institutions were created for this purpose or had this as one of their major goals. Among these institutions were the Federal Housing Administration, the Veteran's Administration, and the Federal Home Loan Banks. Federal tax incentives and mortgage policies represented a substantial subsidy to homeowners. This distortion of the housing market stimulated urban sprawl and has indirectly discouraged higher-density residential development. Coupled with ready credit made available by the above-mentioned institutions, a new pattern in urban growth was made possible. The individual home and a piece of ground became a national goal and metropolitan areas grew through massive development of suburban areas with the attendant decline of the central cities.

In retrospect, these policies proved to be an environmental disaster. Not because the concepts were bad per se, but because the problems were not anticipated. As development occurred, no consideration was given to the simultaneous development of efficient transportation systems to serve the needs of the multitude of new communities being developed. Instead, almost total reliance was placed upon the automobile. This points up the basic consideration that eventually led to the development of the National Environmental Policy Act; that is, major federal actions have a profound effect on the environment, and this effect must be fully examined and the consequences understood so that decisions can be reached that are truly consistant with national goals and policies. The Housing and Urban Development Act of 1968 made a minor inroad into the suburbanization trend by making available federal assistance for rehabilitation of existing housing. Effective land-use programs for the inner city must upgrade rather than abandon areas as has been done in the past.

As metropolitan areas grow, demands for expanding the infrastructure increase. Water and sewer lines are extended, new water and sewage treatment facilities constructed, and new land areas subdivided and developed. This growth is made possible by large expenditures of federal funds, especially in the case of municipal wastewater treatment. Over the sixteen year period following the passage of the Federal Water Pollution Control Act of 1956, construction grants to states and municipalities for wastewater treatment facilities rapidly increased. Expenditures in 1972 were approximately 466 times greater than those of 1957, with more than $1.8 billion being spent over the entire period (Table V) (*15*). Combined with sewerage systems, these grants represent the largest single item budgeted for pollution abatement during this period. Where these facilities were planned as part of an orderly growth and development plan, uncoordinated sprawl was prevented. However, where they followed haphazard growth and development, they contributed to the problem.

It is essential that programs impacting on the environment be coordinated. The review of individual projects on a case-by-case basis is inadequate to even begin to comprehend their magnitude. Planning must be undertaken at the level where the problem occurs and on a scale large enough to include the entire system. A start was made with the establishment of the A-95 clearing house to coordinate required planning. The Department of Housing and Urban Development gave great impetus to effective planning with the requirement for comprehensive regional planning as a prerequisite for receiving federal funds.

Table V Expenditures for Municipal Wastewater Treatment-Works Construction Grants

Fiscal year	*Expenditures (dollars)*	*Fiscal year*	*Expenditures (dollars)*
1957	844,000	1965	69,755,000
1958	16,884,000	1966	81,479,000
1959	36,429,000	1967	84,176,000
1960	40,295,000	1968	122,109,000
1961	44,085,000	1969	134,530,000
1962	42,103,000	1970	176,377,000
1963	51,738,000	1971	478,366,000
1964	60,432,000	1972	413,407,888
		Total	1,859,309,888

D. Land-Use Planning Policy

Land-use planning must be examined at four levels of government—national, state or provincial, regional, and local—to adequately insure long-term protection of the environment. Broad national land-use policies should be set that will encourage development that will protect the environment. State and provincial growth and population distribution policies should be consistent with both national priorities and local desires. Choices must be made as to where growth should occur; e.g., should there be trade-offs between growth of presently existing medium-size cities and the creation of new towns? Governmental programs usually begin with parochial problems, such as coastal zoning; desert, mountain or other recreational development; power plant siting; mineral extraction; and water resource development. This is generally true because such initial programs attempt to rectify past mistakes in land-use decisions, such as allowing massive building in ecologically fragile areas like flood plans or coastal wetlands.

It has been suggested that national sanctions, such as withholding funds from projects that stimulate growth and development without an effective state or provincial land-use planning program, should be applied. A site-by-site approach cannot adequately deal with the problem. The multiplicity of local governments in a given region cannot effectively solve regional environmental problems. In contrast, significant local problems pertaining to land use are generally overlooked by large regional planning agencies. Large agencies can evaluate environmental problems unique to the region and hence determine environmental impact at an early stage in the planning process. Regional agencies can coordinate land development and density patterns by supervising the spatial arrangement and location of new construction, land use, and transportation facilities.

The disaggregation of the metropolitan region through development of independent and rival jurisdictions compounds problems in effecting regional solutions. For example, in the United States, the Saint Louis, Missouri, standard metropolitan statistical area contains approximately 500 local governments, of which about 93% have taxing power (*16*). Local governments have generally failed to correct regional air pollution problems because of the lack of a unified approach. In the United States, land-use policies formed in the 1920s still shape local land-use decisions. Regions have the geographic coverage to aggregate area-wide problems. However, in the United States, they generally do not have the regulatory authority to effectively implement their plans and policies, and there is no unit of government that can use a coordinated approach to reach optimum solutions for common regional problems. Regional councils of government and planning agencies have planning responsibility in some areas of regional public interest, such as transportation, highways, and housing, but very rarely do they have the authority to implement and enforce their plans through effective regulation. Special-purpose districts, such as air pollution control districts, generally have specialized functions and limited authority.

Land-use and transportation planning must be interrelated with enforcement of regulations to maintain desired air quality. Area-wide political processes must be developed for elected officials to pursue common goals extending beyond the bounds of local jurisdictions. Air pollution control should be planned for by the same agency that does land-use planning because of the overlapping externalities that are associated with both problems. Planning without the decision-making power contributes little to effective air pollution control.

A change in the institutional makeup of planning regions to incorporate decision as an integral part of the process would aid in solving the problem. A regional agency can best coordinate physical, social, economic, and environmental considerations within the region's projected growth and land-use patterns. This analysis will determine the region's capacity for growth. Generally United States courts rule in favor of actions in the land-use area where these actions are in accord with established land-use plans. Therefore, the inclusion of air quality objectives in land-use plans would clearly define the region's desires in this area. The result would be more effective action and a stronger legal base for achieving air quality goals, which, through the plan, would be placed in proper perspective with other land-use desires, such as space allocation; and activity, intensity, and magnitude of development.

Related land-use decisions, environmental goals, and regulatory actions should be made at the lowest level of government that can effectively work with these interrelated problems, the region. Regional land-use plans are reviewed by local agencies with differing needs. Therefore, their reviews collectively may not work for the best interests of the entire region. Sound development–environment decisions are difficult for local governments to make if they rely on the real estate tax, because it places them in competition for new development. Local land-use actions are also subject to pressures from special interest groups. It is evident that a modification of some traditional governmental structures will have to be made to resolve regional air quality problems.

E. United States Federal Legislation

Land-use planning is significant in maintaining air quality standards. Through federal regulation, states are directed to identify those areas (counties, urbanized areas, standard metropolitan statistical areas) that, due to air quality and/or projected growth rate, may have the potential for exceeding any national standard. After extensive analysis and public comment, the states were required to submit an evaluation of the impact on air quality of projected growth in each potential problem area. If the evaluation indicates that standards may be exceeded during a 10 year period, a plan is required to insure maintenance of the standards. Land-use considerations, such as growth in population, energy, waste and development of residential areas, commerce and industry are part of the evaluation.

States are required to obtain legal authority to ". . . prevent construction, modification, or operation of a facility, building, structure or installation, or combination thereof, which directly or indirectly results or may result in emissions of any air pollutant at any location which will prevent the attainment or maintenance of a national standard" (*17*). Maintenance regulations returned to the states, by federal edict, some land-use controls previously delegated to local governments. State and local interaction is necessary if air quality implementation plans are to be successful. Local desires often go unnoticed or unheeded at the state level. As part of the 10 year maintenance program, long-term changes in land use can be made to optimize community desires for growth, development, and mobility with desired air quality levels.

Land-use planning is a means of placing competing environmental, economic, and social requirements in balance. Traditionally, this process has been delegated by states to local governments. This focus is too narrow. To be truly responsive to needs, the planning process must be broader in geographic scope. A return of some of the land-use planning functions to the state will help. National policies and goals for land use will set the direction for future actions by state and local governments. Without them, national interests, such as environmental protection, will not be met because of local conflicts. Ineffective land-use regulation can develop through misinterpreted or inferred federal policy. On the other hand, land-use regulation works best at the lowest level of effective government due to the wide diversity in land resources and local use choices.

Most states have studied, proposed, or enacted a variety of land-use legislation. Development of state-prepared land-use plans is of primary concern. Land-use controls are essential if the plans are to work. Therefore, enabling laws for state review of local plans or direct state control for critical areas or large-scale development are necessary. State land-use legislation initially concerned problems peculiar to each state. Coastal zone management, wetland development, surface mining, and power plant siting are a few of the special concerns of direct interest to specific states.

Land-use control for air quality and environmental improvement has been brought about by a variety of federal policies and legislation. State and local programs can produce a workable environmental–land-use planning process that will enable all environmental goals to be met.

REFERENCES

1. "Demographic Yearbook, 1962," p. 316. United Nations, New York, New York, 1962.
2. "Demographic Yearbook, 1972," p. 353. United Nations, New York, New York, 1972.
3. "1972 National Transportation Report," p. 247. U.S. Govt. Printing Office, Washington, D.C., 1972.
4. Institute of Public Administration, Teknekron, Inc., and TRW, Inc., "Evaluating Transportation Controls to Reduce Motor Vehicle Emissions in Major Metropolitan Areas," NTIS No. PB 213 374. Nat. Tech. Inform. Serv., U.S. Dept. of Commerce, Springfield, Virginia, 1972.
5. TRW, Inc., "Prediction of the Effects of Transportation Controls on Air Quality in Major Metropolitan Areas," NTIS No. PB 214 176. Nat. Tech. Inform. Serv., U.S. Dept. of Commerce, Springfield, Virginia, 1972.
6. Alan M. Voorhees & Associates, Inc., and Ryckman, Edgerley, Tomlinson & Associates, "A Guide for Reducing Automotive Air Pollution," NTIS No. PB 204 870. Nat. Tech. Inform. Serv., U.S. Dept. of Commerce, Springfield, Virginia, 1971.
7. *Fed. Regist.* **38,** 15194 (1973).
8. J. M. Thompson, "Methods of Traffic Limitations in Urban Areas." Organisation for Economic Co-Operation and Development, Paris, France, 1972.
9. GCA Corp. and TRW, Inc., "Transportation Controls to Reduce Motor Vehicle Emissions in Major Metropolitan Areas," NTIS No. PB 218 938. Nat. Tech. Inform. Serv., U.S. Dept. of Commerce, Springfield, Virginia, 1972.
10. *Fed. Regist.* **38,** 15834 (1973).
11. G. M. Smerk, *Land Econ.* **41,** 64 (1965).
12. U.S. Department of Commerce, "The Automobile and Air Pollution: A Program for Progress–Part II." U.S. Government Printing Office, Washington, D.C., 1967.
13. W. Vickrey, *in* "Contemporary Economic Issues" (N. W. Chamberlain, ed.), p. 205. Richard D. Irwin, Inc., Homewood, Illinois, 1969.
14. W. Vickrey, *in* "Contemporary Economic Issues" (N. W. Chamberlain, ed.), p. 209. Richard D. Irwin, Inc., Homewood, Illinois, 1969.
15. R. H. Sullivan, "Proceedings of the Joint Conference on Recycling Municipal Sludges and Effluents on the Land," p. 1. National Association of State Universities and Land-Grant Colleges, Washington, D.C., 1973.
16. "Census of Governments." U.S. Department of Commerce, Bureau of the Census, Washington, D.C., 1962.
17. *Fed. Regist.* "Code of Federal Regulations, Protection of the Environment," pp. 213–259 (1973).

3

Episode Control Planning

Darryl D. Tyler

I. Introduction

This chapter discusses three different types of air pollution episodes: (a) the atmospherically induced episode, (b) the single-source episode, and (c) the accidental episode. Because each of these episodes has different causes and effects, each must be controlled in a different manner.

The atmospherically induced episode occurs when meteorological conditions, generally temperature inversions, reduce the effective volume of air that is available to dilute pollutant emissions. Pollutant concentrations rise to a level that acutely affects public health. Pollutant emissions during an episode will generally remain constant. What has changed is the volume of air available to disperse these pollutants. Since the weather cannot be controlled, it is necessary to reduce emissions. Short-term emission reduction is thus the objective of the atmospherically induced episode control plan.

The single-source episode develops when a pollution source is physically located so that its emission presents a threat to receptors in the immediate area. Such situations develop when a source in effect overwhelms the ability of the atmosphere to disperse its effluent before it reaches the receptors. The source may be located too close to the receptors; it may have too short a stack; it may have recently increased its production and thus increased its emissions; or it may be experiencing a control equipment breakdown resulting in increased emissions. The single-source episode may be independent of adverse meteorological conditions, and it generally involves only a small receptor area, in contrast to atmospherically induced episodes, which generally cover large receptor areas. Since the point-source episode is an almost continuous threat, long-term control measures or plant closure are the only solutions. Unlike control programs for most sources, it may be necessary to curtail or close the source until long-term controls can be implemented because of its substantial impact on public health.

Accidental episodes are of two kinds—real accidents, where a release has occurred accidentally and where safety measures are taken after the fact, and potential episodes, where an accidental release is possible and safety measures are taken in advance to counter its potential effects. Such events are generally independent of adverse meteorological conditions and are generally due to a prior transportation accident, a process malfunction, or a natural disaster. The air pollution control agency must perform a number of support functions during an accidental episode, even though prime responsibility may be in the hands of another agency, e.g., the civil defense agency.

Although other governmental agencies may play vital supporting roles, the air pollution control agency is the one primarily responsible for developing the control plans and coordinating the implementation of those plans during atmospherically induced and single-source episodes.

In accidental episodes, the air pollution control agency will generally play a supportive role, vital as that role may be. In all three cases, the air pollution control agency's role will be (a) monitoring air quality to define the problem and to confirm dispersion predictions; (b) predicting pollutant concentration to estimate the potential extent and severity of the problem; (c) reducing such emissions as are under its control and informing the public on these matters and on the degree of hazard and the precautions that can be taken by the public to protect themselves.

II. Planning for the Atmospherically Induced Episode

Atmospherically induced episodes can occur anywhere when atmospheric dispersion conditions are insufficient to adequately disperse the quantity of pollutants being emitted. There are areas in any country that are more susceptible than others. For instance, in the United States, the maximum number of air pollution potential days occur along the Appalachian Mountains and in central California (Fig. 1).

However, just because an area has the meteorological potential for episodes does not mean it will have episodes; it must also have the emission potential. Episode-prone areas are areas with both significant sources of pollution and poor meteorological dispersion characteristics. From the information presented in Figure 1 and knowledge of the location of heavily polluted cities, a list of episode-prone United States cities such as Los Angeles, California; Chattanooga, Tennessee; Birmingham, Alabama; Charleston, West Virginia; and Pittsburgh, Pennsylvania can be developed.

A. Objective

There are two guidelines for the design of an episode control plan. First, and most important, the plan must be adequate to protect public health. Second, as more sources are brought under control by long-term control or enforcement efforts, an area's episode potential, hence its need for an episode control plan, will decrease. Due to what appears to be a long-term energy crisis and other economic, social, and political pressures to slow the implementation of long-term pollution control strategies, it is likely that episode control programs will be needed longer than originally envisioned.

B. Plan Requirements

In order to be effective an episode control plan must contain at least the following provisions: (a) legal authority, (b) criteria, (c) an emission reduction plan, (d) a surveillance system, and (e) a communications system.

1. Legal Authority

The need for legal authority cannot be overemphasized. The entire plan is meaningless if the legal authority for its implementation is in-

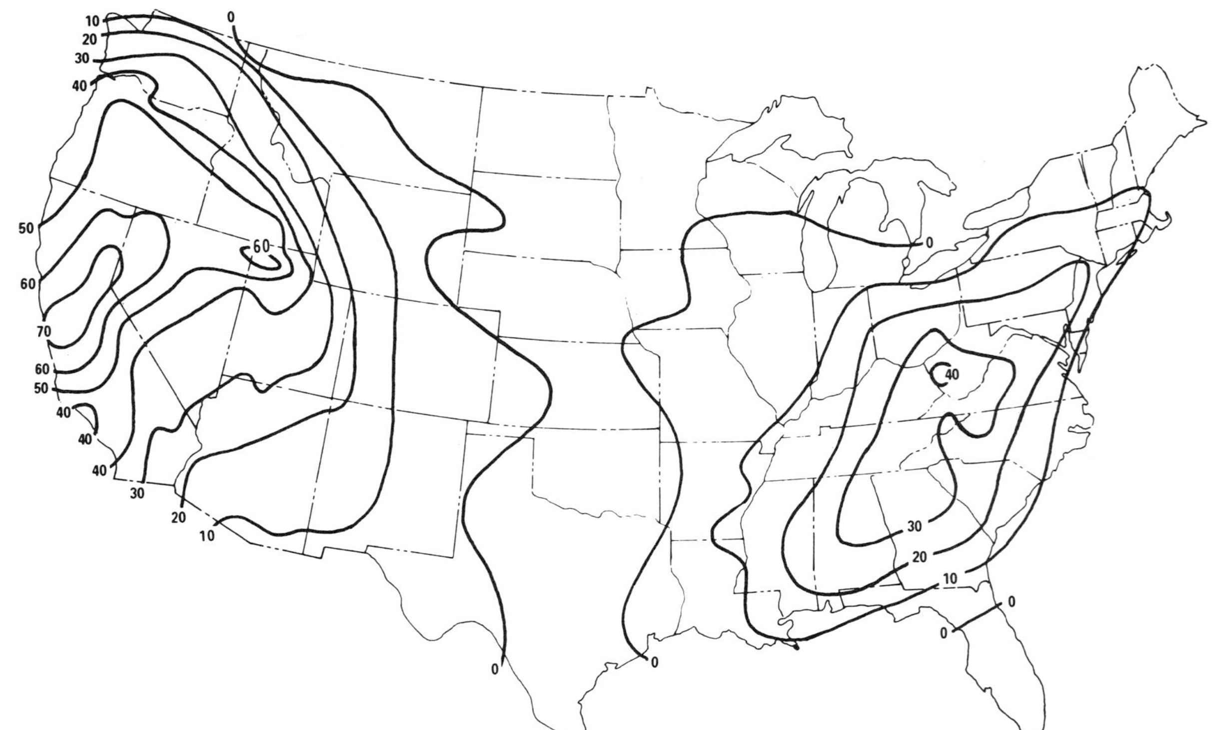

Figure 1. Isopleths of total number of forecast days of high metero logical potential for air pollution in a 5 year period. Data are based on forecasts issued since the program began, 1 August 1960 and 1 October 1963 for eastern and western parts of the United States, respectively, through 3 April 1970. G. C. Holzworth "Mixing Heights, Wind Speeds and Potential for Urban Air Pollution Throughout the Contiguous United States," AP–101, United States Environmental Protection Agency, Research Triangle Park, North Carolina, 1972.

adequate. If the legal authority presently available for implementing an episode control plan is inadequate, special legislation to provide adequate authority must be sought. Legislation should be prepared at the state level permitting the declaration of an air pollution episode by the Commissioner of Public Health, Director of Environmental Protection, or other official with the approval of the governor. The authority should provide the necessary legal power to fully implement an emergency air pollution control plan. The following items should be included in the legislation:

a. Statement of the intent of the legislation to prevent and minimize air pollution disasters that could affect the health of people within the community.

b. Identification of the area of jurisdiction.

c. Identification of enforcement authority; for example, designation of the Department of Health, Department of Air Pollution Control, or other agency as the prime agency responsible for carrying out emergency functions.

d. Specification of the basis upon which the action will be initiated.

e. Specification of the manner in which the specific orders are to be issued and the time period they remain in force.

f. Assignment of responsibilities for development of an episode control action plan, prepared in advance, indicating the specific measures to be taken by owners and operators of major emission sources.

g. Specification of the responsibilities for evaluation of the preventive and abatement measures taken during an episode.

The legislation should clearly stipulate emergency powers, including the authority to fine or take court action in the event noncompliance is encountered. Voluntary compliance schemes should be encouraged as an interim measure but are not adequate substitutes for enforceable episode control programs.

2. Criteria

A system designed to protect the public against the acute effects of high concentration of pollutants must consider: (a) methods of identifying situations where corrective actions should be taken, and (b) the capabilities and strategies for control when action is necessary to prevent further deterioration of air quality. The objective of the episode control program is to prevent levels of pollution from being reached that present an imminent and substantial threat to public health (Table I).

Table I Significant Harm Levels[a]

Sulfur dioxide—2620 μg/m³ (1.0 ppm), 24 hour average.
Particulate matter—1000 μg/m³ or 8.0 COH[b] values, 24 hour average.
Sulfur dioxide and particulate matter combined—Product of sulfur dioxide in micrograms per cubic meter, 24 hour average, and particulate matter in micrograms per cubic meter, 24 hour average, equal to 490×10^3, or product of sulfur dioxide in parts per million, 24 hour average, and COH values, 24 hour average, equal to 1.5.
Carbon monoxide—57.5 mg/m³ (50 ppm), 8 hour average. 86.3 mg/m³ (75 ppm), 4 hour average. 144 mg/m³ (125 ppm), 1 hour average.
Photochemical oxidants—1200 μg/m³ (0.6 ppm), 1 hour average.
Nitrogen dioxide—3750 μg/m³ (2.0 ppm), 1 hour average. 938 μg/m³ (0.5 ppm), 24 hour average.

[a] *Federal Register* **39,** No. 50, Title 40, Part 52, p. 9672 (1974).
[b] COH, coefficient of haze.

Because the emphasis is on avoiding an emergency rather than reacting to one, a series of episode criteria is recommended. A four-stage sequence has been used by most air pollution control agencies, with each stage corresponding to a different degree of severity. The first step is generally activated by an adverse meteorological forecast and serves only to warn the control agency and the public of a potential problem. The next three stages are activated by deteriorating air quality and the prediction of continued poor atmospheric dispersion, with each stage prompting more stringent control measures. The four stages of increasing severity and control are generally designated (a) forecast, (b) alert, (c) warning, and (d) emergency (Table II).

Using a persistence model, Horie (*1*) revealed flaws in currently used episode management criteria like those presented in Table II. Horie found that such criteria tend to delay reaction to an on-going episode to the point that control actions are too late to affect the course of the episode. He stated that since higher concentration levels have shorter persistence, it is not wise to use criteria with the same minimum duration times for each of the episode stages.

Each control step should be designed with the intent of preventing the next concentration level from being reached. The four-stage sequence provides protection to the public and also provides for an orderly reduction of emissions. Its purpose is to avoid overkill or unnecessarily severe abatement but yet provide adequate protection of public health.

The selection of criteria for officially declaring the existence of an episode and activating the chain of events that ensue is a subjective decision involving considerations other than health. It finally comes down to a determination of the number of episode declarations that can be

Table II Episode Criteria[a]

a. Air pollution forecast

An internal watch by the department of air pollution control shall be actuated by a national weather service advisory that atmospheric stagnation advisory is in effect or the equivalent local forecast of stagnant atmospheric condition.

b. Alert

The alert level is that concentration of pollutants at which first stage control actions are to begin. An alert will be declared when any one of the following levels is reached at any monitoring site:

Sulfur dioxide	800 $\mu g/m^3$ (0.3 ppm), 24 hour average.
Particulate matter	3.0 COHs or 375 $\mu g/m^3$, 24-hour average.
Sulfur dioxide and particulate matter combined	Product of sulfur dioxide ppm, 24-hour average, and COHs equal to 0.2; or product of sulfur dioxide $\mu g/m^3$ 24 hour average, and particulate matter $\mu g/m^3$, 24 hour average equal to 65×10^3.
Carbon monoxide	17 mg/m^3 (15 ppm), 8 hour average.
Oxidant (ozone)	200 $\mu g/m^3$ (0.1 ppm), 1 hour average.
Nitrogen dioxide	1130 $\mu g/m^3$ (0.6 ppm), 1 hour average; 282 $\mu g/m^3$ (0.15 ppm), 24 hour average.

and meteorological conditions are such that pollutant concentrations can be expected to remain at the above levels for 12 or more hours or increase, or in the case of oxidants, that the situation is likely to recur within the next 24 hours unless control actions are taken.

c. Warning

The warning level indicates that air quality is continuing to degrade and that additional control actions are necessary. A warning will be declared when any one of the following levels is reached at any monitoring site:

Sulfur dioxide	1600 $\mu g/m^3$ (0.6 ppm), 24 hour average.
Particulate matter	5.0 COHs or 625 $\mu g/m^3$, 24 hour average.
Sulfur dioxide and particulate matter combined	Product of sulfur dioxide ppm, 24 hour average and COHs equal to 0.8; or product of sulfur dioxide $\mu g/m^3$, 24 hour average and particulate matter $\mu g/m^3$, 24 hour average equal to 261×10^3.
Carbon monoxide	34 mg/m^3 (30 ppm), 8 hour average.
Oxidant (ozone)	800 $\mu g/m^3$ (0.4 ppm), 1 hour average.
Nitrogen dioxide	2260 $\mu g/m^3$ (1.2 ppm), 1 hour average; 565 $\mu g/m^3$ (0.3 ppm), 24 hour average.

and meteorological conditions are such that pollutant concentrations can be expected to remain at the above levels for 12 or more hours or increase, or in the case of oxidants, that the situation is likely to recur within the next 24 hours unless control actions are taken.

d. Emergency

The emergency level indicates that air quality is continuing to degrade toward a level of significant harm to the health of persons and that the most stringent control actions

Table II ***(Continued)***

are necessary. An emergency will be declared when any one of the following levels is reached at any monitoring site:

Sulfur dioxide	2100 $\mu g/m^3$ (0.8 ppm), 24 hour average.
Particulate matter	7.0 COHs or 875 $\mu g/m^3$, 24 hour average.
Sulfur dioxide and particulate matter combined	Product of sulfur dioxide ppm, 24 hour average and COHs equal to 1.2; or product of sulfur dioxide $\mu g/m^3$, 24 hour average and particulate matter $\mu g/m^3$, 24 hour average equal to 393×10^3.
Carbon monoxide	46 mg/m^3 (40 ppm), 8 hour average.
Oxidants (ozone)	1000 $\mu g/m^3$ (0.5 ppm), 1 hour average.
Nitrogen dioxide	3000 $\mu g/m^3$ (1.6 ppm), 1 hour average; 750 $\mu g/m^3$ (0.4 ppm), 24 hour average.

and meteorological conditions are such that pollutant concentrations can be expected to remain at the above levels for 12 or more hours or increase, or in the case of oxidants, that the situation is likely to recur within the next 24 hours unless control actions are taken.

[a] *Federal Register* **39,** No. 50, Title 40, Part 52, p. 9672 (1974).

socially and economically tolerated versus the health threat that can be tolerated by the community. Since there does not appear to be a pollution threshold level at which persons suddenly begin to suffer from acute adverse health effects, the levels at which episodes are declared would have to be set at a very low level if we want to protect every person. Such a system would result in an untold number of episode declarations during the period before long-term emission controls become adequate to eliminate the potential for such declarations. Such a large number of episode declarations would bring social and economic chaos. We are therefore in a position of having to select episode declaration levels that will protect as many persons as possible without resulting in episodes being declared at an economically catastrophic frequency.

Los Angeles County, California, has declared as many as fifteen ozone alerts in one 4 month summer-autumn period (*2*). This seems to be a high number, but since the procedures at the alert stage were only to warn the public and request a voluntary curtailment of driving, such alerts were not very disruptive. If, however, each alert had triggered significant mandatory restrictions in vehicle use that would have been necessary if ozone levels were to be reduced during the alert, the community would have been greatly disrupted during the 4 month period.

3. Emission Reduction Plan

The heart of the episode control plan is a set of strategies for achieving rapid stepwise reductions of pollutant emissions. Progressively more stringent control strategies must be specified for each episode stage as greater emission reduction is required.

a. Sources and Control Problems. To develop realistic and effective abatement strategies of the type that will accomplish maximum emission reduction with a minimum disruption of city routine, it is necessary first to identify sources contributing to the problem. A good emission inventory provides the basic direction for the control plan. An inventory that gives only annual emission rates from each source can be used, but one with daily and hourly rates is of far greater value. In general, the more accurate the inventory information, the better control can be tailored to the situation. Before prescribing abatement strategies, careful consideration must be given to the relative contribution of each source category (Fig. 2).

Inasmuch as fuel combustion generates 81% of the sulfur oxides and 70% of the particulates in the example region presented in Figure 2, emergency control efforts must focus on this category if significant reductions are to be achieved. The major fuel burning sources in the midwestern or eastern United States are assumed to be steam electric power plants, industrial sources, and residential space heating, in that order.

Switching power plants to low-sulfur and low-ash fuels and the transfer of load to plants outside the affected region and to more efficient plants are possible methods for reducing emissions during episodes. There are a number of problems that may develop when trying to implement these strategies: (a) The power plant site may not have sufficient space for storing low-sulfur-content fuel. (b) If natural gas is the alternate fuel supply, it may not be available during cold weather when residential demand is high. (c) If the episode covers a large area, it may be difficult to find sufficient power sources outside the affected area. In

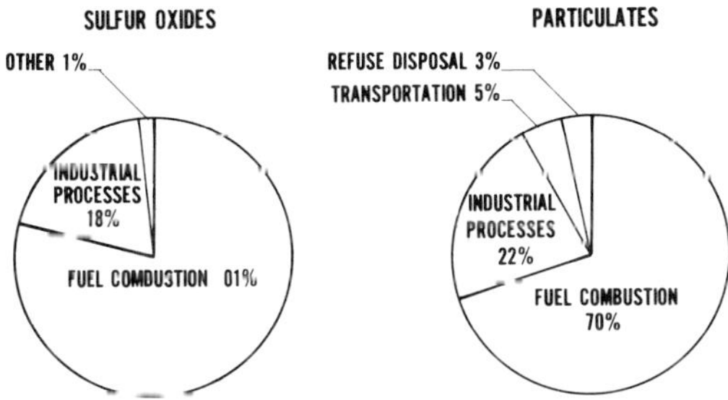

Figure 2. A typical breakdown of sources of pollution in an eastern or midwestern United States city.

many areas, the bulk of the electric power consumed is produced locally, and there isn't sufficient capacity outside the area to cover the demand. In any event, it is important that the most efficient power units be used to the maximum degree during an episode so that production at inefficient (high pollution per kilowatt) units can be curtailed. If the transfer of power and related maneuvering would result in a blackout, the related adverse effects could be greater than those associated with an uncontrolled episode, e.g., the adverse effects on health from the loss of air conditioning during a hot spell or of heating during a cold period.

Significant pollution reduction from fuel combustion for industrial process and space heating can be obtained by the use of low-sulfur and -ash content fuels and by reducing steam loads and production rates to the greatest extent possible. As with power plants, the availability of high-grade fuel may be a problem.

The pattern of residential fuel usage allows some but not a great deal of flexibility. In an emergency, those units heated with natural gas or distillate fuel oil would be largely ignored. Curtailment efforts should focus on large multiple-unit residential buildings where coal or residual oil is burned. The only alternatives are a standby supply of low-sulfur-content fuel for each user and/or reduction in room temperatures. The ability to implement either of these strategies is limited.

While refuse disposal accounts for only a small percentage of the total sulfur oxides and particulate emissions, it is easy to control for short periods. Thus, during an episode, open burning and incineration should be stopped at a very early stage.

While emissions from automobiles and other forms of transportation usually account for only a small percentage of particulates and sulfur oxides emissions, they are the major source of carbon monoxide and the hydrocarbons and nitrogen oxides that create photochemical oxidants. The control of the automobile is the most difficult task the air pollution control officer faces during an episode. Some of the problems are:

a. The necessity of curtailing auto emissions in the morning hours if one hopes to keep the production of oxidant down in the afternoon. This means that people will have to be told the day before that certain restrictions are needed. The major problem in implementing this strategy is that dependable methods for forecasting oxidant concentrations and dosages a day in advance have been developed for only a few areas.

b. Even if oxidant episodes could be accurately predicted in advance, our ability to reduce emissions in even the most episode prone cities without immobilizing the population is very limited. Those cities with the greatest oxidant problem are the same cities that rely almost solely on motor vehicles for their transportation needs. It is one thing to reduce the use of automobiles in New York, New York, or Chicago, Illinois, with their good mass transit systems, but it is quite another situation in Los Angeles, California, with its very limited public transit system.

The key to success of any oxidant episode control system will undoubtedly be advance planning. The most likely method for the control of automobile use will be to curtail those activities that attract automobiles. In other words to close places of employment, shopping, entertainment, and recreation. If we assume that people do not drive just for the sake of driving, but rather that they drive because they have to go some place, if we eliminate the places for them to go to, they will not drive. Thus, emissions will be reduced. In such a system, it will not be necessary to establish complicated systems for exempting those persons, such as doctors and public officials, who need to drive during an episode. It would not be necessary to enforce a driving ban, as the assumption would be that no one would be driving without a reason. It would only be necessary to establish a procedure to check that establishments ordered to close were closed. It is important that places of enjoyment as well as places of employment be closed, otherwise everyone may drive down to the beach for the day.

Short-term reduction of emissions from industrial process sources will assume widely different forms, depending upon the characteristics of the processes involved. Some can be curtailed or shut down readily with little economic hardship and with little hazard. In other cases, sharp changes in production rates are not only costly but also are likely to generate markedly greater pollution releases than during normal operation. This requires extensive preplanning for each significant process source. Under such a program responsible parties recommend definitive curtailment procedures to the control authority stating the actions that should be taken at each stage of an episode. If the authority concurs, these are the procedures officially required during a declared episode.

When abatement strategies are developed, careful consideration should be given to the degree of expense involved in curtailing an operation. Weight should be given to both the cost of reducing the operation and the emissions from the operation. Thus, an operation with low emissions and a high cost involved in stopping the operation might not be shut down, whereas a source with high emissions and an inexpensive shutdown cost would be a candidate for early shutdown. A selective strategy

Table III Episode Control Actions[a]

Category	Source type	Curtailment plan desirable[g]			Switch to low sulfur fuel			Load shifting			Limit cleaning and start up operations			Planned process modification			Partial shutdown			Start no new batches			Reduce to standby			Shutdown or ban			Selective shutdown			Increase activity		
	Alert level	1	2	3	1	2	3	1	2	3	1	2	3	1	2	3	1	2	3	1	2	3	1	2	3	1	2	3	1	2	3	1	2	3
Power generation		X	X	X	X	X	X	X	X	X	X	X	X		X	X																		
Other fuel	Industrial	X	X	X	X	X	X				X	X		X	X	X		X	X		X	X		X	X			X						
Burning source	Commercial	X	X	X	X	X					X	X						X	X		X	X		X	X			X						
	Processing	X	X	X	X	X					X	X						X	X		X	X		X	X			X						
	Residential				X	X	X				X	X	X					X	X															
Incineration	Municipal	X	X	X													X							X				X						
	Commercial	X	X	X													X							X				X						
	Residential																									X	X	X						
	Open																									X	X	X						
Manufacturing[b]	Continuous	X	X	X	X	X	X				X	X	X	X	X	X		X	X						X			X						
	Batch process	X	X	X	X	X	X				X	X	X	X	X	X					X	X		X	X			X						
Commercial[c]	Entertainment	X	X	X	X	X					X	X						X										X						
	Office work	X	X	X	X	X					X	X						X										X						
	Business	X	X	X	X	X					X	X						X										X						
Processing[d]		X	X	X	X	X					X	X						X										X						
Agricultural	Processing[e]	X	X	X	X	X					X	X		X	X	X		X	X						X			X						
	Field operations[f]																						X	X	X	X	X	X						
Governmental	Schools																									X	X	X						
	General office																									X	X	X						
	Public safety																															X	X	X
Construction																					X	X					X	X						

Table III (Continued)

Essential Facilities	Food distribution											
	Hospital & medical											
	Pharmacies											
	Public safety											
	Communications											
	News media											

[a] An excerpt from "Guide For Air Pollution Episode Avoidance," AP-76. United States Environmental Protection Agency, Office of Air Programs, Research Triangle Park, North Carolina, 1971.
[b] Manufacturing includes metallurgical, chemical, petroleum, mineral, paper, mining, etc.
[c] Commercial includes financial, stores, entertainment, offices, services, wholesalers, restaurants, etc.
[d] Processing includes laundries, dry cleaners, garages, service stations, food preparation, etc.
[e] Agricultural processing includes ginning, milling, feed and feed supplement processing.
[f] Agricultural field operations include spraying, dusting, field burning, grading, plowing.
[g] Size and emissions determine which commercial, processing, and agricultural plants must file curtailment plans.

of this sort may present legal problems in that it may not be considered equitable to all sources.

b. Preplanned Abatement Strategies. To assist the air pollution control agency in the development of abatement strategies, all major sources of pollution must be required to prepare and submit preplanned strategies for reducing their emissions during declared air pollution alert, warning, or emergency. Each of the chapters in Part C of Volume IV, Process Emissions and Their Control, includes detailed information on the cutback and shutdown of industrial processes during episodes. Their curtailment strategies should be designed to reduce or eliminate emissions in accordance with the general abatement strategies outlined in Table III. In most cases, plans will be required only from specific sources. However, it may also be necessary to require plans from government agencies that control the operation of multiple sources.

The strategy submitted by each source should identify the source and the emission reduction it can achieve. It should describe the manner in which emissions will be reduced during declared alert, warning, and emergency stages and the amount of time needed to institute these reductions. The time factor is very important in that some batch process operations can be shut down very quickly, whereas continuous operations, such as steel mills and petroleum refineries, may require 2 or 3 days for complete shutdown (Table IV).

It is important that preplanned abatement strategies be updated frequently to include process changes or improvements in pollution control equipment. This latter point is of special importance in view of the fact that the large uncontrolled sources will be the first asked to apply their emission reduction plans.

All preplanned strategies should be reviewed by the air pollution control agency. If the responsible official determines that a given plan is not effective, he will disapprove the plan and order that it be resubmitted. Hearings may be required by a proper review board to resolve differences in opinion. If the source does not resubmit a plan, the air pollution control agency will make the necessary revisions. The source will be responsible under law for complying with this plan during a declared episode.

The major advantage of preplanned strategies is that they help establish an orderly plan of operation that would surely prove to be chaotic and ineffective if the emission-reduction plans have to be developed and implemented during the actual episode. When preplanned strategies are developed, the source knows what is expected of it during a declared episode, and the control agency knows what kind of control it can ob-

Table IV Preplanned Abatement Strategy for Iron and Steel Industry Emissions

Summary of the steps that should be taken by integrated iron and steel mills during each stage of air pollution episode.

Forecast
1. Prepare to switch to low-sulfur fuel where necessary.
2. Prepare for an internal watch to insure minimum emission operation, especially in coke-oven operation.
3. Prepare to stop operations that can be postponed:
    a. Slag quenching.
    b. Incineration.
    c. Scarfing.
4. Prepare to operate coke oven for minimum emissions.
5. Prepare to shut down all steel-refining furnaces without high-efficiency air pollution controls.

Alert
1. Switch to low-sulfur fuel.
2. Reduce sintering operation by 30%.
3. Stop operations that can be postponed.
4. Increase coking time.
5. Start shutting down of all steel-refining furnaces without high-efficiency air pollution controls. Do not add more iron or scrap.

Warning
1. Continue all steps taken in alert period.
2. Stop sintering plant entirely.
3. Shut down all furnaces without high-efficiency controls.
4. Prepare to shut down all furnaces. (Furnaces may be kept banked for protection.)
5. Reduce coke operations to a minimum.

Emergency
1. Continue all steps taken in warning period.
2. Shut down all furnaces. (Furnaces may be kept banked for protection.)

tain in advance. During the actual episode, it is only necessary for the agency to declare the episode stage and not to devise individual source control plans.

Preplanned strategies are also of great value if a source does not follow the curtailment orders issued by the control agency during an episode and it becomes necessary to obtain a court injunction to enforce the order. If preplanned strategies have not been developed, the control agency will in effect have to develop one before it can seek an injunction. Courts are generally not willing to issue vague curtailment orders such as one requiring a facility to reduce emissions to the greatest extent possible or even one that specifies a percentage reduction. Such orders are very difficult to monitor and enforce because they lack specifics. When the control agency petitions a court for an injunction, it must specify exactly what the source should be ordered to do in order to curtail its emissions. During the November 1971 episode in Birmingham, Alabama, it took federal and local officials 6 hours to develop the specific curtailment steps required of approximately twenty-five major sources before they could petition the court for the injunctions. If preplanned strategies had been available, most of this valuable time could have been saved. The court simply would have been asked to enforce the preplanned strategies.

The court is also likely to ask questions concerning the ability of a source to implement the requested order without damaging equipment or endangering employees. Such ill effects are far less likely to occur when a carefully developed preplanned strategy is implemented rather than a strategy developed at a moment's notice during an episode. Thus, the court is more likely to grant an order that is based on the preplanned strategy.

c. Development of the Emission-Reduction Strategy. The two techniques most often used for determining the amount of emission reduction required are simple rollback, in which all sources in the affected area are ordered to reduce emissions a specified percentage, and simulation techniques. Simulation techniques utilize emission inventory data to predict pollutant isopleths for a given set of meteorological parameters, i.e., wind speed and direction, mixing height, and atmospheric stability. It is possible to demonstrate the impact of an individual source on area air quality and thus which sources should be curtailed if air quality is to be improved. The effect of preplanned abatement strategies can be predicted by the model. In this way, it is possible to determine (a) if the preplanned abatement strategies can achieve the desired reduction in concentrations, (b) whether other strategies can be developed that will be more effective, and (c) the least costly set of strategies that will satisfy the desired goal (see also Chapter 1, this volume).

While the simulation technique is a very effective tool, the development of such a model requires a sophisticated analysis of an excellent emission inventory along with an extensive amount of air quality and meteorological data. If resources are not available for utilizing this approach, a less sophisticated strategy of controlling the largest sources first can be employed. The largest sources and uniform rollback strategies have been used in the control of most episodes to date, even though these strategies may be neither the most effective nor the least costly.

Only with the simulation model can one hope to approach an optimum solution for a specific episode.

4. Surveillance System

Episode avoidance requires that accurate air quality, meteorological, and source control information be available at all times for rapid decision making. The episode plan must, therefore, include a system for surveillance of air quality and meteorological conditions and the fast reporting of that data in a useable format. Air quality data describe present conditions, i.e., the effect of past control actions. Present and forecast meteorological conditions indicate the air pollution potential for the immediate future. As the episode plan is implemented, source surveillance is needed to verify that source curtailment is indeed being done as directed (see also Parts B and C, Volume III).

a. Air Quality Monitoring. For episode avoidance purposes, air quality data are needed quickly—in no less than a few hours after the pollutant has been detected and concentrations determined. Monitoring techniques employed in obtaining data on long-term changes in air quality may not be suitable for use in episode situations.

Statistics such as averages, ranges, and deviations, which form the basis for long-term control plans, can be developed using averaging times of up to 1 year, and such data can be obtained by random, continuous, or intermittent sampling. For episode avoidance, however, it is necessary to utilize continuous air monitors, because an episode lasts only a few days, and control actions must be taken quickly.

One hour averaging times are adequate for surveillance of episodic or potentially episodic conditions. Shorter averaging times provide more information (better statistical estimates) but increase the need for automation because of the bulk of data. Averaging times longer than 6 hours entail too much delay in the response of the control program.

Since episodes require control to protect human health, monitoring sites should be located where health is most threatened, such as areas of high population density, near large stationary pollutant sources, near areas of high traffic density, or wherever the anticipated pollution impact might be greatest.

One of the most difficult air quality monitoring problems encountered during an air pollution episode is determining suspended particulate matter concentrations in real time. The most common method of measuring suspended particulate matter is the high-volume sampler. The sample is collected over a 24 hour period, desiccated for a period up to 24 hours, and then weighed. This method does not give rapid enough response during an episode. This is in contrast to the essentially instantaneous results that are available when continuous air monitoring instruments are used to monitor gaseous pollutants.

Since the high-volume technique is the one that can be related to the health effect studies that serve as the basis for taking action during an episode, it is important that any short-term particulate monitoring techniques be correlated with high-volume measurements. This relationship is the source of the problem, because most episode plans have been developed around tape samplers as the source of suspended particulate matter data. There is no universal correlation between tape sampler and high-volume sampler results. The relationship between the two methods varies from place to place and to some degree from time to time, depending upon the composition of the particulate matter. When tape samplers are to be used to monitor particulate matter, extensive correlation data should be gathered at each monitoring site.

Short-term techniques for using the high-volume sampler include placing two to four samplers at each site so that a 24 hour sample is available every 6 to 12 hours. This requires a lot of sampling equipment. Another alternative is to use only one sampler but take shorter-term samples (generally 6 to 12 hours). Shorter sampling time does not present a measurement problem when particulate matter concentrations are high, as would be the case during an episode, but does require around-the-clock servicing of equipment and weighing by agency personnel. When the multiple-sampler or short-term sample approach is used, the samples are desiccated or dried for only a short period before they are weighed. The main problem with these procedures is that the sample is rushed through the drying procedure prior to weighing. Cutting the drying time by increasing oven temperature may not be significant with respect to moisture equilibration, but it is possible that other volatile matter may be driven off, thus reducing the weight of the sample.

Other short-term particulate sampling techniques being tested include nephelometer, vibrating reed, and beta gauging techniques. None of these techniques is yet considered an acceptable replacement for the high-volume sampler. Thus, obtaining accurate short-term particulate matter samples is expected to present a problem to the air pollution control officer for some time into the future.

b. Meteorological Observations. Current and forecast meteorological conditions are probably the most important information for determining if control actions are needed, for how long, and to what degree. These initial meteorological forecasts alert the control agency and the public. In the United States, the National Weather Service is responsible for providing forecasts of air pollution potential on both local and national scales (Fig. 3).

Regular observations made by the National Weather Service are not generally made at a location representative of urban conditions; therefore, some cities are equipped with a dedicated low-level sounding unit that takes soundings in the urban area. Because some sections of the country have a more serious air pollution problem than others, specially trained air pollution meteorologists (focal point meteorologists) are assigned to these areas. The focal point meteorologist, after analyzing the overall weather pattern, issues a daily dispersion outlook (DO), and if atmospheric dispersion conditions are poor, may issue an air stagnation advisory (ASA). The DO is issued only to the control agency while the ASA is publicly disseminated.

Dispersion outlooks are issued to control agencies to provide forecasts of general dispersion conditions for periods of up to 36 hours. Format and content of the DOs are determined by the needs of the control agency. For this reason, a local weather service official should be included at all planning meetings pertaining to episode avoidance procedures. By understanding the objectives and procedures of the episode control program, the forecaster can tailor his forecasts to provide the maximum support to the control agency. Weather service forecast offices (WSFO) with a focal point meteorologist are required to issue a DO Monday through Friday. Other forecasts offices issue them as needed.

In addition to the services available from the weather service, larger control agencies in episode-prone areas may wish to acquire and evaluate more meteorological data than is available from the National Weather Service. Such control agencies should have a staff meteorologist in charge of collecting and evaluating the supplementary data. The staff meteorologist should also play a key role in the development and implementation of any simulation modeling that the agency does in conjunction with the development of an episode control strategy. If a full-time meteorologist is not warranted, private meteorological consultants are available.

c. Data Handling. Making air quality measurements and meteorological observations quickly available for rapid decision making is accomplished by computerizing the data processing system; i.e., the data is telemetered to the control center, processed by computer, and automatically graphically displayed. The advantage of a computerized system is its ability (a) to process large quantities of input data in a short time, (b) to store large quantities of data with rapid retrieval, (c) to process multivariable data and conveniently display such data, (d) to perform simulations using predictive models, and (e) to lower the man-

Figure 3. Location of National Weather Service offices; ★ = weather offices with air stagnation advisory responsibility, APFP = air pollution focal point, APFP/O = APFP and dedicated observer.

power requirement during an episode. The major disadvantage of a computerized system is its cost, both initial equipment costs and operating costs. Another factor in considering a computerized system is that some of the currently available monitoring equipment is not equal in reliability to the telemetering and data processing system to which it will feed data. Therefore, it is likely that a technician will be required to visit the monitoring sites often during an episode if valid monitoring is to be assured. No matter how sophisticated the data processing system, it cannot upgrade poor quality data.

In many areas, manual data processing will be adequate. There are few areas that can justify an automatic data retrieval and processing system for episode reasons alone. Episode control is only one of many things that must be considered when the decision is being made to use automatic rather than manual data handling techniques. When the data are processed manually, appreciable planning is necessary to insure the required processed data will be available in time to assist the control agency with decisions. Dry runs actually simulating episode conditions should be practiced until the system functions smoothly.

d. Source Surveillance. An equally important surveillance tool during an air pollution episode is the air pollution inspector. While the uncovering and reporting of violations is important to the long-term control program, during an episode this function may be vital to the protection of public health. During an episode, all available inspectors should be assigned to around-the-clock surveillance. The inspector should have the power to enforce all emission reduction or shutdown orders issued during the episode. This should include the authority to enter without court order any potential source of pollution other than a private residence.

The inspector should have a copy of the source's approved preplanned abatement strategy when making an inspection. It will be necessary for him to verify only those curtailment procedures that are specified in the strategy. The vehicles used by the inspector should be radio equipped in order that violations can be reported quickly to the control agency and the inspector can receive orders directing him to verify other reported violations.

5. Communications

To successfully carry out the various control activities during an air pollution episode will require joint participation by all segments of the community. Communications, then, is another important element in the implementation of an episode plan. A major function of the communications network is to transmit status reports and control actions to noncontrol agency personnel. Individuals or groups needing such information include:

Public officials who may be called upon to participate in the decision-making process.

Major emission sources, which require latest status reports to effect source control plans designed to reduce emissions.

Public safety, civil defense, or emergency agencies, which are assigned tasks relative to maintaining public safety during an episode and which need to be kept current on the status of events.

The news media, which provide the direct communication link to the general public.

a. Episode Control Center. To assure that air quality and meteorological data is quickly assembled, analyzed, and disseminated to the persons or groups just mentioned, it is nece ry to establish an episode control center to handle all incoming and outgoing communications. The center should be located at the source of authority. This means that the center should be readily available to the offices of the local air pollution authority. The center will generally be located in the offices of the control agency.

b. Operations. The communication operations required during an episode include the following:

i. *Data collection.* During an episode, all data from the monitoring network should be reported to the control center by telephone or teletype on a preplanned schedule and format. Reporting pollutant levels every 1 to 6 hours should be adequate.

ii. *Notification of pollution sources.* Each major emission source must be alerted at each stage of the episode to initiate its emission reduction plan for that stage. This can be accomplished by telephone calls to appropriate plant officials. The appropriate plant officials should be determined in advance of the episode. Some agencies have used a callback system by which the plant official has a predesignated number to call in order to verify that the call requiring him to initiate control actions was official. Methods other than the telephone have been used. For example, the Los Angeles, California, Air Pollution Control District maintains a radio transmitting system that continuously broadcasts and is received by most major industrial sources in Los Angeles County during all hours of plant operation. Public announcements have also been used in some areas. General notification to small emission sources can be made through the news media.

iii. *Notification of the news media.* The press release can be the primary means of communicating with the news media. In order to save valuable time during an episode, a previously preplanned press release for each episode stage should be kept on file and revised before release as the specific situation dictates.

iv. *Notification of sensitive persons.* Another group that must be kept advised throughout the episode are persons who are believed to be most susceptible to acute health problems during an episode. These include persons suffering from respiratory ailments such as bronchitis and emphysema and persons with heart conditions. In many cities, this may represent 15% or more of the population. In order that this group be kept advised, it will be necessary to have a coordinated effort between the local air pollution control officials, public health officials, physicians, hospitals, and the public. A medical advisory committee on air pollution episodes should be established. This committee should include the commissioner of health, the director of the air pollution control agency, and knowledgeable physicians. The medical advisory committee should provide advice on what precautions sensitive persons should take and when and by what means they should be notified.

It may be ill-advised to notify sensitive people via the public media, as it is important that the physician be able to detect that a particular patient is reacting physiologically to an episode and not psychologically. Communicating information through the local health agency or medical society may be more advisable. Implementing such a procedure is time consuming and difficult to coordinate. In most cases it will be necessary to use news media in spite of the aforementioned problem. Advice to sensitive persons may include (a) remain indoors with the windows closed, (b) use a fan or air conditioner to circulate the air, (c) do not smoke, and avoid rooms where others are smoking, (d) avoid unnecessary physical exertion, (e) call your physician if you have a cough or difficulty in breathing.

The Los Angeles County, California, Air Pollution Control District has created "A School and Health Smog Warning System" (*3*) to provide notices to school children and susceptible individuals in order that they may take measures for their own protection. Based on the prediction of high ozone level, the Los Angeles County Medical Association has set forth protective measures that may be taken. Based on the same prediction, Los Angeles County schools have initiated voluntary procedures to minimize the physical activity of school children.

v. *Contact with legal authority.* Persons who may be called upon to participate in determining what course of action is to be taken must be kept informed of the latest episode status. The ideal situation is to have all officials with the authority to initiate action assembled at the episode control center from which the action can best be directed. Higher authorities, such as the governor, mayor, city council, or health commissioner, must be kept informed of all developments. It is likely that if the episode goes into the emergency state, action will be required that only they have authority to initiate.

vi. *Contact with other air pollution control agencies.* It is suggested that an episode coordinating committee be established in air quality control regions encompassing more than one air pollution control agency. Coordination is an absolute necessity in any kind of episode plan. If the committee does not assemble in one place, it is advisable that a conference-call telephone system be installed. The episode stage and the corresponding abatement strategies used in each neighboring governmental jurisdiction must be compatible.

vii. *Contact with other local governmental agencies.* If the police, fire, civil defense, and public health departments are assigned definite tasks during an episode, they must be kept informed of the episode status. In some locations, existing civil defense, disaster, or emergency networks and procedures can be used. The telephone company, though not a public agency, has a similar need to be informed, mainly because of possible overloading of facilities.

viii. *Obtaining weather information.* Air pollution potential forecasts will be received at the episode control center. The advisory will be received directly from the focal point meteorologist. The National Weather Service should be recontacted at regular intervals for forecast updates.

ix. *Contact with field inspectors.* During an emergency, two-way radio contact should be maintained with all field inspectors. The inspectors can be dispatched to suspected trouble spots. They can likewise keep the control center advised of violations and obtain instructions on how to handle the situation.

c. Communications and Operations Manual—Standard Operating Procedure. The communication operations described above should be

Table V Communications Checklist of Air Pollution Episode Emergency Notifications[a]

EMERGENCY declared: Time ______________ Date ______________

The following contacts must be made from the control center to communicate the emergency declaration statement:

	Telephone number	Called by (initials)	Time
Director, Illinois EPA	________	________	____
Commissioner, Chicago DEC	________	________	____
Illinois Pollution Control Board	________	________	____
USEPA-APCO Chicago Office	________	________	____
Chicago, Office of Mayor	________	________	____
Chicago, Health Dept.	________	________	____
Chicago, Dept. of Police	________	________	____
Chicago Transit Authority	________	________	____
Chicago Public Library	________	________	____
Commonwealth Edison Co.—Dispatcher through representative in control center	________	________	____
Peoples Gas—Dispatcher	________	________	____
City News Bureau (will return call) (send any copy of news release)	________	________	____

Select action plans for sources in the affected regions and make required contacts including:

State of Indiana
Cook County
Local jurisdictions in affected region in Illinois and Indiana
City of Chicago Dept. of Streets and Sanitation, Board of Education, and Chicago Housing Authority
Metropolitan Sanitary District
Northern Illinois Gas Co.
Federal Sign Co.

[a] Extracted from: "Air Pollution Episode Control Communications Manual." Chicago Metropolitan Air Quality Control Region, State of Illinois; State of Indiana; City of Chicago, Illinois; Cook County, Illinois; Village of Bedford Park, Illinois; Village of McCook, Illinois; Village of Morton Grove, Illinois; Argonne National Laboratory, Argonne, Illinois.

written and assembled into a communications and operations manual. The receipt and processing of data and information and the action advisories that are developed from these data and information must proceed according to a predetermined plan. The communications and operations manual should specify in detail, operation by operation, how contacts are made or received by the episode control center. The manual should be designed with a set of predetermined procedures for each episode stage. It is advisable that checklists be developed for each operation at each episode stage (Table V). When the operation has been completed, the checklist is filled in in the episode logbook.

d. Episode Logbook. An episode logbook is kept by the control center. Entries are made whenever high pollutant levels are reported or whenever a meteorological potential forecast is received. Whenever any episode stage is declared, the episode logbook becomes a full-time log, and all pertinent developments are noted in it. All checklists form a part of the permanent record of the control center. They are the objective evidence of actions taken during the course of an episode. When the episode is over, the logbook will serve as the primary source of information for the analysis of the episode and a determination of the effectiveness of control actions.

III. Planning for the Single-Source Episode

A. Objective

As progress is made toward attainment of desired air quality goals, the likelihood that a single source will create a local episode condition will be reduced. In the meantime, however, single-source episode conditions do exist and/or will develop, and the air pollution control officer must develop a plan for responding to these situations.

Many single-source episodes are not technically episodes, in that they may have existed over an extended period of time but only recently have been identified. The factors that make these situations similar to episodes are that the threat to public health and welfare is substantial, and the condition, once identified, must be abated quickly by means similar to those used in an atmospherically induced episode, i.e., production cutbacks or cessation of operations.

An example of a single-source situation is one with a short stack located in a narrow valley with homes located immediately downwind of it in the most prevalent downwind direction. The source could also be located in the heart of a city's residential area with homes in all directions. The situation may have existed for years, but due to some change in the source emission, some change in the vegetation of the area, or some change in the density of affected population, the source is identified by the air pollution control agency as a problem. The investigation that the agency must conduct to determine if the source is presenting a significant threat to health and welfare of the citizens in the area would be defined by the episode plan.

A second example involves a potential single source in that a large volume of demolition waste are about to be burned immediately adjacent to a large residential area. A home for the elderly is also located in the area. Again the procedure for determining if the burning of this material would present a significant health threat, and thus should be prevented, would be defined by the episode plan.

B. Plan Requirements

The single-source episode plan must include at least the following elements: (a) legal authority, (b) criteria, (c) surveillance system, and (d) source–receptor simulation modeling capability.

1. Legal Authority

As with the atmospherically induced episode, adequate legal authority is needed to control the single-source episode. The same enabling legislation that authorizes action during the atmospherically induced episode should be sufficient for the single-source episode. In both cases, the legislation must be written to allow the control agency to take the necessary action to prevent episode conditions from developing when they are imminent and not be so worded as to authorize action only after an episode condition has developed. Without preventive authority, the control agency would be unable to take effective action during the situation described in the second example above.

2. Criteria

As was the case in the atmospherically induced episode, the objective will be to prevent or alleviate air pollution situations that pose a significant threat to public health. Therefore, the same pollutant levels that define significant harm for the atmospherically induced episode (Table I) can be used here. The alert, warning, and emergency criteria levels and the corresponding staged emission reductions used for the atmospherically induced episode will have little use in the single-source episode, because the controls required are needed continuously and are generally not dependent on an adverse meteorological condition.

Some single-source episodes may result from a pollutant not defined by a national, state, or provincial ambient air quality standard. In fact, there may be very little known about the health effects of the pollutant.

In such instances, the medical advisory committee, discussed in the preceding section on atmospherically induced episodes, should be asked to evaluate the situation and establish a significant harm level for the pollutant. The committee would arrive at this level after reviewing air pollution and industrial hygiene literature on the pollutant and/or consulting with known authorities on the pollutant. The air pollution control officer should have sufficient authority to utilize the committee's recommendations to effect the necessary control. Having the commissioner of health serve on the committee can also provide this authority.

3. Surveillance System

In most episode or suspected episode situations, the impact on public health will not be so obvious that action can be taken without a period of surveillance. Surveillance may include plant inspection, source testing, and ambient air quality monitoring.

a. Plant Inspections and Source Testing. Since the source's episode potential is being investigated, it is important that the source be observed during its worst condition from an emission standpoint. Observing the source during its least favorable condition will require repeated investigation during all phases and hours of operation. An element of surprise is often needed when conducting such investigations.

b. Mobile Air Quality Monitoring. Since episodes can potentially occur at any location where a source is located in the proximity of people, it is unlikely that a fixed ambient air quality network operated by the air pollution control agency will provide sufficient coverage. A portable or mobile ambient air quality monitoring capability will be needed. Ideally monitoring equipment placed in mobile vans not constrained by the need to be located in an existing structure or near a source of electrical power should be available. The need to place monitoring sites in existing buildings or near electrical power not only restricts the monitoring station but also results in a substantial delay in establishing the station while all the details of obtaining permission to use the building and establishing electrical power are resolved. Where the community perceives a source as a threat to its health, it can be very helpful in providing assistance in the location of sites to monitor the source.

If high pollutant concentrations over short time periods are the problem, continuous monitoring will be needed to detect pollutant excursions. For many of the more common air pollutants, continuous instruments are readily available. For the more exotic pollutants, continuous field instruments are not available. In such cases, it is generally necessary to collect samples in the field over a period of time and return to a laboratory for analysis. When this is necessary, the sampling period should be kept short enough for high-concentration excursions to be detected and not lost due to averaging over a longer time period. Since an enforcement case is likely, chain of custody procedures must be established for handling all samples.

c. MOBILE METEOROLOGICAL MONITORING. During any surveillance effort, it is important to prove that the pollutant concentrations recorded at a monitoring site are a direct result of the pollutant discharged from the source. Collection of meteorological data (wind speed and direction) during the ambient sampling program will greatly aid in making this association.

The initial location of air monitoring sites should be based on meteorological information that is available from the nearest weather station. Monitors should be concentrated downwind of the source in the most prevalent wind direction for that time of the year and at a distance consistent with the source's stack height and the wind speeds that are most prevalent for that time of year. As on-site meteorological data is collected, it may indicate that a monitoring site should be moved so as to be in a better receptor location.

Most portable wind systems can be powered by mechanical means (spring winding) or by batteries, and the recorder may be exposed to the elements. This eliminates many of the siting problems that are faced with air sampling equipment. However, as with air sampling equipment, vandalism can be a problem.

d. SOURCE–RECEPTOR SIMULATION MODELING. Simulation or meteorological modeling can be used to determine the ambient air pollution potential of a source and the best location of air monitoring sites for detecting high ambient concentrations. If good source emissions data are available, they can be combined with meteorological information and stack height characteristics to predict ambient concentration downwind of the source. From such an analysis, one can predict if the source has the potential for creating an episode situation. This screening technique can be used before ambient monitoring is initiated. The modeling technique is the only one available for determining if a potential, but as yet nonexistent, source will create an episode situation. This would include the proposed open burning situation discussed in the second introductory example above.

The major problem in using simulation modeling is the need for good emissions and meteorological data. It is important that emissions be either estimated or tested for under the source's worst operating conditions. Even if precise emission data are not known, modeling can be used to predict relative concentration and thus be used to select the most critical areas as sampling site locations.

The United States Environmental Protection Agency, with the assistance of the Argonne National Laboratory, developed a time-sharing isopleth–area program (*4*). This program, when supplied meteorological and source information, predicts the downwind maximum concentration and defines the area within a specific concentration.

IV. Planning for the Accidental Episode

A. Objective

An accidental air pollution episode is an unexpected event that caused or may cause contaminants to be released into the atmosphere in amounts that are potentially hazardous to public health and welfare. Such situations are independent of prolonged and widespread meteorological stagnation conditions. An accidental episode may be due to a transportation accident, a process malfunction, or be the secondary result of a natural disaster.

Responsibilities for action to protect citizens during such an event normally rests with local police, fire, and civil defense departments and, in a more serious crisis, with mayors and governors. Air pollution control officers are often asked to support these groups or persons by providing advice on the nature and behavior of the atmospheric contaminants involved.

There are a number of episode situations that may be encountered:

a. Pollutant releases that have already occurred when notification is received (fast release or explosion). In such situations very little can be done to alleviate the problem. If time permits, the air pollution control agency can predict where the contaminant is likely to be transported, what the pollutant concentration may be, how toxic it will be, and the precautions that can be taken by citizens to protect themselves.

b. Pollutant releases that are taking place at the time of notification (slow, uncontrolled release). The response is similar to (a) above.

c. Pollutant releases that are imminent, such as an industrial process out of control or a natural disaster approaching an area. The air pollution control agency can advise on what area is likely to be affected, what the pollutant concentration may be, how toxic it will be, and the precautions that can be taken, and it can monitor air quality to confirm the pollutant concentrations.

d. Pollutant releases that may occur if a critical salvage or containment operation fails. The control agency can perform the same operations as noted in (c) above, with more time allowed for planning.

B. Plan Requirements

Four specific tasks can be performed by the air pollution control agency in order to assess an episode threat to health and develop recommendations on what and where protective action should be taken by the public:

a. *Air quality monitoring* to define the problem and confirm dispersion predictions.

b. *Meteorological monitoring* in order to make dispersion predictions.

c. *Meteorological dispersion predictions* to project the extent and severity (pollutant concentration) of the episode.

d. *Effects consultation* on the degree of hazard and precaution that can be taken by the public to protect themselves and their property.

The air pollution control agency should confine its efforts to the above areas and should not become involved with directing operations designed to remove the source of the episode or to evacuate citizens. The expertise to do these things generally does not exist within air pollution control agencies. The implementation of evacuation efforts is the responsibility of police and civil defense agencies. The air pollution control agency's responsibility is to define the problem as to its effects or likely effects on the public and to make recommendations to local officials as to what actions can be taken to alleviate this threat.

In order that the above mentioned tasks can be performed, the air pollution control agency must have the following capabilities: (a) ambient air monitoring capability for hazardous pollutant, (b) meteorological monitoring capability, (c) access to effects information on hazardous materials and information on what precautions the public can take to protect themselves and their property, (d) ability to calculate the hazard area by diffusion analysis, and (e) protective equipment for all agency personnel that may be potentially exposed to the hazardous materials.

1. Ambient Air Monitoring for Hazardous Pollutants

The development and implementation of a hazardous pollutant air monitoring capability is very difficult for the following reasons:

a. There are so many potential pollutants, that it is difficult to have a capability for monitoring all of them.

b. There may be a significant hazard from being exposed to the pollutant while taking the air samples.

c. During episodes that result in a sudden release of material, it is difficult to monitor in a timely fashion.

d. For a potential episode situation, it is difficult to know where to place the monitors due to the variability of the wind direction.

e. In some episodes, the pollutant will not be known and thus there is an inability to know what to monitor for.

In spite of these and many other problems, there are some minimum monitoring capabilities that the air pollution control agency can and should develop. The agency should have one or more multiple-purpose detection kits. Kits of this type are available prepackaged from several manufacturers. A separate reagent-filled tube is used for the detection of each pollutant. One manufacturer lists eighty gaseous or vaporous chemicals for which detector tubes are available. These kits are hand operated in the field, have no power requirement, and provide a direct reading of pollutant concentration.

In some instances, the pollutants released, either directly or as a combustion product of a hazardous pollutant fire, may be the same, although in a higher concentration, as those routinely monitored by the control agency. In such cases, the same equipment can be used after it is recalibrated for higher concentrations. In some instances, it may be possible to modify an existing monitor to be sensitive to a pollutant other than what it is routinely used for. During an episode involving chlorine, the United States Environmental Protection Agency found that sulfur dioxide analyzers using the coulometric principle could be modified for monitoring chlorine. The control agency should know the multipollutant capability of their monitoring instruments. The agency's monitoring capability may be greater than it was originally thought to be.

Since it is not possible to have a monitoring capability for all hazardous pollutants, it is necessary to know which pollutants are most likely to be encountered. Resources Research, Inc., under a contract with the United States Environmental Protection Agency, developed a rating system that enabled them to establish priorities for hazardous pollutants most likely to be encountered during an accidental episode (*5*). The accidental episode potential for stationary sources was based only on accidental case statistics, whereas the potential for mobile sources was based on case statistics; production figures; and physical, chemical, and toxicological properties of transported materials. Their combined final list of fifteen pollutants in order of their priority is:

Hydrogen sulfide	Acetaldehyde	Formaldehyde
Vinyl chloride	Hydrogen cyanide	Acetone
Ethylene oxide	Acrylonitrile	Butane(s)
Chlorine	Methanol	Propane
Anhydrous ammonia	Benzene	Ethyl chloride

In order to further define their monitoring needs, the control agency should become familiar with hazardous materials that are routinely stored, handled, manufactured, or transported through its area.

One of the problems noted above was the potential hazard that the pollutant may present to the person collecting the samples. In no event should a monitoring technician enter the area without proper protective equipment. During a potential episode where it is possible to use continuous monitoring equipment, it may be desirable to telemeter the data or establish some other remote reading capability. During a March 1972 potential episode in Louisville, Kentucky, involving a wrecked chlorine barge, the United States Environmental Protection Agency used a high-power reflecting telescope to read a monitoring instrument some 800 to 1000 feet (240–300 m) away.

2. Meteorological Monitoring

Before any meteorological monitoring is initiated, the control agency should first contact its national weather service. In many instances, arrangements can be made to obtain special forecasts during an episode.

The air pollution control agency should have at least one battery-operated wind system with a portable tripod reserved for episodes. The agency should also have the capability to make low-level pilot balloon observations. Some types of accidents can produce large and/or rising plumes that will be largely under the influence of upper winds. If the cloud is dispersing near the surface, pilot balloons are useful for predicting wind flow. Since topography is very important in determining how a pollutant will be dispersed through an area, the control agency should have a set of topographic maps for its area of responsibility.

During a potential episode situation, one that may occur if a critical salvage or containment operation fails, it will be necessary to have reports (24 hours a day) on current wind speed and direction and updated forecasts. This information will be used to continuously revise the hazardous area forecasts. If the episode occurs, a hazardous area forecast must be current so that protective measures can be taken immediately in the specified area.

3. Health Effects Information

In order that the threat to public health can be determined, it is necessary to know the concentration of the particular substance that is hazardous. "Threshold Limit Values" (TLV) (*6*) are probably the best source of this information. Other sources are the "Chemical Safety Data Sheets" of the Manufacturing Chemists Association (*7*), the "Industrial Safety Data Sheets" of the National Safety Council (*8*), "Dangerous Properties of Industrial Materials" by N. Irving Sax (*9*), the National Fire Protection Associations' "Fire Protection Guide on Hazardous Materials" (*10*) and "Fire Protection for Chemicals" (*11*), the National Safety Council's "Chemical Safety Slide Rule" (*12*), and the "Accidental-Episode Manual" developed for the United States Environmental Protection Agency (*13*).

The Manufacturing Chemists Association, Washington, D.C., has established a Chemical Transportation Emergency Center (CHEMTREC), which operates a 24 hour hot line that can be reached toll free for information on the hazardous potential of substances released as a result of a transportation accident. CHEMTREC can also be used to locate the manufacturer of the problem material who can provide assistance on its control and containment.

The Division of Oil and Hazardous Materials, United States Environmental Protection Agency, has developed the Oil and Hazardous Materials Technical Assistance Data System (OHM-TADS). Through the use of a remote computer terminal available at each of the United States Environmental Protection Agency regional offices, OHM-TADS can provide information on a substance's TLV, irritation level, odor threshold, safety precautions, and general air pollution hazards, as well as voluminous information on the substance's water pollution potential.

It is important that the air pollution control agency establish a medical advisory committee. By using such sources as the health effects information noted above and other sources, including direct contacts with known authorities on the health effects of hazardous materials, the medical advisory committee can determine what constitutes a hazardous concentration. This information can be used along with dispersion modeling to determine the hazard area. The medical advisory committee can also develop information on what precautions citizens can take to protect themselves and advise officials on the desirability of evacuation.

4. Hazard Area Estimation

During an actual or potential episode, the most important task is to protect people who are being or may be exposed to the hazardous materials. It is therefore necessary to define the area downwind of the accident (or potential accident) where ambient concentrations may endanger the health and safety of the population. Through the use of the meteorological diffusion model and information on source emission rate, plume height, wind speed and direction, and meteorological stability, it is possible to predict ambient air concentrations of the involved substance downwind of the accident (or potential accident) site. By comparing these concentrations with health effects information on the substance, it is possible to determine the areas where a health threat exists (*13*).

In most cases, the hazard area will be within a 45° sector downwind of the source. The length of the sector is determined by the factors noted above. During an episode, the hazard area sector map should be updated every hour as meteorological and/or source conditions change. For potential accidents, it will be possible to prepare hazard area sector maps in advance for various meteorological and source strength conditions.

The significant inputs needed to make hazard area estimates are the meteorological parameters affecting transport and diffusion and source strength or emission rates. In an accident, accurate source strength estimation is an almost impossible task. The best source of this information is generally those persons most closely associated with the cause of the accident, e.g., the plant engineer, the tank manufacturer, the trade association, the chemical producer. These people should assume responsibility for the control or containment of the accident situation, not only because of their legal responsibility, but also because of their familiarity with the source.

During a potential episode situation, it may be necessary to make hazard area estimates based on several estimates of source strength. For instance, during the chlorine barge accident noted above, it was necessary to make hazard area estimates for many different source conditions including a continuous slow leak from a control valve, a continuous fast leak from a ruptured pipe, and the near instantaneous release from the rupture of from one to four tanks on the barge. Depending on the situation that actually developed, the appropriate estimate would have been used.

5. Protective Equipment

The choice of protective equipment depends on the job to be performed by those wearing the equipment. In most cases, air pollution control personnel will not be expected to take part in source control or containment operations and therefore should not expect to work in a contaminated area for extended periods. The purpose of this equipment will be to allow the control official to leave the contaminated area in safety if for some reason he must enter or is caught in the contaminated area. For instance, respiratory protective devices of no more than a 1 hour capacity would be quite adequate for most air pollution control agency personnel needs.

Protective equipment, like monitoring equipment, must be routinely checked. There are few things worse than needing oxygen and then finding the tank empty or the control valve malfunctioning. In cases where a respiratory device will not be used frequently, a self-contained chemical oxygen generator may be preferable to a tank device. It has the advantages of requiring little maintenance and having a long shelf life.

Most protective equipment manufacturers will provide details on the advantages and disadvantages of the various types of protective equipment and will provide training in the use of the equipment selected. These services should be used in order that the best equipment for the task to be performed be selected and the use of the equipment be fully understood.

C. Social, Political, and Administrative Considerations

There are a number of social, political, and administrative considerations during an accidental episode.

1. Community Contingency Plan

A community air pollution disaster contingency plan should be developed. The air pollution control agency's response to accidental episodes will be most effective when coordinated by a community-wide contingency plan. The contingency plan should specify responsibilities, those to be alerted and in what sequence, how communications are to be handled, how the public is to be notified, and if necessary, evacuated, and when and how to terminate the emergency.

2. Manpower

The amount of manpower needed in an extensive long-term operation can be staggering. It may be necessary to man air quality monitoring, meteorological, and dispersion forecast posts 24 hours a day by professionals or technicians.

3. Operational and Communication Problems

The operational and communication problems that must be solved at a disaster site include:

a. A command center must be established where all information relative to the accident will be received, at which all decisions are made, and from which such decisions will be disseminated to all involved parties. While it is probably desirable that the command center be located relatively close to the disaster site, precautions must be taken to assure that the command center is not itself in a danger zone. Alternate centers should be selected and provisions made for moving operations to them if earlier ones become endangered.

b. Communications must be established between the disaster site and the command center so that activities at the site can be monitored.

c. A wind monitoring system must be established near the disaster site and communications established between the wind monitoring system and the meteorologist making the dispersion forecasts.

d. If the dispersion forecaster is not at the command center, communication must be established between him and the center.

e. Communications must be established between the dispersion forecaster and the nearest National Weather Service office in order that the forecaster may be kept abreast of the general weather situation.

f. If ambient air quality monitoring from a ground site or an aircraft is being conducted, communication will have to be established from the monitoring site or aircraft to the command center.

g. If key public officials will not be at the command center during critical operations, communications must be established to them. In any event, communication procedures will have to be established for reaching them when they are not at the command center.

h. Communication procedures will have to be established between the public officials and the general public. This will include radio and television hookups, as well as the use of police and fire communications systems. A public information office is needed to relieve technical persons from that responsibility. Good public information is the best way to counteract rumors.

4. Personnel at Disaster Site

The first rule is only send those persons to a disaster site who are absolutely necessary to organize the tasks the agency may be called on to perform. Do not send only technical experts, because there are many administrative problems that must be dealt with if the operation is to be effective. The persons first sent to the site must have authority to call in the agency's technical experts as needed. Do not send a large group of unnecessary observers who serve no useful purpose other than to eat left-over Red Cross food. Surplus personnel create protection problems if they do not bring their own protective equipment with them. During a critical operation in the removal of a chlorine barge that had collided with a railroad bridge at Morgan City, Louisiana, in January 1973, there were twice as many observers at the control center as there were necessary personnel. While sufficient breathing devices had been assembled for the necessary personnel, there were not nearly enough devices for all the observers. If the control center had become contaminated, chaos would have developed, and it is likely that necessary forecasts and warnings would not have been issued to the public. A complete operation could have failed as a result of having too many observers at the control center.

At an actual disaster, public officials may be near the state of panic looking for someone who can advise them on the severity of the problem and on what action they can take to protect the citizens. Since it is very likely that the air pollution control agency's advice will be sought and relied upon heavily, it is important that those persons sent to the site have adequate knowledge. Do not send observers. Such persons, because they have air pollution control credentials, will be forced to provide advice even if they are not equipped to do so. This could not only prove embarrassing for these persons and to the agency, but it also could be dangerous. This same need to provide expertise also applies to providing assistance over the telephone. No advice is better than bad advice. With no advice, there is at least a 50–50 chance of doing the right thing.

5. Acquiring Necessary Information

The air pollution control agency must know where a disaster involving release of a pollutant to the atmosphere has occurred, the nature of the pollutant, the emission rate, the plume size, the meteorological conditions, what is being done to alleviate the situation, if residents have been evacuated, who is on the scene, etc. To assist the control agency in obtaining this information, a copy of a telephone questionnaire is presented in Table VI.

6. Evacuation

Resist the temptation to advise a large scale evacuation. The control agency should only define the problem for local officials. They, in turn,

Table VI Report of Accidental Episode

Date ____________

1. Time of notification ____________
2. Name of caller ____________
   Title of caller ____________
3. Nature of emergency (i.e. fire, spill, explosion) ____________
4. What pollutant(s) is/are causing emergency situation ____________
5. Owner of material ____________
6. Has owner been notified ? ____________
7. Does owner plan to do anything? ____________
8. Location of emergency ____________
9. Owner of property where emergency is located ____________
10. Has owner of property been notified? ____________
11. How long has material been released? ____________
    a. Emission rate ____________
    b. Plume rise ____________
12. How long is material likely to be released in the future? ____________
13. Is any kind of monitoring now being done? ____________
14. Is any monitoring planned for the future? ____________
15. Any idea of what ambient air concentrations are? ____________
16. Any idea of toxic effects of released material? ____________
17. What is being done to alleviate the situation? ____________
18. Is caller knowledgable or does he have access to information on combating the emergency? ____________
19. Have residents of area been alerted or evacuated? ____________
    a. Number of people evacuated ____________
    b. Size of area evacuated ____________
20. Who is on the scene? ____________
    a. Local air pollution agency ____________
    b. State air pollution agency ____________
    c. State police ____________
    d. Local police ____________
    e. Fire department ____________
    f. Coast Guard ____________
    g. EPA Office of Water Programs ____________
    h. DHEW Division of Emergency Health Service ____________
    i. Other ____________
    j. Is Office of Emergency Preparedness involved? ____________
21. Description of emergency area
    a. Rural or urban ____________
    b. Topography ____________
    c. Local weather condition (e.g., wind speed, wind direction) ____________
22. Who to call for additional information ____________
23. Will caller have additional information later? ____________
24. Where can caller be reached? ____________
25. If caller will have additional information, when will it be available? ____________
26. Who else has been notified? ____________

should decide on what actions to take in order to reduce the threat to the public. Evacuation is itself a dangerous experience for certain segments of the population, and an air pollution control agency usually does not have an appreciation of these problems. The air pollution control agency should be able to define the health threat, but other officials will have to consider this, along with other factors, in deciding whether evacuation efforts should be made.

7. Subsequent Safety

After the disaster has passed the air pollution control agency has the responsibility to help determine whether areas are once again safe to occupy. This would include, for instance, monitoring in low-lying areas after a chlorine release.

8. Environmental Impact

The agency would be expected to investigate the environmental impact of the disaster and file a report on it.

9. Extent and Limitations of Responsibilities

In the case of the potential disaster involving a potential release of a pollutant into the atmosphere, one of two situations may develop:

a. Local officials may not be aware of the potential severity of the situation and may not provide adequate assistance to the air pollution control agency by establishing communications and operational procedures by which to take advantage of the agency's services.

b. Local officials may be aware of the severity of the problem, and may attempt to assign all disaster control responsibilities to the air pollution control agency.

These can both be bad situations in that the former will not allow the agency to adequately provide its services and the latter may get it involved in a situation that it is unequipped to handle.

During the March 1972 Louisville, Kentucky, chlorine barge salvage operation previously mentioned, local elected officials were unaware of the potential severity of the operation until it was well under way. Until these officials became aware of the problem and then took charge of the operation, it was very hard for federal air pollution control officials to provide their services in an effective manner. For instance, operations were first established at the United States Coast Guard operation center, and dispersion forecasts were made there. It was then learned that the Coast Guard had no responsibility for warning the public. All they were going to do in the event of a chlorine release was call the Federal Office of Emergency Preparedness and local civil defense officials and report that a release had occurred. It was later learned that the office of Emergency Preparedness had no communication procedures established, so forecast operations were established at the Louisville civil defense headquarters. It was then found that communications had not been established between Louisville civil defense officials and their counterparts across the Ohio River in Indiana. It was necessary for air pollution control officials to participate in establishing this communication link. Many of these problems could have been eliminated if local officials had understood the potential danger and the steps necessary to alleviate the threat.

In a situation near West Lafayette, Ohio, where a railroad car of phosphorous pentasulfide wrecked and burned, local officials were aware of the potential threat but were in a near panic as to what could or should be done to protect the public. When an air pollution control official arrived only as an observer, they forced him to make recommendations concerning evacuation that he was totally unequipped to give. As stated earlier, do not send observers to a potential or actual disaster situation or they are likely to find themselves in a similar situation.

If air pollution control agency officials arrive at a potential disaster site and local officials attempt to assign all responsibility to them, including containment or control operations, they must explain that this is beyond their technical capabilities.

In conclusion, while the air pollution control agency should be prepared to perform support functions during an accidental episode, it must also be aware of the limitations in assuming full responsibility for the control operation.

REFERENCES

1. Y. Horie, *Proc. 67th Ann. Meet.* Air Pollut. Contr. Assoc., Pittsburgh, Pennsylvania, *1974* Paper No. 74-209 (1974).
2. J. C. Mosher, E. L. Fisher, and M. F. Brunelle, "Ozone Alerts in Los Angeles County 1955–1971." Air Pollution Control District, County of Los Angeles, Los Angeles, California, 1972.
3. R. L. Chass, J. N. Birakos, M. F. Brunelle, and J. Mosher, *Proc. 65th Ann.* Meet. Air Pollut. Contr. Assoc., Paper No. 72-59 (1972).
4. R. F. King, "Time-Sharing Isopleth-Area Program." Center for Environmental Studies, Argonne National Laboratory, Argonne, Illinois, 1972.
5. U.S. Environmental Protection Agency, "Accidental-Episode Manual," No.

APTD-1114, pp. 3–4. Office of Technical Information and Publications, Research Triangle Park, North Carolina, 1972.
6. American Conference of Governmental Industrial Hygienists, "Threshold Limit Values for Chemical Substances and Physical Agents in the Workroom Environment with Intended Changes for 1976." Cincinnati, Ohio, 1976.
7. Manufacturing Chemists Association, "Chemical Safety Data Sheets." Washington, D.C.
8. National Safety Council, "Industrial Safety Data Sheets." Chicago, Illinois.
9. N. I. Sax, "Dangerous Properties of Industrial Materials," 2nd ed. Van Nostrand-Reinhold, Princeton, New Jersey, 1963.
10. National Fire Protection Association, "Fire Protection Guide on Hazardous Materials," 4th ed., Boston, Massachusetts, 1972.
11. C. W. Bahme, "Fire Protection for Chemicals," National Fire Protection Association International, Boston, Massachusetts, 1961.
12. National Safety Council, "Chemical Safety Slide Rule." Chicago, Illinois, 1967.
13. U.S. Environmental Protection Agency, "Accidental-Episode Manual," No. APTD-1114. Office of Technical Information and Publications, Research Triangle Park, North Carolina, 1972.

4

Organization and Operation of Air Pollution Control Agencies

Jean J. Schueneman

I. Philosophy of Air Pollution Control

A. Objectives of Control Programs

The basic objective of control programs is to preserve the health and welfare of man, both now and in the distant future. Other objectives are protection of and prevention of damage to plant and animal life, prevention of damage to physical property and interference with their normal use and enjoyment, provision of visibility required for safe air and ground transportation, ensuring continued economic growth and development, and maintenance of an aesthetically acceptable and enjoyable environment.

Those planning or operating air pollution control programs must consider the collective needs and desires of their community and make every effort to translate them into a plan of action. The program should be designed to provide for air management in a broad sense and encompass both short- and long-range goals.

To accomplish long-range objectives, a series of shorter-range objectives must be established if an orderly, planned program is to result. These include collecting fundamental facts, determining conditions existing in the field, analyzing information, developing a plan of action, and putting the plan into effect.

In the field of air pollution control in major urban centers, there are usually a wide variety and great number of activities that could be undertaken to improve present air quality or to prevent degradation of air quality in the future. There is a temptation to move ahead on all fronts simultaneously and to engage in so many activities that none receive the degree of emphasis and effort necessary to assure success. Such overextension should be avoided in favor of a more concentrated attack on a manageable group of activities that can be promptly and successfully completed with the resources available. Unless this is done, the program administrator will find himself and his program in the midst of a confusing, meaningless series of efforts and without any completed program segments to show (*1*).

B. Public Policy in Air Pollution Control

Public policy may be defined as the central ideas and purposes that guide the public actions of individuals and social institutions, both public and private. There are four central ideas that should be accepted as axioms of public policy in the air pollution field.

First, action should not be postponed until the air is so polluted that a "crash" type corrective program of control is necessary. Drastic actions taken on short notice in response to a sudden awakening to a problem that has been worsening for years are far more costly and disruptive of normal economic and social patterns than control measures planned in advance and gradually applied. It is possible and prudent to continually look far into the future and identify trends in community growth and activity patterns that are adverse to air conservation and to take steps, in an evolutionary manner, to head off or redirect situations to the end that serious problems are not allowed to develop.

Second, the air cannot be purified after it is polluted. Prevention rather than cure must be the aim. The control or prevention of air pollutants at their sources, therefore, is of fundamental importance.

Third, air pollution control cannot be accomplished without close co-operation among all levels of government, industry, and the public. In the United States, the primary source of the police power on which regulatory programs for air pollution control are based rests with the states. In other countries, provinces usually have similar powers, either inherently or by delegation. The need for states and provinces to exercise this power in the air pollution field, either directly or by delegation to appropriate sublevels of government, is therefore obvious, as is participation by the national government. Preferentially, commercial, industrial, municipal, and individual participation should be voluntary, in recognition that their long-term interests are best served by clean air.

Finally, the entire structure of any air resource management program rests on enlightened public opinion. The air resource of any area will be managed—or neglected—to the extent its citizens desire and demand.

The establishment of precise statements of public policy in the air resource field is deterred by several factors. The relationships between adverse effects of air pollutant concentrations have not been unequivocally delineated. Policy makers and the public are having difficulty grasping the present and future significance of the population, urbanization, and energy use explosions. The jurisdictional area of policy makers in local government is often too small to encompass the region for which policy needs to be made. Disparity exists between those who pay for the costs of pollution control and those who benefit from it. There are formidable obstacles, but they are not insurmountable.

C. The Air Resource Management Concept

The basic use of our air resource is to sustain life. All other uses must yield to the maintenance of air quality that will not degrade, either acutely or chronically, the health or well-being of man. There remain but two major areas of possible compromise. These are the aesthetics and the economic impact of air pollution and its control. The cost borne by society to achieve a desired quality of air should be in balance with the benefits to be attained (*2*).

Until recent years, the control of air pollution was based upon the correction of existing problems. Little or no consideration has been given to long-range planning or how proposed community or regional master plans might influence future ambient air quality. This lack of coordinated planning has tended to maximize the creation of air pollution problems. As long as this situation is allowed to exist, air pollution problems will be created faster than they can be resolved. The use of the air resource management concept should improve this condition materially.

Early air pollution control programs were oriented along two major lines: (a) control of smoke and particulates from combustion sources and (b) control of other discrete emissions into the atmosphere that were deemed a "nuisance" and were capable of technical and economic control. In both cases, it was hoped that the desired air quality would be achieved. What this "desired" air quality was remained rather ill defined. In more recent years, ambient air quality standards have come into existence, and these provided specific objectives toward which air pollution control efforts could be directed. Further, the air has been recognized as a limited resource that must be preserved and protected. But at the same time, the air must be used as a source of oxygen and as a medium for disposal of the residual materials resulting from human activity that cannot otherwise be disposed of. Thus, especially in major urban centers, but on a global basis as well, it has become necessary to make plans and decisions as to how the air is to be used and protected. The approach to doing so has been called "air resource management" (*2*). The elements of an air resource management program may be listed as follows (*2*):

a. Development of a public policy on air conservation.

b. An organizational framework and staff capable of operating along functional lines (e.g., engineering, technical services, field services) and its funding support.

c. Delineation of realistic short-range goals that can be effectively met in a reasonable time period, say 5 years.

d. Continual assessment of existing air quality and preparation of estimates of the future situation.

e. Assessment in depth on a continuing basis of the emissions from all existing pollution sources and those expected to exist in the future.

f. Development of the necessary information about factors that influence the transport of air pollutants.

g. Assessment of the effects of air pollution on man and his environment.

h. Establishment of ambient air quality standards (referred to by some as objectives or goals).

i. Design of measures and programs calculated to bring about the air quality desired.

j. Development of long-range air use plans, fully integrated with other plans for energy supplies, land use, transportation, recreation, refuse disposal, etc., to cope effectively with projected changes.

k. Development of an understanding of the broad impact of changing science and technology on air resources and their potential effect on the social and economic character of modern society.

l. Development of an effective information and educational program to inform the public of the need to solve air pollution problems promptly and effectively.

The mechanisms and organization for reaching decisions in a community (state or province) are of key importance in the success of an air resource management program. They must provide for the assimilation of information into goals and policies and the subsequent enunciation of a public policy on the management of a community's air resource. Appropriate community involvement in the decision-making process must be encouraged to ensure a base of public support for action. The degree of success achieved in making decisions and development of policy will depend upon the clear delegation of responsibility to a key individual. It follows that full support by top management (including chief elected executives) is needed to develop and implement a set of program goals.

Long-range planning must be an integral part of an air resource management program. It helps smooth the way for anticipated social and technological changes in the urban environment and makes it possible to consider a much broader scope of alternate programs to achieve or maintain high quality air. For example, in a 25-year plan, consideration could be given to the elimination of certain existing combustion operations and substitution of electrical energy or district heating as one means of reducing pollutant emissions—a program too complicated to consider on a short-range basis. Planning forces program directors to think ahead about technological and other changes and predisposes them toward changes to which they can agree and support. Advance planning, with its integral assembly of information about the future, makes it possible to identify what may happen if a given trend were to continue and to challenge the often-made assumption that it is "good" to have more of everything and to use more materials, fuels, and services. With proper long-range planning, such assumptions can be evaluated in all of their ramifications, and instead of being restricted to finding ways to accomodate unbridled growth, alternatives can also be examined and steps can be taken to guide and direct future trends in advantageous ways. Without planning, the director will find himself reacting to changing events rather than influencing them as he should.

The air resource management concept provides a base for organization of tasks so that program directors can perform them systematically, purposefully, with understanding, and with a reasonable probability of accomplishment. It endeavors to develop a point of view, identifies approaches for finding what should be done, and describes how to go about it. In this way, it is more likely that the steps required to achieve and maintain desired air quality will be taken expeditiously and in an equitable manner.

The general air resource management approach described above is not entirely applicable in all situations. In certain areas where air quality is already very good, it may be desirable also to incorporate measures that prevent any future significant deterioration of air quality rather than to allow development to occur up to the point where generally applicable air quality standards would be just barely met. Such areas might be scenic vistas or recreational areas. In other cases, it may not be socially and economically desirable or technically possible to achieve ambient air quality in keeping with generally applicable air quality standards. Such areas might include dry, arid regions where wind-blown dust makes it impossible to achieve particulate matter standards and areas where oxidant levels are high because of naturally occurring ozone. In such cases, the usual air resource management approach may be modified to incorporate measures to reduce pollution levels to the lowest practicable levels, even though air quality standards are not met.

II. Governmental Air Pollution Control Programs in the United States

A. Local Governmental Programs

1. Extent of Programs and Budgets

There were about 183 full-time local governmental air pollution control programs in the United States in 1973 (Table I) (*3*). There are quite a number of other governmental agencies that conduct some, but very limited, air pollution control activities. The number of full-time agencies has increased over the years since 1952, with marked increases in the 1961–1965 and 1965–1973 periods (Table II). There were 49 agencies that existed in 1965 that did not exist in 1973. They had a total budget, including Federal grants, in 1965, of about $1.8 million. Of these, 26 were city agencies that in 1973, had been incorporated into county or regional programs; the remainder were mostly small agencies that discontinued operations. The *median* budget for all local agencies was about 38 cents per capita in 1973, an increase over the 15 cents per capita in 1965 and the 10 cents of the 1952–1961 period.

The funds available to local agencies have increased markedly over the years. Local governments have increased their appropriations from $2.5 million in 1952 to $10.5 million in 1965 to $34.0 million in 1973. Added to these funds were federal program grants of $3.6 million in 1965 and $20.0 million in 1973 (Table III). This has resulted in large increases in *average* budget per capita served from 8 cents in 1952 to 21 cents in 1965 to 55 cents in 1973 for all agencies collectively. However, increased agency operating cost, as indicated by the average salaries of municipal employees, has also risen markedly, and thus the real increase in agency capability has been much less than the total budget and expenditures per capita would indicate (Table III). Another factor has been the continuing increase in urban population, which tends to further reduce the apparent increase in agency capability as compared to the magnitude of the job to be done. The real increase in per capita budget based on the total urban population, adjusted for increases in operating cost, has been from 6 cents per capita in 1961 to 9 cents in 1965 to 17 cents in 1973. However, some of the urban population now receives air pollution control services from state governments rather than local governments, and thus the situation in 1973 is somewhat better than these statistics indicate.

The population served by local air pollution control agencies has increased from 31 million in 1952 to 68 million in 1965 to 99 million in 1973. This represents 30%, 50%, and 65%, respectively, of the total urban population of the nation. Between the local programs listed in Table I and the state programs described later, there is now complete coverage of the entire population of the country with an air pollution control program of more or less adequate capability.

A few major local air pollution control programs spend a large portion of the total funds spent by local governments. The 9 largest agencies (spending $1.0 million per year or more) account for 47% of the total; the 20 largest agencies (spending $500,000 or more) account for 63% of the total local governmental expenditures. The four largest agencies

Table I Local Air Pollution Programs and Budgets—United States[i] (Agencies with Federal Grant in 1973)[a]

City or area	Population[b] 1970 (×1000)	1973 budget ($1000)[c] Non-federal	Federal	Total	1973 per capita budget (cents)	1965 total budget ($1000)
Alabama						
Huntsville	138	44	44	88	63.8	15
Jefferson County (Birmingham)	645	49	147	196	30.4	163
Mobile County	317	36	108	144	45.4	29
Tri-County (Lawrence, Limestone, Morgan)	146	23	50	73	50.0	—
Alaska						
Cook Inlet (Anchorage)	116	50	50	100	86.2	—
Fairbanks	31	78	20	98	316.1	—
Arizona						
Maricopa County	968	167	92	259	26.8	129
Pima County	352	68	59	127	36.1	33
California						
Fresno County	413	76	91	167	40.4	—
Humboldt County	100	35	53	88	88.0	7
Kern County	329	100	87	187	56.8	—
Los Angeles County	7,032	5,192	805	5,997	85.3	3,663
Monterey and Santa Cruz County	374	110	85	195	52.1	—
Orange County	1,420	221	72	293	20.6	126
Riverside County	459	320	150	470	102.4	86
Sacramento County	631	82	81	163	25.8	28
San Bernardino County	684	700	86	786	114.9	284
San Diego County	1,358	500	432	932	68.6	68
San Francisco (Bay Area)	4,278	3,223	498	3,721	87.0	1,048
Santa Barbara County	264	60	60	120	45.5	—
Stanislaus County	195	21	62	83	42.6	—
Ventura County	376	179	99	278	74.0	30
Yolo and Solano County	136	77	41	118	86.8	—

Colorado						
Boulder City and County	132	11	16	27	20.5	9
Colorado Springs	229	25	20	45	19.7	—
Denver City and County	1,105	72	197	269	24.3	151
Mesa County	54	7	9	16	29.6	—
Pueblo City	97	26	31	57	58.8	—
Tri-County (Adams, Arapahoe, Douglas)	356	46	87	133	37.4	48
Connecticut						
Bridgeport	157	40	80	120	76.4	34
Bristol	55	14	26	40	72.7	—
Fairfield	56	16	4	20	35.7	23
Greenwich	60	20	11	31	51.7	—
Meriden	56	19	20	39	69.6	—
Middletown	37	21	34	55	148.6	19
Milford	51	18	10	28	54.9	—
New Haven	138	65	59	124	89.9	—
Norwalk	79	40	20	60	75.9	—
Stamford	109	55	80	135	123.9	—
Stratford	50	18	4	22	44.0	22
Washington, D.C.	757	143	234	377	49.8	78
Metropolitan Washington Council of Governments[d]	2,847	5	100	105	3.9	23
Florida						
Broward County	620	54	85	139	22.4	—
Hillsborough County	490	111	81	192	39.2	10
Jacksonville City	529	105	140	245	46.3	—
Palm Beach County	349	73	75	148	42.4	52
Georgia						
Fulton County	608	79	127	206	33.9	24
Illinois						
Chicago	3,367	1,850[e]	770	2,620	77.8	1,163
Cook County	2,124[f]	944	350	1,294	60.9	150
Granite City	40	23	30	53	132.5	—

Table I (*Continued*)

City or area	*Population*[b] *1970* (×*1000*)	*1973 budget ($1000)*[c] *Non-federal*	*Federal*	*Total*	*1973 per capita budget (cents)*	*1965 total budget ($1000)*
Indiana						
Anderson	71	27	31	58	81.7	—
East Chicago	47	69	69	138	293.6	34
Evansville	139	50	33	83	59.7	25
Gary	175	104	100	204	116.6	106
Hammond	108	60	30	90	83.3	—
Indianapolis (Marion County)	792	185	100	285	36.0	40
Michigan City	39	25	15	40	102.6	—
St. Joseph County	245	38	35	73	29.8	—
Vigo County	115	27	23	50	43.5	—
Iowa						
Cedar Rapids, (Linn County)	163	33	30	63	38.7	16
Des Moines, (Polk County)	286	38	76	114	39.9	—
Kansas						
Kansas City (Wyandotte County)	186	55	82	137	73.7	32
Topeka (Shawnee County)	155	9	24	33	21.3	—
Wichita (Sedgwick County)	351	27	79	106	30.2	—
Kentucky						
Jefferson County	695	232	164	396	57.0	108
Maryland						
Allegany County	84	15	17	32	38.1	—
Anne Arundel County	298	88	112	200	67.1	27
Baltimore City	906	206	206	412	45.5	77
Baltimore County	621	180	180	360	58.0	27
Frederick County	85	15	18	33	38.8	—
Montgomery County	523	130	140	270	51.6	—
Prince Georges County	661	85	87	172	26.0	—

Massachusetts[g]						
Berkshire County	149	10	29	39	26.2	—
Boston Metro (Plymouth, Essex, Norfolk, Suffolk, and Middlesex Counties)	2,754	193	300	493	17.9	219
Central Air Pollution Control District (Worcester County)	346	28	65	93	26.9	—
Fitchburg	43	12	10	22	51.2	—
Merrimack Valley Air Pollution Control District (parts of Essex and Middlesex Counties)	434	29	86	115	26.5	—
Pioneer Valley Air Pollution Control District (Franklin, Hampshire, and Hampden Counties)	642	61	117	178	27.7	44
Southeastern Air Pollution Control District (Plymouth, Dukes, Norfolk, Nantucket, and Barnstable Counties)	500	28	85	113	22.6	—
Worcester City	177	17	35	52	29.4	35
Michigan						
Flint	193	49	33	82	42.5	—
Grand Rapids (Kent County)	393	45	47	92	23.4	10
Macomb County	625	37	53	90	14.4	—
Muskegon County	157	20	34	54	34.4	21
Wayne County (Detroit)	2,667	847	1,000	1,847	69.3	133
Minnesota						
Olsted County	84	13	20	33	39.3	—
St. Cloud County	64	15	28	43	67.2	—
St. Louis County	221	30	36	66	29.9	—
St. Paul	310	148	148	296	95.5	69
Missouri						
Independence	153	38	17	55	35.9	—
Kansas City	112	24	21	45	40.2	—
Greene County (Springfield)	507	147	110	257	50.7	—
St. Louis	622	226	323	549	88.3	177
St. Louis County	951	258	256	514	54.0	174
Montana						
Great Falls	60	10	17	27	45.0	—
Missoula County	58	20	9	29	50.0	—
Yellowstone County	87	15	22	37	42.5	—

Table I (Continued)

City or area	Population[b] 1970 (×1000)	1973 budget ($1000)[c] Non-federal	Federal	Total	1973 per capita budget (cents)	1965 total budget ($1000)
Nebraska						
Lincoln and Lancaster Counties	198	77	44	121	61.1	—
Omaha	347	127	60	187	53.9	8
Nevada						
Clark County	273	85	127	212	77.7	50
Washoe County	121	39	58	97	80.2	41
New Jersey						
Central (Linden, Perth Amboy, Rahway, Sayreville, South Amboy, and Woodbridge)	248	39	118	157	63.3	—
Elizabeth	113	56	45	101	89.4	10
Hudson Municipal (Hudson County)	609	36	108	144	23.6	—
Suburban (Parts of Essex and Union Counties)	500	49	78	127	25.4	—
New Mexico						
Albuquerque	328	82	63	145	44.2	15
New York						
Erie County	1,113	629	211	840	75.5	114
Monroe County	712	149	87	236	33.1	—
Nassau County	1,423	555	155	710	49.9	150
New York City	7,868	6,110	999	7,109	90.4	1,220
Niagara County	236	96	96	192	81.4	47
Onondaga County	472	98	24	122	25.8	—
Rockland County	230	60	60	120	52.2	—
Suffolk County	1,117	115	76	191	17.1	31
Westchester County	891	80	50	130	14.6	—
North Carolina						
Cleveland County	73	25	19	44	60.3	—
Durham County	133	19	24	43	32.3	18
Forsyth County	214	34	63	97	45.3	—

Gaston County	148	21	16	37	25.0	26
Guilford County	289	22	13	35	12.1	24
Mecklenburg County	355	102	109	211	59.4	—
Rowan County	90	13	14	27	30.0	17
Western (Buncombe, and Haywood Counties)	187	36	72	108	57.8	33
Unifour APCP (Alexander, Burke, Caldwell, and Catawba Counties)	227	20	60	80	35.2	—
Ohio						
Akron (Medina, Summit, and Portage Counties)	755	151	179	330	43.7	130
Canton (Stark County)	369	64	64	128	34.7	—
Cincinnati	1,421	321	565	886	62.4	97
Cleveland (Cuyahoga County)	1,702	620	403	1,023	60.1	277
Lake County	197	28	39	67	34.0	—
Lorain	78	24	20	44	56.4	29
Mansfield (Richland County)	130	19	24	43	33.1	—
Montgomery, Darke, Miami, Clark, Preble, and Greene Counties	1,045	257	368	625	59.8	—
Portsmouth (Adams, Lawrence, and Scioto Counties)	153	36	67	103	67.3	—
Steubenville (Columbiana, Jefferson, Belmont, and Monroe Counties)	301	41	65	106	35.2	—
Toledo (Lucas County, City of Rossford, and City of Oregon)	479	161	128	289	60.3	63
Youngstown (Mahoning, and Trumbull Counties)	536	44	88	132	24.6	18
Oklahoma						
Oklahoma City (Oklahoma County)	511	82	85	167	32.7	13
Tulsa City–County	402	100	100	200	49.8	33
Oregon						
Columbia-Willamette (Clackamus, Columbia, Multnomah, and Washington Counties)	910	222	314	536	58.9	—
Lane County	213	86	88	174	81.7	—
Mid-Willamette (Benton, Linn, Marion, Polk, and Yamhill Counties)	208	74	90	164	78.8	52
Pennsylvania						
Allegheny County	1,605	440	660	1,100	68.5	370
Lehigh Valley	255	12	12	24	9.4	15
Philadelphia	1,950	516	766	1,282	65.7	366
Greater York	115	17	8	25	21.7	—

Table I (Continued)

City or area	Population[b] 1970 (×1000)	1973 budget ($1000)[c] Non-federal	Federal	Total	1973 per capita budget (cents)	1965 total budget ($1000)
South Carolina						
Charleston	248	48	58	106	42.7	—
Columbia	284	11	5	16	5.6	—
Greenville	241	27	37	64	26.6	—
Spartanburg	174	15	30	45	25.9	19
Tennessee						
Chattanooga (Hamilton Ccunty)	254	137	142	279	109.8	37
Knox County	276	46	60	106	38.4	—
Nashville (Davidson County)	448	85	134	219	48.9	31
Memphis (Shelby County)	722	88	92	180	24.9	—
Texas						
Corpus Christi (Nueces County)	245	17	34	51	20.8	—
Dallas	845	143	123	266	31.5	—
El Paso City–County	322	47	95	142	44.1	—
Ft. Worth	393	72	64	136	34.6	—
Galveston County	170	62	61	123	72.4	—
Houston	2,133	426	316	742	60.2	—
Jefferson	245	29	42	71	29.0	—
Laredo (Webb County)	220	16	24	40	18.2	—
Lubbock City–County	179	30	33	63	35.2	—
San Antonio	835	68	65	133	15.9	—
Virginia						
Alexandria	111	42	42	84	75.7	
Fairfax County	488	79	61	140	28.7	—
Richmond	250	74	35	109	43.6	29
Roanoke	89	21	21	42	47.2	14

Washington						
Benton, Franklin, and Walla Walla Counties	118	9	23	32	27.1	—
Northwest Air Pollution Control District (Skagit, Whatcom, Island, and San Juan Counties)	165	61	46	107	64.8	—
Olympic Air Pollution Control District (Thurston, Pacific, Grays Harbor, Mason, Jefferson, and Callam Counties)	219	46	63	109	49.8	—
Puget Sound Air Pollution Control District (King, Kitsap, Pierce, and Snohomish Counties)	1,675	450	457	907	54.1	—
Southwest Air Pollution Control District (Clark, Cowlitz, Lewis, Skamania, and Wahkiakum Counties)	252	78	75	153	60.7	—
Spokane County	287	24	49	73	25.4	—
Yakima County	145	16	27	43	29.7	—
West Virginia						
Wheeling	48	14	10	24	50.0	6
Wisconsin						
Milwaukee	1,054	319	158	477	45.3	186
Racine	95	15	45	60	63.2	—
Total	98,548	34,033	20,033	54,066		12,483[h]

[a] Data from Schueneman (3) and United States Environmental Protection Agency reports.
[b] Source: Department of Commerce, Bureau of the Census; 1970 Census of Population Reports; and population as given in United States Environmental Protection Agency grant award information for population served by agency.
[c] Budget as given by United States Environmental Protection Agency Program Grant, unless otherwise stated.
[d] Planning grant given to Metropolitan Washington Council of Governments.
[e] Chicago also had $2.5 million available for air pollution control (motor vehicles) in 1973.
[f] Cook County agency serves population outside city of Chicago. Population is approximate.
[g] Local programs operated by state, except Fitchburg and Worcester City.
[h] Does not include $1,788,000 budgets of forty-nine agencies that existed in 1965 but that did not exist in 1973, except for those for which data were not available in 1973.
[i] Data not available for significant programs known to exist in 1973 in Manatee County, Florida; Dade County, Florida; Boston, Massachusetts; Minneapolis, Minnesota; The Interstate Sanitation Commission (New York City, New York Area); and Harris County, Texas.

Table II Per Capita Expenditures of Local Air Pollution Control Agencies—United States[a]

Annual expenditures per capita (cents)	*1952*		*1955*		*1961*		*1965*		*1973*	
	A[b]	*B*[c]	*A*	*B*	*A*	*B*	*A*	*B*	*A*	*B*
1–9	22	49	36	47	36	42	38	29	3	2
10–19	16	35	27	36	28	33	46	35	8	4
20–29	3	7	7	9	14	17	23	18	29	16
30–39	3	7	2	3	3	4	8	6	31	17
40–49	0	0	4	5	2	2	5	4	23	13
50–59	0	0	0	0	2	2	5	4	25	14
60–69	1[d]	2[d]	0	0	0	0	5[d]	4[d]	21	12
70–79	—	—	—	—	—	—	—	—	13	7
80–89	—	—	—	—	—	—	—	—	11	6
90–99	—	—	—	—	—	—	—	—	3	2
100–109	—	—	—	—	—	—	—	—	2	1
110–119	—	—	—	—	—	—	—	—	3	2
120 or more	—	—	—	—	—	—	—	—	5	3
Total	45	100	76	100	85	100	130	100	177	100
Median (cents per capita)	10.0		10.0		10.8		15.2		37.8	

[a] In 1952–1965, this is agencies spending $5000 or more per annum. In 1973, it is all agencies with federal grants. About six significant agencies had no grant. Data are from Schueneman (*3*) and United States Environmental Protection Agency files.
[b] Number of agencies reporting.
[c] Percent of agencies reporting.
[d] Agencies with a per capita budget of 60 cents or more.

(New York, New York; Los Angeles County, California; San Francisco Bay Area, California; and Chicago, Illinois) alone spent over $19 million in 1973 (Table I). The total manpower available to local agencies in 1973 was about 3000, incuding all types of personnel.

2. Organizational Patterns

Local governmental agencies (cities, counties,* etc.) have traditionally carried out the major portion of the regulatory and other parts of air pollution control programs. This was true because the most severe air pollution problems were first recognized and addressed in the major urban centers. Until the late 1940s, cities operated nearly all of the air pollution control programs. As recently as 1961, three-fourths of local

* By "counties" is meant the principal administrative subdivisions of states or provinces.

control agencies were in city governments (Table IV) (*3, 4*). But since that time, in recognition of the regional nature of air pollution problems and the difficulty and poor economy of operating several separate agencies engaged in the same work in the same metropolitan area, various region-wide organizations were created. In regional areas where there

Table III Total Expenditures and Population Served by Local and State Air Pollution Control Agencies—United States[a]

Item	*1952*	*1955*	*1961*	*1965*	*1973*
Number of local programs	44	70	72	117	177
Population served (millions)	31	36	43	68	99
Local funds (millions of dollars)	2.5	4.9	7.9	10.5	34.0
Federal grant funds to local agencies (millions of dollars)	0	0	0	3.6	20.0
Total local and federal funds (millions of dollars)	2.5	4.9	7.9	14.1	54.0
Per capita budget, for population served, local agencies (cents)	8	14	18	21	55
Operating cost index used to do cost adjustment[b]	—	—	1.00	1.18	2.04
Per capita budget, for population served, local, agencies, cost adjusted (cents)	—	—	18	18	27
Total urban population (millions)	102	111	127	135	153
Per capita budget, for total urban population, local agencies, cost adjusted (cents)	—	—	6	9	17
Number of state and territorial programs	1	3	18	35	55
Population of states with programs	NA[c]	NA	10	154	210
State funds (millions of dollars)	NA	NA	2.0	4.4	26.8
Federal grant funds to states (millions of dollars)	NA	NA	0	1.7	27.0
Total state and Federal funds	NA	NA	2.0	6.1	53.8
Per capita state budget for population served (cents)	NA	NA	20	4	24
Per capita state budget for population served, cost adjusted (cents)	NA	NA	20	3	12
Total United States population (millions)	NA	NA	183	194	210
Per capita state budgets; total United States population; cost adjusted (cents)	NA	NA	1	3	13
Total state and local funds	NA	NA	9.9	14.9	60.8
Total federal grant funds	NA	NA	0	5.3	47.0
Total state, local, and federal grant funds	NA	NA	9.9	20.2	107.8
Per capita budget, all funds, total United States population (cents)	NA	NA	5	10	51
Per capita budget, all funds, total United States population, cost adjusted (cents)	NA	NA	5	9	25

[a] Data from reference (*3*) and other sources.
[b] Average salary of municipal employees in the United States used as an index with 1961 equal 1.00. Agencies spend 70–80% of their budgets for salaries.
[c] NA means not applicable or not available.

Table IV Jurisdictional Areas of Local Air Pollution Control Agencies—United States[a]

Kind of jurisdiction	*1961*	*1965*	*1973*
City	64	61	49
County	15	46	76
City–county	1	11	28
Multicounty	2	6	16
Multicity	3	6	4
Others	0	0	15
Total	85	130	188

[a] Data from Schueneman (*3*) and the Air Pollution Control Association (*4*).

is no major city but where there are a number of smaller towns and various industries, it has not been economically feasible to operate a separate control agency for each city or county. In some such regions, a number of counties have joined together to form a single agency large enough to function effectively.

In 1973, only 26% of the local agencies were in city governments, while 40% were at the county level and 34% were in a city–county, multicounty, multicity or some other broader coverage kind of organizational arrangement (Table IV). Yet, many major metropolitan areas do not have any formal regional organizational setting for air pollution control work. Included in this group of areas are New York, New York; Chicago, Illinois; Philadelphia, Pennsylvania; St. Louis, Missouri; Houston, Texas; Washington, D.C.; and Minneapolis–St. Paul, Minnesota. Some major cities that are covered by regional organizational structures include Los Angeles, California; San Francisco, California; Detroit, Michigan; Boston, Massachusetts; Buffalo, New York; Nashville, Tennessee; Birmingham, Alabama; Seattle, Washington; Phoenix, Arizona; Jacksonville, Florida; Miami, Florida; and all of the major cities in Ohio.

Some of the more unusual organizational arrangements have come into being during the past few years. Many city–county agencies have been formed in places where an urban center is unable to annex territory but major growth is occurring in the surrounding territory where governmental services may be weak or nonexistent. In such cases, the central city provides services throughout the area surrounding the city and in the county in which the city is located. In Massachusetts, the state has authority to establish regional air pollution control districts and to operate programs therein. The local governments included in the district pay the state for the services. Five such districts are presently operating. In New Jersey, arrangements have been made to facilitate the joint operation of air pollution control programs by combinations of local governments. Three such agencies are operating. In Ohio, the state has made arrangements for the existing major city air pollution control agencies to provide air pollution control services to surrounding counties, acting as agents for the state. The state, with the aid of federal grants, provides the necessary funds. About ten such service areas were in operation in 1973. In the state of Washington, the state is empowered to designate air pollution control regions made up of appropriate combinations of cities and counties. When an area is designated, the local governments set up the air pollution control agency and operate it. There were seven such regions in 1973. In the Washington, D.C., region, the chief executives of Maryland, Virginia, and the District of Columbia have designated the Metropolitan Washington Council of Governments as the agency to develop plans for control of air pollution in the region, but the council does not have any regulatory powers nor are the plans mandatory on the participating entities. However, the Council's recommendations do have substantial effect. In the Minneapolis–St. Paul, Minnesota, area the Twin Cities Metropolitan Council, by agreement with the state, does the air quality planning for the region. The council cooperates with the state in this work and also approves or disapproves applications for construction of any major facility before it is acted on by the state. In the New York City, New York, region, the Interstate Sanitation Commission provides certain collaborative and coordinative services for the constituent States of New York, New Jersey, and Connecticut.

In some states, the legislature has concluded that air pollution control activities are more appropriately carried on at the state level of government than at the local level. Thus, in Louisiana, Arkansas, and Rhode Island, the local governments are given no role in the air pollution field, and in South Carolina, no new air pollution control agencies may be set up by local government. In addition to these four states, there are ten other states in which there are no local governmental air pollution control programs (Delaware, Hawaii, Idaho, Maine, Mississippi, New Hampshire, North and South Dakota, Utah, and Vermont). In these states, all necessary actions must be undertaken by the state government.

In 1956, local air pollution control agencies were widely dispersed in various departments of local government, with health and building departments being most common (Table V). In those times, much of the air pollution control work involved inspection of fuel burning plants (new construction and existing plants), and in many cities, such work was done by a building or safety department. Safety was involved be-

Table V Organizational Placement of Local Air Pollution Control Programs—United States[a]

Kind of agency	*1956* Number	*1956* Percent	*1965* Number	*1965* Percent	*1973* Number	*1973* Percent
Health department	23	28	68	52	100	56
Separate agency	16	20	27	21	61	34
Environmental agency	0	0	0	0	14	8
Building department	19	23	10	8	0	0
Safety department	10	12	10	8	2	1
All other departments	14	17	15	11	2	1
All agencies	82	100	130	100	179	100

[a] Substantial agencies with more than one employee assigned full-time to air pollution work. Data from Schueneman (*3*) and the Air Pollution Control Association (*4*).

cause of the pressure boilers used in some fuel burning plants. As the scope of air pollution work expanded, health departments became more involved because of their work in various aspects of environmental health and the concern about the adverse health effects of air pollution. In some areas, air pollution control work was of sufficient scope and importance to warrant a separate agency, reporting directly to the executive head of the local government or to a board or commission. By 1973, there were more air pollution control agencies in health departments (56) than in any other department. The agencies in building, safety, and other departments had all but disappeared. Separate agencies were next most prevalent. And by 1973, a new kind of major department had evolved in some local governments, namely, a Department of Environmental Control or some similar title. Such departments may be charged with responsibility for such activities as water pollution control, solid waste, water sanitation, radiological health, and noise abatement, as well as air pollution control. Fourteen local air pollution control organizations were in such environmental agencies in 1973 (Table V). There is little difference in the distribution of agencies among the various departments by virtue of the size of the agency, with the exception that more of the very large agencies are separate entities or in a department other than health than is true for the smaller agencies.

Depending on the nature of the air pollution problem in a community and the kinds of activities being carried on, the internal organization may take many forms. A fairly common organization is shown in Figure 1. About 25–35% of the staff will tend to be in the field services division, about 15–20% in technical services, and the remainder in other ac-

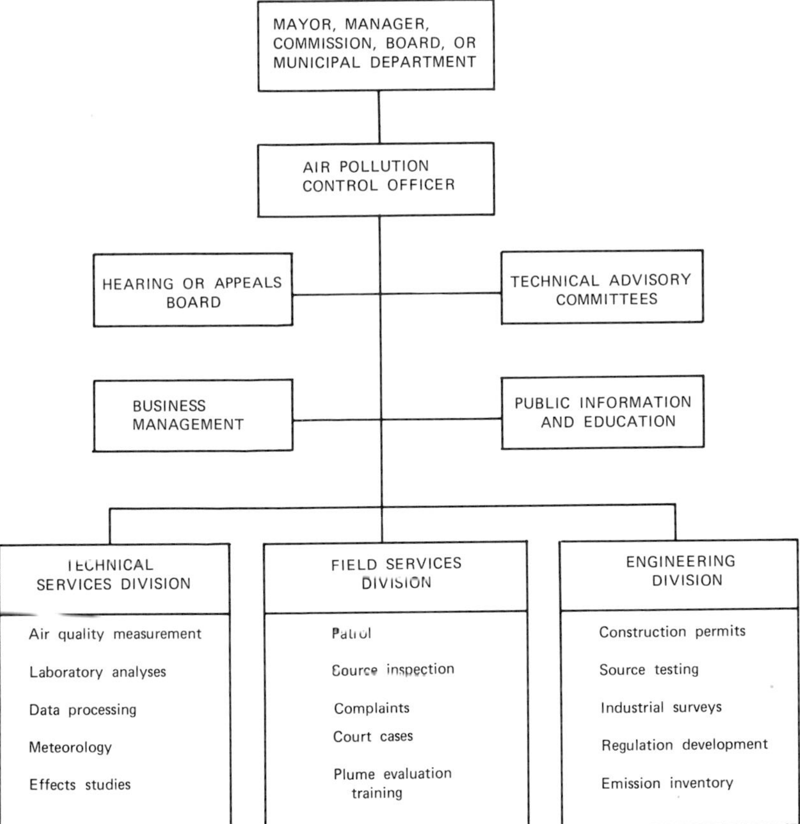

Figure 1. Typical oganization chart for a local governmental air pollution control agency (*3*).

tivities, varying widely. Technical committees provide advice and consultation to the air pollution control officer on a variety of matters, such as air quality standards, emission control regulations, construction plan approval criteria, air pollution emergency action, public relations, and complex scientific questions. Some committees are maintained on a continuing basis, while others may be called together only for a limited time to help with a certain problem. The hearing or appeals board decides on appeals of persons regulated from actions taken by the air pollution control officer—a means of trying to achieve an equitable settlement of points at issue without time-consuming, costly, and inconvenient court action. Such boards are usually appointed by the mayor or other administrative entity and consist of three or five people.

3. Activities of Local Agencies

Activities of local air pollution control programs are principally directed toward bringing about actual control of pollution sources through enforcement of laws, rules, and regulations. The nature and extent of program activities varies tremendously, depending on problems existing in the area and money and personnel available for agency operations. Very small programs having limited staff and problems, and usually serving areas with a population less than one or two hundred thousand, may only seek abatement of nuisance conditions called to their attention by public complaints and perhaps operate air sampling stations and report on new source construction activity as a service to the state agency. This may require only the part-time services of one or more persons to accept and investigate complaints and to negotiate with responsible parties and secure abatement. In complex cases, assistance may be secured from the state agency or consultants.

Somewhat more extensive programs confronting more pervasive or complex problems may also make self-initiated observations and cite violations of antinuisance regulations, issue permits for relatively less complex sources, maintain a registration system for sources of pollution, issue operating permits, and conduct a limited air pollution measurement program. The agency will have the power and capability to prosecute violators and will likely have a close relationship with the state air pollution control agency.

Moderate sized agencies, employing in the range of ten to twenty-five people will carry on all the activities mentioned in the previous paragraph and will likely also be more involved in relationships with other governmental agencies, enforce a broader range of emission regulations, issue permits to more kinds of facilities, maintain an air pollution episode plan in readiness, conduct limited public information activities and operate a relatively complete air quality monitoring program. Modest laboratory services may be included, or they may be provided by the state or through contractual arrangements.

Comprehensive agencies serving larger cities or metropolitan areas will engage in a full line of air pollution control activities in the fields of management, planning and evaluation, field enforcement, engineering, technical services, and public information. A complete list of these activities is provided in Section III,B of this chapter, and many of them are described in some detail in Section IV of this chapter.

The distribution of agency manpower into the various types of activities is generally about 20–25% in management functions; 25–35% in enforcement; 20–30% in engineering and 15–20% in technical services

(*5*). For example, the San Francisco Bay Area, California, agency with a staff of 220 has 89 in enforcement (40%), 26 in engineering (12%), 64 in technical services (29%), and 41 in all other functions (19%), including administration, planning, public information, legal counsel, and others. Wayne County, Michigan, including Detroit, with a staff of about 100 people allocates 33% to enforcement, 13% to engineering, 22% to technical services, and 32% to all other functions, including administrative services, legal counsel, data processing, public information, and training.

B. State Government Programs

1. Extent of Programs and Budgets

The state of California adopted a law in 1947 permitting the formation of county air pollution control districts. This was the first specific state law concerning air pollution, although previously several states had passed laws enabling cities to control smoke, and nearly all states have long had general laws dealing with nuisances. In 1951, Oregon adopted the first statue creating an air pollution control program at the state government level and shortly thereafter began operation with an annual budget of $60,000. By 1956, the states of New Jersey and California had also established identifiable full-time air pollution programs and, for these three states, annual expenditures were about $250,000. In 1961, there were nineteen states operating air pollution programs with a budget of $5000 per year or more (Table VI). These programs served a total of 103 million people and had a total budget of a little more than $2 million. However, in only about five of them were the budgets and programs of any consequence when compared with the job to be done (*6*).

By 1965, thirty-one states had specific laws that called for some form of air pollution program at the state level. Thirty-four states and Puerto Rico had an air pollution program with a budget of $5000 or more (Table VI). Of these, thirteen were newly created within the year, at least in part because of the stimulatory effect of federal program grants that were first made available in 1965. These thirty-five agencies had budgeted state funds of $4.46 million. Added to the increase in state budgets was $1.7 million in federal program and survey grant funds, making a total of $6.3 million available to state agencies, more than three times the amount in 1961.

In 1973, all states and territories except American Samoa had an air pollution control program. Total state appropriations were $26.8 million. Federal grants added $27 million to make a total of $53.8 million avail-

Table VI Budget of State and Territorial Air Pollution Control Programs—United States (1961, 1965, and 1973)[a]

State or territory	Population 1970 (1000)	Budget ($1,000)[b] State or territory	Federal	Total	1973 Per capita budget (cents)	Total budget 1965 ($1,000)	Total budget 1961 ($1,000)
Alabama	3,444	272	362	634	18.4	15	0
Alaska	300	40	70	110	36.7	0	0
American Samoa	28	—	—	0	00.0	0	0
Arizona	1,771	322	227	549	31.0	0	0
Arkansas	1,923	120	202	322	16.7	58	0
California	19,953	5,000[c]	1,058	6,058	30.4	2,437[d]	1,161[d]
Colorado	2,207	455	457	912	41.3	52	10
Connecticut	3,032	495	1,007	1,502	49.5	148	7
Delaware	548	180	193	373	68.1	19	18
District of Columbia	757	143	234	377	49.8	78	42
Florida	6,789	592	700	1,292	19.0	167	63
Georgia	4,590	288	500	788	17.2	78	0
Guam	87	20	55	75	86.2	0	0
Hawaii	769	234	175	409	53.2	74	10
Idaho	713	108	178	286	40.1	10	0
Illinois	11,114	1,253	1,748	3,001	27.0	124	0
Indiana	5,194	200	343	543	10.4	75	0
Iowa	2,825	261	540	801	28.3	0	0
Kansas	2,247	316	412	728	32.4	0	0
Kentucky	3,219	555	493	1,048	32.6	64	0
Louisiana	3,641	230	350	580	15.9	57	0
Maine	992	93	192	285	28.7	0	0
Maryland	3,922	657	606	1,263	32.2	280	38
Massachusetts	5,689	223	566	789	13.8	303[e]	72[e]
Michigan	8,875	479	833	1,312	14.8	117	20
Minnesota	3,805	237	429	666	17.5	5	5
Mississippi	2,217	164	420	584	26.3	0	0
Missouri	4,677	179	254	433	09.2	69	0
Montana	694	110	220	330	47.6	25	0
Nebraska	1,483	36	109	145	09.7	0	0
Nevada	489	23	45	68	13.9	0	0
New Hampshire	738	78	186	264	35.8	19	0
New Jersey	7,168	1,309	1,913	3,222	44.9	523	104
New Mexico	1,016	162	400	562	55.3	21	0
New York	18,237	2,600	2,515	5,115	28.0	529	228
North Carolina	5,082	322	800	1,122	22.0	0	0
North Dakota	618	45	45	90	14.6	15	0
Ohio	10,652	1,805	405	2,210	20.7	55	95
Oklahoma	2,559	115	231	346	13.5	12	0

Table VI (*Continued*)

State or territory	*Population 1970 (1000)*	*Budget ($1000)*[b]			*1973 Per capita budget (cents)*	*Total budget 1965 ($1000)*	*Total budget 1961 ($1000)*
		State or territory	*Federal*	*Total*			
Oregon	2,091	580	62	642	30.7	104	63
Pennsylvania	11,794	958	1,100	2,058	17.4	248	60
Puerto Rico	2,712	342	419	761	28.1	171	0
Rhode Island	947	89	134	223	23.5	0	0
South Carolina	2,591	394	300	694	26.8	68	0
South Dakota	666	11	32	43	06.4	0	0
Tennessee	3,924	476	500	976	24.9	25	0
Texas	11,197	2,094	2,157	4,251	38.0	44	23
Utah	1,059	145	160	305	28.8	0	0
Vermont	444	97	155	252	56.8	0	0
Virgin Islands	63	50	100	150	238.1	0	0
Virginia	4,648	460	833	1,293	27.8	0	0
Washington	3,409	622	345	967	28.4	115	24
West Virginia	1,744	422	497	919	52.7	151	0
Wisconsin	4,418	293	638	931	21.1	0	0
Wyoming	332	32	68	00	27.1	0	0
Total	206,103	26,786	26,973	53,749		6,355	2,043

[a] Data from Schueneman (*3*) and United States Environmental Protection Agency reports.

[b] Budgets for 1973 taken from United States Environmental Protection Agency grant awards. May not match with other data because of variable fiscal years.

[c] Does not include approximately $2,700,000 available as research funds in state of California. Approximately $3,500,000 of state subvention funds to local agencies are included in local agency nonfederal amounts given on Table I.

[d] Includes funds for research and technical assistant administered by Department of Public Health and regulation of emissions for motor vehicles administered by Motor Vehicle Pollution Control Bureau and Department of Highway Patrol.

[e] Includes funds for state in general, metropolitan Boston area, and metropolitan Springfield area.

able, an increase of a factor of more than eight since 1965 (Table VI). Five states (New York, California, Illinois, New Jersey, and Texas) had budgets greater than $3 million per year. The seven states with budgets in excess of $2 million accounted for 48% of the total expenditures, while the fourteen agencies with budgets greater than $1 million accounted for about 62% of the total. Per capita expenditures ranged from a low of 6 cents in South Dakota to a high of $2.38 in the Virgin Islands. The *median* for the 55 agencies was 28 cents per capita. The *average* per capita expenditure, based on the total budgets of all states and the total population of the nation was 1 cent in 1961, and 4 cents in 1965 and 26 cents in 1973.

While the growth in funds available to state air pollution control agencies has been very substantial from 1961 to 1965 to 1973, there are two other factors to consider to put the budget increases into proper perspective, namely, agency operating cost (as reflected by the average salary of employees) and the increase in population to be served. Average salaries have about doubled in the 1961–1973 period and population to be served has increased by about 15% in the same period (Table III). After considering these factors, the real growth in agency capability from 1965 to 1973 is a factor of about 4.3 rather than the 8.8 indicated by available budgets alone (Table III).

Considering the total funds available to state and local agencies, including federal grants, there has been an increase from about 10 to 20 to 108 millions of dollars from 1961 to 1965 to 1973, respectively, which has been offset by increasing population and monetary inflation so that the real increase in capability is only about half that indicated by the raw monetary figures (Table III).

2. Organizational Patterns

State air pollution control programs in the United States generally began in health departments. In 1968, of the fifty-three states and territories having air pollution control programs, forty-two were in health departments. The other eleven operated as separate agencies reporting to a board or commission or the governor. At that time, most agencies were small, and the health department was considered to be a good place to house the air pollution activity in view of the interests of health departments in environmental health generally and the adverse health effects of air pollution. In the 1968–1975 period however, there has been a great expansion of interest in environmental quality and a recognition of the need to integrate and coordinate all environmental quality programs (7). In late 1970, the federal government formed the Environmental Protection Agency, reporting directly to the president. States acted similarly, and by 1974, a total of thirty states had created separate environmental agencies of one kind or another and had placed the state's air quality work in that agency (Table VII). Other activities typically conducted by environmental quality agencies are water pollution control, solid waste control, noise control, radiological protection, pesticide regulation, and toxic substances control. Some environmental agencies also include shoreline management, water use control, fish, shellfish, game protection, and natural resources management.

Table VII Organizational Location of State Air Pollution Control Agencies United States[a]

Kind of agency	*1968*		*1970*		*1972*		*1974*	
	Number	*Percent*	*Number*	*Percent*	*Number*	*Percent*	*Number*	*Percent*
Health departments	42	79	36	68	23	43	18	33
Separate air or air and water	11	21	8	15	6	11	6	11
Environmental agencies	0	0	9	17	25	46	30	56
Total	53	100	53	100	54	100	54	100

[a] Data from Schueneman (*3*), the Air Pollution Control Association (*4*), and Smith *et al.* (*5*).

Most state air pollution control activities are carried out under the cognizance of a board or commission, usually appointed by the governor. In forty-four states and territories such is the case (*8*). Some of the boards and commissions are concerned only with air quality, while others are more broadly concerned with various environmental quality issues. The boards and commissions establish broad agency policy and usually formally promulgate the rules and regulations under which the operating agency conducts its program. Nearly all boards and commissions are made up of part-time people, and thus, all staff work is done by the regular line employees. In eleven states and territories, there is neither a board nor a commission, or the body is only advisory to the administrative agency. In such cases, the agency rules and regulations are promulgated by the head of the agency. This is also the case for the United States Environmental Protection Agency.

Most states in which local governments also operate air pollution programs have organized their activities to augment and support the local programs and to conduct such other direct activities as may be necessary to get the total job done. This includes provision of a full line of services in areas where there is no local program and provision of certain services in areas served by local programs where the local agency is unable to do so. The state and local agencies have generally organized their activities to avoid duplication and omission. In three states (Louisana, Arkansas, and Rhode Island) the state organization covers all activities, and no local programs are authorized. In South Carolina, no new local programs are authorized. In many states, district offices have been established at strategic locations throughout the state so that program activities can be carried on without severe inefficiency of personnel utilization caused by long travel times from the state capital to the site of field work. New York, for example has one hundred fifty of its staff in the central office in Albany and a total of about one hundred people in nine district offices around the state. Texas has twelve district offices and Ohio has five.

The internal organization of state air pollution programs varies widely, depending on whether the agency is in an environmental quality agency, a health department or is a separate entity. Further, the basic structure may be either functional (e.g., enforcement, laboratory, planning) or programmatic (e.g., air pollution, water pollution, solid waste), depending on the desires of the agency administrator. Environmental quality agencies have often gone to the functional type structure to a full or partial degree, but many are more programmatic than functional. The Oregon Department of Environmental Quality (Fig. 2) is largely programmatic but partly functional. The California organization (Fig. 3) is fairly typical of a comprehensive air pollution control agency working

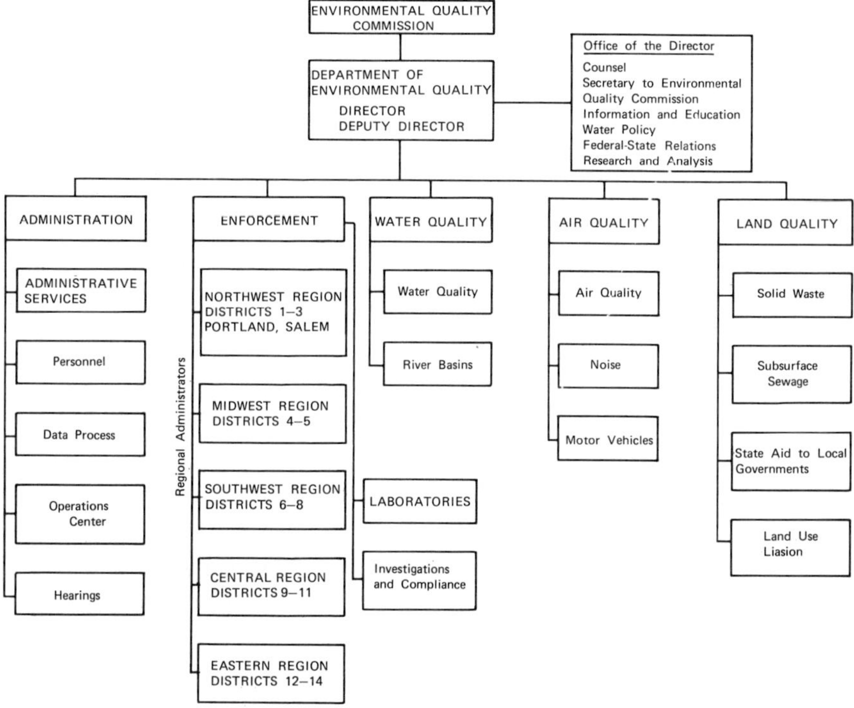

Figure 2. Organization chart of the Oregon Department of Environmental Quality (February 1974).

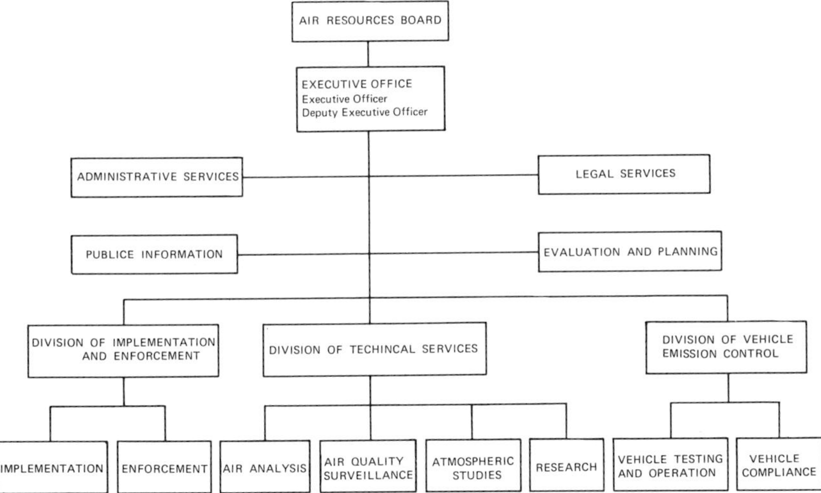

Figure 3. Organization chart of California's air pollution control program (October 1973).

under an air pollution control board. A similar structure would be expected for a self-sufficient unit located in a health department or environmental quality agency except that the board might be attached to the head of the parent agency rather than in a direct position over the air pollution unit director. A less comprehensive air pollution organization within a state might also be organized in a fashion similar to that shown for a local agency (Fig. 1) except for the structure above the air pollution control officer. District offices might also be added.

3. Activities of State Agencies

While activities of state agencies in 1965 were rather limited, by 1973 most states had rather extensive programs in air pollution control (*8*). These varied in magnitude depending on the population and nature of the air pollution problem in the state, the extent of local governmental activities, and the budget available to the state agency. All states had the necessary basic laws to establish the powers and duties of the air pollution control agency. All states had legal authority to conduct all necessary activities, partly as a result of the Federal Clean Air Act, which required states to adopt such laws. If they do not, the federal government is empowered to conduct air pollution activities on a direct action basis in the state. In fifteen states and territories, there is no local governmental air pollution programs. In these areas, the state-level agency conducts all of the air pollution control work.

The activities and role of state agencies are described in general terms in Section III,A of this chapter, and specific program elements are described in some detail in Section IV. However, there are some particular activities that are so unique or unusual that they are worthy of special mention. One such activity is the provision of state funds to local governments for operation of air pollution control programs. A number of states, including Maryland, New York, and California, do so. In Maryland, the state generally pays for about half of local program costs. In California in 1973, the state provided about $3.7 million to those local agencies that adequately coordinated their activities with other local agencies in the same air basin. All state agencies are required by federal law and regulations to review the applications of local agencies for federal control program grant funds. This procedure seeks to aid in coordination of state and local governmental programs. In New York, 5 cents is set aside from each motor vehicle registration fee to provide funds for the state to conduct studies and other activities relating to air pollution from motor vehicles. A total of $350,000 per year is made available to the state air pollution control agency in this manner. In California, the state highway patrol operates twenty mobile vehicle emission testing units. Vehicles are stopped at the roadside and an emission test is made. If emissions are excessive, the vehicle owner is required to have repairs and adjustments made. In Texas, New York, West Virginia, Ohio, Pennsylvania, North Carolina, Illinois, New Jersey, Washington, and other states, district offices are established around the state to provide services in the local area and to coordinate state and local governmental programs. In many states (see Section IV,C,8), it is necessary to review applications filed by representatives of pollution sources for tax relief in connection with pollution control expenditures. All states are involved in review of environmental impact statements that are prepared and circulated for review on all projects supported by federal funds or undertaken by the federal government, in accordance with the National Environmental Policy Act. In California, the state agency is responsible for regulating emission rates from new motor vehicles manufactured for sale in that state. The program includes setting of emission standards, testing of vehicles as they come off the assembly line and certification of prototype vehicles for future model year production (*9*). In New York, a procedure has been established that requires pollution sources that are in violation of an emission standard to post a performance bond to guarantee that an agreed upon compliance schedule will be adhered to. If the source fails to proceed with abatement action in a timely manner, the bond is forfeited. More than $600,000 in such bonds were posted in 1972. New York also enacted a $150 million environmental improvement bond issue. The money was used to abate air pollution arising from state-owned facilities and buildings owned by local governments, such as schools, incinerators, and hospitals. Illinois has formed an Institute for Environmental Quality, which provides centralized management for a large number of state agencies concerned with research in environmental fields, including air quality control (*10*).

C. Interstate Programs

Even though many major urban centers and smaller communities are located astride state boundary lines, there has been very little formal action to ensure uniformity of action among the two or more states concerned with the common, region-wide air pollution problem in such areas. While the Federal Clean Air Act of 1977 [Section 102(c)] specifically grants the consent of the United States Congress for two or more states to enter into agreements or compacts for cooperative efforts and the establishment of special joint agencies, only three states have done so, and no states have entered into any agreement concerning mutual or joint regulatory or enforcement action. Several interstate compacts have been proposed over the years, and some have been provided for in the laws of the concerned states, but only one, involving three states, has been approved by the United States Congress. The general opinion is that such interstate arrangements are so cumbersome and legally complex that more can be accomplished by informal cooperation among operating control agencies than by more formal, legally binding mechanisms (*11*).

The one interstate compact agency, the Interstate Sanitation Commission, made up of New York, New Jersey, and Connecticut, was originally established in 1936 to deal with water pollution problems. In 1956, the United States Congress granted the commission authority to extend its activities into the air pollution field (*12*). In 1961, the states of New York and New Jersey adopted laws implementing the authority granted by Congress. Connecticut did not join in this action until a later date. The Commission's principal work involves air pollution measurements, assisting with study of interstate flow of air pollutants, acting as a clearing house for public complaints involving interstate pollution and participation in design of area-wide comprehensive studies and coordination of activities of the various state and local agencies during air pollution episodes (*13*).

The Metropolitan Washington Council of Governments provides coordinative and service functions in behalf of the various state and local governments concerned about air pollution in the Washington, D.C., area. The Council is made up of local members of the general assemblies of Maryland and Virginia and of the United States Congress and members of the governing bodies of three counties in Virginia, two counties in Maryland, several cities in these counties, and the District of Columbia. The council has been delegated responsibility, by the governors of Maryland and Virginia and the mayor–commissioner of Washington, D.C., for developing plans for attainment and maintenance of air quality in the region. The council also provides services to an air pollution control agency coordinating committee, assembles and distributes information and acts as the coordinating agency during air pollution episodes.

Many state and local agencies with common interests in an area including more than one state have created informal committees to help coordinate activities throughout the interstate area. The groups are usually made up of representatives of the operating state and local air pollution control agencies. They concern themselves with design of air monitoring systems, sharing of data, development of coordinated program plans, uniform emission control requirements, air pollution episode plans, etc.

III. Role of Various Levels of Government in Air Pollution Control

A. State and Provincial Government

State and provincial governments have been active in the air pollution field only in recent years. State and provincial programs around the world expanded markedly during the 1965–1975 decade.

In the United States, key features of the 1970 federal law were the requirements that states develop comprehensive plans for attainment and maintenance of nationally promulgated ambient air quality standards, that the states have adequate legal authority to take necessary actions in the air pollution field, and that the states actually carry out and enforce the implementation plan. The states could, of course, rely upon local governments to carry out parts of the plan under provisions of state laws, charters, and constitutions that empowered local governments to engage in air pollution control programs, but even then, the basic responsibility remained with the states. Many states have, in fact, relied heavily upon local governments in the development and accomplishment of implementation plans, but some states have assumed the entire responsibility. In some seventeen states and territories there, is essentially no activity in the air pollution field at the local government level. Between these two extremes, there is a full spectrum of variation from one state to another as to the respective roles of state and local government.

It is impossible to describe in a broadly applicable way the role of state and provincial governments in the air pollution field. However, all states and provinces have roles, to one extent or another, in the categories of (a) leadership, (b) coordination, (c) evaluation, (d) service, and (e) operations.

Leadership is the primary role. The principal mechanism for exercising this role is through the development of a comprehensive plan for air resource management throughout the state or province. A major element of such a plan will be an implementation plan for attainment and maintenance of national ambient air quality standards, where they have been enacted. Such plans should embody both short- and long-range goals and should be updated from time to time. The plan should spell out the duties and responsibilities of various units of government, reasonable in the light of existing or obtainable resources. The plan should be developed jointly with any local governments that are involved. Once adopted, it should be given wide distribution and should be used as an operating guide by all concerned. The state or province should take the lead in seeking adoption or revision of applicable state or provincial laws and the powers and duties delegated to local governments. The state or province should provide leadership to air pollution programs in cities, towns, and counties as may be appropriate. Some state and provincial agencies have established a special office within their own programs to provide for liason with local governmental programs and have issued guidelines as to the proper role of local agencies (*14*). Promulgation of model ordinances and pollutant emission limitations for use by local agencies is another useful role. The state or province may also conduct demonstrations, perhaps in cooperation with one or more local agencies or others, of new techniques and equipment, new organizational patterns, better methods for operating control programs, etc. Through such demonstrations local programs can be accelerated, the cost of duplicate or ill-conceived experimentation avoided, and uniformity promoted.

In the field of coordination, the state or province has an important role in seeing that the activities of local agencies along with those of the state or province augment each other and are not unnecessarily duplicative. Uniformity of policies and procedures, enforcement policy, codes, and reporting should be achieved to the extent possible. The state or province can foster uniformity and coordination through publication of operational and technical guidelines and through conduct of regular periodic meetings of representatives of local agencies. Agreements should be developed as to the respective roles of the state, provincial, and local agencies. These should be reduced to writing and be updated periodically. The state or provincial air pollution agency should also seek coordination of its activities with others within its parent organization (e.g., department of health or environmental control) to ensure that conflicts and inefficiencies are avoided. Further, the state or provincial air pollution agency should maintain contact with activities of state and provincial offices concerned with industrial development, land-use planning, agriculture, transportation planning, housing and urban renewal, education, building regulation, public works, motor vehicle inspection, and others. Efforts should be made to coordinate appropriate activities of these agencies with those of the air pollution agency and also to ensure that policies and programs of these other agencies will augment rather than deter activities of local air pollution control programs. It is often helpful to arrange for regular periodic meetings of representatives of the various key agencies and to create a formal committee to enhance cooperation and coordination.

In the field of evaluation, the most important function is to periodically prepare and publish an overall appraisal of the air pollution situation throughout the state or province. The evaluation should consider trends in air quality, emissions from pollution sources, the extent of source control, staffs and budgets of control programs, and effects of air pollution. Consideration should be given to the influence of growth of population and pollution sources, meteorology, governmental structure, and control program operations. From the evaluation, recommendations should be made for future actions and priorities. Another function in the evaluation field is the appraisal of local air pollution programs. The states and provinces should determine, in cooperation with the local programs, whether proper objectives have been identified and whether adequate progress is being made toward achievement of such objectives through suitable program operations. The evaluation should seek to improve the operation of the local program through specific helpful recommendations. A particular element of such evaluations should be a continuing cooperative program to assure the quality of field and laboratory measurements made by the local agencies.

States and provinces may provide a variety of services to local agencies. Training is one activity particularly suited to the state and provincial level. The training should be designed to supplement the more basic training available in universities and at national facilities. Likely subjects include orientation in the broad aspects of air pollution, visible emission evaluation, operation of field air sampling equipment, and local laws and regulations. Appropriate use should be made of self-instructional and packaged training materials available from the national government and elsewhere. Another vital service is the establishment of pollutant emission standards and equipment design criteria. Their establishment requires the skills of a variety of professional disciplines that may not be available at the local government level. States and provinces may also provide a variety of specialized technical assistance and financial support to local agencies. The acquisition, storage, and retrieval of air pollution publications from many sources is a library function beyond the capability of most local agencies and private organizations, but the state or province should maintain a specialized air pollution library and provide information from it to local agencies and others. States and provinces should assist local agencies in connection with review of plans for construction of facilities that may cause air pollution. In the case of industrial process equipment and indirect sources (those sources that generate motor vehicle travel and its attendant pollution), local agencies often do not receive a sufficient number of plans to maintain competence to review them. In such cases, the state or province should assume full responsibility or provide consultative services, as appropriate. States and provinces may also provide laboratory services for making those relatively infrequent, more complex analyses that are beyond the capability of local agencies and, similarly, provide services for some kinds of source emission testing. In many cases, a valuable service that can best be provided by the state or provincial agency is the computerized storage, retrieval, analysis, and periodic reporting of air quality data. This service at the state and provincial level can often save a great deal of local agency staff time and costly duplication of computer hardware and staff capability.

In the direct operations area, the activities of states and provinces will vary widely, depending on the division of responsibilities between state, provincial and local agencies. The state or province will usually have direct operational responsibilities and programs in regard to motor vehicle emission inspection and control, development of comprehensive plans for implementation of national ambient air quality standards, a basic minimal air quality monitoring network, administration of laws relating to tax relief for air pollution control expenditures by industry, power plant siting, consolidation and reporting of air quality and emissions data, research, management of problems and programs involving a neighboring state or province, control of air pollution from state- or province-owned facilities, and the full gamut of program operations in areas not served by a local agency. Beyond these activities, and those described in the preceding paragraphs, states and provinces, may also engage in many of the same activities as local programs, as described in the next section. In many states and provinces, district offices are strategically located in various parts of the state or province to reduce travel time and improve effectiveness.

B. Local Government

Air pollution programs of cities, counties, and other local agencies are principally directed toward regulatory control of air pollution. The nature of the local program will depend on the nature of the problems, funds, and personnel available to operate the program; the nature of national, state and provincial laws setting out the powers and duties of local agencies; and the activities and policies of the national, state and provincial air pollution agencies. Local program elements include:

1. Development of a comprehensive air resource management plan, including short- and long-range goals.
2. Selection of air quality goals and objectives.
3. Assistance to the state or provincial agency in the development of plans for implementation of national ambient air quality standards.
4. Development of a register of air pollution sources, emissions from them, degree of emission control, and related data.
5. Measurement of air quality, including operation of a chemical laboratory.
6. Development of rules and regulations concerning pollutant emissions, operating practices, design of equipment, fuel composition, etc.
7. Evaluation of effects of pollution.
8. Investigation of public complaints.
9. Agency-initiated detection of violations of emission control and other rules and regulations applicable to stationary sources and motor vehicles.
10. Negotiation and abatement of violations of rules and regulations.
11. Abatement of pollution from sources owned by local government.
12. Provision of leadership and motivation for development and implementation of plans to reduce motor vehicle travel and pollution from individual vehicles.
13. Annual or other periodic inspection of pollution sources.
14. Issuance of operating permits.
15. Development of industry group abatement plans.
16. Measurement of pollutant emissions from stacks.
17. Prevention of new sources of pollution through review of construction plans.

18. Provision of consultative serves to industry and equipment purveyors on emission control means.

19. Establishment and operation of a fee system.

20. Conduct of office and formal hearings.

21. Court prosecution of chronic and flagrant violators of the law and regulations.

22. To the extent possible, prevention and, as necessary, management of air pollution episodes and pollution from industrial and other accidents.

23. Provision for cooperation among local governmental agencies, such as those concerned with planning, zoning, transportation, solid waste disposal, water pollution control, building inspection, and boiler inspection, and for cooperation with neighboring local governments.

24. Design and operation of systems for storage and retrieval of data.

25. Preparation of regular activity reports and special topical reports to the public and higher levels in the local government.

26. Conduct of public information program.

27. Review of environmental impact statements prepared pursuant to national, state, provincial, or local law with regard to air quality implications.

28. Provision of on-the-job training, continuing education, and specialized air pollution training to agency staff.

29. Maintenance of liaison with the provision of information to citizen groups interested in air conservation.

30. Maintenance of a library of technical literature, national regulations, operational guidelines, and other materials.

C. State–Provincial–Local Relationships

It is generally accepted that responsibility for control of air pollution should be the at the lowest level of government capable of dealing effectively with the problem in its entirety. With locally controlled action, the program can be better tailored to local needs, economy of operation can be improved, and the program can be more accessible and responsive to local citizens. However, there are problems associated with local control. Cities and counties with small populations are not often in a financial position to hire the needed technical personnel nor to acquire the necessary equipment to run an adequate program. The multiplicity of local governments within metropolitan areas, often in excess of a hundred, cannot be expected to operate separate programs in an effective and efficient manner. County and multicounty jurisdictional areas are necessary. Industrial and commercial interests sometimes control the actions of local government. If such interests control major pollution sources and also do not want to abate pollution from them, the local citizens are unable to have their desires for clean air satisfied. In such cases, the necessity for the state, provincial, or national government to provide for needs of the citizens may become evident.

In the United States, responsibility for determining the relationships between state and local governmental agencies rests with the state, since local governments have only those powers and duties specifically given to them by the state. The general nature and scope of such arrangements are given in broad state laws and in constitutions. It remains for the operating air pollution control agencies to work out the details of state–local relationships in control program operations. Some general criteria for determining the role of local governments include the size of the state, the location and size of cities and metropolitan areas, the population of the state and its distribution, the resources available to local agencies, the nature and magnitude of the air pollution problems and the interests and desires of local governments (*5*). The following general principles apply to the distribution of activities and responsibilities between state or provincial and local agencies:

a. States and provinces should be responsible for certain kinds of activities that, by law or because of operational practicalities, cannot reasonably and effectively be carried on by local agencies. Included in these actions would be development of a basic state–province-wide air pollution control program plan; promulgation of operating procedures and data recording forms; generalized training; provision of highly specialized skilled employees; review of local programs; conduct of specialized, complex laboratory analyses; and storage, retrieval, and reporting of data on state–province-wide basis. In order to foster regional approaches to control program operation and to encourage local cooperation in state–province-wide programs it may be desirable for the state or province to provide some of the funds for local program operations under specified terms and conditions. The state or province is responsible for all activities in areas not served by a local program.

b. Independent, comprehensive, local control programs should be encouraged in those areas that include a concentrated population great enough to provide adequate financial support for a control program, i.e., at least 250,000 or more people. The area should include all of an air quality control region, where such regions exist, and desirably coincide with the boundaries of existing units of general government, e.g., a county.

c. In metropolitan areas, it is desirable to have a single agency responsible for air pollution control throughout the area. If this cannot be achieved, the best alternative is to have as few responsible agencies as possible within the area.

d. Jurisdictional areas having a population of less than 250,000 can be expected to handle less complex matters, such as control of smoke from small combustion operations, open burning, and abatement of nuisances from small industrial and commercial establishments. These smaller jurisdictions can also assist by carrying out routine surveillance of sources and air quality monitoring programs.

e. Except as noted above, states and provinces should retain and exercise responsibility for control of the larger and more complex air pollution sources. The state or province cannot be expected to effectively regulate a multiplicity of small sources. However, effective county, multicounty, or district programs may largely eliminate the need for a state or provincial regulatory program.

f. In small, highly urbanized states or provinces a state–province-operated, state–province-wide control program may be the best approach. Conduct of certain operations of the state–province program on a district basis will probably be advisable.

g. States and provinces may appropriately retain state–province-wide responsibility for control of pollution from specified sources, e.g., motor vehicles and electricity generating plants. The state or province should also retain the right to assume jurisdiction over major single sources of pollution within areas having comprehensive local programs when pollutants from such sources adversely affect areas outside the area in which the source is located and the local agency is not effectively dealing with the source.

h. State or provincial laws should clearly define those matters that are the responsibility of local jurisdictions and those that are retained by the state or province.

i. State and provincial laws should provide readily usable authority and clear, comprehensive, detailed guidelines for establishing, operating, and financing air pollution control districts—based on location of sources, characteristics of air quality control regions (when they exist), distribution of population, local desires, and existing governmental structure.

j. The state or province should retain the right to reassume rights and responsibilities it delegates to local jurisdictions when, after a hearing, it is found that the local jurisdiction is not effectively controlling air pollution. To prevent local agencies from abdicating their responsibilities as a means of avoiding the expense of operating an air pollution program, the local jurisdiction should be obligated to pay for the costs of operating the program for which the local agency was rightfully responsible.

D. Interjurisdictional Relationships

The need for changing the geographical jurisdictions and boundaries of local government in many areas, particularly in the major metropolitan areas, arises because of the growing maladjustment between what local governments are called on to do and their ability to perform many functions independently, including air pollution control, which are area-wide in nature. There are four general reasons why local governments as now constituted are unable to perform area-wide functions effectively: (a) fragmentation and overlapping of governmental units, (b) disparities between tax and service boundaries, (c) state or provincial constitutional and statutory restrictions, and (d) overlapping of state and provincial lines by a metropolitan area.

There are a variety of ways for metropolitan areas to adjust their areas and powers to be able to meet the needs of their citizens. However, the initation and operation of air pollution programs cannot wait until major reforms in metropolitan government occur. It becomes necessary, then, to make the best of existing conditions. The preferred organization to handle air pollution control is a unit of general government with a jurisdictional area equal to or larger than the air quality control region. In many cases, this will be a county with the county commissioners acting as an air pollution control board or an agency operated by the county such as the health department. In many cases, however, county governments have little authority within incorporated areas, although city–county health or environmental departments having jurisdiction throughout the county are becoming more common. If the county agency does not operate in the central city, the situation often will not be too unsatisfactory, since the central city and the balance of the county may each have populations large enough to support a good program. The two programs could be operated on a parallel basis by simple informal coordination. An acceptable, and perhaps politically attainable approach to establishing this type of program organization would be to adopt state or provincial enabling legislation permitting county governments to operate air pollution programs in all parts of the county or, on local option, in all parts of the county except in incorporated cities larger than some minimum size—e.g., 250,000 population. In metropolitan areas consisting of more than one county, a multicounty air pollution control district is desirable and may be attainable. Failing this, the best choice would be separate county programs coordinated by informal agreements among the counties.

Intergovernmental agreements under which a governmental unit conducts an activity jointly with one or more other governmental units, or contracts for its performance by another governmental unit are the most widely used means of broadening the geographical base for handling common functions in metropolitan areas. The greatest weakness of this approach is that a community that has a number of troublesome pollution sources may not agree to enter into such an arrangement.

In spite of all the difficulties that may exist in governmental structure and operation, it is still possible for operating agencies to arrange cooperation and mutual assistance through informal mechanisms. There is rarely any obstacle to prevent agencies in a metropoltian area from sharing certain facilities and equipment, pooling air quality and pollutant emission information, agreeing on air quality and emission standards, and conducting coordinated simultaneous projects of all kinds. This sort of cooperation is the least that should occur in any case.

International situations must be handled at the national level through diplomatic channels or international agencies. Such dealings are slow and time consuming. On the Canada–United States boundary, the International Joint Commission has been involved in transboundary flow of pollution for more than 50 years and is continuing to function in this field. In the United States, the Clean Air Act of 1977 (*15*) provides a formal procedure for handling international flow of pollution, provided the other country affords the United States similar arrangements. Beyond these formal arrangements, the state, provincial, and local agencies responsible for air pollution control on both sides of an international boundary can resolve many problems and provide for cooperation through informal working arrangements. Such actions do in fact occur in most situations.

E. Relationship with Transportation and Land-Use Planning Agencies

The problem of air resource management is intimately related to many other community problems. Communities have grown and developed haphazardly, and as a result, existing cities have many undesirable features, including air pollution. While it may be possible, through traditional approaches of pollutant emission reduction, to achieve relatively clean air in most of the cities of today, it is increasingly evident that this will not be possible in future years as major urban centers continue to grow in size and as the rate of energy use per capita and other pollution-causing activities continue to rise. If clean air is to be provided in the future, better provision must be made in community development and transportation plans for air resource conservation, and even more importantly, such plans must be brought to fruition. It has become obvious that the air resource of an area does have a limited carrying capacity. Only a certain burden of pollutants can be discharged into the air and still meet ambient air quality standards. It follows that there is a finite limitation on the number of people that can be accomodated on a particular block of land, given a certain rate of pollutant release per capita. Of course, more people can be accommodated if the rate of pollutant emission per capita can be reduced, but in some major urban centers, doing so is becoming so difficult and costly as to be impractical. It is therefore essential that air quality considerations be incorporated into land-use and transportation plans.

There are a number of activities in the land-use and transportation planning process in which air quality matters may be considered (*16, 17*). These include long-range planning and transportation needs studies on a regional scale; policy planning, in which attainment of air quality goals is established as a specific objective; design of specific transportation systems; urban design, in which the location and form of structures are determined; new town planning; and transportation engineering to achieve smooth traffic flow.

While the desirability of cooperation between air pollution control agencies and land-use, zoning, and transportation agencies has been recognized for at least 30 years, there still remains a regrettable lack of communication and coordination. Only a very few air pollution control agencies have actively sought to cooperate with these agencies, and even fewer have established routine working relationships. Similarly, only about 20% of the planning agencies are actually engaged in consideration of air quality, although about 40% do recognize environmental aspects as an element in their overall planning programs. The need for closer cooperation is urgent.

There are some signs of progress. In the United States, noteworthy action has been brought about by the Federal Aid Highway Act [23 U.S.C. 109(j)], which requires that urban transportation plans and specific plans for highway sections be reviewed by cognizant air pollution control agencies and that these plans be made consistent with the air quality implementation plan. Similarly, environmental impact statements required by the National Environmental Policy Act and some similar laws at the state level have caused designers of federally aided and other projects to consider the air quality (and other environmental) consequences of their plans. Through formally established procedures, air quality control agencies become involved in review of the projects and associated impact statements. A few regional and local planning agencies have undertaken important air quality appraisals of urban and transportation plans they develop.

Some of the specific activities that air pollution control agencies can take or that they can foster in planning and transportation agencies are:

a. Establish a unit within the air pollution control agency to specialize in land-use, transportation, zoning, and other community planning. Include professional land-use and transportation planners in the unit.

b. Foster establishment of an environmental planning committee or working party made up of representatives of various agencies whose activities guide or influence community growth and development.

c. Establish a formal working relationship and project cross-referral system with zoning agencies.

d. Get involved in the review of environmental impact statements and in their preparation.

e. Study plans in existence or being made for transportation, land use, water supply, sewage disposal, open spaces, etc. for the purpose of coordinating these plans with the air resource management plan. Inform others in the planning field of the work being done on air resources.

f. Make projections of pollutant emissions from present and future sources, considering several possible emission control programs and geographical distributions of pollution sources. Consider population growth and transportation, including substitution of mass transit for private automobiles, industrialization, technological changes, etc.

g. Develop and apply mathematical models for estimating present and future air pollution levels based on meteorological and pollutant emission data and land-use patterns. Study pollution levels that would result on the basis of several possible pollution emission control programs and land-use patterns.

h. Prepare a series of estimates of the impact of alternatives for control of pollutant emissions and land use that will result in various levels of air quality. Consider such things as cost of control, influence on transportation and social patterns, need for industrial expansion, availability of fuels, community desires and purpose, and effects of various pollution levels.

i. Design air resource management activities in the light of such community planning matters as allocation of land uses and performance standards for new land uses.

j. Encourage use of district heating to minimize pollution caused by a multitude of smaller heating plants, and encourage public refuse collection to eliminate the need for on-site incineration of refuse and its resulting air pollution problems.

k. Encourage and support development patterns that are compatible with use of mass transit as the appropriate transportation mode.

i. Oppose development plans that do not require provision of paved streets with curbs to avoid the dust from unpaved roads and driving on unpaved shoulders.

F. Environmental Impact Reviews

In the United States, the National Environmental Policy Act of 1969 (*18*) sets forth a strong national policy directed toward encouraging harmony between man and his environment and enrichment of the understanding of the ecological systems and natural resources important to the nation. A key element in the act is a requirement that an environmental impact statement be prepared on every proposal for federal legislation and other proposed major federal actions that significantly affect the quality of the human environment. Included in this requirement are actions by any nonfederal agency proposing a project supported in whole or in part by federal funds. These environmental impact statements are reviewed by state and local agencies, particularly those responsible for enforcement of environmental standards, and are available for review and comment by the general public.

The National Environmental Policy Act requires that the environmental impact statement on any proposed action include a detailed review of the following:

a. The environmental impact of the proposed action.

b. Any adverse environmental effects that cannot be avoided should the proposal be implemented.

c. Alternatives to the proposed action.

d. The relationship between local short-term uses of man's environment and the maintenance and enhancement of long-term productivity.

e. Any irreversible and irretrievable commitments of resources that would be involved in the proposed action should it be implemented.

While the environmental impact statement review process does not, as such, include provision for any entity to directly stop the conduct of an action, it does require all concerned to incorporate environmental considerations into their planning process along with economics, production, transportation, and other factors that had usually been considered in the past. It has forced the public and public agencies to knowingly consider the broader implications of action that were only narrowly considered in the past and to accept more responsibility for the protection of the environment and of natural resources, for the benefit of future generations of people.

Air pollution control agencies are concerned with a wide variety of actions and projects upon which environmental impact statements are prepared. Each agency should ensure that the procedures for notification of the agencies of proposed projects are such that all relevant projects do in fact come to the attention of the agency. The United States Office of Management and Budget has issued guidelines and requirements in this field to help insure proper circulation of environmental impact statements (and to implement certain other laws) in its Circular A-95. The circular provides for state, regional, and metropolitan clearinghouses, which aid in the securing of comments and views of state and local agencies that are authorized to develop and enforce environmental laws, including air pollution control laws. Each state and local air pollution control agency should work closely with the cognizant clearinghouse in these matters.

Environmental impact statements are prepared under the regulations issued by the United States Council on Environmental Quality, pursuant to the above-noted National Environmental Policy Act (*19*). In addition, each federal agency has published regulations establishing its own formal procedures for complying with the environmental impact statement requirements. These have been published in the United States *Federal Register*.

The air pollution control agency can use the environmental impact statement process for several purposes, including seeking more adequate provisions for air conservation in connection with individual projects; exerting pressure to provide broader and longer-range consideration of air quality in overall planning of such things as highway systems, transit systems, and urban redevelopment projects; mobilization of opposition to poorly conceived projects that do not give adequate consideration to air quality; encouragement of basic changes in design criteria to the end that air quality is more fully considered in future projects; and fostering consideration of air quality matters in various kinds of long-range planning activities, since many of these are supported by federal funds. In addition to activities under the federal law, in the United States, state and local agencies may become involved in similar activities provided under state laws, since at least eleven states now have laws similar to the federal law that apply to state-conducted or -supported activities and, in a few cases, that apply to actions of local governments and even certain private entities.

IV. Air Pollution Control Program Elements

A. Finding and Evaluating Problems

1. Emission Inventory (see also Volume III, Chapter 17)

Knowledge of what pollution is emitted to the atmosphere, where, when, and by whom, is basic to an air pollution control program. Such data, in conjunction with air quality data, indicate the degree of emission control needed to achieve air quality goals and help establish the priority schedule for abatement action. The most elementary inventory consists of a listing of pollution sources of various kinds such as major coal, oil, and refuse-burning plants; pyrometallurgical plants; chemical factories; and cement mills. A more detailed inventory would include the amount and kind of fuel and refuse burned, the amount of various process materials used, the kinds of process and combustion equipment, the nature of emission control equipment or methods, and the variation of operations with time. A rough source inventory can be prepared on the basis of published information. Improvements can be made by use of mailed questionnaires. To achieve better accuracy and completeness, field inspection and evaluation are necessary. In some cases, stack emission measurements are needed.

Emission estimates are calculated on the basis of some index of the rate of operation (e.g., amount of fuel used, amount of raw material used), composition of fuels or raw material, nature of process and emission control equipment in use, and an "emission factor," which relates the operating parameters to the rate of emission of various pollutants. Emission factors are based on actual stack emission measurements, process information, materials balances, or theoretical considerations. At least one compilation of such factors has been published (*20*). Data for keeping a source inventory up to date can be secured in the course of other control program operations, including annual inspections, issuance of operating permits, and review of construction plans.

In an emission inventory, data are usually grouped in two ways—(a) by pollutants and (b) by source categories. Pollutants most often included are particulate matter, sulfur dioxide, nitrogen oxides, carbon monoxide, and hydrocarbons. Particulate matter may be subdivided into "fine" (e.g., less than 5 μm in size) and "coarse." They may also be subdivided on the basis of their chemical composition. Hydrocarbons may be subdivided into methane and nonmethane, into photochemical reactivity classes, or into various groupings by particular species. Other pollutants sometimes inventoried include organic acids, aldehydes, ketones, and ammonia. Source categories include fuel combustion (stationary sources), refuse combustion, transportation, solvent evaporation, gasoline handling, and industrial processes. The latter may be subdivided into chemical, metallurgical, mineral, and petroleum processing, for example. Procedures for conducting a rapid preliminary emission inventory have been described by Ozolins (*21*). Inventory record systems may be prepared manually or by use of a computer (*22*).

In the United States, the federal government maintains a national data bank of pollutant source information. The system is known as the National Emission Data System. State and local emission data systems should be consistent with this system, since all state and local agencies are required to submit data to the system. Procedures for operating in conformity with the NEDS system have been published (*23*). A compatible system for state use is also available. It is the emission inventory subsystem of a comprehensive data handling system developed by the United States Environmental Protection Agency (*24, 25*). It is also possible to use these basic pollution source files for programming agency work on inspections, permits, and routine surveillance through use of a computerized enforcement management system (*26*).

2. Air Quality Measurements

Every air pollution control agency should have continuous access to data indicating the quality of its atmosphere in order to determine current and future emission control requirements, evaluate trends in air quality over the years as a measure of control program effectiveness, identify sources of pollution, warn of oncoming problem situations, establish standards or goals for air quality, forecast and identify air pollution episode occurrences, and publish daily air quality information for public information. At intervals of 5 years or so, relatively intensive studies of air quality are needed to explore variation of concentration of important pollutants with time and geographic location, atmospheric reactions, meteorological relationships, effects of pollution, etc. Between such intensive studies, less intensive continuous monitoring will provide as much information as can be effectively used. The nature of appropriate measurement activities will vary from area to area and from time to time. The agency director must give attention to the costs and manpower involved in operating field sampling stations, conducting laboratory analyses, and analyzing and publishing data. Techniques for estimating costs are available (*27, 28*). Monitoring activities are expensive and use substantial manpower. Without proper allocation of agency resources, monitoring work tends to absorb more resources than it should, to the detriment of other priority activities. Many details concerning operation and management of air monitoring systems have been published (*28*).

Since air quality data are used as a basis for emission control programs that may be very costly, it is essential that all reasonable measures be taken to ensure the quality of the data. Guidelines and procedures for conducting a quality assurance program extending across every significant step in the system from selection of methodology, through reagents, standards, and systems, to operational technique should be a part of every air quality monitoring program. Guidelines and procedures for conducting a quality assurance program have been described (*28*) and specific information concerning quality assurance in using reference methods for measurement of sulfur dioxide, carbon monoxide, suspended particulate matter, and oxidant have been published (*29*).

Air quality measurements are needed to appraise the impact of particular sources of pollution on their immediate neighborhoods. While general atmospheric air quality and emission control procedures will take care of the air pollution situation as a whole, localized problems may still develop. In this type of situation, a small area is studied intensively for a few days or weeks, until enough data are available upon which to base a control plan.

In the United States, minimum requirements for air monitoring have been prescribed by the federal government (*30*). Appropriate arrangements should be made to incorporate data collected by local agencies into data systems operated by state and national agencies and for publication of data, with interpretation, for the information of all concerned. The federal government requires that all air quality data of certain kinds be reported into the SAROAD system (Storage and Retrieval of Air Data) (*31*). A computerized air quality data handling system (AQDHS), which is compatible with SAROAD and which is part of a comprehensive data handling system, has been made available for state and local agency use (*32*). All air quality data should be carefully screened and edited before it is permanently entered into the data bank to avoid future false conclusions. Procedures for doing so are available (*33*).

3. Monitoring Air Pollution Effects

Most of the work on determining relationships between concentrations and effects of pollutants is done by national governments and universities. However, state, provincial, and local agencies should conduct special studies to verify the validity of application of general relationships to specific situations in their jurisdictions.

Some monitoring of effects of pollution can be done rather easily. For example, local weather bureau records provide data on change in visibility over the years. Special visibility observations and photographs taken by air pollution control agency personnel at a regular time and place over the years will provide indices of visible pollution.

Effects on vegetation can be kept under surveillance by horticulturalists in the area, with only brief training and nominal effort. With a bit more work and training, test specimens of selected species of vegetation can be set out in various parts of an area to indicate possible pollutants causing vegetation damage and the day of occurrence of injurious concentrations. Procedures are also available for measuring effects of pollutants on metals, fabrics, and rubber.

Mortality data, analyzed for relationships between pollution levels and time and cause of death and place of residence may yield useful information. Studies of morbidity can be made, using data recorded by hospitals, industries, and others. Panels of volunteers can be used to develop information on the prevalence of eye, nose, and throat irritation; odor, and decreased visibility. When repeated from time to time, such studies may yield information on trends toward improved or worsened conditions.

Public complaints about odors and dustfall, while difficult to interpret because of extraneous influences, can sometimes yield indications of trends in these two kinds of pollutant effects.

4. Public Complaints

Public complaints about air pollution provide a measure of the kinds of air pollution people are able to detect with their senses. Unless matters complained about are corrected, the public will be displeased with the air pollution control program. When analyzed with respect to location, time, and weather, public complaints can help locate troublesome sources, guide pollution control efforts, and contribute toward selection of air quality goals. If the control agency fails to act on complaints, there will be a loss of public support, allegations of favoritism and "politics" in the enforcement effort, and occasionally a time-consuming mandamus suit. However, compared to other control program activities, complaint handling is an inefficient way to obtain a given measure of progress toward control. Efficiency in handling complaints can be enhanced by requiring the complainant to identify himself or by requiring that all complaints be in writing and signed. Every effort should be made to have the complainant identify the source of trouble and to record as much information as possible about the dates, times, and circumstances when the condition complained of exists. Since many of the conditions complained about are transient in nature, the employment of inspectors with radio-equipped automobiles is an advantage, permitting inspection soon after a complaint is made.

Investigation of public complaints requires considerable skill on the part of the enforcement officer. Cases can become quite involved and time-consuming. However, following well-founded procedures can minimize problems and use of resources and improve the prospects for a satisfactory conclusion (*34*). Complaint cases fall into three general categories. One type involves a complaint about air pollution that is a violation of a rule or regulation. If the enforcement officer verifies the violation, a notice of violation is issued and actions are taken to get the violation abated. A second type involves a situation that is not a violation of a specific rule or regulation, and yet an unsatisfactory condition exists. In such cases, the enforcement officer attempts to secure a reconciliation through persuasion and appeals to the good will of the pollution source. If such efforts are not successful, the case may go one of two ways. If the condition complained about only affects a very few people, it is a private dispute and the control agency can take no further action. But if a considerable number of persons are affected by the pollution, the situation may constitute a public nuisance subject to abatement action by the control agency. If such is believed to be the case, a notice of violation of an antinuisance regulation can be issued to the responsible source and be followed through to final abatement. The third general kind of complaint is one in which the agency has no legal grounds for action. These cases are simply closed after an explanation of the agency's position to the complainant.

5. Public Interest and Concern

It is generally recognized that all problems of the environment must be considered as a whole. At the same time, the definition of "health" has been tending toward the one used by the World Health Organization, "a state of complete physical, mental, and social well-being and not merely the absence of disease or infirmity." Thus, increasing attention is being given to sociological and economic matters associated with environmental health problems. Studies directed toward evaluation of public perception of air pollution and the influence such perception has on attitudes and actions (*35–38*) therefore assume increasing importance. Information obtained from such studies can be used in design of program elements concerned with public information, interjurisdictional relationships, air quality goals, regulations, time schedules, and community organization.

6. Visible Plume Abatement

One of the most important segments of an air pollution regulatory program is the detection and abatement of visible emissions. Such work is a vital part of the overall program to reduce emission of various kinds of particulate matter including smoke from combustion of fuels or refuse, metallurgical fumes, dusts from a variety of processes and operations, and mists from atomization or condensation of materials. The prohibition of smoke emissions darker than a specified shade of gray or black, as measured by use of the Ringelmann chart (*39*), has been practiced in many parts of the world since about 1910. In the past several decades, visible emissions other than black or gray smoke have increasingly been regulated by limiting the maximum opacity that a stack plume may have, opacity being the degree to which light transmitted through the plume is obscured (*34*).

Nearly all regulations allow short periods of plumes of higher percentage opacity for a few minutes at a time and do not limit the opacity if it is due to uncombined water (e.g., condensed steam). The opacity concept has been found to be workable and legally acceptable (*40, 41*). While it is not possible to quantitatively determine the amount of particulate matter emission reduction that will be achieved in a community through enforcement of a given limitation on the opacity of stack plumes, it is possible to make rough estimates of the extent of reduction achievable for particular sources.

Control of emissions of particulate matter through enforcement of a limitation on the Ringelmann density of smoke and the opacity of other emissions has many practical advantages as compared to programs based only on limitations on the allowable mass rate of particulate emissions. It is possible for a field inspector to keep a great many sources under observation at a low cost. The use of a helicopter for broad area surveillance is practical (*42*). The skills required of an inspector are nominal and easily acquired through training. The need for expensive stack tests to determine mass emission rates is reduced, thus saving money for the sources and the public agency. No expensive equipment is needed to measure visible emissions, and it is easily possible for pollution source operating personnel to determine for themselves the opacity of their emissions and thus prevent violations through proper operation and maintenance of equipment. The procedure makes it possible to bring about control of particulate emissions that escape through windows and roof vents, which cannot usually be measured so that a mass emission limit would be possible. There are, however, some limitations to opacity regulatory programs, namely, that emissions cannot be readily measured at night, water droplets in the effluent may make it impossible to measure opacity, and opacity limitations are not a convenient basis for design of emission control equipment. Therefore, mass emission limitations are usually imposed concurrently with limitations on the visual shade and opacity of emissions.

Vendors of emission control equipment, regulatory officials, and pollution source owners, as well as the general public, often desire installation of control equipment that will prevent any visible emission as a means of avoiding public complaints and ensuring minimum emissions. However, the verbal and visual goal of "no visible emissions" cannot be used in control equipment design calculations. But, if a mass emission rate that has been found to result in a "clear stack" is known, then design can proceed. Data on emission rates (in grams per standard or actual cubic meter, for dry and wet type emission control equipment) that will result in a "clear stack" have been published (*43*), and therefore, it is often possible to design for a clear stack.

Certain procedures must be followed by field inspectors in making visible emission measurements and they must be initially trained and be periodically rechecked to ensure that proper measurements are made. Initial training usually takes three or four days and is accomplished through use of a plume generator, which emits a black or white plume, the opacity of which is measured photoelectrically. The trainee is graded on his ability to report results in keeping with the instrumental measurements (*34*). The Ringelmann chart is the basic field measurement device.

In the United States, the federal government (then the United States Public Health Service) has published specifications for film strips having optical densities of 20, 40, 60, and 80%, which can be used to prepare a handy pocket-size smoke inspection guide. These specifications are published in the United States *Federal Register* (42 Codified Federal Regulations, Part 75).

Every state in the United States has adopted regulations limiting the opacity of emissions. Most states limit the allowable opacity to 20%, while a few have specified 10%, and two have prohibited all visible plumes, except for the few sources that can show that there is no feasible means of eliminating visible emissions.

7. Routine Periodic Inspection of Pollution Sources

Many air pollution programs provide for periodic agency-initiated inspections of installations that may or do cause air pollution. The frequency of inspection may be the same for all kinds of units (e.g., once per year), or more frequent inspections may be made of important sources that have a tendency to cause excessive emissions. An objective of such periodic inspections is to detect faulty equipment and arrange for its repair, thus preventing excessive emissions due to defective equipment. Such periodic inspections are also useful in continually updating pollutant emission inventory data and source registers, keeping in touch with developments as to means being used to control emissions, maintaining interest of operating personnel in air pollution control matters and motivating them to maintain minimum pollutant emissions, and fostering exchange of information among those having similar source control problems. Also, new practices or equipment, which may require modifications to existing regulations or enactment of new ones, are discovered at an early date, making it possible to curb undesirable situations before they become too widespread.

For effective utilization of a periodic inspection program, procedures for scheduling inspections, making field inspections, and recording and reporting data should be documented in agency operating guidelines. Helpful material for agency use in doing so is available (*34*). Scheduling may be done manually in a small agency or if only few major sources are to be included in the periodic inspection program. But in most large agencies, a formal management system, probably including use of a computer, will be needed to schedule the regular inspections and any follow-ups and to maintain the data gathered in usable form. One such enforcement management system is available in packaged form (*26*). Inspections involving a great many kinds of sources may be scheduled on an area-by-area basis to minimize travel time and costs of field personnel. Inspections of larger, more complex sources may be assigned to specialists to ensure quality and uniformity of the field work. Technical reference material (*44–46*) should be made readily available to inspection personnel.

8. Stack Sampling

Actual measurement of emission of pollutants is the most accurate and sometimes the only way of determining exactly the emissions from a particular source. However, the procedure is time consuming and expensive, and therefore its use is limited, with various estimating procedures being used much of the time. The purposes of stack sampling are (a) to determine emissions from particular sources to detect violations of emission regulations, (b) to develop factors for estimating emissions from classes of sources, (c) to obtain information on efficiency of emission control means for reference use in reviewing plans for new installations, and (d) to maintain and improve accuracy of emission inventory data.

Although tests are usually made by the control agency, some laws provide that those responsible for the pollutant source make or have made such emission tests as are reasonably required by the control agency and that the agency will be permitted to prescribe or approve testing methods and witness the testing. Even so, every agency should be capable of doing its own tests or having them done for it by contract. In providing for a stack testing capability in the control agency, consideration must be given to the number and qualifications of field and laboratory personnel required, provision of laboratory services, equipment cost and selection, and other administrative matters (*28, 47*).

The results of stack tests may ultimately be used as evidence in a court case relating to violation of an emission control regulation. Therefore, it is important that prescribed or standard methods be used, since unorthodox or new procedures are difficult to defend under an attack in court. Features of the procedure that must be given particular attention are location of the testing point in the exhaust system, number of sampling points in the duct, sampling equipment used, methods of determining stack gas velocity, maintenance of isokinetic conditions, handling of the collected sample and recording of results. Since the results of stack tests may be used in court, careful attention must be given to documentation of the chain of custody of the samples, identification of persons involved in doing field and laboratory work, and recording of results throughout the procedure so that the information will satisfy the legal rules for evidence admissible in a court proceeding (*28*).

9. Continuous Emission Monitoring

Because of high costs involved in making manual stack emission measurements of the kind described in the preceeding section, it is not feasible to make such tests very frequently. Thus, such tests do not provide a fully adequate body of information to indicate the effect of variation of source operating parameters on emissions, to alert operating personnel to existence of excessive emissions, nor to serve as a sole basis for enforcement of emission control regulations. Since this is the case, there is an increasing use of instruments to continuously measure emissions from sources. One type of instrument, which has been in rather general use for years to measure smoke emissions from fuel-burning plants, uses detectors of changes in light transmittance across a section of a duct or stack. In many cases, these are installed voluntarily by source owners as a means of aiding boiler operators in maintaining efficient combustion conditions and conserving fuel. In some jurisdictions, governmental control agencies require the use of such instruments on specified sources as a means of helping plant operators to avoid excessive smoke emissions. In more recent times, as technology for making measurements has become available, continuous emission monitors for other contaminants have come into wider use. Pollutants often measured include particulate matter, sulfur dioxide, nitrogen oxides, and a number of specific materials associated with chemical processes. In the past, only a few governmental regulations have required such measurements, but the practice is increasing. In even fewer cases have data from continuous emission monitors been used as a basis for determining compliance with an emission regulation. This is because of the lack of assurance of the accuracy of the measurements and because the instruments are under the control of the source owner. However, it is likely that emission monitors will be more widely used as the basis for law enforcement as more information and experience with the instruments is accumulated. The problem of source owner control of the instruments can be overcome by periodic, unannounced independent calibration of the instruments by the control agency and by providing for the inspection, at any time, of the instruments and the data obtained from them (including calibration data) by the control agency.

The cost of installing and operating continuous emission monitors is substantial and may be an unreasonable burden to place on smaller pollution source owners when total emissions are small or where the possibility of exceeding emission limitations is not great. Thus, requirements for continuous emission monitors should be restricted to relatively large sources, to those that have a tendency to exceed emission control limits because of processes or operations that require unusual care in operation to maintain minimum practicable emission rates, and to those sources that may emit relatively highly toxic or noxious pollutants.

B. Correcting Problems

1. Adopting Air Quality Objectives or Standards

Ambient air quality objectives, standards, or goals are the keystone to air resource management. They describe the air quality to be achieved or preserved—the end to which control measures are directed. While it is possible to operate air pollution programs without having air quality goals, it is more desirable and sounder technically to have them. Their key impact is in long-range planning as a tool in design of preventive actions. Without air quality standards, the program will be largely directed toward correction of existing problems whose detrimental effect is so great as to be almost overwhelming. With air quality standards, it is much more feasible to provide for a sound control program for all sources of pollution, even though they may not be, individually, creating a specific nuisance.

In the United States, the federal government, under provisions of the Clean Air Act, has promulgated air quality standards that are applicable throughout the entire country (*48*). Each state government is required to prepare and submit, for federal approval, a plan for attainment and maintenance of the national standards. State and local governments may, if they wish, establish standards that are more stringent than the national standards and may adopt standards for pollutants for which there are no national standards. Many states have elected to exercise this option.

2. Adopting Emission Control Regulations

Regulations that limit emission of pollutants to the atmosphere or otherwise provide for reduction of air pollution levels are the heart of air pollution control programs. The regulations may embody a number of general approaches, including (a) emission limitations that prescribe the allowable concentration, mass, or visual appearance of stack effluents; (b) equipment design specifications and requirements that certain items of equipment be used; (c) prohibitions on the use of certain equipment or practices; (d) limitations on the content of certain elements or compounds in fuels or raw materials; (e) restrictions on the location of specified sources; and (f) requirements for extraordinary emission control procedures during times of poor atmospheric dispersion.

The nature and extent of emission control regulations are governed by desired air quality, the kinds of pollution sources in the area, dispersive characteristics of the atmosphere, and estimates of the total amount of pollutant emissions that can be tolerated. Emission control requirements can be determined by one of five methods: (a) establishing an air quality goal, measuring present air quality, inventorying current pollutant emissions, and calculating emission reduction needed to meet the air quality goal (*49*); (b) the rollback approach—based on an assumption that at some time in the past, air quality was satisfactory—estimating what emissions were at that time, and calculating the degree of reduction of present emissions needed to reach the level existing at the earlier time (*50–52*); (c) calculating emission reductions that can be achieved by use of available emission control methods and equipment and estimating improvement in air quality that would result (*53*); (d) selecting a "source fence line" air quality standard and, on the basis of stack gas atmospheric dispersion formulas, calculating the amount of pollutant that may be emitted from a single source with various stack heights and not exceed the air quality standard; and (e) determining minimum emission rates attainable by application of the best practicable means of emission control and limiting emissions to such rates.

Preparation of regulations involves extensive technical knowledge and exercise of judgment in a field unfamiliar to many legislators. Therefore, air pollution laws and ordinances often do not contain specific emission control requirements, but rather contain general policy statements and provide for administrative machinery to carry out the intent of the law. Authority to adopt regulations is usually given to a board or commission or a designated executive. Nearly always, public hearings are required before a regulation may be adopted. The board, commission, or executive, in drafting regulations, is limited by the language of the basic law granting regulation-making power and must be careful to remain within delegated powers lest the regulation be found invalid by a court. In addition, the regulations must be adopted in the manner required by law. For example, the regulations must be formally adopted at a proper meeting. A written record must be kept of the vote, and the regulation must be filed with a specified official and perhaps published in a designated place. A record should be kept of all these actions, and the preamble to the regulations should state that the required steps have been taken. The regulation should specify upon whom it is binding, who are excepted, and what actions are required or prohibited, so that those regulated will know what is expected of them.

In preparing regulations, care must be taken to ensure that a workable procedure for enforcement is possible. Regulations that require expensive, time-consuming, or complex procedures for enforcement or that cannot be clearly understood and acted upon by the parties regulated will not likely be successful in practice. Consideration should also be given to any undesirable side effects the regulation may have (e.g., creation of a solid waste or water pollution problem) and to the impact the regulation may have on the rate of use of fuels and raw materials to ensure that such resources are not needlessly wasted. Expected general growth of the community and pollutant emissions must be taken into account as regulations are prepared to ensure that they will be adequate for a sufficient period of time to avoid overly frequent, disruptive, and costly changes in the future.

After regulations are adopted, the agency should evaluate to see that desired air quality improvement is achieved and what the effects of compliance are on the community. If the regulation is found to be improper for any reason, it should be revised or repealed.

3. The Enforcement Process

Once requirements of various kinds have been established in laws, ordinances, and regulations, it becomes the agency's job to see that the public complies. This is done through a combination of programs of education to advise the public what is expected; follow-through on public complaints; finding illegal conditions; securing abatement of illegal conditions through conference, conciliation, and persuasion; and prosecution of those few violators who refuse to cooperate. A vigorous public education program concerning actions expected and prohibited can do much to secure voluntary compliance, thus saving much agency effort in finding and correcting situations one by one.

The most common actions in securing abatement are to bring about compliance with specific regulations. The first action is to find the violation. This done by field inspectors on patrol, area surveys, and investigation of public complaints. The effectiveness of field inspection can be enhanced by use of radios by field men, use of aircraft for surveillance, and occasional concentration of many inspectors into a limited geographic area (*42*). It is necessary to conduct surveillance activities during all hours of the day and on weekends as well as weekdays. Some agencies prefer to have field enforcement people wear uniforms and drive distinctively marked cars. When a violation is found, an investigation is made that includes inspection of the offending source, interrogation of responsible parties, and interview of complainants or other witnesses. From this, the necessary abatement action is determined, i.e., a decision is made as to whether corrective or punitive action is required. Punitive action is usually taken if the offender has repeatedly violated requirements or there seems to be no good reason why the violation occur. Such cases include open burning of refuse, smoke emissions due to negligent operation or maintenance, and excessive smoke from vehicles. In such punitive actions, a citation to appear in court is issued, or in some cases, a "ticket" is issued, requiring payment of a fine in the same manner as for tickets for illegal vehicle parking. If corrective action is indicated, the responsible party is given a notice specifying the violation that has occurred and indicating the action to be taken. A time for compliance commensurate with the corrective action needed is specified, and follow-up action is instituted as dictated by the enforcement policy of the control agency (*34*). The control agency may offer advice to the offending party as to the means that may be used to abate the violation and may provide the names of a number of consulting firms that can assist the violator with development of detailed abatement plans. In the United States and Canada, lists of consulting firms are published annually by the Air Pollution Control Association (*54*).

In matters for which there is no specific law or regulation, particularly nuisance abatement, the procedure is somewhat modified. Usually an agency inspector investigates the situation, which may involve odors, dustfall, soiling, or some other condition. If he is of the opinion that a nuisance exists, he attempts to secure abatement by negotiation with the party responsible for the condition. If this fails to bring about correction, a hearing may be held before the pollution control officer or a hearing board. Based on the views of the hearing body, an order to abate the nuisance may be issued or the case may be dismissed. If the order to abate is not obeyed, the matter may be referred to a court for adjudication. Hearings may also be held when agency inspectors do not believe a nuisance exists but some citizens do. This allows the hearing body to get all the facts and opinions upon which to base a decision and gives the citizens an opportunity to make their views publicly known.

Other enforcement procedures involve failure to obtain a construction permit, to secure or renew an operating permit, to provide a required report, or to make a required test. In such cases, it is common practice to allow the responsible party an opportunity to correct his failure before filing court action. However, if the person is a repeated violator or seems to have violated intentionally, the matter may be referred to court immediately upon discovery.

When the abatement action required is one that will take a long time to correct, be very costly, or involve a large segment of a community's industry or public, it is common practice to arrange a long-term abatement schedule. The schedule will require that certain actions be taken at various times over a period of years. Periodic progress reports may be required. It is essential that each increment of progress in the schedule be enforceable, so that any failure of the source to proceed with each step of the schedule is a cause for legal action. If this safeguard is not provided, the agency may find that the entire time for abatement has elapsed before enforcement action can be taken, with the resulting undue suffering of the public because of continuing pollution. In a few jurisdictions, the source is required to post a performance bond to guarantee that the source will proceed in a timely fashion with the promised abatement action. If the source proceeds in accordance with the approved abatement schedule, the source suffers no penalty, but if the source fails to perform as agreed, the bond is forfeited.

Since it is possible that any action taken by the agency may result in a court action, it is important that all work be accurately done and meticulously recorded (*34*). The actions must be in accordance with the law, regulations, and precedents developed in previous court actions.

The enforcement process involved in adjudication of odor problems is essentially the same as used in other matters. However, there is no objective method for measuring odors, making it necessary to use the human nose and human reaction as a means for determining presence and objectionableness of odorous pollutants. Methods of detecting presence of odors and locating their origin have been described (*34, 55*). These may be applied in enforcement proceedings. The reaction of the public to odors may be determined by house-to-house surveys, student odor surveys, and public opinion polls (*56*).

4. Variances, Hearings, and Appeals

A variance is permission granted by the control agency, board, or commission to build, maintain, or operate an installation that does not comply with regulations. When a new law is put into effect, unless a period of grace is written into the law, variances will be needed to allow time to make corrections. Variances may also be needed to allow time to secure equipment, conduct research and pilot plant studies, or budget expenditures. If no satisfactory way to correct a problem is at hand, a variance to give time to develop one may be better than installing unsatisfactory available controls forthwith. Variances should expire on a set date and be renewable only after reconsideration of the case.

In many jurisdictions, the law provides for an administrative hearing process whereby the agency is empowered to hold hearings associated with enforcement of air pollution laws and regulations. The procedures are usually relatively informal, with the objective being to develop the facts in the case without the heavy expense of formal court actions. The agency is usually given subpoena powers and testimony is under oath, but the parties are not bound by the strict rules of evidence that prevail in courts. Both parties (the agency and the source) are allowed to present their positions and evidence and to cross-examine witnesses. Depending on the facts in the case and the agency's statutory powers, the agency hearing officer may issue a corrective order, dismiss the case, levy a civil penalty, or take other appropriate action. If the civil penalty is not paid or compliance with the order does not ensue, the matter is referred to court. The administrative hearing procedure offers many advantages and flexibility in handling air pollution cases to both the control agency and the pollution source. Clear-cut rules of procedure are necessary to ensure that cases are handled in a fair and orderly manner.

Appeal and hearing boards provide a quasijudicial means for individuals aggrieved by actions of the agency administrator to have their views heard without the expense of a court case. The board will ordinarily be appointed by the mayor, governor, legislative body, or the air pollution control board or commission. There may be three to five members, made up of such persons as lawyers, scientists, engineers, or representatives of industry and the general public, the idea being to form a group that will objectively represent the total public interest. One board may serve the purposes of both an appeal and a hearing body. Appeal and hearing boards can usually grant variances and reject or modify actions taken by the agency administrator, and in some cases, issue corrective orders to be complied with by pollution sources. Hearing boards may also consider cases in which the administrator has not been able to secure abatement and will try to find some means, short of prosecution, to get needed correction of violations.

A few laws require that before the air pollution program administrator can issue an order to correct a violation, a hearing be held before a board or commission. This procedure is unworkable, since boards are usually made up of unpaid people who cannot spend the time necessary to hold the required number of hearings, and it is unnecessary, since most cases are clear-cut violations that should be corrected. It is preferable to hold hearings only when requested by a regulated person or the program administrator and to authorize the administrator to issue, as he sees fit, orders and notices authorized by law and regulations. Even when hearings are held only on request, the number of hearings may be substantial. For example, in a populous jurisdiction the hearing board may hold more than a hundred meetings per year.

5. Control of New Installations

A key element in an air pollution program is the regulation of new installations. Some means should be provided that will ensure that all newly constructed facilities will be built in such a way that they can be operated without causing pollutant emissions in excess of those allowed by law or regulations and that the new facility will not cause ambient air quality standards to be exceeded. The most effective way to do this is to require that approval be obtained from the control agency of construction plans before installation begins. The procedure should apply not only to the usual stationary sources (e.g., power plants, cement mills, foundries) but also to sources that are associated with or induce motor vehicle traffic (e.g., shopping centers, roads, sports stadiums). The latter kind of facilities are sometimes called indirect sources. Through the new source review procedure, installations are avoided that are poorly designed, would cause excessive air pollution, or that would be placed in an inappropriate location and thus require costly modification or reconstruction. The procedure also makes it possible for the control agency to keep its emission inventory up to date and to place specific additional requirements on sources as conditions to obtaining authority to proceed with construction. The new source review procedure, however, does impose on the air pollution control agency a requirement that it have available necessary skilled personnel to determine from sets of plans and specifications whether a proposed installation will meet emission limitations and not cause ambient air quality standards to be exceeded. A possible disadvantage of a permit system is that innovative techniques for control will be discouraged, since source owners and control agencies tend to select previously tried methods of control for which experience is available rather than risk disapproval of an application for a permit to construct or modify an installation that may actually be an improvement over past practice.

Operation of a construction permit program requires the establishment of guidelines, procedures, and systems for accomplishing a number of tasks (*52*). These include:

a. Notification of source owners and operators, contractors, engineering firms, and others of the requirement to obtain a permit prior to construction. Arrangements should also be made with other government agencies to alert the pollution control agency of proposed facilities that come to their attention. Direct communications, announcements in the mass media, and letters distributed through trade and professional associations may be used.

b. Preparation of rules or regulations that identify the kinds of facilities included and excluded; set forth procedures for making application, processing the application, and appealing agency decisions; specify the authority of the agency to deny applications and to impose permit conditions; describe variance conditions and procedures; spell out testing requirements; etc.

c. Provision for distribution of application forms, copies of regulations, test procedures, instructions for completing applications, etc.

d. Receipt and evaluation of applications with regard to emissions as compared to regulations and impact on ambient air quality.

e. Conduct of inspections as construction proceeds to ensure that the approved plans are followed.

f. Provision of guidelines and procedures for conferences with applicants, hearings on appeals of denials of applications, placing conditions on permits, granting of variances, cancellation of a permit if construction does not proceed or deviates from approved plans, preparation of court cases, etc.

g. Establishment and operation of a permit data and work management system.

h. Provision for coordination with other agencies that issue permits, review environmental impact statements, and others.

In reviewing applications, the agency should make reference to available publications that indicate the nature and extent of emissions that may be expected from installations and the efficacy of various types of emission reduction measures (*44–46*). Such publications should also be made known to parties engaged in design of facilities. Similarly, the agency should use available publications for evaluating the impact on air quality of indirect sources such as roads (*57–61*) and airports (*62, 63*). If the designers of facilities and the control agency use the same evaluative approaches, much disagreement and redesign can be avoided.

If an applicant and the agency disagree on whether a proposed installation will perform satisfactorily, a conditional permit can be issued, even though the agency feels the installation will not perform adequately. The applicant should acknowledge in writing the fact that the agency believes the installation will not be satisfactory and agree that changes or replacements will be made if necessary. After a permit to construct a facility has been issued and the installation has been built, it must still meet emission control requirements. To ensure that this is the case, each newly built installation should be inspected and emission measurements should be made. If emissions are satisfactory, a permit to operate is issued. If not, corrections must be made and additional emission measurements conducted. These tests not only ensure that legal requirements are met but also provide information useful in future plan reviews.

To reduce the number of plans that must be reviewed, it is well to exempt from construction permit requirements all installations that are habitually designed so as to cause no problems or are of such nature that plan review and approval is unlikely to bring about any significant reduction in pollution emissions. Such sources include gas-fired units, all but very large units fired with light distillate fuel oil, stationary internal combustion engines, small roads and parking lots, and other small and insignificant facilities.

If, for some reason, a permit system cannot be established, an alternate is to establish a registration system. In a registration system, those planning to build a new facility are required to submit only information on the location where the unit will be built, a general description of what it will do, and information on the nature and location of expected pollutant emissions. The agency does not approve or disapprove the installation, but merely issues a registration form or certificate. Reliance is placed on the applicant for meeting emission regulations after the unit goes into operation. The registration procedure does, however, give the agency knowledge of the installation so that it can be checked after construction and an opportunity to call attention of the applicant to emission requirements and to offer him advisory services.

In the United States, the federal government has promulgated emission limitations for certain major new sources (*64*). These apply throughout the nation. Also, national emission standards for hazardous air pollutants that apply nationwide have been promulgated (*65*). Standards have been set for asbestos, beryllium, mercury, and vinyl chloride. In regard to both of these federal standards, enforcement authority may be delegated to states. If this is not done, enforcement is carried out by the federal government.

6. Sealing Equipment

Authority to seal equipment and thus prevent its use and continued violation of emission regulations is a powerful enforcement tool. It should be used sparingly and cautiously. Application of this punitive and preventive power is often more damaging to the offender than fines levied by a court. Some agencies use this power to ensure correction of defects in space heating equipment by sealing units at the end of the heating season. Thus, the owner is not deprived of its use when needed, but he knows he must get it corrected during warm weather or face either prosecution for breaking the seal or getting cold. It is also a useful tool in preventing further construction on a unit being built in a manner different from that prescribed in approved plans.

7. Disaster Prevention

There are two general kinds of air pollution disaster, namely, (a) those associated with stagnant meteorological dispersion conditions and emissions from all sources, and (b) those associated with accidental release of air pollutants from a single industrial, material storage, or other source. Each air pollution control agency should be prepared to discharge its responsibilities, in cooperation with other agencies, with regard to both kinds of disasters.

Air pollution episodes associated with stagnant meteorological dispersion conditions can occur in any area where pollutant emissions are great enough to cause a build-up of concentrations, since stagnant dispersion conditions occur almost everywhere, although their frequency varies from place to place. Valley locations, where atmospheric ventilation is restricted by topography, are particularly prone to generalized pollution episodes. Therefore, most air pollution control agencies should have an action plan for management of such episodes. The plan should include meteorological and pollutant level criteria for determining that an episode may occur or is occurring; procedures for reducing pollutant emissions from stationary and mobile sources; means for securing real-time quality and meteorological data; a preplanned system of communications with all concerned agencies, the public, and pollution sources; predetermined legal authorities for taking various actions required; and a procedure for setting up and operating an episode operations control center. Agencies and organizations that will likely be involved, along with the air pollution control agency, include the police department, communications media, school authorities, industrial groups, health and hospital groups, and the civil defense agency. Guidelines for use in large, medium-sized, and small urban areas are available (*66–68*). The episode plan must be simple, effective, preplanned, predisseminated, and understandable by those who must take action during an episode (*69*). It is desirable to conduct drills of the episode plan in order to work out defects at a noncritical time and to test the effectiveness of the plan (*70*). In interstate and interprovincial areas, it is very desirable to have a single plan that applies throughout the region to prevent confusion and promote effectiveness. In the New York City, New York, area, the Interstate Sanitation Commission acts as the coordinating group for the three states involved (*71*). In the United States, the air pollution levels representing the degree of severity of air pollution and denoting the several stages of an episode used by most agencies are those published by the federal government (*72*). However, some state and local agencies have adopted somewhat different levels (*71, 73*).

While the main activity associated with air pollution episodes is to reduce pollutant emissions, action may also be taken to protect those members of the public who are particularly subject to injury by high pollution levels. In some hospitals, special air cleaning systems are activated to remove pollutants by use of filters and activated carbon from the air in rooms occupied by people with serious respiration or cardiac problems. In most cities, special advisories are broadcast over the radio during episodes advising sensitive people to take self-protective actions. In Los Angeles County, California, at certain pollutant levels, strenuous activities are avoided in the schools and special news bulletins are issued to advise people with cardiac and respiratory ailments of actions they should take. The control agency also provides special advisories to such people by use of a recorded telephone message which is kept current in relationship to air quality conditions (*42*).

Pollutant releases due to industrial accidents, spills, rail and highway accidents, fires, windstorms, etc. are also of concern to air pollution control agencies, along with other public safety organizations. The concerned agency should develop contingency plans that include procedures for identification of polluting materials, estimating impacted areas, protecting the public, and restoration of normal conditions. Industrial facilities should be required to develop contingency plans and to provide back-up equipment as needed to minimize the impact of such situations as power failures, cooling water failures, equipment malfunctions, pipe and tank ruptures, fires, and flooding. Safeguards should be provided in advance to prevent release of air contaminants to the extent practicable (*74*). In the United States, the federal government has published a manual to assist air pollution control agencies in such activities (*75*).

8. Motor Vehicle Emission Control

There are two basic parts to motor vehicle emission control programs, namely, (a) ensuring that new motor vehicles are constructed in such a way that emissions are minimized to the extent feasible (and necessary to achieve clean air) within available technology and (b) seeing that engines and pollution control devices and systems are kept in place and properly maintained. In most nations, the job of seeing that new motor vehicles are properly constructed will be done by the national government. The job of policing in-use vehicles can, in most countries, be done best by state, provincial, and local governments.

A national program for control of pollution from new motor vehicles has been in effect in the United States since 1968. Other countries, notably Japan, have also instituted programs. The program in the United States is based on the Federal Clean Air Act, which authorizes the federal government to prescribe pollutant emission standards for new motor vehicles. Such regulations have been promulgated (*76*). Once these standards, and a suitable test procedure, have been established, the automobile manufacturers are required to produce vehicles that meet the standards. Prototype models of future year vehicles and engines are tested by the manufacturers in accordance with federally prescribed procedures. The data from the testing is provided to the federal government, and the test vehicles are delivered to the federal government at prescribed stages of the testing procedure so that the federal government can verify the results. Once a prototype is found to meet the standards, mass production can proceed. Provisions are made for testing production vehicles to ensure that they too meet prescribed standards. The federal law also provides for warranty, by the manufacturers, of the emission control systems and devices; a prohibition against tampering with emission control systems by new car marketers; and provision for recall and repair or modification by the manufacturer of any class of vehicles that fails to meet prescribed standards when the vehicle is in public use at no cost to the vehicle owner.

Control of air pollution from in-use vehicles began in the 1960s with programs to control smoke from vehicles, particularly diesel-powered trucks and buses. There are three general kinds of vehicle smoke control programs, (a) establishment and enforcement of limitations on visible emissions from vehicles operating on the road; (b) smoke emission limitations based on visual observations of "excessive smoke" enforced as part of a vehicle safety inspection program; and (c) limitations on smoke emissions from vehicles using instrumental measurement of smoke emissions under prescribed engine operating conditions, often also as a part of a vehicle safety inspection program. On-the-road vehicle smoke control programs are based on regulations that limit the density of smoke, as measured by reference to the Ringelmann chart, or limit the time period or travel distance over which any visible smoke may be emitted. The regulations based on time or travel distance of the vehicle are generally preferred, since they avoid the need for training inspectors to use the Ringelmann chart and the problems of the inspector in positioning himself in a proper location for observing the smoke density of emissions from a moving vehicle. All visible emissions from gasoline-powered vehicles can reasonably be avoided by proper maintenance, and therefore a complete prohibition of visible emissions from such vehicles is appropriate. With diesel-powered vehicles, even a well operated and maintained engine will make some smoke for short periods during acceleration from a standstill and when changing gears. Therefore, an allowance for some visible emissions during such events should be provided. Enforcement of on-the-road smoke emission limitations is accomplished by inspectors patrolling in automobiles. It is generally desirable for the inspectors to be in uniform and that the vehicles be equipped with lights and sirens in the manner of regular police vehicles (*46*). Such programs can cause many smoking vehicles to be taken off the road and repaired. For example, eight inspectors in Los Angeles County, California, cite about 4500 violations per year (*42*).

National, state, provincial, and local governments can operate vehicle safety inspection programs. In some of them, a visual inspection is also made to determine whether excessive smoke is emitted. The standard is not usually precise but more often is a verbal description as to what shall be considered excessive. If a vehicle is found to be emitting excessive smoke, the owner is required to have the vehicle repaired and to have it inspected again. Such programs make it possible to abate part of the vehicle smoke problem. For example, in the city of Cincinnati, Ohio, about 0.2% of all vehicles passing through their safety inspection lanes are rejected because of excessive smoke emissions (*77*).

The third type of vehicle smoke program uses an instrument that measures the light obstruction of the exhaust stream or measures the extent of the blackening of a piece of filter paper through which some exhaust gases have been drawn. The vehicles may be tested as part of a regular complete safety and roadworthiness inspection, as a special periodic smoke abatement inspection, or in a roadside spot-check program. Offending vehicles are required to have repairs and adjustments made and to again be inspected. Fines may be imposed, and serious offenders may be ordered out of service until repairs are made.

In areas where motor vehicle pollution is severe, it may be necessary to conduct a program to ensure that gaseous emissions are kept to a minimum. There are a variety of possible programs, including:

a. Visual inspection to see that pollution controls are still in place. This is of limited value since several factors that influence emissions cannot be observed by visual inspection (e.g., carburetor adjustment, engine tuning).

b. Engine adjustment checks in which idle speed, air:fuel ratio, and spark timing are measured. If found to be out of adjustment, corrections are required of the owner.

c. Engine tune-up programs that require vehicle owners to have their engines tuned and their spark plugs and breaker points replaced at intervals representing expected service life.

d. Emission inspection programs involving measurement of emissions with the engine at idle and a requirement that those having excessive emissions be adjusted and repaired.

e. Emission inspection programs as in (d), but with testing of the engine under load on a dynamometer.

f. Emission inspection programs as in (e), but also including a determination of the adjustments or repairs needed for the information of the owner.

Each type of program has a number of advantages and disadvantages (*78*). The selection of the program to be used should be preceeded by careful evaluation of several options. The states of California and New Jersey and the city of Chicago, Illinois, have established inspection–maintenance programs.

Vehicle inspection maintenance programs may be operated in a variety of ways, including spot checks done at roadside by mobile testing units, conduct of mandatory periodic inspections at private garages licensed by the government, mandatory periodic inspections at government operated inspection stations (possibly in conjunction with safety inspection), and incorporation of emission inspection into general roadworthiness inspections of trucks and buses as part of a public service commission (or similar agency) inspection. Each kind of program has advantages and disadvantages to be evaluated before selecting an approach for a given area (*79*). Any inspection system does entail possible abuses by the automotive service industry that must be guarded against. They also increase vehicle maintenance costs, although that is offset in part by improved gas mileage. They also impose on the public a loss of time in having inspections made and do induce vehicle trips needed to get to and from inspection facilities and, if needed, repair facilities (*80*). The emission reductions to be expected from an inspection–maintenance program depend on the frequency of inspection, the performance levels required (or, expressed another way, the percentage of vehicles rejected on inspection), and whether an engine idle mode or loaded mode test cycle is used. For a loaded mode test, reduction of total fleet hydrocarbon emissions range from 8%, if 10% of the vehicles are rejected, to 15%, if 50% of the vehicles fail. The carbon monoxide emission reductions are 3–4% less. With the idle mode test, total fleet emission reductions are in the range of 1–4% less than with the loaded mode test for hydrocarbons and carbon monoxide at the various rejection rates (*72*). The emission reductions achievable are generally a percentage of the emissions. Therefore, the absolute benefit (in terms of reduction of the mass rate of pollutant emissions) becomes smaller as the mass rate of emissions from vehicles is reduced. Thus, in the United States, where the average vehicle emission is declining because of production of cleaner new cars, the benefits of inspection–maintenance are likely to decline as the years go by.

Another approach to control of air pollution from in-use vehicles, which may be applied at a time when the fleet of vehicles in public use is not made up of vehicles that were constructed in accordance with presently known means for control of pollution, is to retrofit recently developed technology to older vehicles. Such measures include lean idle air/fuel ratio, vacuum spark advance disconnect, catalytic exhaust gas oxidizers, air bleed into the intake manifold, and exhaust gas recirculation. Emission reductions per individual vehicle of carbon monoxide, hydrocarbons, and nitrogen oxides range up to 60% (*72*). Whether such retrofits will find widespread application is unclear. Many major urban centers have such programs under consideration, and the state of California has a program of this type in operation.

Governmental expenses in operating programs for controlling air pollution from in-use vehicles are substantial. To provide funds for such programs, governments may increase gasoline taxes or automobile license fees, charge fees for inspections, increase the price of driver's licenses, or rely on other sources of revenue.

9. Fuel Regulation

Early fuel regulations limited the volatile content of coal because high-volatile content coal tends to produce excessive smoke when used in hand-fired furnaces. It has been found to be essentially impossible to get firemen to use sufficient care in hand firing to be able to burn high-volatile coal without causing smoke. The only practicable answer is to prohibit use of high-volatile coal in hand-fired furnaces.

In many countries, regulations limiting the sulfur content of fuels are used to reduce sulfur dioxide emissions and thus reduce sulfur dioxide concentrations in the atmosphere. Such regulations became feasible as technology for removing sulfur from petroleum products became practical. In the case of coal, compliance is achieved by using naturally occurring low-sulfur coals and by washing coals to remove part of the native sulfur. More recently, procedures for removing sulfur dioxide from flue gases have become available, and for plants equipped with such equipment, sulfur in fuel limitations are replaced by limitations on sulfur dioxide emissions.

In the United States, the lead and phosphorous content of gasoline has been regulated as a means of limiting lead levels in the ambient air and protecting catalytic exhaust gas control devices on automobiles, which are rendered ineffective by lead and phosphorous (*81*). Regulations also provide for registration of motor vehicle fuels and additives to motor vehicle fuels and crankcase oil and testing by the manufacturer to determine whether there would be any adverse effects from use of a fuel or additive on health or welfare or on the emission control performance of any motor vehicle or vehicle engine (*82*).

This type of regulation is basic in nature, is very effective, and can be enforced readily. An enforcement program involves checking fuel composition at its source (mine or refinery), in transport vehicles, and at distribution depots. Lists of approved fuels are prepared and made public at frequent intervals. Inspections and fuel analyses at the point of use are necessary to substantiate violations. Complaints that illegal fuel is being sold must be checked out. Standards of measurement and procedures for testing must be developed and made public. In some cases, fuels that do not meet required characteristics are permitted, provided it can be shown that they can be used and that they emit no more pollutant than a fuel that does meet the required characteristics. If this is the case, methods of demonstrating emission performance must be established so that "nonstandard" fuels can be tested.

10. Operating Permits

Some control agencies use operating permits (sometimes called certificates) as one part of the overall source emission control program. There are two general kinds of operating permit. One kind is issued after a newly constructed source is put in operation and found to be in compliance. If the source does not comply, the operating permit may be withheld until changes are made to ensure compliance. Such permits usually apply to any source for which a permit to construct is required. The permit will usually remain in effect as long as the source complies with all applicable regulations. Failure to comply is a cause to revoke the operating permit and, if operations continue, to seek court action to force discontinuance of operations and penalties. Another kind of operating permit is one that must be renewed at intervals, such as a year or two. The agency decides which kinds of sources must secure permits and how often they must be renewed. Major sources are usually included. Smaller sources that have a tendency to violate emission regulations or cause increased emissions because of poor operation or maintenance are included. The general rule is that if an annual (or other periodic) inspection will help assure minimum emissions, the class of sources is included in the operating permit program. In conducting the program, it is necessary to maintain a file of sources that have operating permits and to arrange for the annual (or other periodic) inspection and permit renewal. If violations of emission regulations are found, appropriate action is taken to secure abatement. If equipment is found to be in a poor state of repair to the extent that excessive pollutant emissions would be expected, action is taken to require that repairs be made.

During the course of periodic inspections for renewal of operating permits, the field inspector should update information in the agency's files on pollutant sources and control equipment on each premise, check emission inventory data, check compliance with fuel quality regulations (and perhaps collect a fuel sample), and verify ownership data.

If violations of regulations are found, notices and corrective orders can be issued and the operating permit can be revoked or not be renewed. If corrections are not promptly made, the owner can be prosecuted both for failing to have a valid operating permit and for any other violation.

Since conduct of an annual (or other periodic) inspection for operating permit renewal and other purposes is costly to the agency, many of them charge a fee for annual permit renewal that is large enough to defray the agency's costs of making the inspections and doing other related work.

11. Citizen Suits

Private citizens have long held the right to seek relief and compensation for damages they suffer due to pollution discharged to the atmosphere by another party. Over the years, people have gone to court to seek abatement of air pollution, usually in the form of foul odors or deposition of dust and soot. A number of cases have been filed by people who have had their crops damaged by sulfur dioxide and by those who had experienced injury to their cattle or crops (usually fruit trees) caused by fluorides discharged from aluminum smelters and phosphate fertilizer plants. In such cases, the courts have usually awarded monetary damages and have often ordered discontinuation of the offending discharges. Such law suits are difficult for private citizens to undertake because of the technical difficulties of collecting evidence and the costs of court proceedings. In the past, such suits could only be undertaken by a party who had suffered personally.

In the United States, the United States Clean Air Act and a number of state laws now provide a broader range of means by which the public, either as private citizens or through their organizations, can seek, through court action, to bring an end to air pollution. These laws generally make it easier for citizens or organizations to take court action against a pollution source, especially those violating any pollution control law or regulation. Some laws also enable citizens and organizations to pursue court actions to require governmental agencies to faithfully and diligently carry out the provisions of an air pollution control law. In the United States, air conservation organizations have filed many law suits in recent years seeking court orders directed at government agencies to require them to promulgate regulations, to strengthen regulations, and to promptly enforce laws and regulations. Many of these suits have been successful in securing activity by governments to properly implement air pollution control laws.

12. Minimizing Motor Vehicle Travel

In some major urban centers, pollution from motor vehicles is so severe that, even with application of all available means to reduce emissions from individual vehicles, it will not be possible to achieve ambient air quality standards. In such cities, attainment and maintenance of air quality standards will be accomplished only by reducing the amount of motor vehicle travel and improving traffic flow. This involves, in the short-term, revisions in urban transportation systems and, in the long-term, major revisions in the overall community land-use development and transportation plans.

Measures to reduce motor vehicle travel in the context of existing communities involve a variety of programs to increase the use of less polluting mass transit and decrease the use of personal automobiles. Incentives for encouraging use of mass transit include provision of more desirable vehicles, more frequent service, lower fares; provision of parking areas in the fringe areas of the city, improving passenger waiting areas to provide more comfort and protection against foul weather, provision of feeder buses into major transit routes, designation of lanes on streets for exclusive use of buses to shorten travel times, and better availability of schedule information. At the same time that incentives are provided to use mass transit, it is usually necessary to impose penalties on the use of private autos in order to get more people to switch to mass transit. Such penalties may include reduction of the number of parking places in traffic-congested areas, increased taxes or rates for parking spaces, stricter enforcement of on-street parking laws, higher tolls on routes of entry into the central city, and taxes on cars entering the city through use of a special license or windshield sticker. Some combination of measures should lead to some reduction in use of personal vehicles.

In addition to measures for reducing use of personal vehicles, it may be necessary to apply measures to improve or reroute traffic flow in areas of heavy vehicle concentration. Improved traffic flow reduces stop and go driving patterns and engine idling in traffic. Among the measures that may be employed are automated traffic control systems, widening of traffic lanes, exclusion of vehicles from some streets or areas, providing additional lanes at intersections, prohibiting on-street parking, street straightening, pedestrian movement controls, and provision of inner and outer loop roadways to route vehicles around rather than through congested areas.

Other procedures for reducing motor vehicle travel include a variety of ways for increasing use of carpools, such as exclusive lanes on streets for cars with two or more riders; reduced charges or preferential location of parking for those using carpools; carpool locator systems; and educational campaigns. Improved taxi service and taxi sharing also are of value. Staggering of work hours and a four day work week instead of five will also help reduce travel and traffic congestion. Higher vehicle fuel and license taxes also reduce personal vehicle use but may unduly impact on people with limited incomes.

The air pollution control program manager in a city requiring transportation controls to reduce air pollution will need to evaluate the extent to which air pollution levels may be reduced by application of the various measures described above. An evaluation must also be made of the costs, feasibility, public acceptance, and enforceability of the measures. Procedures for doing so have been published *(78)*. Evaluations have been made for a number of these measures in a variety of cities and these are available *(83, 84)*.

Longer-term approaches to reduction of motor vehicle travel involve major modifications in the overall community transportation system (roads, rail travel, bus systems, etc.) and changes in land-use patterns. This involves the evaluation of the air quality aspects of various possible generalized urban forms, land-use plans, and land-use densities and the amount and mode of travel expected to be associated with each alternative plan. Various land-use and transportation alternatives are translated into expected pollutant emissions for designated small subareas, and then ambient air concentrations of pollutants are calculated by use of atmospheric dispersion models. At least one set of procedures and models for this purpose has been described *(85)*. Trade-offs must be made between high-density land use, which minimizes travel but concentrates air pollutants from stationary sources, and lower-density development, which disperses emissions from stationary sources but results in more vehicular pollution because of the greater travel associated with low-density development.

C. Operating the Agency

1. Initiating an Air Pollution Control Program

Whatever the reason for concern about air pollution in a community or the persons or groups interested in doing something about air pollution, the basic necessity in getting a control program initiated is to develop enough interest among the people of the jurisdiction to support enactment of an air pollution control law and establishment of a full-time governmental control agency. One way to develop interest is to form a clean air committee, preferably appointed by the mayor or governor. It is helpful for a staff person from government to be assigned to work with the committee and to act as its executive secretary, since voluntary committee members seldom have the time to carry out all the needed work. Legislative groups often use the study committee approach to develop a basis for formulation of laws. These committees may be made up of members of the legislative body or of others from private walks of life. Another commonly used initial step is the conduct of a preliminary survey by an existing governmental agency. For any of these approaches, the general procedure is as follows:

a. Draw in additional people to serve as technical resource people or advisors.

b. Reach preliminary agreement on what the problems are, and form subgroups to study various aspects of the problem (e.g., legislation, governmental organization, sources of pollution, effects of pollution, meteorological and topographical aspects).

c. Secure facts from existing information in the community, library research, public meetings, outside consultants, and special studies.

d. Develop possible approaches to solution of air pollution problems.

e. Place the report on the problem and the proposed plan of action before the community through mass media, meetings, and individual contacts.

f. Study reactions to the proposal and revise the proposed plan as warranted.

g. Present the plan to appropriate legislators or legislative bodies and support actions to implement the proposal.

An effective way to bring together interested and knowledgeable people is the governor's or mayor's conference. The head of the government calls the meeting and arranges a program. The purpose is to present as much information as possible about the problem to as large a group of concerned people as can be conveniently assembled. The conference will attract top-flight speakers and listeners and will usually get much coverage by the mass media. After the conference, the head of government will have a better idea of how and where to go and whom to call upon to assist him, and a foundation for public support will have been laid.

Once the program comes into being, it is better to provide for a few program elements in a more concentrated manner than to move on all program aspects at once. It will take time to hire and train personnel, and therefore, the relatively simple things should be done first, gradually building up to more complex activities. It may be wise to omit some controversial aspects from initial plans in order to get a start on more generally accepted aspects and take on the more difficult problems as the staff develops know-how. Initial successes on a smaller segment of the problem will do more to create strong public support for an enlarged program than will a scattering of effort, with no completed successes on many facets of the program.

2. Intragovernmental Relationships

Many governmental agencies engage in activities peripheral to air pollution control program operations, such as issuance of building, zoning, and ventilating system permits; industrial inspection; boiler inspection; nuisance abatement; land-use and transportation planning; civil defense; and refuse disposal. The air pollution program must be coordinated with these other programs if waste and confusion are to be avoided. An example of such coordination is agreement with the building department that all construction plans be filed there, but that a building permit will not be issued until the air pollution agency has reviewed and approved the plans. Another example is air pollution program representation in activities of planning and zoning agencies.

3. Public Relationships

Community recognition of an air pollution problem and an action program to resolve the problem will come about much more readily when the public is convinced that a problem exists, concerns them, and can be solved. A public information program is needed to provide the public with the information they need to arrive at proper decisions. Some specific elements of a public information program are:

a. Inform the people of the nature, extent, causes, and effects of air pollution in their community.

b. Describe the steps that need to be taken to control air pollution. Explain the economics of these measures, which, in most cases, will indicate that the costs of control will be less than the costs of tolerating pollution.

c. Instill the idea that clean air is a desirable and attainable goal and that a well-planned program will not disrupt community life or industrial progress.

d. Make known the results of successful control efforts undertaken elsewhere.

e. Explain the transport of pollutants from one jurisdiction to another for the purpose of securing control action throughout an air basin.

f. Inform people about things they should do to minimize pollution (e.g., cease open burning of refuse, have automotive emission control systems serviced, keep home heating plants properly adjusted).

g. Describe the activities of the control agency and the status of compliance with regulations by major sources.

Other aspects of public relations activities of air pollution control agencies include work with conservation, civic, and professional groups; arranging for public seminars and workshops; arranging for public participation in public hearings on regulatory and other actions; publication of monthly and annual reports and information bulletins; awarding of citations for outstanding control work; use of public advisory boards; and activities associated with annual Cleaner Air Week and Earth Week. In the United States and Canada, the Air Pollution Control Association has published a handbook on public relations for the control agency (*86*), and the United States Government has published a list of organizations interested in air conservation (*87*). Public information and education activities are discussed in detail in Volume V, Chapter 5 of this book.

4. Air Pollution Control Boards and Commissions

There are several general kinds of laws under which the requirements to be met by pollution sources may be established. In one pattern, the law or ordinance adopted by the legislative body will contain the specific limitations on emissions and other requirements on the conduct of the governed sources. Provision may or may not be made for an advisory board to assist the responsible administrative agency in implementing the law. In another pattern, the basic law establishes only broad legislative policy and purposes and specifies that a designated administrative agency shall establish the necessary detailed requirements for controlling pollution by promulgation of rules and regulations consistent with the general intent of the legislation. The law may or may not provide for an advisory board or commission to review the agency's regulations and to provide advice thereon. In the third pattern, the basic law establishes only broad legislative policy and purposes and specifies that a designated board or commission shall establish the necessary detailed requirements for controlling air pollution by promulgation of rules and regulations. The rules and regulations are then implemented by an administrative air pollution control agency, which may work under the supervision of a governmental executive (such as the mayor or governor) or under the supervision of the board or commission.

The first approach, the one of having all requirements in the law itself, is favored by legislative groups that wish to retain strong control over the control program. Such is the case in a considerable number of city governments. However, as the air pollution problem increases in complexity, as it does at the state, provincial, or national level, legislative bodies have increasingly chosen to put only broad policies into the law, leaving the details to an administrative agency or to a board or commission. A disadvantage of incorporating details into the basic law is the difficulty of making changes, should the need arise, as it often does. In the United States, the Federal Clean Air Act leaves most of the detailed rule- and regulation-making activity to the executive branch of government; even though the basic law does specify detailed requirements for emissions from new motor vehicles.

The use of boards and commissions is viewed as a desirable means of broadening the base of responsibility, particularly since many decisions are based on limited information and use of judgment and have impact on various community interests. However, boards and commissions are not always responsive to public desires for clean air, particularly if their membership includes a number of industrial representatives. They also detract from the ability of chief executives (mayors and governors) to meet their responsibilities to the public, since the terms of office of board and commission members are usually staggered in such a way that a newly elected executive may not be able to control membership or policies. The existence of a board or commission may also cloud and diffuse responsibility between the board itself and the operating agency, thus making it difficult for the chief executive to fix responsibility for inadequate performance and to take remedial action. It is not uncommon for at least some board members to be uninformed or only marginally concerned about air pollution matters, thus reducing their effectiveness. These various defects of boards and commissions have prompted some people to favor placing authority and responsibility for air pollution control squarely on the chief executive or the head of the operating air pollution control agency, and to give him all necessary powers, including rule- and regulation-making to implement general policies set out in the basic law. It is reasoned that adequate protection against arbitrary or ill-advised actions by the chief executive or the air pollution control administrator are provided by hearings, recourse to the courts, and the ballot box.

Most boards and commissions serve without pay, although a few receive a nominal per diem allowance or honorarium, and nearly all are reimbursed for any travel expenses incurred in attending meetings. Membership on boards varies considerably from one place to another. Those included may be governmental officials whose work is related to air pollution and its control serving in an ex officio capacity, industrial representatives, physicians, lawyers, engineers, etc. The appointing official (governor or mayor) may be allowed to appoint any person he desires or the law may specify certain general qualifications the appointees must have, such as occupation or professional qualifications. There has been a trend in recent years for laws to exclude from membership on boards or commissions any persons who have strong ties with any enterprise that will be regulated by the board's rules.

The air pollution control board may also serve as the hearing and appeal board. This arrangement is considered inappropriate by some, since the board judges the reasonableness of its own actions. The advantages of this dual role for the control board are that fewer people are needed for boards and the fact that the control board is best informed on its own regulations and is thus better able to interpret them than would be a separate appeal board. Since the board has basic responsibility to the public for protection of air resources, it may be held that an independent appeal board should not be put in a position of being able to obstruct the control board by granting exceptions to the control board's actions if the appeal board should be so inclined. Protection against arbitrary actions by the control board, in any case, is available to those regulated, by resort to the courts.

5. Data Storage and Retrieval

Air pollution control agencies acquire, record, store, retrieve, analyze, and publish large volumes of data. The data result from many of the agency's operations and involve such things as air quality data, pollutant emissions, pollutant sources, emission control equipment, fuels and materials used, source tests made, abatements effected, permits issued, inspections made, hearings held, and court actions undertaken. The data may be used to keep case histories, monitor program operations, schedule activities, and to make various statistical tabulations and analyses. The degree of sophistication of the data system and the extent to which the electronic computer is used will vary from one air pollution control agency to another. The smaller agencies may operate most efficiently with only manual systems. In other agencies, a partially computerized system using a time-sharing remote computer terminal may be most advantageous (*88*). In large agencies, modern systems using punched cards, magnetic tapes, and data processing machines are nearly always needed.

In choosing a data system, consideration must be given to initial and operating costs, the needs and uses of the data involved, and the accessibility of the data for program purposes. There is, of course, no point in storing data for which there is no foreseeable use. Coding should be as simple as possible to minimize errors and reading and coding time. The system should be designed so that transfers of data from one record to another is minimized. Therefore, it is desirable, for example, to punch data record cards directly from field or laboratory reports. The basic record can be overprinted with card column numbers and coding information to reduce the need to refer to coding manuals or rely on memory.

The total data system should be so designed that information from various decks of data cards can be related in any desired manner. For example, one might want to relate air quality measurements at certain locations to pollutant emissions from sources within 1 mile (1.6 km) of the air sampling station. Geographic location coding is particularly necessary not only to interrelate data within the agency, but also to make use of data sorted by other agencies such as the land-use planning agency. Thus, particular attention should be given to location coding, using the method (grid system, census tracts, city block, etc.) of greatest overall utility.

In the United States, the federal government operates a national data system and bank for air quality data and pollutant emissions. The federal government encourages state and local agencies to use the same system or a compatible system so that all data can be analyzed by computer on a national basis. Manuals and procedures for using such systems have been distributed (*23, 30*). The federal government has also developed a comprehensive data handling system for use by state and local agencies and is encouraging the use of the system. It includes systems for storage, retrieval, and use of air quality data and emissions inventory data and a system for managing the various kinds of information needed to schedule and operate an enforcement program. The system has been described in a series of publications (*24–26, 31, 89*).

6. Fee Systems

Most air pollution control agencies collect fees for such functions as periodic inspection of pollutant sources, review and approval of construction plans, and issuance of annual or one-time operating permits. Fees are charged on the premise that those who do (or may) pollute the atmosphere should pay a larger share of the cost of operating the control program than the community as a whole. Fee schedules may be based on a flat fee for all permits, a filing fee, fees based on capacity or size of equipment, annual fees, and fees based on equipment costs (*52*). In Michigan, a recent innovation has been the imposition of a surveillance fee. The annual fee for each source is related to the magnitude of pollutant emissions and the costs incurred by the agency for maintaining surveillance over the various sources of pollution. The use of fee systems to help pay for agency operating costs does seem attractive in that they reduce the amount of money that must be collected by general taxation to support the program. However, the reliance on fee systems to cover all or part of the agency's costs can lead to undesirable situations. The agency may tend to concentrate its efforts on activities that produce a large amount of revenue, regardless of the efficacy of the activity in the control of air pollution. Important program elements that do not produce fees may be inappropriately deemphasized. Further, agency personnel may tend to rush through their work in order to produce more fees, and thus the quality of the work may suffer.

In years gone by, most local air pollution control agencies recovered less than 25% of their operating expenses from fees, with the median being about 20% (*90, 91*). However, 20% of the agencies recovered more than half of their operating costs in fees. No more recent data have been published. The Michigan program, noted above, recovers nearly all control program costs.

There has been much discussion of the use of emission charges or taxes as a means of both raising revenues to cover the cost of control program operation and as a means of encouraging sources to install pollution control equipment. In such schemes, the source would pay a fee or tax based on the magnitude of its emissions, or on the basis of the extent to which available controls had been applied, or on some other basis that would result in lower fees if emissions were low rather than high. Several such schemes have been described (*92, 93*). To date, such schemes have not been widely used.

7. Development of Guides for Compliance

The operation of an air pollution abatement program depends upon the ability to get the owners and operators of pollution sources to apply the necessary measures to reduce emissions to the atmosphere and to properly operate and maintain both basic combustion and processing equipment and air pollution control equipment. Some facets of the control program may depend upon an ability to get many members of the public to do certain things, such as to have their heating plants adjusted annually, or to refrain from certain activity, such as the open burning of refuse. The simple promulgation of regulations and the case-by-case enforcement of the regulations would be very time consuming and costly. Effectiveness can be improved and enforcement costs can be reduced if efforts are made to inform those who are to take action as to what the requirements of the regulations are and how they may comply. This can be done by use of various public informational channels and through the preparation and distribution of special guidelines for compliance to the persons who are expected to take abatement actions. These approaches apply to those situations wherein a substantial number of persons will be taking the same or similar actions. They would not apply to those relatively few sources of a particular class that must be handled on a case-by-case basis.

A guide for compliance is a publication that describes the requirements of a regulation and the various specific alternative actions that persons may take to comply. It may include the names of appropriate equipment vendors and consultants who can assist the regulated party. Cost estimates of taking control actions may be given for typical installations. Any benefits that will accrue to the regulated person because of the control action should be pointed out. The required time schedule for taking action would be specified. These guides can often be prepared best in consultation with those who will have to take control action so that the issuing control agency will be better able to provide needed information in a manner that the users will find to be most helpful. Also, the useful experiences of various persons and organizations can be identified and incorporated so that all concerned can use it. After the guideline is prepared, it is given wide distribution so that it gets into the hands of those regulated. Trade associations, fuel and equipment suppliers, and others can often assist with such distribution. Further, the guideline should be given to each regulated person at the time the agency issues any formal request for compliance action.

8. Tax Relief and Financial Aid

Because of the substantial expenditures that must be made by industrial and governmental institutions for control of air pollution, a number of programs have been set up by various governmental agencies and private groups to ease such financial burdens. Included among these are preferential consideration of loans for pollution control expenditures by small businesses; reduced interest rate loans for pollution control expenditures offered by banks; state, provincial, and local governmental low-interest-rate, tax-exempt industrial development bonds; and various tax exemptions for pollution control expenditures (*94*).

National, state, provincial, and local tax exemptions for pollution control expenditures began to appear in about 1963, usually in the form of property tax and sales tax exemptions for pollution control equipment. Franchise and income tax credits and deductions and an accelerated amortization schedule tax allowance for pollution control facilities are among the forms of incentives provided.

Many state, provincial, and local governments can issue industrial revenue bonds to help industry purchase pollution control equipment. In the United States, such bonds are exempt from federal income taxes, and thus, the industrial establishment is able to borrow money at a rate of about 2% below that which would have to be paid on taxable corporate bonds (*95, 96*). The saving to industry is substantial. For example, on a 20 year, $10 million dollar issue, the company would save about $3 million in interest charges. The procedure is not useful, however, for expenditures of less than about one-half million dollars because of the costs of legal fees, negotiations, underwriting, printing, etc. In such revenue bond arrangements, the government agency uses the proceeds from the sale of bonds to pay for the pollution control equipment that is then leased to the company. The company then makes lease payments to the government, which in turn uses the funds to pay interest on the bonds and to retire them. When the bonds are retired, the government gives title to the equipment to the company. Other similar arrangements can be made in the form of installment purchase of the pollution control equipment. The company (rather than the government) pledges its resources to pay back the bonds and, therefore, must have a good enough credit rating to insure repayment. In another format, the government agency may set up a tax-exempt bond program and pledge its own credit to repay the bonds and then use the proceeds of bond sales to help industry finance pollution control costs. As before, the industry pays the government back for any funds provided to it, with interest.

In a few states and provinces, bond issues have been made to provide direct payments and loans to local governments to cover the expenses of abating pollution at state, provincial, and local government-owned facilities. The bonds are retired over the years by use of regularly collected tax moneys.

9. Plans for Achieving and Maintaining Air Quality Standards

Air pollution control programs of the past have concentrated largely on abatement of known sources of pollution through application of reasonably available control technology. Such programs have resulted in some rather dramatic improvements in air quality in many areas. Now, however, there is increasing recognition of the need for long-range plans for conservation of air resources. Such plans seek to incorporate consideration of air quality goals into comprehensive community plans and particularly into the land-use and transportation elements of such plans. Such plans also seek to consider all other aspects of environmental quality, along with air resources.

As a result of the United States Federal Clean Air Act of 1970, each state prepared a plan for attainment and maintenance of national ambient air quality standards. These plans included such elements as legal authority, growth projections, emission limitations, staffing and funding of the control agency, review of new sources prior to construction, air quality and source emission monitoring, intergovernmental cooperation, and transportation and land-use controls where needed. The plans were prepared by the states in accordance with national requirements (*72*) and generally provided for attainment and maintenance of air quality standards through 1975 or 1977 (*97*). These plans incorporated some consideration of the future growth and development of communities and the impact of such growth on air quality. But most of them did not address the more complex issues and problems associated with growth of communities over the following 10 to 20 years.

Development of a long-range plan for conservation of air resources requires that each urban community develop a set of policies on urban growth, land-use development, transportation systems, etc., and a set of goals and objectives that the community desires to achieve. Air quality factors must be included in each of these policy elements along with other environmental quality factors. Once general policy statements are available, various studies can be made of future conditions, and evaluations can be made of various alternative plans. The results of these studies can be used to modify the general policies, as needed, and to select a particular set of rules and guidelines for the future. In the air quality element of such planning, there are some procedures for projecting future air pollutant emissions (*98, 99*), for translating such emissions into ambient air quality by use of atmospheric diffusion models (*100–102*), and for evaluating the cost–effectiveness of various approaches to control (*103–106*). Methodologies for the evaluation of the air quality implications of various alternative land-use development and transportation plans have been described (*16, 85, 107*), and further developments are continuing. Future plans for air quality control will, in major urban centers, include both the traditional measures for control of emissions from sources and more innovative measures such as total regional emission ceilings, emission allocation on a subregional basis, limitations on emission density for classes of land use, changes in urban form, requirements for use of district heating, and energy conservation measures. These approaches will begin finding application in the mid-1970s and beyond. The important need is for advance planning and phased orderly implementation to replace short-range corrective measures on a crash basis.

10. Power Plant Siting

Power plants are a major source of air pollution, and with the rapid growth in use of electricity, the location of new plants has become a matter of substantial concern because of their impact on environmental quality. Sites for new plants that are acceptable to all concerned have become increasingly difficult to find as nations have become more densely settled. Power companies have been confronted with an ever increasing array of conditions to be met, citizen opposition, and time delays. The balancing of various public and private interests associated with new power plants has become very difficult in regard to plant sites, fuel supply sources, power transmission lines, environmental impacts, costs, etc.

In the United States, some states, notably Maryland and New York, have adopted special laws and programs to manage these problems. The programs may involve a variety of elements, including preparation of environmental impact reports, designation and review of alternative sites, consolidation of governmental review and approval mechanisms, raising of revenues to finance studies and perhaps site acquisition, and formalized arrangements for public participation in site considerations. The programs are usually concerned with all environmental factors and incorporate the regulatory measures and agencies that have traditionally worked with power plants, such as public service commissions. These programs should help insure that power plants are put in the most advantageous location in the total public interest and help make the overall proposal, review, and construction activity proceed in a more orderly fashion.

11. Staffing and Manpower Utilization

The number and type of personnel needed to operate an air pollution control program will vary with many factors, including the nature and number of pollution sources to be regulated, the extent of restrictions on atmospheric ventilation, the age of the agency and the extent to which various problems have been solved, the size of the agency, the population served, the number and type of regulations to be enforced, and the size of the geographic area served. This being the case, it is not possible to give a simple estimate of the staffing needs for a given agency. However, a model has been developed that considers several variables and predictors and makes it possible to estimate the number of people needed to accomplish the major functions of a control agency (*108*). The model was applied to the United States to estimate the total number of people required to staff the aggregate of state and local agencies. A number of assumptions were involved, and therefore the full report should be consulted before making any detailed use of the data developed and displayed in Table VIII. Comprehensive local programs are those that conduct all necessary activities without assistance from the state or other

Table VIII Predictive Model Output for Required Man-Years by Function and Agency Type—United States[a] (1974)

Function	Local						State		Combined		
	Comprehensive (agencies)		Noncomprehensive (agencies)		Minimal regulatory						
	Man-years	Percent	Man-years	Percent	Man-years	Percent	Man-years	Percent	Man-years	Percent of total	Percent of subtotal
Office of the director											
Policy, public relations, intergovernmental relations, and systems analysis	403.5	8.8	28.8	9.8	64.4	20.0	268.2	9.5	764.9	9.5	23.8
Administrative and clerical support	1085.2	23.8	77.5	26.4	64.4	20.0	720.4	25.4	1947.5	24.3	61.0
Staff training and development	202.0	4.8	15.3	5.2	—	—	145.9	5.1	381.2	4.8	12.0
Statewide training operations	—	—	—	—	—	—	47.3	1.7	47.3	.6	1.4
AQCR[b] planning and evaluation	—	—	—	—	—	—	63.0	2.2	63.0	.8	1.8
Subtotal	1708.7	37.4	121.6	41.5	128.8	40.0	1244.8	43.9	3203.9	40.0	100.0
Technical services											
Laboratory operations	187.8	4.1	11.3	3.8	—	—	133.5	4.7	332.6	4.2	26.2
Operation and monitoring network	214.0	4.7	20.3	6.9	64.4	20.0	—	—	298.7	3.7	23.8
Data reduction and processing	159.4	3.5	3.7	1.3	—	—	143.5	5.1	306.6	3.8	24.4
Special field studies	110.1	2.4	7.4	2.5	—	—	73.3	2.6	190.8	2.4	15.2
Instrument maintenance and calibration	90.5	2.0	7.2	2.4	—	—	44.4	1.6	142.1	1.8	10.4
Subtotal	761.8	16.7	49.9	17.0	64.4	20.0	394.7	13.9	1270.8	15.9	100.0

Field services											
Scheduled inspections for permit renewal	556.9	12.2	46.2	15.8	64.4	20.0	366.5	12.9	1034.0	12.9	62.0
Complaint handling and field patrol	375.2	8.2	65.5	22.4	64.4	20.0	—	—	505.1	6.3	30.0
Preparation for legal actions	74.8	1.6	9.8	3.3	—	—	43.2	1.5	127.8	1.6	8.0
Subtotal	1006.9	22.1	121.5	41.5	128.8	40.0	409.7	14.4	1666.9	20.8	100.0
Engineering services											
Calculation of emission estimates	91.4	2.0	—	—	—	—	60.8	2.1	152.2	1.9	8.3
Operation of permit system	684.9	15.0	—	—	—	—	500.9	17.7	1185.8	14.8	63.7
Source testing	161.8	3.5	—	—	—	—	124.9	4.4	286.7	3.6	15.3
New regulations, engineering reports, and emergency procedures	146.5	3.2	—	—	—	—	97.7	3.4	244.2	3.0	12.7
Subtotal	1084.6	23.8	—	—	—	—	784.3	27.7	1868.9	23.3	100.0
Grand total[c]	4562.0	100.0	293.0	100.0	322.0	100.0	2833.5	99.9	8010.5	100.0	—

[a] Data from reference (*108*).

[b] AQCR = Air Quality Control Region.

[c] Differences in man-years and number of positions reported for State and local agencies due to rounding of numbers.

group. Noncomprehensive local agencies conduct most program activities but require specialized support from the state or elsewhere for such functions as engineering and meteorology. Minimal local agencies conduct only limited functions, such as operation of sampling equipment, answering of complaints, and conduct of annual inspections on relatively less complex sources. The total of 8010 people required by state and local agencies was estimated to be composed of the occupational categories tabulated below. The above-noted report *(108)* indicated a vacancy rate

Occupational category	*Percent of total manpower*
Engineers	28
Chemists	6
Meteorologists	2
Sanitarians	5
Other professions	4
Inspectors	19
Technicians	12
Administrative–clerical	24
	100

of about 10–20% in air pollution control agency staff positions and an annual turnover rate of about 4% for technical and professional people and 16% for administrative and clerical people.

Air pollution control agencies normally do not have all the people they need to accomplish all of the assignments they have. Therefore, it is essential that the best possible use be made of available staff. Measures for relieving manpower shortages include *(109)*:

a. State, provincial, and local sharing of specialized personnel.

b. Use of task groups drawn from several agencies to accomplish complex activities.

c. Provision of highly specialized personnel by the state or province to serve all agencies in the state or province.

d. Provision of services by local agencies in their jurisdictions to eliminate the need for state–provincial personnel to travel to the area to do the job.

e. Use of highly skilled people for only complex work and their relief from routine chores.

f. Use of people having only the minimum required skills to do routine work.

g. Establishment of a special group to handle court cases.

h. Use of consultants to do complex jobs that are beyond the capabilities of agency staff.

i. Elimination of permits and inspections of minor sources.

j. Handling of as many complaints as possible by telephone rather than by personal visit.

k. Elimination of nonessential air quality monitoring.

l. Management of abatement by classes of sources in which there are many sources as a group rather than individually.

m. Requirement for sources to make emission tests rather than have the agency do so. The agency can monitor the work.

n. Having people report for work directly in the field rather than passing through the office.

o. Programming of work on a geographic basis to reduce travel time.

p. Publication of informational brochures to avoid the need for individually prepared letters in response to common inquiries.

q. Provision of two-way radios to field men.

r. Use of standard enclosures, letters, and forms to supplement individually prepared communications.

s. Use of existing educational institutions to conduct training rather than having the agency do so, as is done in New Jersey (*14*).

t. Use of university students to augment the work of professionals, such as lawyers (*110*).

12. Approaches to Program Evaluation

As in any management situation, evaluation of air pollution control programs and the evaluation of the effectiveness of air pollution control agencies is an important element in overall air resource management and air pollution control agency operation. While all agree with this concept, there are as yet no generally agreed upon criteria or procedures for making such evaluations. Many "experts" seem able to generally agree on whether a program or an agency in a given case is appropriate or effective, but they do so on the basis of broad, subjective opinion, with few being able to agree on objective criteria for making such judgments. This situation prevails because of the many facets of an air pollution control program, the wide variety of the nature and causes of air pollution problems in different areas; the multiplicity of general approaches to management of air pollution problems that can be made to work, the strong influence of factors beyond the immediate control of agency administrators that influence the kind of air pollution control program he may operate; and the changing of circumstances, attitudes, and control efforts with time and public priorities.

There are at least thirty elements in the overall program of an air pollution control agency that should be given relative priorities for assignment of available funds and manpower. No one distribution is right for all jurisdictions. It has not been possible to identify factual and objective criteria that would make it possible to determine the correct distribution of agency manpower among such agency activities as source surveillance, enforcement, ambient air measurements, data reporting, public education, etc. The distribution is usually the result of the subjective judgment of the agency administrator. Air pollution problems obviously vary from one jurisdictional area to another because of the nature of the sources involved, climate, topography, land-use patterns, etc. The kind of laws, regulations,. and programs needed to best attack and solve the problem in one city will not necessarily be the same as in another. Some control agency administrators work in areas where legislative bodies have freely adopted all of the laws needed to attack air pollution problems, but in other areas, necessary laws have not been adopted. Similarly, funds to operate programs have been more or less liberally appropriated. Since such circumstances prevail, it is difficult to evaluate whether a particular air pollution control agency has been as effective and efficient as it should have been, based on such end-result criteria as improvement in air quality or reduction in emissions.

A major influence on progress in air pollution control has been the ordering of public priorities. In areas such as St. Louis, Missouri, and Pittsburgh, Pennsylvania, in the 1940s, great priority was placed on abatement of black smoke, and a great deal was accomplished. These priorities and public desires change markedly from time to time and from one area to another. They have a great impact on what a control agency can and cannot do. Thus, the evaluation of a given agency and a comparison of the effectiveness and efficiency of one agency as compared to another is very complex indeed.

In spite of these complexities and problems, it is nevertheless necessary to perform evaluations of three general kinds: (a) Is the control agency properly organized and operated to accomplish the most it can with available resources and legal authorities? (b) Is the overall control program properly established and directed toward the right objectives? (c) Is the total air pollution control effort bringing about the desired end results?

Evaluations of control-agency organization and operations involve relatively well-known approaches used in the management science field. Included would be analysis of lines of authority and communications, productivity per manhour, extent of emission reductions achieved per dollar spent by the agency, adequacy of recordkeeping and reporting procedures, distribution of resources in accordance with needs to accomplish identified priority tasks, arrangements for cooperation with other governmental entities, means for assuring the adequacy and accuracy of work done, adequacy of program planning and review procedures, appropriateness of operational rules and guidelines used by employees, extent of conformity of the program with governing policies of higher authorities or others having a controlling impact on the program, qualifications of staff members in relationship to desired or needed qualifications, and suitability of the organizational structure in relationship to the agency's assigned mission.

Evaluations of overall control programs and their objectives involve consideration and appraisal of the adequacy of broad program goals (e.g., reaching or maintaining certain air quality standards, securing emission controls of a specified level of particular classes of sources); the extent of funds and manpower available to the control agency; the basic laws and administrative rules and regulations in relationship to those needed to achieve program goals; the placement of the control agency in the overall governmental structure and its relationships with other agencies; the general philosophy and attitude of the agency (e.g., whether it is as aggressive as it should be in enforcing laws, whether it works properly with other agencies); the composition and functions and operations of air pollution control advisory and appeals boards; the opinions of elected officials, the mass media, public interest groups, and others as to the overall nature of the control program; and the extent of consistency of the program with mandated national guidelines.

Evaluation of whether the overall air pollution control effort is bringing about the desired end results is the truly basic evaluation. For example, if one is basically concerned about air quality and that certain air quality standards be achieved, then a comparison over a series of years between existing air quality and these standards is the one prime measure of program effectiveness. But many factors influence progress toward achievement of air quality standards (or the lack thereof), and use of such simple criteria does not tell the whole story. Other indirect indications of overall progress that might be used would include the extent to which existing regulations are complied with, the ratio between existing emissions and those that would exist in the absence of controls, the relationship between existing levels of emission control and that which might be achieved by use of the best practicable means for emission control, and the change in emission rates and totals over a number of years of control program operation. Each of these indices are useful but also are subject to various influences that may detract from their validity and usefulness in specific situations.

Several technical papers have been published that provide useful information to those interested in control-program evaluation (*111–114*). A procedure for evaluating the effectiveness of an air pollution control agency that can be used by public interest groups has been distributed by the United States National Air Conservation Commission (*115*). While evaluation procedures are difficult and uncertain, it is possible to conduct such evaluations. The results will be useful if they are properly interpreted and if the evaluation leads to the identification of problems and means for their solution and if the evaluation in fact does evaluate agency performance and not some other external influencing factor. A key element to remember is that activity does not necessarily reflect effectiveness.

13. Air Pollution Indexes

There are several situations in which it may be desirable to have a means of describing "air pollution" as one overall entity, rather than in a series of concentrations of several individual pollutants. It is also sometimes desirable to express pollutant concentrations in terms more familiar to the public or in terms that incorporate consideration of the significance of the concentration in terms of effects. Such indirect means of expressing pollutant levels are referred to as air pollution indexes. One use of such indexes is in informing the public about air pollution. Some believe that a series of numbers expressing concentrations of several pollutants in parts per million or micrograms per cubic meter and presenting the air quality standards for such pollutants in similar terms is too confusing and complex for the general public to understand. There is some feeling that a simple index number, ranging from 0 to 100, as in the familiar school grading system, or a series of words, ranging from "good" to "bad" or "clean" to "polluted," would be a more useful way to convey information to the public. Another use for an air pollution index would be in comparing the "air pollution" situation in several communities, each having different levels of several principal pollutants. One community may have relatively high concentrations of sulfur oxides and particulates and relatively low concentrations of other pollutants. Another community may have relatively low concentrations of sulfur oxides and particulates and relatively high concentrations of other pollutants. Comparison of the severity of the "total air pollution problem" in these two communities would be facilitated if there were an agreed upon procedure for reducing the several concentrations to a single value representing an appropriate weighting of the significance of the levels of the various pollutants. A similar need for an index arises in the development of systems for avoidance or management of severe episodes of air pollution. Adverse effects could occur during a great variety of combinations of possible concentrations of the several pollutants. It would be helpful to have a system that would make it possible to put the various pollutant levels into a formula that would consider the effects of the various pollutants both singly and in combination with the other pollutants present and produce an appropriately weighted single value that would indicate whether episode prevention or control actions were needed at a particular time.

Several such air pollution indexes have been developed (*73, 116–120*). Some of these have been tailored to the special needs of the jurisdictional areas for which they were developed, and others are said to be generally applicable. Several general approaches are involved. One is to relate existing pollution levels to ambient air quality standards, with the standard providing the reference baseline for each pollutant, and then converting existing concentrations into a percentage of the standard. The percentages for several pollutants can then be added together and perhaps be divided by a factor to produce a single index value. Other systems seek to relate existing pollution levels to the array of values experienced over the past to give an indication of whether pollution levels are better or worse than usually encountered. Others are based on the probability that some adverse effect, such as decreased visibility, will be experienced. Some indexes include only sulfur dioxide and particulate matter, while others include consideration of all major pollutants routinely measured. The rating scales usually define "clean air" as that which is substantially below the air quality standard or that which represents the pollution level on the few cleanest days of the year in the area, whichever coincides with the lowest pollution level. "Severe" air pollution or "very polluted" air is often related to the level of pollution calling for emergency abatement action in the episode control plan that applies in the community. In some systems, a pollutant index is calculated for each pollutant on a uniform basic principle and also for the sum of all pollutants. The information reported to the public may then be only for the pollutant having the highest index number or perhaps that number and the sum will be reported. The reporting of the "worst pollutant" index is deemed necessary to reveal the occurrence of high values of a single pollutant at times when all other pollutants are at low levels. In such cases, the overall index would be low and would tend to be misleading because of the impact of the one pollutant present in high concentrations.

In general, indexes have some value. But at the same time, they have a high potential for giving false impressions. It is important to be sure that those who prepare and use indexes fully understand the basis for their preparation and the significance and limitations that should be attached to them.

REFERENCES

1. J. J. Hanlon, "Principles of Public Health Administration," 2nd ed., p. 128. Mosby, St. Louis, Missouri, 1955.
2. A. N. Heller, J. J. Schueneman, and J. D. Williams, *J. Air Pollut. Contr. Ass.* **16,** 307 (1966).
3. J. J. Schueneman, *in* "Air Pollution" (A. C. Stern, ed.), 2nd ed., Vol. 3, pp. 719–796. Academic Press, New York, 1968.
4. Air Pollution Control Association, "1975–1976 Directory—Governmental Air Pollution Control Agencies." Pittsburgh, Pennsylvania, 1975.
5. D. R. Smith, N. G. Edmisten, and D. J. deRoeck, *J. Air Pollut. Contr. Ass.* **22,** 943 (1972).
6. J. J. Schueneman, *J. Air Pollut. Contr. Ass.* **13,** 116 (1963).
7. E. H. Haskell, *Environ. Sci. Technol.* **5,** 1092 (1971).
8. United States Environmental Protection Agency, "State Air Programs Digest—Fiscal Year 1972." Washington, D.C., 1972.
9. California Air Resources Board, "Air Pollution in California—Annual Report 1973." Sacramento, California, 1974.
10. I. Feller, A. J. Engle, R. S. Friedman, D. C. Menzel, and J. T. Sacco, "Intergovernmental Relations in the Administration and Performance of Research on Air Pollution." Pennsylvania State University, University Park, Pennsylvania, 1972.
11. F. K. Kleinman, *J. Air Pollut. Contr. Ass.* **21,** 71 (1971).
12. Public Law 946, 84th United States Congress, 2nd Session, Aug. 3, 1956.
13. T. R. Glenn, *J. Air Pollut. Contr. Ass.* **22,** 444 (1972).
14. L. P. Goldshore and J. J. Tozzi, *J. Air Pollut. Contr. Ass.* **21,** 115 (1971).
15. Clean Air Act of 1970 (42 U.S.C. 1857 et seq.) includes the Clean Air Act of 1963 (P.L. 88-206), and amendments made by the Motor Vehicle Air Pollution Control Act (P.L. 89-272 of Oct. 20, 1965), the Clean Air Act Amendments of 1966 (P.L. 89-675 of Oct. 15, 1966), the Air Quality Act of 1967 (P.L. 90-148 of Nov. 21, 1967 the Clean Air Act amendments of 1970 (P.L. 91-604 of Dec. 31, 1970), and Clean Air Amendments of 1977 (P.L. 95-95 of Aug. 7, 1977).
16. J. C. Fensterstock, J. A. Kurtzweg, and G. Ozolins, *J. Air Pollut. Contr. Ass.* **21,** 395 (1971).
17. E. W. Hauser, L. W. West, and A. R. Schleicher, *Traffic Quart.* **20,** 71 (1972).
18. The National Environmental Policy Act of 1969, P.L. 91-190, 42 U.S.C. 4321 et seq.
19. Code of Federal Regulations, Title 40, Chapter V, Part 1500, "Preparation of Environmental Impact Statements: Guidelines." United States Government Printing Office, Washington, D.C., 1973.
20. United States Environmental Protection Agency, "Compilation of Air Pollutant Emission Factors," 2nd ed., Publ. No. AP-42. Research Triangle Park, North Carolina, 1973.

21. G. Ozolins and R. Smith, "A Rapid Survey Technique for Estimating Community Air Pollution Emissions," Pub. Health Serv. Publ. 999-AP-29. United States Department of Health, Education, and Welfare, Cincinnati, Ohio 1966.
22. S. G. Shanks, W. A. Cote, A. Boyer, and D. G. Cooper, *J. Air Pollut. Contr. Ass.* **22,** 85 (1972).
23. United States Environmental Protection Agency, "Guide for Compiling a Comprehensive Emission Inventory" (revised), Publ. No. APTD-1135. Research Triangle Park, North Carolina, 1973.
24. United States Environmental Protection Agency, "Comprehensive Data Handling System—Emissions Inventory Subsystem User's Guide," Publ. No. APTD-1550. Research Triangle Park, North Carolina, 1973.
25. United States Environmental Protection Agency, "Comprehensive Data Handling System, Emissions Inventory Subsystem Program Documentation," Publ. No. APTD-1551. Research Triangle Park, North Carolina, 1973.
26. United States Environmental Protection Agency, "Enforcement Management System User's Guide," Publ. No. APTD-1237. Research Triangle Park, North Carolina, 1972.
27. H. R. Hickey, W. D. Rose, and F. Skinner, *J. Air Pollut. Contr. Ass.* **21,** 689 (1971).
28. United States Environmental Protection Agency, "Guidelines for Technical Services of a State Air Pollution Control Agency," Publ No. APTD-1347. Research Triangle Park, North Carolina, 1972.
29. United States Environmental Protection Agency, "Guidelines for Development of a Quality Assurance Program," Publ. No. EPA-RA-73-028 (Four Parts: (a) Carbon Monoxide; (b) Total Suspended Particulate Matter; (c) Photochemical Oxidant; (d) Sulfur Oxides). Research Triangle Park, North Carolina, 1974.
30. Code of Federal Regulations, Title 40, Chapter I, Subchapter C, Part 51, "Requirements for Preparation, Adoption, and Submittal of Implementation Plans." United States Government Printing Office, Washington, D.C., 1973.
31. United States Environmental Protection Agency, "User's Manual: SAROAD—Storage and Retrieval of Aerometric Data," Publ. No. APTD-0663. Research Triangle Park, North Carolina, 1971.
32. United States Environmental Protection Agency, "Air Quality Data Handling System User's Manual," Publ. No. APTD-1086. Research Triangle Park, North Carolina, 1972.
33. G. J. Nehls and G. G. Ackland, *J. Air Pollut. Contr. Ass.* **23,** 180 (1973).
34. M. I. Weisburd, "Field Operations Manual for Air Pollution Control—Vol. I, Organization and Basic Procedures," Publ. No. APTD-1100. United States Environmental Protection Agency, Research Triangle Park, North Carolina, 1972.
35. California Department of Public Health, "Air Pollution Effects Reported by California Residents." Berkeley, California, ca. 1958.
36. W. S. Smith, J. J Schueneman, and L. D. Zeidberg, *J. Air Pollut. Contr. Ass.* **14,** 418 (1964).
37. J. Schusky, "Public Awareness and Concern with Air Pollution in St. Louis Metropolitan Area." United States Department of Health, Education, and Welfare, Public Health Service, Washington, D.C., 1965.
38. N. Z. Medalia and A. L. Finkner, *U.S. Pub. Health Serv., Publ.* 999-AP-10 (1965).
39. R Kudlich, *U.S., Bur. Mines, Inform. Circ.* 8333 (1967).

40. Appellate Department, Superior Court, Los Angeles, California, People vs. International Steel Corp. (1951) California App. 2d Supp. 935, 226 P. 2d 587, at 938-9 and 1956; 137 C. A. 2nd (Suppl) 859; 291 Pac. 2nd 587.
41. United States Supreme Court (351 U.S. 990, 100L. Ed. 1503). Appellate Department Superior Court, Los Angeles, California, People vs. Plywood Manufacturers of California; People vs. Shell Oil Co., People vs. Union Oil Co.; People vs. Southern California Edison Co. (137 C.A. 2d Supp. 859; 291 P. 2d 587) (1955).
42. County of Los Angeles, California, Air Pollution Control District, "Biennial Report, Fiscal Years 1971–72 and 1972–73." Los Angeles, California, 1973.
43. Industrial Gas Cleaning Institute, *J. Air Pollut. Contr. Ass.* **23,** 608 (1973).
44. M. I. Weisburd, "Field Operations and Enforcement Manual for Air Pollution Control. Vol. III: Inspection Procedures for Specific Industries," Publ. No. APTD-1102. United States Environmental Protection Agency, Research Triangle Park, North Carolina, 1972.
45. J. A. Danielson, ed., "Air Pollution Engineering Manual," Publ. No. AP-40. United States Environmental Protection Agency, Research Triangle Park, North Carolina, 1973.
46. M. I. Weisburd, "Field Operations and Enforcement Manual for Air Pollution Control. Vol. II: Control Technology and General Source Inspection," Publ. No. APTD-1101. United States Environmental Protection Agency, Research Triangle Park, North Carolina, 1972.
47. United States Environmental Protection Agency, "Administrative and Technical Aspects of Source Sampling for Particulates," Publ. No. APTD-0754. Research Triangle Park, North Carolina, 1971.
48. Code of Federal Regulations, Title 40, Chapter I, Subchapter C, Part 50, "National Primary and Secondary Ambient Air Quality Standards." United States Government Printing Office, Washington, D.C., 1973.
49. J. J. Schueneman, J. D. Williams, and N. G. Edmisten, *Atmos. Environ.* **2,** 353 (1968).
50. R. I. Larsen, *J. Air Pollut. Contr. Ass.* **11,** 71 (1961).
51. California Department of Public Health, "Technical Report on California Standards for Ambient Air Quality and Motor Vehicle Exhaust." Berkeley, California, 1960.
52. A. Stein, "Guide to Engineering Permit Processing," Publ. No. APTD-1164. United States Environmental Protection Agency, Research Triangle Park, North Carolina, 1972.
53. L. A. Winkelman, *J. Air Pollut. Contr. Ass.* **14,** 441 (1964).
54. Air Pollution Control Association, *J. Air Pollut. Contr. Ass.* **23,** 635 (1973).
55. N. A. Huey, L. C. Broering, G. A. Jutze, and C. W. Gruber, *J. Air Pollut. Contr. Ass.* **10,** 441 (1960).
56. S. W. Horstman, R. F. Wromble, and A. N. Heller, *J. Air Pollut. Contr. Ass.* **15,** 261 (1965).
57. E. M. Darling, "Computer Modeling of Transportation Generated Air Pollution," Rept. No. DOT-TSC-OST-72-20. United States Department of Transportation, Washington, D.C., 1972.
58. J. L. Beaton, A. J. Ranzieri, and J. B. Skag, "Motor Vehicle Emission Factors for Estimates of Highway Impact on Air Quality," Rept. No. FHWA-RD-72-34. Federal Highway Administration, Washington, D.C., 1972.
59. J. L. Beaton, A. J. Ranzieri, E. C. Shirley, and J. B. Skag, "Mathematical

Approach to Estimating Highway Impact on Air Quality," Rept. Nos. FHWA-RD-72-36 and 37. Federal Highway Administration, Washington, D.C., 1972.
60. J. L. Beaton, A. J. Ranzieri, E. C. Shirley, and J. B. Skag, "Analysis of Ambient Air Quality for Highway Projects," Rept. No. FHWA-RD-72-38. Federal Highway Administration, Washington, D.C., 1972.
61. J. L. Beaton, E. C. Shirley, and J. B. Skag, "A Method for Analyzing and Reporting Highway Impact on Air Quality," Rept. No. FHWA-RD-72-39. Federal Highway Administration, Washington, D.C., 1972.
62. J. E. Norco, R. R. Cirillo, T. E. Baldwin, and J. W. Gudenas, "An Air Pollution Impact Methodology for Airports," Phase I, Publ. No. APTD-1470, United States Environmental Protection Agency, Research Triangle Park, North Carolina, 1973.
63. United States Department of Transportation, "Airports and Their Environment: A Guide to Environmental Planning." Washington, D.C., 1972.
64. Code of Federal Regulations, Title 40, Chapter I, Subchapter C, Part 60, "Standards of Performance for New Stationary Sources." United States Government Printing Office, Washington, D.C., 1973.
65. Code of Federal Regulations, Title 40, Chapter I, Subchapter C, Part 61, "National Emission Standards for Hazardous Air Pollutants." United States Government Printing Office, Washington, D.C., 1973.
66. United States Environmental Protection Agency, "Guide for Air Pollution Episode Avoidance," Publ. No. AP-76. Research Triangle Park, North Carolina, 1971.
67. United States Environmental Protection Agency, "Guide for Control of Air Pollution Episodes in Medium-Sized Urban Areas," Publ. No. AP-77. Research Triangle Park, North Carolina, 1971.
68. United States Environmental Protection Agency, "Guide for Control of Air Pollution Episodes in Small Urban Areas," Publ. No. AP-78. Research Triangle Park, North Carolina, 1971.
69. A. D. Rossin and J. J. Roberts, *J. Air Pollut. Contr. Ass.* **22**, 254 (1972).
70. E. J. Croke and S. G. Booras, *J. Air Pollut. Contr. Ass.* **20**, 129 (1970).
71. A. I. Mytelka, *J. Air Pollut. Contr. Ass.* **22**, 447 (1972).
72. Code of Federal Regulations, Title 40, Chapter I, Subchapter C, Part 51, "Requirements for Preparation, Adoption, and Submittal of Implementation Plans." United States Government Printing Office, Washington, D.C., 1973.
73. L. Shenfeld, *J. Air Pollut. Contr. Ass.* **20**, 612 (1970).
74. B. B. Crocker, *Chem. Eng.* **70**, 97 (1970).
75. United States Environmental Protection Agency, "Accidental Episode Manual," Publ. No. APTD-1114, Research Triangle Park, North Carolina, 1972.
76. Code of Federal Regulations, Title 40, Chapter I, Subchapter C, Part 85, "Control of Air Pollution From New Motor Vehicles and New Motor Vehicle Engines." United States Government Printing Office, Washington, D.C., 1973.
77. City of Cincinnati, Division of Air Pollution Control, "Annual Report, 1973." Cincinnati, Ohio, 1974.
78. United States Environmental Protection Agency, "Evaluating Transportation Controls to Reduce Motor Vehicle Emissions in Major Metropolitan Areas," Publ. No. APTD-1364. Research Triangle Park, North Carolina, 1972.
79. Northrop Corporation, "Mandatory Vehicle Emission Inspection and Maintenance," Final Rept, Vol. 1. Anaheim, California, 1971.
80. S. I. Schwartz, *J. Air. Pollut. Contr. Ass.* **23**, 615 (1973).

81. Code of Federal Regulations, Title 40, Chapter I, Subchapter C, Part 80, "Regulation of Fuels and Fuel Additives." United States Government Printing Office, Washington, D.C., 1973.
82. Code of Federal Regulations, Title 40, Chapter I, Subchapter C, Part 79, "Registration of Fuel Additives." United States Government Printing Office, Washington, D.C., 1973.
83. United States Environmental Protection Agency, "Transportation Controls to Reduce Motor Vehicle Emissions in Major Metropolitan Areas," Publ. No. APTD-1462. Research Triangle Park, North Carolina, 1972.
84. United States Environmental Protection Agency, "Prediction of the Effects of Transportation Controls on Air Quality in Major Metropolitan Areas," Publ. No. APTD-1363. Research Triangle Park, North Carolina, 1972.
85. S. D. Berwager and G. V. Wickstrom, "Estimating Auto Emissions of Alternative Transportation Systems," Rept. No. DOT-OS-20004. United States Department of Transportation, Washington, D.C., 1972.
86. Air Pollution Control Association, "How To Tell the Air Pollution Story." Pittsburgh, Pennsylvania, ca 1970.
87. United States Environmental Protection Agency, "Groups that Can Help—A Directory of Environmental Organizations." Washington, D.C., 1972.
88. W. Dittrich, W. Brown, G. Hallman, and E. Nelson, *J. Air Pollut. Contr. Ass.* **21,** 555 (1971).
89. L. Hedgepeth, "Comprehensive Data Handling System, Emission Inventory Subsystem, Installation and Test Guide." United States Environmental Protection Agency, Research Triangle Park, North Carolina, 1972.
90. Air Pollution Control Association, "Statistical Tabulation of Air Pollution Control Bureaus, United States and Canada." Pittsburgh, Pennsylvania, 1952.
91. Air Pollution Control Association, "Statistical Tabulation—Part I, Air Pollution Control Governmental Agencies, United States and Canada." Pittsburgh, Pennsylvania, 1956.
92. G. Fishelson, *J. Air Pollut. Contr. Ass.* **24,** 44 (1974).
93. A. M. Schneider, *J. Air Pollut. Contr. Ass.* **23,** 486 (1973).
94. S. K. Mencher and D. S. Jones, "Guide to Technical and Financial Assistance for Air Pollution Control," Publ. No. APTD-1119. United States Environmental Protection Agency, Research Triangle Park, North Carolina, 1972.
95. B. H. Dickens, *J. Air Pollut. Contr. Ass.* **22,** 13 (1972).
96. F. X. Reilly, *Ind. Develop. Mfr. Rec.* **141,** 16 (1972).
97. L. J. Duncan, "Analysis of Final State Implementation Plans—Rules and Regulations," Publ. No. APTD-1334. United States Environmental Protection Agency, Research Triangle Park, North Carolina, 1972.
98. Research Triangle Institute, "The Development of Methodology to Permit Projection of Air Pollutant Emissions for Geographic Areas." United States Environmental Protection Agency, Research Triangle Park, North Carolina, 1973.
99. E. W. Peterson, *J. Air Pollut. Contr. Ass.* **23,** 11 (1973).
100. TRW Inc., "Air Quality Display Model." United States Department of Health, Education, and Welfare, Public Health Service, Washington, D.C., 1969.
101. A. D. Busse and J. R. Zimmerman, "User's Guide for the Climatological Dispersion Model," Publ. No. R4-73-024. United States Environmental Protection Agency, Research Triangle Park, North Carolina, 1973.
102. R. I. Larsen, "A Mathematical Model for Relating Air Quality Measurements

to Air Quality Standards," Publ. No. AP-89. United States Environmental Protection Agency, Research Triangle Park, North Carolina, 1971.

103. Ernst and Ernst Inc., "A Cost Effectiveness Study of Air Pollution Abatement in the National Capital Area." United States Department of Health, Education, and Welfare, Public Health Service, Washington, D.C., 1969.
104. United States Environmental Protection Agency, "Air Quality Implementation Planning Program, (IPP), Vol. I. Operator's Manual." Washington, D.C., 1970.
105. P. Morgenstern and K. A. Hagg, *J. Air Pollut. Contr. Ass.* **22,** 774 (1974).
106. F. C. Hamburg and F. L. Cross, *J. Air Pollut. Contr. Ass.* **21,** 66 (1971).
107. United States Environmental Protection Agency, "National Conference on Managing the Environment," *Proc.* Washington, D.C., 1973.
108. United States Department of Health, Education, and Welfare, "Manpower and Training Needs for Air Pollution Control" (A report to the Congress of the United States, 91st Congress, 2nd Session), Doc. No. 91-08. Washington, D.C. (1970).
109. J. J. Schueneman, *J. Air Pollut. Contr. Ass.* **17,** 670 (1967).
110. E. F. Wilson, W. Reilly, J. C. Donlevie, and F. Voigt, *J. Air Pollut. Contr. Ass.* **22,** 801 (1972).
111. S. Wiel, *J. Air Pollut. Contr. Ass.* **22,** 437 (1972).
112. P. H. Arbesman, W. S. Baker, K. L. Johnson, and S. N. Roth, Paper #50, *65th Annu. Meet.* Air Pollut. Contr. Ass., Pittsburgh, Pennsylvania, *1972.*
113. S. Weil, Paper #51, *65th Annu.* Meet. Air Pollut. Contr. Ass., Pittsburgh, Pennsylvania, *1972* Paper No. 72.
114. S. B. Baruch and S. N. Roth, Paper #52, *65th Annu. Meet.* Air Pollut. Contr. Ass., Pittsburgh, Pennsylvania, *1972.*
115. National Air Conservation Commission, "Appraisal: Guidelines for Evaluating the Effectiveness of Your Official Air Pollution Control Program." National Tuberculosis and Respiratory Disease Association, New York, New York, 1972.
116. L. R. Babcock, *J. Air Pollut. Contr. Ass.* **20,** 653 (1970).
117. M. H. Green, *J. Air Pollut. Contr. Ass.* **16,** 703 (1966).
118. San Francisco Bay Area Air Pollution Control District, "Combined Pollutant Indexes for the San Francisco Bay Area," Inform. Bull. No. 10-68. San Francisco, California, 1968.
119. W. A. Thomas, L. R. Babcock, and W. D. Shults, "Oak Ridge Air Quality Index," Rept. ORNL-NSF-EP-8. Oak Ridge National Laboratory, Oak Ridge, Tennessee, 1971.
120. Anonymous, *Air Eng.* **10,** 28. (1968).

5

Public Information and Educational Activities

Edward W. McHugh

I. Public Relations, Public Information, and Public Education

It is useful to distinguish between the various kinds of informational activities that are usually lumped under the heading "public relations" or, simply, "PR." In the broadest sense, the term includes all activities that are designed to influence public perception of any product, activity, idea, corporation, or government agency. Descriptive titles of the PR practitioner vary with the segment of the public that is to be influenced or the techniques employed. For example, a person who works at securing product or personal publicity is a PR man or a press agent; if the publicity is to be paid for, it is accomplished through an advertising man. Other related activities are suggested by other titles—you are in "government relations" if you work with local, state, or federal agencies to attain corporate goals; "employee relations" if you work to get along with the workers; and in "corporate relations" or "community relations" if you work to make your company look good to the general public. There are also people in customer relations, stockholder relations, alumni relations, and press relations—all terms descriptive of the segments of the public targeted by the activity.

Regardless of what title the practitioner uses, the basic techniques used to influence public attitudes and policy are the same, whether the employer or client is a government agency or private corporation. The difference is in the goals and objectives of the agency, and these vary considerably from group to group and may vary over any given time span.

In this chapter, we will review public relations techniques used by federal, state, and local governmental agencies and private corporations and suggest how the public information and educational activities of such agencies and corporations might be organized. For illustrative purposes, we will cover the internal activities of a public agency—the Bay Area Air Pollution Control District (of San Francisco, California)—in developing and executing programs to carry out the objectives of the agency.

A. Press Relations

In its earlier manifestations, public relations activities were mostly confined to getting a client's name in the papers. This was important to politicians, people in show business—sports, musical and theatrical enterprises—and to resort promoters. The practitioner was called a press agent, and he was usually a former newspaperman who retained some connections with his ex-colleagues. The occasion for client mention was frequently provided through a publicity stunt—a pseudoevent designed to attract press notice. Jim Moran, a famous practitioner, provided a classic example of this when he sought to dramatize the fact that changing horses in the middle of a stream was entirely possible by doing so as newsreel cameras whirred, during the 1944 presidential campaign. Since the slogan of the Democratic candidate, Franklin D. Roosevelt, was, "Don't change horses in the middle of a stream!" it was apparent that Moran was seeking to help his client, presidential aspirant Thomas E. Dewey. In spite of Moran's heroics, Dewey lost.*

In general, press-agentry is held in the lowest esteem by those who seek to "professionalize" public relations. As the editors of *Effective Public Relations* put it, "In the orchestra of public relations, press-agentry represents the brass section" (*1*). Although the term "press agent" may have fallen into disfavor, the function continues. While still used, the publicity stunt has given way to more prosaic means of getting free advertising. Pretty girls are always interesting enough to photograph, and their pictures are interesting enough to get into the newspapers, even if the girl is holding a sausage to promote National Sausage Week or being Miss Rheingold or wearing a lab coat to promote Cleaner Air Week.

The people who used to be press agents are now styling themselves public relations counsel, community affairs directors, or similarly impressive-sounding functionaries. But among newspapermen, they are known merely as "flaks." This term, defined in the "Dictionary of American Slang" as "a professional publicity worker or press agent," is a mildly derogatory word. Apparently it was derived from an older expression, "puffery," which suggested that flattering comments about the PR man's clients, which he sought to insert in the paper were "puffs." Since the explosion of antiaircraft shells made puffs, and since antiaircraft fire was known as "flak," a new application of the term was coined. As a matter of fact, newspapers and electronic media (radio and television) find much of the material developed by PR men useful, particularly on a slow news day. There is a given amount of time and space to fill, and as a result, informational material, even if produced by flaks, is often welcomed.

There is an important theoretical difference between governmental information work and public relations in the private sector. The government employee is a public servant and therefore has an obligation to be

* Moran was famous for other feats—selling iceboxes to Eskimos and finding a needle in a haystack—that frequently enabled him to get newspaper coverage. The question might be asked as to whether this kind of press is informative or simply the kind of fun that ingratiates the press agent with some city editors.

honest with the press. His role is as much to help reporters get a story as to make his agency look good. Of course, there are constraints on how much information he can provide, such as national security and the like, but the government public information officer should promote free flow of information, not obstruct it.

B. News Releases

Securing public awareness of an event, person, or product through newspaper coverage may be done by summoning reporters to a press conference, sending them a news release, or having lunch with the publisher. The most frequently used technique is the press, or news, release. The Environmental Protection Agency's excellent booklet, "Don't Leave It All to the Experts" (*2*), starts the section on the press with the statement, "The press is in the news business." This point should be borne in mind by any agency or business when considering making a public announcement about an event that the Board of Directors or the boss thinks is earthshaking. What might be important to that agency could be of small consequence to the public—at least in the judgment of a city editor. And it is good to keep in mind that the city editor makes such judgments dozens of times every day, and he owns his job because he makes few mistakes. In other words, the cardinal rule governing the issuance of a press release is that it must contain news. There are a few simple rules that make press releases more useable:

a. Put the name, address, and phone number of the releasing agency at the top of the release, together with the name and extension of a person who can be called by a reporter or rewrite man for further information.

b. Indicate at the top whether the material is for immediate release or to be released at a certain future time. If the material is "embargoed," or prohibited from being used before a certain time, it should be for a sound reason. For example, an announcement might be made in the course of a speech that is to be delivered at 10 A.M. on Wednesday that has important, newsworthy implications. A press release covering the announcement might be issued the day before, but the announcement would be embargoed for release at 10 A.M. Wednesday, because otherwise the audience for whom the announcement was intended would read about it in the morning papers before the speech. Even if the release is marked for immediate release, the date of issuance should appear so that the editor can judge how current the material is.

c. The release should be double-spaced (or even triple-spaced) with generous $1\frac{1}{2}$ inch margins on both sides of the page, so that an editor can make notes and indicate changes in the copy when he sends it to rewrite.

The usual routing of a release is from the city desk to rewrite, then back to city desk for checking. When approved, the story goes to the news editor, who determines where the story will be placed in the paper and how much space will be allotted. It then goes to the copy desk, where a headline is roughed in. Finally, it is sent to the "back shop," where it is typeset and put into the page makeup.

d. Editors are flooded with releases every day, so it is good practice to put the news at the top. In other words, devote the first paragraph (or "lead") to the substance of the story and the subsequent paragraphs to explanatory material. The editor can make a quick determination of whether or not to use the story if the release is done in this form. If not, he may not have the patience to wade through a lot of copy to find out what it's all about.*

e. Use simple declarative sentences and avoid technical jargon. Where technical terms are essential, make sure they are explained.

f. If the release runs more than one page, it is customary to put the word "more" at the bottom of all pages prior to the last and end the release with asterisks, cross-hatch signs (# # #), or the old telegrapher's signal for ending transmission, "30."

g. Make it a point to find out the name of the reporter who specializes in air pollution—the environmental reporter—for each paper, so the release can be addressed to him personally. But make sure the release is sent to the city desk as well, because the reporter might be on assignment, out of town, on vacation, or sick. If there is no regularly assigned reporter, just send it to the city desk.

C. Radio, Television, and Wire Services

The press release is the most widely used technique to secure newspaper coverage. It is customarily sent to radio and television stations, as well as magazines and trade publications, but it is not as successful when employed in this manner. In order to understand radio news in particular, you have to understand how commercial newswire services operate.

The function of a newswire service is to transmit foreign, national, regional, and local news and feature stories to its client newspapers and radio and television stations. The client may purchase the newspaper

* Most news stories are written in that way for several reasons—when a reporter writes a story, it is expected that the news editor will cut it back from the bottom thus the least essential details come last. Newswire services always transmit copy in the expectation that the story will be given different weight by different papers, depending on the nature of their circulation area and the preference of various editors.

wire, which gives full news and features, or the radio wire, which gives very short versions of news and feature stories. The client may also buy special services, such as the financial wire, which gives, in addition to business stories, securities and commodities market quotations. The sports wire is another special service with stories, box scores, race results, and such sporting statistics.

Wire service offices are called bureaus, and the means by which news is collected varies from bureau to bureau, depending on its location and number of employees. In many, cases, the stories are rewrites of stories already printed in nearby newspapers. Bureau reporters may cover stories themselves, or they may take stories over the phone from "stringers"—part-time employees who are usually reporters from local papers.

Wire service editors at smaller bureaus will be more likely to use a press release than those at larger bureaus. This is because more things are happening in Chicago than in Helena, Montana. The editor in Chicago must feel that the press release contains a news story with at least regional if not national importance before he allows it to take up valuable wire time. The editor in Helena is faced with the problem of filling his "split" of wire time.

While it is true that the radio wire carries shortened versions of stories that run on the newspaper wire, there are many cases when the radio wire puts out stories that do not appear on the newspaper wire. This is occasioned by the publication in one of the client newspapers of a story that wasn't used by the editor on the bureau's newspaper desk. It should be understood that it is very tough to sort out stories that will be run, from those that will be "spiked" when you have many clients with various shades of editorial opinion serving various submarkets within a given region. Your inclination is to run everything, but you are dealing with a finite amount of time.

The radio wire transmits more stories than the average radio station uses, so it is up to the "newsroom" editor to exercise his news judgment on which stories will be read. But essentially, radio news reports are merely segments of teleprint rolls that have been torn off and pasted into a sequence that can be read on the air, allowing suitable pauses for commercials. Of course, many radio stations, particularly FM, have no news programs at all. Others, particularly network-owned, have very extensive local newsgathering staffs.

D. The Television Newsroom

Television also relies on newswire stories, but to a lesser extent than the average radio station. Television stations under contract to or owned by the networks, receive news items from the network's national and international news staff. Stories in local papers provide leads that can be followed up with a filmed interview or other visual treatment. In the United States, television news is classified as "public service" by the Federal Communications Commission, but unlike other kinds of public service programming, television news makes money. As a result, television newsgathering is usually adequately staffed with capable newsmen. They operate with a cameraman who normally shoots 16 mm color film, which is edited and processed into a news show.* Where the reporter and the cameraman go depends on the assignment editor. He watches the news service wire, which gives a daily schedule of press conferences each morning, and he checks local papers for any stories that might lend themselves to visual treatment. Fully staffed radio and television newsrooms, as well as newspapers, also tune in to police radio broadcasts for leads on stories.

If television coverage is given to any news event, it is because the newsroom assignment editor has disptached a reporter and a cameraman. Frequently, the assignment editor will consult with the news producer before doing so, because there are likely to be many story options on any given day, and the producer is ultimately responsible for what goes on the air. Since both of these people are usually trained newsmen, their judgments are seasoned by experience. The cameraman shoots the event in 16 mm color with a magnetic sound stripe. He will often take "cutaways"—shots of the newspaper reporters writing, of the audience if there is one, and a close-up of his reporter diligently noting what is being said—on a smaller camera without sound. The film will be returned to the station for processing and review by the reporter, editor, and producer. After seeing all the film for the day, the producer will send a list to the film room of the stories he wants to use and the sequence. The film room puts the edited film on program reels—an "A" reel for the main story and a "B" reel for the cutaways—then sends them to the control room for programming. The director runs the program from there, giving instructions to the cameramen in the studio as to placement, cuing the anchormen via the hand signals of the floor manager, and running the rolls of news film.

The remarkable part of television news production is that these programs are put together so rapidly. From the time film arrives in the station to the time it is on the air could be as little as one hour, but that is the barest minimum. Producers can work comfortably within a two-

* Substantial technical breakthroughs have been made recently in the use of videotape in portable cameras. Within a few years, 16 mm film for news is likely to disappear.

hour time frame, will get the job done in an hour and a half, and will do it in an hour only if they are dealing with a really big story.

The newest technical development in television news dissemination has made "remote-live" broadcasts possible. The portable electronic camera, popularly known as a "minicam," can be carried to the scene of a news story and either transmit sound and pictures back to the studio for live broadcasts or video-tape the event for later use. As mimicams and video tape come into wider use, the time constraints on television news production will shrink because less film processing will need to be done.

E. The Press Conference

The press release is effective in getting coverage in a newspaper, and if the newspaper uses the story, there is a good chance the radio wire will transmit it, and if that happens, the story will be put on the air, and television might use it as well. With this in mind, there are a few more admonitions in order concerning press releases:

a. Always include wire service bureaus on your press list. They may not use the story as frequently, but when they do, they would rather work from your release than from a newspaper clip.

b. Don't bother to send a release to a radio station that does simple "cut and paste" news broadcasts.

c. Mail to the active, well-staffed radio stations for the same reason you mail to wire services.

d. Send to the television stations for a slightly different reason—television news editors might see visual possibilities in stories that a newspaperman would spike. If the visual possibilities are so obvious that *you* see them, you might consider holding a press conference instead of using a straight release.

It is sensible to hold a press conference if the story you want to tell is important enough to prompt reporters to come. There is nothing sadder in the life of a PR man than to set up a press conference, bring in his client or boss, then sit and wait for reporters who don't show up. They will certainly show up if the subject of the press conference is newsworthy; they may show up anyway if it is a slow news day. Commitments or promises by media people to cover any given event are never to be relied upon, because they can be turned aside at any time by a late-breaking hot news story.

Press conferences are favored by politicians and government officers, because they are most likely to get full coverage and can't be accused of favoritism or leaking a story. In general, they are disliked by newspaper reporters because they have difficulty following up on questions and their competitors all can hear what line of questioning is being pursued. Further, the newspaper people resent the front-row treatment usually given radio and television. But one must recognize that television is the most important of the media, and television likes press conferences, so there will always be press conferences.

As to timing, it would be foolish to schedule a press conference at 5 P.M. if your target is the 6 P.M. television news—or even at 4 P.M. The time must be set with a view toward the deadlines of the major newspapers as well. Only radio has enough flexibility in programming to be unconcerned about time. The day of the week is also important. For some reason, most press conferences are set for Monday mornings, and as a result, not all get covered. Ideally, a press conference will be set when no one else is having one and when there are no big stories breaking that would divert reporters. Weekend press conferences work well in areas where television and newspapers maintain an adequate weekend staff. Some political leaders in Washington are finding that Saturday morning press conferences pay dividends in that the wire services pick up the story (Saturday is usually a slow news day) and the politicians' message rattles around radio newsrooms all weekend.

A few considerations that should be kept in mind when holding a press conference:

a. Select a location that is convenient to the reporters you hope will cover.

b. Notify the media about one week prior to the date, then follow up by telephone. In the notice—set up in the same form as a release—indicate the subject of the meeting.

c. Build the press conference around a "star" or spokesman who will be present to answer questions.

d. Provide background material, including the text of any speech that will be part of the proceedings.

e. Although it isn't necessary, it's nice to have coffee and donuts, or some such, available for the reporters. It's appreciated.

f. If the wire service has a daily press conference bulletin—which is customary in the larger metropolitan areas—make sure your event is listed. City desks and assignment editors pay close attention to this service, which usually goes over the wire at about 7:30 A.M., because they are then at the point where the decisions must be made. The day's schedule of press conferences is before them, and they've got to send out the reporters.

F. Press Inquiries

Press releases and press conferences are outgoing press relations activities; that is, they seek to influence the public through stories in the media. A paradox is that the media resent the idea of being "used" by anybody to communicate with the public, in spite of the fact that they are and they know it. Without the input of flaks and flakkery, newspapers would be thinner and television time would be harder to fill.

Perhaps as a result of this feeling, newspaper people are easier to get along with when they are seeking news from you on their own volition. This comes about in several ways. There may be a wire story out of Washington, D.C., indicating a change in the regulations of the United States Environmental Protection Agency (EPA). The story is interesting only to those who know all about EPA and what it is doing, and this isn't enough for the editor. He wants to get the "local angle"—what will the change mean to local agencies and industries? So he assigns a reporter to find out. If the reporter digs up a little illustrative material, it will be written into the wire service story. If he gets a lot, a separate story will be written as a "side bar"—a story on the same subject usually placed on the same page close to the main story.

Another occasion for a reporter-initiated contact is a feature story. These are newspaper (or television) essays on a given subject that are timeless—not related to the news on a date certain. If a scientist makes a notable technological breakthrough in air pollution control, a story on the subject is a news story. If a scientist has made many such discoveries and is now enjoying his retirement by raising roses in Petaluma, a story about him is a feature story.

In the late sixties and early seventies, the news media discovered the environment. Many feature stories were written on air pollution, open space, resource conservation, and similar topics. Often, the writer had little background in the subject and had to do much research. The first person a newswriter contacts for general background is the PR man for any public agency dealing with the problem he is interested in. From there, he might go to the technical people in the agency and to industrial and environmental sources. It is useful to keep several points in mind when handling a press inquiry:

a. The reporter is after facts. If you don't know the answer, refer him to someone who does. If you aren't sure whom to refer him to, offer to find out and let him know.

b. Don't try to fool a reporter. Be absolutely candid. If he is treading on sensitive ground, tell him so—tell him you cannot answer the question, that you have no comment. Lies are poison, because when they surface, you have destroyed forever your credibility with that reporter and with his colleagues.

c. Try to answer as many questions as possible without referring the reporter to someone else. The PR man is paid to talk to reporters; an engineer isn't. Further, most managements prefer to have one person responsible for what appears in the press. Then they know who to blame.

These admonitions are concerned mostly with relations with newspaper reporters or magazine writers. These are likely to need more help because they do more of an in-depth treatment. Electronic journalists frequently try to have the subject write the story. They ask wide-ranging, general questions to encourage the person being interviewed to talk, then edit the responses into a story. This isn't meant as criticism; it takes considerable skill to do that kind of editing. Further, some television writers approach a feature story in the same way as newspapermen. Only after extensive research do they summon the camera, and they then have very specific ideas about whom they want to talk to and what they expect to be said.

Radio newspeople are inclined to be casual about any given story. The most frequent contact between radio news and PR men is made by telephone and most often is a request for comment on a national news development. In other words, they seek a side bar to a wire story. There are regulations that require that no taping take place unless the subject is aware of the fact, and in most cases, a radio reporter will ask that you identify yourself at the beginning of a tape. However, there are some who will simply call you up and start right in, without advising you that you are being taped and without any beep on the phone that would warn you. It isn't worth it to complain; the best course is to be a little careful when any radio station calls. Be sure of your facts and decline to comment unless you are. Just assume you are being taped unless you are assured you aren't.

G. Meeting the People

People communicate best when they are face to face; the words used are modified by intonation and infiection. Facial expressions and gestures lend emphasis. A smile, a shrug, or a raised eyebrow might carry more meaning than the spoken words. On the telephone, gestures and facial expressions are not part of the conversation, and where written communication is used, the words alone carry the message.

While it is impractical to rely exclusively on face-to-face communication, many PR practitioners seem to shun such contacts instead of encouraging them. The best possible way to get a feel for the community is to get out and talk to people. True, one might be misled—the aseptic public opinion surveys are probably more scientific—but nothing can substitute for give-and-take discussion as a learning experience. The PR man learns whether people care very much about air pollution and, if they do, what ideas (or myths) they might have about its cause and control. The people learn something about the agency or company and are able to humanize an entity that is too often highly impersonal.

The most common way to meet people is through speaking engagements. Schools, luncheon clubs, service groups, and professional associations are usually eager to book outside speakers, particularly if they have a motion picture or slide show that will be part of the program. The most important part of any such presentation is the question-and-answer period. The speaker should encourage not only questions, but comments from the audience. To the extent that audience participation can be generated, the program is a success. Obviously, face-to-face communications is inefficient, but it is rewarding and should be pursued to the limits of available time.

Talking on the telephone with the public may not be as vivid an experience as personal contact, but it is—for better or worse—easier to communicate in this manner than through letters. Conflicting forces are at work on people who use the phone. On the one hand, it is easier to pick up the receiver and dial a number while in the heat of anger or frustration, and it is easier to "speak frankly" when one knows that the other fellow can't hit you. But on the other hand, a person with a modicum of understanding and compassion finds it easier to calm an irate person down on the telephone than through written communications.

People make unsolicited phone calls to public agencies dealing with air pollution control because they usually know so little about the subject that they can't write a letter asking for information. They don't know who to write or what to ask for. As a result, such agencies have a great opportunity to project a favorable—or negative—impression through telephone PR.

Telephone PR starts with the switchboard. Incoming calls are sorted according to who or what the caller wants. Specific questions, requests for general information, and requests to be placed on mailing list should be referred to a designated person or the PR department, because the switchboard operator hasn't the time to take care of the request even if she knows the answers to some questions. Impersonality and brusqueness on the part of the switchboard operator are bad PR; friendliness and helpfulness are good PR. Aside from the above routine calls, a

PR person should take all difficult calls, e.g., a caller who is very upset about what is being done or not being done about air pollution or a caller who is incoherent, confused, or who has vague, unspecific questions on the subject.

In the case of public agencies, it is desirable to at least attempt to handle all calls, even if the subject matter deals with problems over which the agency has no jurisdiction. The public finds it difficult to distinguish between air pollution control of stationary sources versus air pollution control of automotive sources. Similarly, air and water pollution seem to be related, so shouldn't the same governmental agency handle both? And why not noise pollution? In some cases, notably the EPA, the same agency does handle such problems, but where they don't the agency that receives the call should at least be able to refer the caller to the proper agency, including address and telephone number.

H. Written Communication

Lawyers and businessmen have great respect for written communications because a "record" is created that, in the event of later disagreements, can be the subject of extensive discussions as to what exactly was meant. For the same reason, many PR men are admirers of skillfully written letters that can project an affirmative feeling in the reader without carrying the information or the response that the reader is seeking.

Congressman Peter Rodino, Chairman of the United States House of Representatives Committee on the Judiciary, received tens of thousands of communications during the time his committee was conducting hearings on the impeachment of President Nixon. It is axiomatic in the United States Congress (and elsewhere) that *all* communications must be answered if humanly possible. The preparation of custom responses to each of those who wrote was out of the question, so the congressman's staff developed a form letter that would, it was hoped, give a satisfactory response to everybody—Democrats and Republicans, pro and anti Nixon. Moreover, it was essential that the letter be absolutely opaque on the subject of the chairman's personal opinions of the president's guilt or innocence. The result was admirable. It emphasized the solemn responsibility of the committee to carefully weigh available evidence and make its judgment solely on the basis of the facts. This satisfied those who thought the president guilty, because they "knew" the evidence would sustain their position. Similarly, those who were afraid the president would be treated unfairly were reassured by the calm, dispassionate tone of the letter. The last paragraph of Chairman Rodino's letter is particularly impressive: "We are going to work expeditiously and fairly.

And, when we have completed our inquiry, we will make our recommendations to the House. We will do so as soon as we can, consistent with principles of fairness and completeness. We are going to be just, honorable, thorough, and worthy of the public trust."

Written communications need not be instruments of ambiguity, and most are not. Few can project the feelings of the writer as economically and succinctly as former United States Senator Stephen Young's immortal response to an irate constituent, "Some jackass wrote me a letter and signed your name to it," but few have the senator's perspicacity—or temper.

"A soft answer turneth away wrath," is the maxim that most PR people rely upon when faced with hostile correspondence. Three oil refinery operators in the San Francisco, California, Bay Area—Shell, Exxon, and Standard of California—were faced with identical letters recently, consisting of one line, "Why does your oil refinery put out so much air pollution?" followed by the writer's name and address. In each case, the responses were lengthy, detailing technical information about how the refinery operates, the social benefits of the products, and the economic importance of the facility in the community. They also indicated that the refineries were operating in compliance with all applicable state, federal, and local controls, and ended with an invitation to the addressee to visit the refinery.

A short time later, Lester F. Allen, Manager of Public Relations for Shell Oil Company in the Western United States, told the authors that it was only the press of other business that prompted a written response. Normally, he handles such inquiries by calling the writer on the phone and arranging to meet with him personally. Allen is a strong believer in face-to-face communication, bolstered by preprepared collateral materials where they are in point.

Nearly all PR people develop collateral materials that are useful in answering requests for general information. These include folders, brochures, booklets, and similar publications, and they may either be enclosed with letters or sent in lieu of a written response.

I. Lobbying

We have described public relations as activities designed to influence public perception of ideas, events, people, and corporate entities. Since government representatives are "servants of the people," influencing their perception may be regarded as a part of public relations. Certainly, public relations activities can play a substantial role in influencing the decisions of lawmakers and the decrees of administrative agencies. The accomplishments of Carl Byoir and Associates on behalf of its clients during the 1940s and 1950s included making important changes in public policy. "For A & P* it defeated a tax on chain stores in New York State. For the same client it later overcame the threat of a federal tax proposed by U.S. Congressman Wright Patman of Texas. For the Tile Council of America it secured helpful building code changes around the country. For the Eastern Railroad Presidents' Conference it won a veto, in Pennsylvania, of a bill that truckers wanted and the railroads vehemently opposed" (*3*).

Such changes are rarely made through lobbying alone. Normally, major publicity campaigns must be mounted, including newspaper advertising, an active speaker's program, and direct mail solicitations for support. Of course, lobbying is done at the decision-making level—the Congress, state legislatures, or administrative bodies—but lobbying of other public bodies, such as city councils, may also be done to secure resolutions of support for presentation to the decision-makers.

A lobbyist is a person who seeks to influence governmental decisions by a variety of techniques, including testifying at hearings, bringing information and arguments to the attention of public officials, and communicating negative and affirmative pressure. The pressure ranges from letter and telegram campaigns, public relations campaigns and visits by constituents, to campaign contributions, entertainment, and gifts. These techniques are all legitimate, although there has been a move toward comprehensive reporting regulations and laws on reporting of gifts, contributions, and entertaining. However, over the years, these activities have been widely regarded as corrupting influences. Consider these observations on the vocation of an unfortunate gentleman named Gillies, who, many years ago asked a California court to enforce a lobbying contract.

> Men who are paid to influence legislation, and who become acquainted with and cultivate the friendship of members through dinners, wines, cigars, and personal attention are certainly not assisting the state in procuring good legislation. If such men escape public prosecution, it is no reason that the time of the courts should be taken up aiding and assisting them in relation to their nefarious business. Courts will not permit themselves to be used for the purpose of aiding or enforcing such contracts. If such persons escape punishment through a public prosecution, they may consider themselves fortunate. (*Le Tourneux v. Gillies,* 1 Cal. App. 546, 550–551) (*4*)

In the United States, the Congress makes decisions on air pollution control; so does the United States Environmental Protection Agency, the several state legislatures, various state-sanctioned regulatory agencies,

* A & P—The Great Atlantic and Pacific Tea Company—a chain of supermarkets.

and a whole covey of city, county, and regional governing boards and agencies. There are similar arrays of bodies and agencies in other countries. Any corporation, association, or public group that seeks to influence decisions on air pollution will find plenty of scope for lobbying.

The close student of the regulatory structure will find that there are levels of input below the actual decision-makers where testimony can be more effective than at the final hearing. Many agencies have advisory councils that formulate regulations that are in virtually final form by the time they are brought before the legal entity that will enact them. Frequently, these are the points where the real work is done, and these are the points at which the information and arguments of the lobbyists are most likely to be welcome.

Lobbyists are most often lawyers, perhaps because most legislators are also lawyers, or because their subject matter is the body of law. On air pollution matters, however, the most effective spokesmen are usually chemists, chemical engineers and other technical adepts.

In summary, the lobbyist must know who to contact, where his presentation (or the presentation of a specialist on his team) will be most effective, and whether to recommend the mobilization of public opinion to support his client's position.

II. Basic Information Programs—Response and Dissemination

Any organization, public or private, is likely to find that people are curious about its activities. If such an organization is engaged in air pollution control—or air pollution production—that interest is heightened. Such organizations must expect such interest and provide for responding to inquiries. With respect to public agencies, one or more people are designated to handle these questions, and they are frequently professional public relations experts. Many agencies are part of a larger structure—a department of the public health agency, for example—and in those cases, the parent agency provides the public information department. Nevertheless, of the sixty-three state, provincial, and territorial air pollution control agencies in the United States and Canada listed in the 1977 Directory of Governmental Air Pollution Control Agencies, forty-two had separate public information departments or specialists. Of the two hundred thirty-one regional, county and city agencies listed, forty-eight had such departments or personnel. The programs of these agencies usually are passive in nature; that is, they respond to inquiries rather than soliciting them, although opportunities for stimulating interest such as "Cleaner Air Week" may be utilized.

The role of a public information officer or department operating a basic program is to provide the agency staff, the news media, other public and private agencies, and the general public with information concerning agency activities. The most outgoing feature of a basic program is the press release. Other techniques are individual responses by mail and telephone, providing speakers upon request, and the distribution of publications. Many agencies send out a periodic newsletter to help keep recipients abreast of current developments.

Most corporate public relations departments cannot specialize in air pollution matters, because the scope of corporate activities is much wider than generating air pollution, although some environmentalists seem to believe otherwise. For such corporate departments, publications prepared by specialists on the subject can be particularly useful. Several years ago, Owens-Corning Fiberglas experienced adverse public reaction due to emissions from its plants in Barrington, New Jersey, and Santa Clara, California. The major problem was plume opacity, which, due to the nature of the process, required extensive work and much money to correct. The Owens-Corning PR people, working with the technical staff, developed a folder that featured drawings of the plant at Barrington with air and water pollution abatement equipment outlined in red. The folder was designed as a walking tour guide, and citizens from nearby communities were encouraged to visit the plant. The program was so successful that it was adopted for use at Santa Clara. The object was to show that Owens-Corning was spending lots of money to clean up its problems and should be complimented, not condemned.

A. Basic Brochure

For the government agencies, a "basic brochure," or folder that presents general information about the agency and its activities is very important. For many, the agency's annual report is regarded as the basic brochure, but it's a poor substitute. The annual report is usually a statistical compilation of how many people are employed, how much money is spent, and how many violators were nabbed, with a smattering of figures showing how much cleaner the air is getting.

Among the several basic brochures that are informative and readable are "Blueprint for Clean Air," published by the Puget Sound (Washington) Air Pollution Control Agency, and "Sick of Pollution? Be Part of the Solution," a booklet published jointly by the Allegheny County (Pennsylvania) Health Department and the Tuberculosis League of Pittsburgh. "Air Pollution and the San Francisco Bay Area," the annually revised booklet of the Bay Area (California) Air Pollution Control

District, currently runs 44 pages in length and includes a 4-page air pollution glossary. The 84-page Los Angeles, California, Air Pollution Control District book, "Profile of Air Pollution Control," includes fold-out charts, some in color. Both the Los Angeles and Bay Area booklets are complete enough to provide a high school student with enough information to write a term paper on the subject. Since many of the information requests received are from such students, it is no accident that the booklets were designed to fill the need.

Among the more comprehensive basic booklets is "A Citizen's Guide to Clean Air," published by the Conservation Foundation of Washington, D.C., under a contract with the EPA. The booklet is 95 pages long, includes a detailed review of federal law, government organization, and a discussion of implementation plans, the auto control program, and citizen participation. Also included are a glossary and a small but good bibliography.* Two publications of the American Lung Association (formerly the National Tuberculosis and Respiratory Disease Association) are quite readable and very useful. "Air Pollution Primer" includes 84 pages of text plus an index and a glossary. It covers the relationship between weather and air pollution, identifies pollutants, and discusses effects. "Controlling Air Pollution," as the title suggests, describes control techniques and hardware for stationary source abatement. The booklet includes excellent nontechnical drawings and a glossary.

Parenthetically, in the course of securing information for this chapter, the author wrote to ten air pollution control agencies throughout the United States. Bearing in mind the axiom referred to in Section I,H on written communication, which is that all mail must be answered as promptly as possible, we anticipated little difficulty. In fact, we received prompt answers from six: Philadelphia, Pennsylvania; Boston, Massachusetts; Houston, Texas; Puget Sound (Seattle), Washington; Allegheny County (Pittsburgh), Pennsylvania; and Chicago, Illinois. After a second request, we received answers from New York, New York; Los Angeles County, California; and Wayne County (Detroit), Michigan. We have yet to hear from Fulton County (Altanta), Georgia. It should be noted that all ten of the agencies selected have public information departments.

B. Special Purpose Publications

In addition to the basic brochure, it is useful to develop special purpose folders keyed to a new development or event, a special activity,

* Our own excellent compendium was described as follows: "*Air Pollution,* Arthur C. Stern, ed., second edition, 3 vols. The most authoritative, comprehensive work available. Highly technical."

or a target population. Essentially, a special purpose publication deals with a single subject, in contrast to, for example, the basic brochure, which deals with many, and periodic publications on the same subject, such as monthly summaries of air quality readings.

The titles of some Los Angeles County, California, Air Pollution Control District publications are illustrative: "Power Plants in Los Angeles County," "Smog Alerts, School and Health Smog Warnings," "The APCD Enforcement Program," and "The Engineering Division of the Air Pollution Control District." Chicago, Illinois, prepared "Color the Sky Blue" to introduce its incinerator, boiler, and furnace control program, and Wayne County, Michigan, printed a 1-page color leaflet, "Pollution Alert," to warn people that the burning of leaves was banned. Another engaging Wayne County leaflet is entitled, "The Bloodhound's Hand Guide to Air Pollution," and it features a splendid drawing of a dolorous bloodhound. The text asks citizens to report odors to county inspectors and describes the sources and characteristics of six odorous compounds.

The EPA has quite a few special purpose pamphlets "Groups That Can Help," a directory of environmental organizations; "Research and Monitoring"; and a combined air–water booklet entitled "Action for Environmental Quality." As befits an agency financially supported by the United States Government, the federal publications are usually four-color, slick paper jobs, with many stomach-churning pictures of smog and belching smokestacks.

The City of Houston, Texas, has a unique approach to general and special purpose publications. Aside from a few mimeographed folders done by its Department of Public Health, it distributed outside publications. It buys a little cartoon-illustrated folder entitled, "Keep It Clean! (The Air, We Mean)" from Imagination, Inc., of St. Paul, Minnesota. The EPA provides its four-color overview of air and water pollution, solid waste management, and radiation hazards called "Toward a New Environmental Ethic." A general purpose, cartoon-illustrated booklet, "Needed: Clean Air," copyrighted by Channing L. Bete Company of Greenfield, Massachusetts, is also stocked, and these (and other such) are sent out to citizens, "tailored to the needs and capabilities of the requester" (*6*).

Government agencies are not alone in producing special purpose folders. For years, companies and associations have been publishing the views of their officers and members in folders and pamphlets, changing them as new developments are made and new needs arise. An unusual case is the publication of the speech of a public official in an attractive, illustrated booklet by the Automobile Manufacturers Association. The booklet (and speech) is entitled "Automobile Pollution and the Great Credibility Gap," and the text was presented at the Institute on the Environment of the American Society for Public Administration on April 7, 1971, by Alexander Rihm, Jr., Director of Air Pollution Control Programs, Division of Air Resources, New York State Department of Environmental Conservation. In brief, the theme is that automobiles aren't causing as much air pollution as some people think and that they are being controlled as efficiently as possible anyway. One statement from the text, "But let us beware of those cases where facts are twisted to support radical action in the environmental area."

C. Periodicals

People who care about air pollution and air pollution control like to keep abreast of developments as they occur. Companies and government agencies are aware of this and try to provide information through periodic publications. The most common form of such publications is the newsletter.

A newsletter is so termed because of its format—its shape, size, and general makeup. Page size is approximately $8\frac{1}{2} \times 11$ inches (22×29 cm), the same size as letter paper. Newsletters may, in fact, be composed of sheets of letter paper stapled together. The *Air Pollution Quarterly* of San Diego County (California) and the *Clean Air Monitor* of the Bureau of Air Pollution Control, Pittsburgh, Pennsylvania, are examples. They are typewritten, with the *Monitor* using a mimeotype stencil and the *Quarterly* an offset process. The *Quarterly* is printed on both sides of the page, the *Monitor* on one. In both cases, the newsletters use a first sheet with a printed masthead, this being the "front page."

Most newsletters are printed on both sides of 17×11 inch (44×29 cm) sheets, so that four $8\frac{1}{2} \times 11$ inch (22×29 cm) pages are produced with a center fold. Some, such as *Air Pollution Control Progress,* of the Philadelphia, Pennsylvania, Air Management Services, take this a bit further, printing on both sides of $25\frac{1}{2} \times 11$ inch (66×29 cm) sheets, thereby producing six pages. This results in a somewhat awkward problem with the folds, one on the left side for the front page, and one on the right side of page three.

Newsletters that are printed are frequently typeset, and a two- or three-column make-up is employed. Philadelphia's *Progress* uses two columns and the *Bulletin* of the California Air Resources Board uses three columns, as does *Environmental Currents* of the City of Chicago, Illinois and *Skylines* of the Cincinnati and Southwestern Ohio Air Pollution

Control Division. *Air Currents* of the Bay Area Air Pollution Control District (San Francisco, California) uses two columns for the front page and three columns for the other three pages.

With a circulation of 33,000, the *EPA Citizens' Bulletin* of the United States federal government has more readers than any other government periodical in the environmental field. The subject matter includes water quality, noise, and other areas of concern to the EPA in addition to air pollution. The *Bulletin* is unusual in that it is slightly different in page size—$8 \times 10\frac{1}{2}$ inches (21×27 cm)—and is printed on both sides of $16 \times 10\frac{1}{2}$ inch (42×27 cm) sheets. The copy is typewritten, then printed with a photo offset process. Two of the sheets are usually printed, producing eight pages, which are then glued together during the folding process. A drawback to this process is that extra-wide margins must be employed because of uncertainty in how much glue the brush on the machine will strip down the center of the sheets. The EPA uses $1\frac{1}{2}$ inch (3 cm) margins and, as a result of this and the smaller page size, employs a one-column make-up.

An idea of the content of newsletters can be gleaned by perusing the headlines. "Train Supports Temporary Energy Accommodation," says the EPA; "July—Freshest Month in City's History," boasts the Chicago, Illinois, Department of Environmental Control; "Hearings Slated for Air Pollution Emergency Plan," announces the San Diego County, California, Air Pollution Control District. Pictures are used in all newsletters that are printed, mostly of agency personnel, governing board members, and city officials. Charts and graphs are employed, often to show how much pollution is being taken out of the air by the agency's control program.

There are few business periodicals that deal exclusively with environmental control, except for such commercial newsletters as the widely known *Air/Water Pollution Report,* a weekly newsletter published by Business Publishers, Inc., of Silver Springs, Maryland, (current annual subscription rate, $175.)

An exception is *Ecolibrium,* a quarterly publication of the Shell Oil Company, Houston, Texas. This is a magazine, rather than a newsletter, with a cover plus from twelve to sixteen 9×12 inch (23×31 cm) pages, three-column make-up, and typeset with pictures. Among the stories are such matters as Shell's research on oil-spill control, work on alternative energy sources, pollution-free refining, and the declining role of natural gas in the nation's energy picture.

Another publication of an oil company, *Exxon, USA,* is a four-color magazine with production values reminiscent of *National Geographic.* Again, this is published quarterly, and while not addressed specifically to environmental problems, devotes much space to the subject. Unsurprisingly, it is published by a company known as Exxon, USA.

D. Air Quality Information

As people become sensitized to air pollution as a community problem, they tend to perceive it—whether it's there or not. In an earlier age of innocence, the golden glow of autumn haze, spiced with the tang of burning leaves and the sharp bite of sulfur dioxide as oil-burning furnaces are fired up after the long summer vacation, used to quicken the citizen's pulse, because such sights and smells meant the start of a new school term, the football season, new friends, and activities. Now these things are understood as air pollution.

Perhaps as a reaction against the bureaucracy that destroyed public innocence about air pollution, many people call agencies with air monitoring capability and demand to know why there is so much air pollution on any given day. They can't see across the street. A check at the nearest air monitoring station often reveals that there is little air pollution, but quite a bit of haze or fog.

The fact is, most people equate air pollution with reduced visibility, and control agencies have a continuing problem in trying to explain the fact that air pollution can be present without serious visibility impairment and that visibility impairment can be present without much air pollution. The best way to handle this problem is to provide readings of air pollution levels every day throughout the year, whether the air is clean or dirty.

The way such readings are made public presents a difficult problem. How do you explain, for example, that 0.50 parts per million of oxidant is severe air pollution, and ten times that much carbon monoxide—5.0 parts per million—is clean air? Many agencies have tried to deal with this problem by developing a combined pollutant index—adding together various readings of air pollution levels, perhaps adjusting them to comparable levels of toxicity, and placing the resultant number on a scale going from clean air at the bottom to heavy or severe air pollution at the top. Pollutants included might vary, but usually include sulfur dioxide, oxidant, carbon monoxide, particulate matter, and nitrogen dioxide.

The need for such indexes was mentioned in an unsigned article appearing in the January 10, 1972, issue of *Chemical and Engineering News*.

> To many specialists at the American Academy for the Advancement of Science meeting in Philadelphia, an essential step in cutting through the

> uncertainties in pollution is somehow to get numbers into the picture. Although lacing their papers with disclaimers, these scientists have confidently worked up indexes. . . .
>
> In overriding their own uneasiness at the probable errors involved in forming these gross indexes, scientists cite particularly a growing "environmental credibility gap" between themselves and the supporters they need most, the public.

Unfortunately, the "probable errors" referred to by the author became of increasing concern to many environmentalists, and combined indexes have been on the decline because they fail to give specific levels of specific pollutants, with some readily perceived relationship to air quality standards or health effects.

It may be that the simpler the reporting the better. As mentioned earlier, the average person regards air pollution as dirty air that cuts visibility. If so, perhaps the M.U.R.C. system of the Air Pollution Control Division of the Wayne County (Michigan) Department of Health is the answer. As explained in one of the agency's special purpose folders, "M.U.R.C.—pronounced 'murk'—is an acronym which means *M*easure of *U*ndesirable *R*espirable *C*ontaminants and is a report of the amount of airborne, microscopic particles of dirt in the atmosphere." The system utilizes a particulate tape sampler and the spot on the filter tape is measured with a densitometer (spot evaluator), then "converted into the M.U.R.C. Index scale by a mathematical formula." The scale consists of numbers from zero to "121 and over," with values assigned such as "Light Contamination," "Medium Contamination," etc. The folder includes the usual disclaimer, but couples it with an assertion that might be called into question by some scientists, "The M.U.R.C. Index is not an overall pollution indicator. Other contaminants—sulfur dioxide, carbon monoxide, hydrocarbons, etc., are measured by other methods and generally rise and fall with the M.U.R.C. Index levels."

An excellent summary of the various index systems currently used in the United States and Canada has been compiled by Gary C. Thom and Wayne R. Ott. The study was sponsored by The Council on Environmental Quality and the U.S. Environmental Protection Agency.* On the shortcomings of the present systems, the authors say, "The diversity of air pollution indices creates potential confusion, raises questions about their technical validity, and prevents the indices from being used to give a national picture of air pollution problems." The authors recommended that a federal interagency task force be established to explore the feasi-

* Thom, Gary C., and Ott, Wayne R., *Air Pollution Indices. A Compedium and Assessment of Indices Used in the United States and Canada,* U.S. Government Printing Office: 1975 590-314/1680.

bility of establishing a standard air quality index.

After several years of experience with a combined pollutant index, the (San Francisco, California) Bay Area Air Pollution Control District (BAAPCD) adopted a reporting system that simply states the readings for each pollutant except sulfur dioxide (sulfur dioxide is highly localized in the Bay Area) and assigns a value for the level reported—"Clean," "Light Air Pollution," "Significant," "Heavy," and "Severe." The designations are roughly based on federal air quality standards. A feature of this system is that each station—there are twenty-four air monitoring stations in the Bay Area—receives a designation for the day based on the designation of the highest level reached by any pollutant. Thus, if the station in San Jose had "Significant Air Pollution" for oxidant, the designation for the station would be "Significant," even if the other pollutants were designated "Clean Air."

Daily reports on air quality levels are made available to the electronic media and the press, but BAAPCD found that they aren't always used. Some editors aren't interested in clean air; they will only report air pollution when there is air pollution. However, television weathermen are notably cooperative in presenting air quality information, and some daily newspapers use the data regularly on the weather statistics page.

The BAAPCD makes air quality designations available to anyone who telephones 863-7664 (SMOG). The "Smog-phone" responds with a recorded message, which is taped every morning with an air quality forecast and every afternoon with the day's readings. Those who want more detail, including newspapers and electronic reporters, call the agency's regular number.

E. Elements of the Basic Program

To summarize, a basic information program should be designed to provide the public, the news media, and other public and private agencies with information concerning air pollution and the agency's activities. The means used to inform are distribution of the basic brochure and special purpose publications, judicious use of the press release and press conference, the development and circulation of a newsletter, and individual responses via mail and telephone.

A related aspect of the program is the regular transmission of air quality data to the public and news media and the encouragement of its use. Providing interpretation of the meaning of the data is an essential concomitant. Another means used is to provide speakers on air pollution and control to business and conservation groups, service clubs, and schools. It might be argued that this belongs under the "Active" rather than basic program, but it belongs here because it is unavoidable. If the technique is vigorously pursued, then it is certainly active.

III. Active Information Program—Outgoing Solicitation

The basic information program is a creature of public interest; it must be established because such a program is the most efficient way an agency or company can respond to the expressed need for information. The active program has as a major goal the stimulation of public interest in the subject of air pollution and air pollution control where such interest is waning or nonexistent. Aside from purveyors of abatement equipment, commercial environmental newsletters, and an occasional company that seeks to take credit for cleaning up its operations, corporations do not gear their PR toward an active solicitation of public concern about air pollution. Accordingly, our discussion will apply primarily to public agencies.

A. Use of News Media

There are few ideas more repugnant to news people than the notion that their medium is being "used" by PR people. But even if the idea is repugnant, it is nevertheless true. In view of the public service doctrine of the United States Federal Communications Commission (FCC), the electronic media—radio and television—have little choice but to grant access to certain governmental and community groups. The print media—newspapers and magazines—are under no such stricture, and if they are to be "used," the material must have genuine news or feature value.

In brief, the Federal Communications Act provided that commercial broadcasters must be licensed to use the people's airwaves. The license, which is granted by the FCC, established pursuant to the act, grants the broadcaster exclusive right to use the frequency sought for three years. When renewal time rolls around, it is incumbent upon the broadcaster to show that the station has used the frequency to serve the community interest, and one of the most important ways this is done is to aver that nonprofit social or civic groups have been given access to the airwaves. PR people are acutely aware of this and take advantage of it.

There are several ways to reach the stations' audience, depending on the broadcaster's programming and policies.

a. Develop a series of radio and/or television spot announcements for use on public service time. Nearly all stations intersperse public service messages among paid commercials. Their formats are the same as paid commercials; that is, they are of 10, 20, 30, or 60 seconds in length and are designed with three goals: (1) get the audience's attention, (2) inform the audience, and (3) compel the audience to take action.

The simplest and cheapest way to do a spot is to write it in the form specified by the station and ask that station announcers read it on the air. Some will, some won't. A more effective way is to prepare the spot on tape, film, or a record, depending on station requirements. Spots of this nature are more likely to be used. Most of the larger stations designate a person as "public service director," who will answer inquiries about the format of the spot. If there is no such person, the program director is the man to see.

An illustration of two messages—a 60-second and a 30-second radio announcement—both designed to carry the same message, is provided by the USEPA's Mobile Source Pollution Bureau.

60-Second Message

You've probably been hearing a lot of talk lately about gas mileage on new cars.

Let's look at the facts. If you buy a small car, you probably can expect to get about 20 miles on a gallon of gas in city driving. But if you want a bigger car, your mileage will go down to only about 10 miles per gallon.

Add air conditioning and you lose another mile. . .

And automatic transmission costs a little more.

In addition, all new cars come with an emission control system to cut down the exhaust pollutants that are fouling our air. Yes, it too affects gas mileage . . . about the same as automatic transmission. But much more important is how it will improve the quality of our air.

From now on, when you hear talk about gas mileage, remember this.

The emission control system on your car is just one of the many factors that affects your mileage.

But it's the only one that means we'll have cleaner air.

30-Second Message

There's been a lot of talk lately about gas mileage on new cars.

With a small car, you can get about 20 miles on a gallon of gas in city driving. But if you want a bigger car . . . air conditioning and automatic transmission . . . your mileage goes way down.

In addition, all new cars come with an emission control system to cut down exhaust pollutants.

It's just one of the many factors that affects your mileage. . .

But it's the only one that means we'll have cleaner air.

The following is a 20-second television film produced by the San Francisco, California, Bay Area Air Pollution Control District:

20-Second Public Service Spot Announcement

VIDEO	AUDIO	
Vista of morning commuter auto traffic	Voice Over:	There *are* better ways than this to get to work.
Automobile and members of a car pool	Driver:	We like the car pool because it saves us the hassle of driving every day, it's more economical, and it's one step in helping to reduce air pollution.
Traffic moving to Bay Bridge toll gate	Voice Over:	Conserving fuel can also mean cleaner air. This message from the Bay Area Air Pollution Control District.
Logo		

b. Encourage use of air pollution subject matter on talk shows and public service feature programs. The talk show program, wherein the audience is encouraged to telephone the station with on-the-air questions directed to a guest, has become quite popular in many markets. (Radio and television people refer to their broadcast areas as "markets" for obvious commercial reasons.) This is a voracious format, with a new guest needed every day, and since air pollution has community-wide implications, experts from local governmental agencies are welcomed.

Nearly every station has a weekly "community concern" type program—usually dull, turgid, round-table discussions that are aired on Sunday mornings when all except invalids are doing anything except watching television or listening to the radio. Producers of such programs are highly approachable, particularly if a package is offered that sounds interesting. For example, a panel that includes an industrialist, an environmentalist and a government man might generate a little excitement.

c. Use "free speech messages" and editorial responses. Many stations have instituted the practice of permitting spokespeople from community groups to express personal opinions on matters of current general public interest. If the viewpoint stated is at variance with that held by others, equal time can be secured for response. Similarly, many stations take positions on public questions and express them in "editorials." Again, the stations are required by the FCC to give equal time for the statement of divergent opinions.

There are other ways to secure access to the broadcast media, such as "community calendars"—announcements of events of community interest—and through regular news programs, a subject discussed earlier. The best source of information is the broadcast station itself. As long as the station is under the imperatives of federal law to provide service to the community, the stations are ready and willing to help.

The only imperative newspapers and magazines operate under is to maximize circulation so that advertising rates will be high enough to generate profits. In order to do this, the "product" must be readable, interesting, and comprehensive. Also, newspapers like to feel that they serve the community interest, even though there is no law that compels them to do so. The net result of this, so far as PR is concerned, is that attempts to get into the papers must be coupled with attractive ideas or events with solid news value.

The best approach is to suggest feature stories. Many newspaper reporters—particularly those who have attained special status, such as science writers and environmental reporters—are expected to develop their own feature stories, naturally with the concurrence of their editors. They are always on the lookout for story ideas and they welcome suggestions. Magazine writers are even more susceptible, since many of them are free lance; that is, they are paid only for what is printed.

B. Use of Other Media of Communication

Billboards, long a substantial advertising medium in America, have declined in importance as state and federal legislation have restricted their number and location. Nevertheless, they retain some importance in some areas, and most billboard companies will make public service postings on billboards that haven't been rented by commercial clients. Since you are expected to supply the paper, it is wise to secure some kind of commitment from the company before going to the expense of designing and printing the posters. It should be noted that billboard companies have no legal obligation to make public service postings—they simply feel it is good PR for them to do so. They are often attacked by some people for advertising liquor and cigarettes, and they like to be in the position of countering such claims by pointing out the good things they do for the Boy Scouts and the United Crusade. Not surprisingly, the PR manager for the billboard company is the person to see. In the case of smaller companies, the manager or owner has the say.

In many metropolitan areas, car cards are posted inside commuter railroads, busses, subways, and streetcars. The same comments apply to this medium as to billboards, with the exception that most rapid transit companies are publicly owned and their policy toward accepting public service postings is more strongly adhered to than most private companies'.

Another kind of outdoor advertising, favored more in Europe than in the United States, is posters. Usually, the space used for posting is not owned or leased by the poster company; it is simply appropriated—fences, walls, anything that is visible to people passing by. In some American cities, the poster operators follow the practice of pasting over posters put up by rival operators, ultimately resulting in enough mutual disruption to compel a *modus vivendi* that will enable some sort of sharing of the market. There are no public service postings done by such operators, but purchase of this medium might be considered in some areas because it is relatively cheap.

C. Paid Advertising

Paid newspaper advertising has been undertaken by many public agencies, because state and federal laws frequently require that notice of public hearings, for example, be published in a newspaper of general circulation. Traditionally, these ads are placed in the "Legal Notices" category of the classified advertising section of the paper. Some agencies have decided to make a virtue of necessity by placing display ads on the news pages. A modest-sized ad of perhaps six or eight column inches on one of the news pages attains far more visibility than anything in the classified pages. A declaration of publication or affidavit is supplied when legal advertising is done, but not with display advertising, so the use of display advertising must be done in addition to, rather than in lieu of, legal ads. The cost of display ads should be checked carefully. Newspapers have two (or sometimes more) rates, depending on who places the ad. "Retail" rates are quite reasonable because store owners, home builders, and other businessmen who rely heavily on newspapers to attract customers provide much of the newspapers' revenues and the papers are reluctant to kill the goose that lays the golden egg. Most of these advertisers either have their own personnel do the ads or pay an agency for creative help. Ads at retail rates are not "commissionable," i.e., the newspaper doesn't pay a commission (usually 15%) to an advertising agency for placing the ad. "National" rates are considerably higher and are applied to everyone, including government agencies, who do not qualify for retail rates. National rates are commissionable.

Many corporations have used newspaper advertising in a manner that is similar to special purpose publications. Full-page ads appeared during the energy crisis explaining the positions of the various oil companies on the subject, and these were frequently reprinted on single sheets and circulated to government and community leaders. Ads in magazines have been used in the same way.

With respect to air pollution control, advertising has been used to ventilate disagreements between whole industries on how to attain automobile emission control—or how not to. The debate over the catalytic converter pitted two industrial giants—E.I. Du Pont De Nemours & Company and Ethyl Corporation—against General Motors. In order to utilize catalytic converters, lead must be removed from gasoline, and Du Pont and Ethyl put lead into gasoline. Thus, Du Pont and Ethyl argued that the use of catalytic converters shouldn't be required. In an apparent attempt to consolidate the support of the oil industry for this position, Ethyl ran a full-page ad in the May 13, 1974, issue of *Oil and Gas Journal* with the headline, "The Catalytic Converter—Read What The Experts Are Saying." The experts selected for the ad made scary comments about sulfuric acid conversion, particulate sulfates, and the sacrifice of gasoline economy, using such expressions as "may be a Frankenstein monster." The ad's conclusion: "With all these uncertainties, why catalysts now? . . . Let's not rush into the wrong solution. Ethyl urges that emission standards be frozen at 1974 model year levels until the promise of new non-catalytic technologies are proven."

On September 12, 1974, General Motors placed a full-page ad in the major daily newspapers of the United States, headed, "General Motors believes it has an answer to the automotive air pollution problem . . . and the catalytic converter has enabled GM engineers to improve performance and to increase miles per gallon." The ad included four columns of copy and illustrative drawings. Concerning lead in gasoline, GM said, "As the public becomes aware that the use of unleaded gas lowers maintenance costs by greatly increasing the life of spark plugs, engine oil and exhaust system components, the demand for unleaded gas should cause it to become available at most other U.S. gas stations." The ad is silent on the creation of sulfur compounds by the converter.

Paid advertising on air pollution control has been much less frequently used on radio and television except where there is a clear product tie-in. For example, Standard Oil of California promoted its gasoline in 1970–1971 by plugging the "anti-air pollution" properties of its additive F-310. Some oil companies, notably Exxon, have used paid television ads to improve their corporate images through reference to the cleanliness of their refineries.

D. Audiovisual Communication

The use by government and businesses of 16 mm sound color motion pictures has been extremely popular for many years. These films have been mostly aimed at school-age populations, since the showing of motion pictures in classrooms has been a favorite device of many teachers to sneak educational concepts into classes that have been led to think they are being entertained. Among the most entertaining have been the travel films done by travel entrepreneurs, oil companies, and motor clubs and fishing, skiing, and hunting films done by sports equipment manufacturers. The best of these films rarely preach, threaten, or use other "hard sell" techniques; they entertain, and thereby subtly coax and cajole the audience into an affirmative feeling about the product, company, or point of view presented.

The environmental movement has engendered a number of films, including some specifically in air pollution. Some have been sponsored by citizen and government groups to show the dangers of pollution; other films by industrial groups to show that they are protecting the air. All tend to be biased and to get out of date rapidly. Thus, the desirability of previewing a film before showing it publically. The films range from 4 minutes (a cartoon entitled "The Deluge") to 54 minutes ("Air Pollution: Take a Deep Deadly Breath"). The films were produced or sponsored by such disparate entities as the National Film Board of Canada, Caterpillar Tractor Company, University of California, National LP Gas Council, General Motors' Detroit Diesel Engine Division, National Tuberculosis and Respiratory Disease Association, Aetna Life and Casualty, Shell Oil, Ethyl Corporation, Automobile Manufacturers Association, Inc., and the USEPA.

Making a film can be expensive but useful if it reaches the audience it is intended for. The key is distribution. There are many commercial distributors who will take films for sale or distribution to civic groups, schools, television stations, and theaters and in the case of rental, charge the sponsor for each showing. The rate depends on the size and type of audience reached. Some special purpose distributors, The Conservation Foundation of Washington, D.C., for example, might take a sponsored film for distribution at little or no cost, if they feel the film advances their special purpose or cause and is an attractive addition to their library.

Among the largest distributors in the United States are Encyclopedia Britannica Educational Corporation of Chicago, Illinois; Contemporary/McGraw Hill Films of New York, New York; Modern Talking Picture Service, Inc., of New Hyde Park, New York; and Associated-Sterling Films of New York, New York, and the United States Environmental Protection Agency.

In past years, television used quite a lot of 16 mm films, particularly for programming in morning and weekday afternoons. This practice has been changing with increasing rapidity with the advent of "soap operas" and as more and more series, originally produced for prime time showings, are given second, third, fourth, etc. runs in daytime slots. As a result of the decreasing importance of television as an available outlet for 16 mm films, the television format (26–28 minutes with cuttable breaks) is being disregarded by many producers.

Another audiovisual technique in wide use is the film strip—still pictures in sequence run in conjunction with a live narrator working from a script or a taped narration. Film strips were extremely popular in schools in the 1950s and 1960s because of their low cost. However, the increasing availability of sound color motion pictures has worked to diminish the importance of film strips.

Other visual support for verbal presentations continue to be popular and easy to produce. These include color 35 mm slides, material for use with overhead projectors, flip charts, strip signs, and other easel-related graphics.

E. Organization for Action—How It Can Be Done

Having reviewed most of the techniques and materials commonly used for both active and passive information programs, it should be interesting to examine an organizational structure that has been designed to use them in support of a governmental air pollution control agency.

The Bay Area Air Pollution Control District (BAAPCD) is headquartered in San Francisco, California, and has jurisdiction over nonvehicular sources of air pollution located in all or part of the nine counties that touch on San Francisco Bay. This area includes about 5600 square miles (14,500 km^2) of land area and 4.8 million people. The agency, which was established in 1955, operates with a 1977–1978 budget of $6.5 million and 224 employees. BAAPCD is a governmental agency with regional responsibility. It is larger than most air pollution control agencies in the United States, but it is not the largest. Its attraction for illustrative purposes relates to its size, the nature of its programs, and the fact that the author has a fair knowledge of its structure.

The relationship between the Public Information and Education division (PI&E) and the rest of the District is shown in Figure 1. The BAAPCD Board of Directors consists of two local government officials—a county supervisor and a mayor or city councilman from within the county—selected from each of the nine counties that comprise the district. The air pollution control officer serves at the pleasure of the board of directors, and the rest of the staff—except for the counsel and the board's secretary—are hired by and are responsible to the air pollution control officer. Staff members who report directly to the air pollution control officer are the counsel, controller (chief of administration),

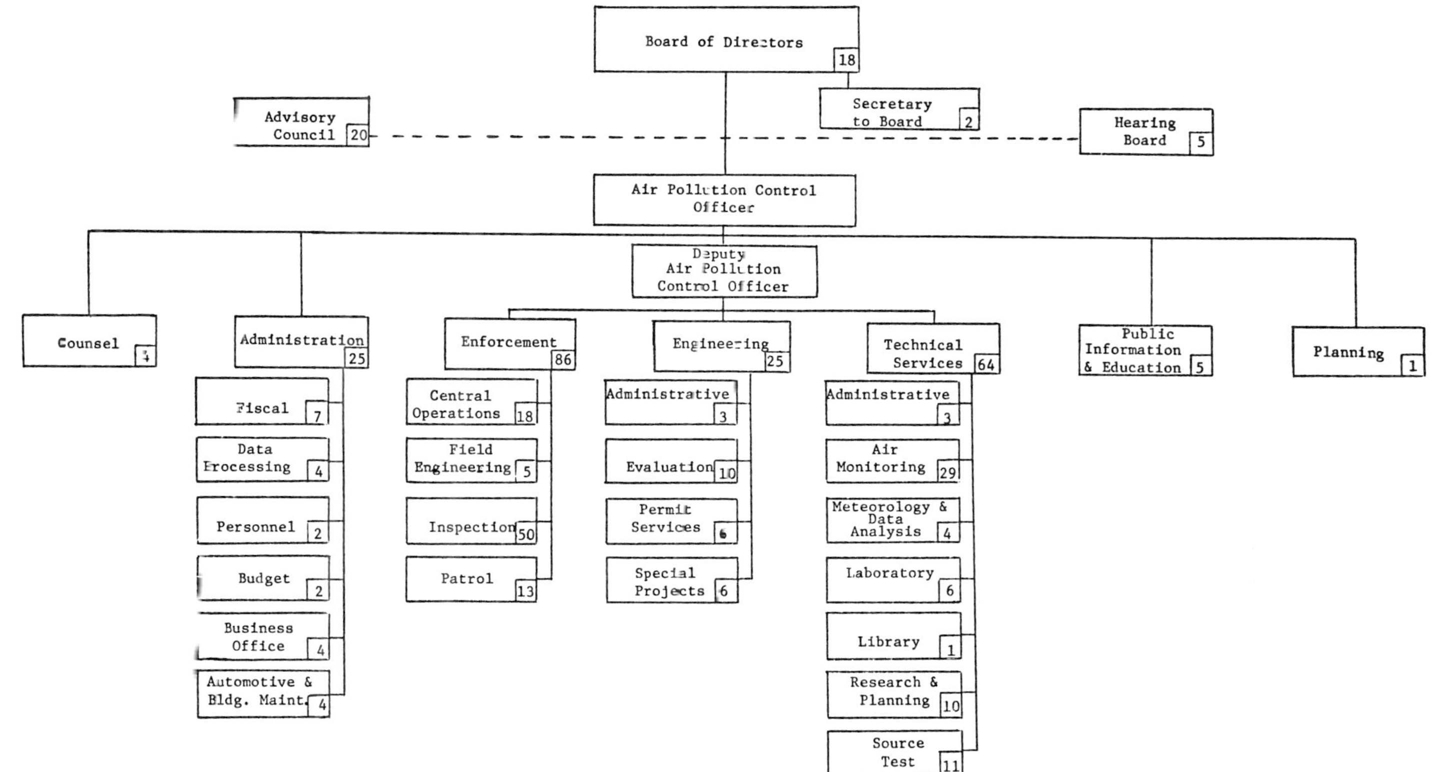

Figure 1. Bay Area Air Pollution Control District Organization Chart.

deputy air pollution control officer, program planning officer, and the director of public information.

The Director of Public Information has both administrative and staff responsibilities. His staff responsibilities to the air pollution control officer (APCO) are several:

a. To evaluate and comment on the district's programs and policies with respect to their effect on public opinion.

b. To prepare and/or edit speeches, statements and reports from the APCO to the board of directors and the public.

c. To represent the district as the need arises before public meetings and legislative hearings, and to act as spokesman to the news media when the air pollution control officer is not available.

The director of PI&E has administrative responsibility for the PI&E division. PI&E consists of two information officers—an "inside" man and an "outside" man—and two clerical employees. The inside man is primarily a writer, with the direct responsibility of producing district publications and handling general correspondence. The outside man works with community groups and schools and handles most of the speaking engagements. In terms of training and skills, both people are equipped to carry out any assignment, and their roles are interchangeable.

The BAAPCD uses most of the techniques and materials discussed earlier, but some are different enough to warrant review. Since much emphasis is placed on contact with the dozens of school districts in the Bay Area, a special "teacher's kit" has been assembled. This is comprised of the following:

a. The district's basic booklet, "Air Pollution and the San Francisco Bay Area," which is revised annually.

b. "Air Pollution Experiments for Junior and Senior High School Science Classes," a publication of the Air Pollution Control Association, 4400 Fifth Avenue, Pittsburgh, Pennsylvania 15213.

c. A folder, "What is Air Pollution?" published by the district.

d. A map of the district.

e. A list of district publications.

f. A list of community and governmental groups that provide information on other environmental concerns.

g. A bibliography of publications suitable for school use.

These materials are packaged in an attractive folder labeled "Air Pollution Information Kit."

The BAAPCD has four copies of its 21-minute, 16 mm sound color motion picture, "It's Your Air," available for loan or use in connection with speeches. Also available is a 60-panel film strip, with commentary in a cassette tape. The outside man often uses a 35 mm slide show—26 slides arranged in sequence in a Carousel type magazine for use in certain kinds of presentations.

Press packets, consisting of all the written material—reports, proposed regulations, and other memoranda—that are provided to the board of directors are prepared and distributed to the news media representatives who cover the twice-monthly BAAPCD Board meetings. Press relations are the particular concern of the director of public information, and he normally handles press inquiries.

"Smog-Phone" is the name given to two serial telephone lines, which respond when 673-SMOG (7664) is dialed. A message is recorded on two Dictaphone Ansafone 30-second tapes each morning to give the outlook for air pollution at each of thirteen locations in the Bay Area for that day. Late in the afternoon, the message is rerecorded with the air quality designation (clean air, heavy air pollution) for the same locations. Transmission of detailed air quality information is undertaken daily through a private wire service, Business Wire, which reaches the major Bay Area news media. The information is sent to Business Wire by means of a Xerox Telecopier 400. This machine uses the telephone to transmit written data. The author intends no endorsement of the products or equipment mentioned in this chapter. We are simply reporting on what is being used.

The clerical employees are a senior clerk–stenographer and a clerk–typist. A recital of the routine duties of these positions in a time frame will help provide a better understanding of how PI&E works.

Senior clerk–stenographer

a. Prepare internal and external correspondence, except for memos personally typed by the Information Officers—daily.

b. Answer telephones and either respond to caller or refer the call within PI&E or elsewhere—daily.

c. Prepare press releases—periodically.

d. Prepare and distribute weekly calendars for all division heads—weekly.

e. Send and receive material on the Telecopier—twice daily.

f. Prepare press packets for board meetings of the advisory council—two or three times monthly.

g. Prepare PI&E's monthly report to the board of directors—monthly.

h. Handle scheduling of district film.

i. Do purchase orders, handle accounting for PI&E, and be responsible for the files—daily.

Clerk–typist

a. Type final copy for the BAAPCD monthly newsletter *Air Currents*—monthly.
b. Type all copy for basic booklet revisions and special purpose publications—periodically.
c. Mail requested materials and keep related records—daily.
d. Keep track of supplies of printed materials—daily.
e. Keep mailing lists updated—twice monthly.
f. Route newspaper clippings of special interest to BAAPCD personnel—weekly.
g. Handle scheduling of the outside man and do related correspondence.
h. Prepare and record Smog-Phone message—twice daily.
i. Provide back-up to senior clerk–stenographer in answering phones—as needed.

Again, these duties are interchangeable, and each person must be able to handle all tasks.

The above tasks delineate the routine patterns that indicate the nature of the BAAPCD information program—provide a basic "response" service to the news media and interested citizens and engage in active solicitation of interest of schools and environmental groups. Although PI&E has used public service spot announcements from time to time, usually indicative of aggressive PR, these have usually been keyed to the introduction of new programs, e.g., the air pollution emergency episode plan.

There have been several new programs undertaken by the BAAPCD of sufficient importance to prompt the development of special purpose materials—the introduction of a three-stage emergency episode plan and the establishment of a vehicle patrol section. PI&E paid for and helped produce a folder entitled, "As a Person with a Breathing Problem, What Should You Do When They Issue a Smog Advisory?" which was imprinted as a publication of the California Lung Association and distributed by its local affiliates. The thinking was that physicians, who would be encouraged to hand the folders to their patients, would give greater credence to the Lung Association than to the BAAPCD.

The vehicle patrol was mandated by the California legislature, in spite of the fact that the BAAPCD had never been involved in controlling auto emissions. The district was required to issue citations to the owners of autos, trucks, and busses with exhaust pipes that emitted smoke of greater opacity than permitted by law (generally, Ringelmann, or 20% opacity). To do this, the district purchased a dozen cars, had them painted and equipped with emergency signals, and trained the personnel who would do the patrolling. PI&E developed a special folder, "Introducing . . . ." (one must turn the page to find out who or what is being introduced). Upon the inception of the program in late 1973, feature stories were done by all major Bay Area newspapers and television stations.

Many other special programs could be mentioned, as indeed other anecdotes and illustrations of PR activities could have been included in this chapter. My aim, however, has been to indicate how public information and education activities might be organized and carried out. If the reader is left with the idea that "there's more to this than I thought," then I have succeeded.

REFERENCES

1. S. M. Cutlip and A. H. Center, "Effective Public Relations," 4th ed., p. 10. Prentice-Hall, Englewood Cliffs, New Jersey, 1971.
2. United States Environmental Protection Agency, "Don't Leave it All to the Experts," Publ. No. 0-478-748. United States Government Printing Office, Washington, D.C., 1972.
3. I. Ross, "The Image Merchants," p. 110. Doubleday, Garden City, Long Island, New York, 1959.
4. R. Carpenter, Executive Director and General Counsel. League of California Cities, "The Role of the Legislative Advocate in State Legislatures" (presented to the Western Political Science Association, April 2–4, 1970), unpublished monograph.
5. "1977 Directory of Governmental Air Pollution Control Agencies," Air Pollution Control Association, in cooperation with the Office Of Air Quality Planning and Standards, United States Environmental Protection Agency, Pittsburgh, Pennsylvania, January 31, 1977.
6. V. N. Howard, Chief, Pollution Control Division, City of Houston, Texas (letter), February 8, 1973.

6

Air Pollution Surveys

August T. Rossano

I. Scope and Objectives of Air Pollution Surveys

An air pollution survey is a critical study of a specified geographical area for the principal purpose of determining the nature, sources, extent, and effects of the air pollution that either exists within the area or that exerts a significant influence upon it. Several different purposes may be served by an air pollution survey. Among these are evaluation of highly localized problems, urban and land use planning, development or improvement of a community air pollution control program, and research. In the context of this discussion, the following types of activities are not considered to fall within the scope of air pollution surveys: brief inspection visits by control officials to investigate complaints or determine the degree of compliance with local regulations; short-term technical investigations limited to the evaluation of combustion, manufacturing, and other polluting operations or performance testing of gas cleaning devices.

II. Survey Types

A. Local Source Problems

Studies of local source problems usually involve specific and identifiable emission sources. An example of this type of survey is a study of the air, water, vegetation, materials of construction, and possibly, animal life in the vicinity of a phosphate fertilizer plant, aluminum reduction plant, or steel mill to determine the extent to which particulate and gaseous pollutants from such a source affect the local ecology. A study of the concentrations of sulfur dioxide and particulates from an isolated coal-burning electric power plant or metal refinery may be conducted to obtain data on the range of influence of such a source or the efficiency of its air pollution control measures (*1, 2*).

B. Site Selection Surveys

A practice that unfortunately is only infrequently followed is the air pollution survey of an area prior to the selection of a site for the construction of a new industrial plant. The proper use of land in an era of expanding technological and population growth requires imaginative and thoughtful community planning and zoning. While such a practice cannot guarantee freedom from future air pollution difficulties, it does provide an opportunity to prevent the unnecessary creation of obviously undesirable situations, and possibly, litigation. In cases where the site has already been selected, a study of environmental conditions existing at the site prior to start-up is highly desirable. This would provide benchmark data that could be used to protect the company's interest from a legal and regulatory standpoint.

C. Environmental Impact Statements

The National Environmental Policy Act of 1969 (Public Law 91-190) requires that all agencies of the federal government include, for each proposed action significantly affecting the quality of the human environment, a detailed statement on the environmental impact, any adverse environmental effects that cannot be avoided, alternatives to the proposed action, the relationship between local short-term uses of man's environment and the maintenance and enhancement of long-term productivity, and any irreversible and irretrievable commitments of resources that would be involved. The review process includes examination and analysis by the appropriate federal, state, and local agencies, as well as by the general public (*3*).

Some of the regional offices of the United States Environmental Protection Agency have prepared guidelines for the preparation of environmental impact statements (*4*). Many states have likewise passed state environmental protection acts requiring the preparation of environmental impact statements for activities that may significantly affect the environment (*5*).

D. Air Pollution Appraisals

An air pollution appraisal is a form of survey that is based on analyses of published and otherwise readily available pertinent technical data and information. The three principal objectives of such appraisals are to determine qualitatively the current status of air pollution in a community or group of communities, to estimate future potential for air pollution intensification, and to offer broad recommendations for prevention or abatement.

The survey findings and recommendations are generally based on data relative to population and population trends, industrial activity, fuel consumption for space heating and power, automotive usage, waste disposal practices, agricultural activity, and patterns of anticipated growth, as well as local and regional topography and climatology. Within recent years, several such appraisals have been made on a community and statewide basis by federal, state, local, and private organizations. Table I lists these as well as other types of air pollution investigations of limited scope, indicating their location, date, scope, and responsible organizations.

In general, community-wide appraisals require one to two man-weeks of field observations with about three or four man-months of either prior or subsequent data collection, analysis, and interpretation. Areas

Table I Selected List of Air Pollution Appraisals and Investigations of Limited Scope

Designation	*Date*	*Responsible organization*	*Scope*	*References*
Air pollution in Poza Rica, Mexico	1950	U.S. Bureau of Mines and U.S. Public Health Service	Investigation following the Poza Rica incident. Review of symptoms, pathological findings, alleged source, and presumed cause.	*14*
Air pollution in London, England	1952	Committee of Departmental Officers and Expert Advisers, Minister of Health	Investigation into the causes of the high mortality and morbidity in London during a 2-week period in December, 1952. Information collected includes meteorological and air pollution factors and medical statistics. Previous air pollution disasters are reviewed.	*15, 16*
Dust deposition in the vicinity of a power station	1952–1954	Central Electricity Authority (Great Britain)	Measurement of dustfall rates in different directions from Little Barford Power Station under various meteorological and plant operating conditions.	*1*
Air pollution in New South Wales	1953	New South Wales Department of Public Health, Australia	Measurement of dustfall by means of station networks in Sydney, Newcastle, and Port Kembla.	*22*
Air pollution in the vicinity of Portland, Oregon	1956	Stanford Research Institute (Menlo Park, California)	The air pollution potential of the Portland area as determined from calculated estimates of emissions under specified meteorological conditions.	*23*
Air pollution in St. Bernard, Ohio	1956	U.S. Public Health Service in cooperation with the Cincinnati Bureau of Air Pollution Control and Heating Inspection	One-month concentrated study of sulfur dioxide, oxides of nitrogen, hydrogen sulfide, suspended particulates, and selected weather factors at one station in a suburb of Cincinnati, Ohio. Results explained in terms of wind direction and speed.	*24*

Atmospheric pollution in Milan, Italy	1956	Milan University Institute of Hygiene with the co-operation of the Health Office of the City Council	Measurements of sulfur dioxide, carbon monoxide, carbon dioxide, hydrogen sulfide, chlorine, fluorine, ammonia, oxides of nitrogen, and suspended particulates with respect to location, season, topography, and other factors.	*25*
An aerometric survey of the city of Honolulu, Hawaii	1956–1957	Truesdail Laboratories, Inc. (Los Angeles, California) at the request of the Board of Health of the territory of Hawaii	Measurement of aldehydes, carbon monoxide, hydrocarbons, oxidants, oxides of nitrogen, surfur dioxide, and particulate matter during 2-week periods of differing weather conditions in September and January. Correlation of pollution with weather, time of day, and location.	*26*
State-wide air pollution appraisals:				
Washington	1956	U.S. Public Health Service in cooperation with the individual state departments of public health and air pollution control boards	Evaluation of the current status and future outlook of air pollution based largely on existing data on such factors as population, fuel usage, industrial activity, waste disposal practices, topography, and meteorology.	*27–41*
Illinois	1956			
Connecticut	1956			
Tennessee	1957			
Texas	1958			
New York	1958			
North Carolina	1959			
Pennsylvania	1959			
Minnesota	1961			
Florida	1961			
Georgia	1962			
South Dakota	1962			
Kansas	1963			
Montana	1963			

Table I (Continued)

Designation	*Date*	*Responsible organization*	*Scope*	*References*
Utah	1962	Utah Legislative Council	Collection and analysis of data on the general character of pollution in the state. Recommendations for a control program are offered to the state legislature.	*42*
Maryland	1962	Maryland State Department of Health	Air quality data on suspended particulates, organic content, beta radioactivity, soiling, and dustfall at Rockville, Maryland.	*43*
Community-wide air pollution appraisals:				
Steubenville, Ohio	1956	U.S. Public Health Service in cooperation with local public health or air pollution agency	Same as for state-wide air pollution appraisals.	*44–54*
Portland, Oregon	1956			
Denver, Colorado	1957			
Birmingham, Alabama	1958			
Elmira, New York	1958			
Charleston, South Carolina	1959			
Hamilton, Ohio	1960			
Lynchburgh, Virginia	1961			
Washington, D.C.	1962			
Chattanooga, Tennessee	1964			
Portland, Oregon	1963	Oregon State Sanitary Authority, Portland Health Bureau, and Associated Oregon Industries		
Bistate study	1957–1959	Illinois Department of Public Health, Illinois.	A study of the sources, nature and extent of air pollution in the Chicago,	*55*

		Indiana State Board of Health, and Purdue University	Illinois, metropolitan area. This study was preceded by an Illinois state study in 1956 by the Illinois Department of Public Health.	
Short-term, intensive studies:				
Fresno, California	1960	U.S. Public Health Service in cooperation with local health and air pollution agencies	Concentrated air quality studies of limited duration at a few selected locations.	*17, 56–65*
Denver, Colorado	1961			
Birmingham, Alabama	1962			
Pittsburgh, Pennsylvania	1963			
Winston-Salem, North Carolina	1963			
Baltimore, Maryland	1964			
Indianapolis, Indiana	1964			
Duquesne, Pennsylvania	1964			
Nassau County, Long Island, New York	1965			
Jacksonville, Florida	1963			
Denver, Colorado	1962	Colorado State University	Measurement of total particulates, soiling, solar radiation, wind, and other weather variables.	*65*
Air pollution in El Paso, Texas	1957–1959	El Paso City–County Health Unit	A 2-year study to obtain data on type, extent, source, and effects of air pollution in the El Paso, Texas, area.	*66*
Air pollution in Phoenix, Arizona	1958	Arizona State Department of Health with the assistance of the U.S. Public Health Service	Diurnal sampling for seven consecutive days at one location for oxidants, nitrogen dioxide, sulfur dioxide, carbon monoxide, and suspended particulates. These data are related to visibility and wind speed.	*67*

Table I (*Continued*)

Designation	*Date*	*Responsible organization*	*Scope*	*References*
Pollutants in Mexico City, Mexico	1959	Laboratory of Industrial Hygiene, Department of Health, Mexico	Measurements at two stations for suspended particulates, sulfur dioxide, and oxidants. Correlation of pollutant data with meteorology, topography, and industrial activity and time of day.	*68*
Air pollution in Tucson, Arizona	1959	Arizona State Department of Health with the assistance of the U.S. Public Health Service	Essentially similar to the Phoenix, Arizona, study.	*69*
Air pollution in Roseburg, Oregon	1960	Oregon State Sanitary Authority	A 6-month study using nine stations gathering data on particle fallout and suspended particulates within or near Roseburg, Oregon.	*70*
Air pollution in Medford, Oregon	1960	Oregon State Sanitary Authority	A 6-month study using a total of ten stations gathering data on dustfall, suspended particulates, oxides of nitrogen, carbon monoxide, sulfur dioxide, and meteorological variables within or near Medford, Oregon.	*71*
The Eureka–Arcata, California, ambient air quality study	1961	California Department of Public Health	Study of nuisance and damage from smoke, cinder fallout, sawdust, and gaseous and particulate pollutants from wood waste incinerators under varying meteorological conditions.	*72*
Air pollution aspects of orchard heating in the Yakima, Washington, area	1961	Washington State Department of Health	Total suspended particulate and soiling data correlated with weather data to determine extent of air pol-	*73*

			lution from orchard heating during period April 1 to June 1.	
Abatement activity (6 reports)	1967–1971	U.S. Public Health Service in cooperation with state and local air pollution control agencies	Investigations conducted in selected areas to assist governmental agencies in their consideration of occurrence of air pollution, adequacy of abatement measures, nature of delay in abating the pollution, and necessary remedial action.	74
Report for consultation on air quality control region (51 reports)	1968–1970	United States Department of Health, Education, and Welfare	Analysis of jurisdictional considerations for selecting suitable air quality regions based on social, economic, and demographic data. Emission inventories and diffusion modeling were utilized to construct maps of current levels of carbon monoxide, sulfur dioxide, and particulate matter. Predictions to the year 2000.	75
Transportation controls to reduce motor vehicle emissions (19 reports)	1972–1973	U.S. Environmental Protection Agency	Analyses of the nature and magnitude of the air pollution problem attributed to motor vehicles, and strategies to attain and maintain national ambient air standards.	76
Real property owned by the federal government (51 reports)	1969–1971	U.S. Environmental Protection Agency and predecessor agency	A summary compilation of the major facilities of real property owned by the U.S. Government that may present air pollution potential. Listings include location, brief description of the installation, operating agency and total floor area.	77

Table I (*Continued*)

Designation	*Date*	*Responsible organization*	*Scope*	*References*
Air pollution reports of federal facilities (33 reports)	1968–1971	U.S. Environmental Protection Agency and predecessor agencies	A listing of federal facilities by state, county, and city by type and amount of fuel used, contribution to air pollution of local region, and the status with respect to implementation of abatement plans.	*78*
Air pollution emission inventory (77 reports)	1968–1974	Environmental Protection Agency	A rapid emission inventory of selected locations using a grid coordinate system. Emission qualities of particulate matter, sulfur dioxide, and carbon monoxide reported in terms of tons per grid unit on an average annual day, and average summer and winter day (Vol. III, Chap. 17)	

as large as a state require about nine man-months of field work and a similar amount of time for questionnaire and data analysis.

E. Short-Term Limited Investigations

Another type of survey, which is less demanding in terms of manpower and time but involves a considerable amount of air sampling and analysis, is the intensive small-scale investigation limited in geographical scope and time. This type seeks to demonstrate the general patterns and levels of pollution by sampling the air continuously for selected gases and solids at a few locations in a community for a period of one to several weeks. Such surveys provide order-of-magnitude figures of levels and fluctuations of pollutant concentrations and serve as a useful device for focusing the attention of the community on its growing problem. This type of survey can frequently be used to advantage in training or orienting new personnel.

F. Research Studies

Though impressive progress has been made in recent years, there still is a considerable lack of understanding of the fundamental mechanisms and processes involved in the formation, intensification, and attenuation of air pollution and its undesirable effects. Past community air pollution studies have subsequently increased our knowledge of the basic concepts of air pollution even though their prime objective was to develop control measures. The Nashville, Tennessee, Community Air Pollution Study provides a rare example of a survey conducted principally in the interest of research (*6*). More such research-oriented air pollution surveys are needed to provide answers and explanations for the many questions inherent in the air pollution phenomenon (*7–10*).

Investigations of the air pollution disasters that occurred in Liege, Belgium (*11, 12*); Donora, Pennsylvania (*13*); Poza Rica, Mexico (*14*); and London, England (*15, 16*) represent special research investigations performed immediately after the episodes in an effort to uncover fundamental facts concerning the nature, extent, and causative factors involved. Other examples of area-wide studies of a research nature are the study of air pollution in the Central Valley of California in 1959 to determine the nature and contribution of pollution from agricultural burning (*17*); the comprehensive health survey in New York, New York, to ascertain the effects of urban pollution on the health of a large group of selected families (*18*); the CHESS studies conducted by the United States Environmental Protection Agency (*19*); the study of the physicochemical characteristics of St. Louis, Missouri, aerosols conducted in 1974 (*20*); and the intensive investigation of the chemical and physical parameters of aerosols in Pasadena, California (*21*).

G. Technical Investigations under the United States Clean Air Act

The United States Clean Air Act of 1963 (Public Law 88-206) empowered the federal government to intervene in interstate air pollution problems that could not be solved by the states involved. Specifically, the Secretary of the Department of Health, Education, and Welfare was authorized under specified conditions to initiate procedures to secure abatement of the pollution. One of the initial steps was the convening, by the secretary, of a public conference the purpose of which was to provide information needed for developing rational recommendations for abatement in accordance with the provisions of the act. At such conferences, data were presented from various sources of information including air monitoring programs and surveys that had been conducted previously. In addition, the United States Public Health Service and others introduced the results of their own studies conducted expressly for the abatement action in question.

Several such technical investigations were made since these provisions of the Clean Air Act went into effect in 1963 (*74*). In general, these intensive studies were carried out over a period of several months and included such aspects as meteorology, air quality, inventory of emissions, interstate travel of pollutants, adverse effects on the community, and progress toward finding a solution to the problem. The results of such investigations led to conclusions that formed the basis upon which the United States Public Health Service presented its recommendations for abatement.

The United States Clean Air Act 1967 amendments (Public Law 90-148) authorized the secretary, under certain circumstances, to intervene in both interstate and intrastate air pollution problems. The United States Clean Air Acts of 1970 and 1977 likewise authorize the administrator of the United States Environmental Protection Agency to intervene in such cases.

Fifty consultations on air quality were performed by the National Air Pollution Control Administration between 1968 and 1970 for the purpose of designating the boundaries of air quality control regions necessary to provide adequate implementation of air quality standards (*75*). Air pollution surveys conducted by the federal government between 1968 and 1973 include eighteen transportation control, transportation strategy, and vehicle emission surveys (*76*). During 1968 and 1971, two series of studies were conducted at government installations. These are described in fifty-one publications entitled "Real Property Owned by the Federal Government" (*77*) (one for each state and the District of Columbia) and thirty-three survey reports of such property in major metropolitan areas (*78*). The seventy-seven air pollution emission inventory reports published between 1966 and 1974 are referenced in the chapter on emission inventory (Volume III, Chapter 17).

III. The Comprehensive Air Pollution Survey

A. Air Resource Management Concept

The ultimate objective of an area-wide air pollution study is generally the abatement of the pollution. Theoretical solutions to the problem may appear simple—eliminate or control all significant pollution sources. However, the practical application of such a solution is most difficult, since the economic cost of radical changes in fuel usage, energy conversion systems, waste disposal methods, and modes of transportation may be prohibitive. The atmosphere must continue to serve as a receiving body for the many tons of pollutants that must be disposed of daily in each community. The fact that this same body of air is also the sole source of the air needed for the basic biological processes of living, ventilation, preparation of food, and other essential functions, presents a dilemma. Continued indiscriminate emission of pollutants into the atmosphere can in time render it unfit for these critical purposes.

What is needed, therefore, is a rational basis for determining the degree of constraints necessary to achieve satisfactory air quality. This involves an estimate of the maximum degree of pollutant discharge consistent with the natural self-purification capacity of the air resource, air quality standards, and regulatory provisions intended to prevent significant air quality deterioration. Obviously, some form of compromise must be reached in which the aforementioned dual role of the atmosphere is optimized. The desirable end product of the area-wide study should be the development of an air resource management plan. Such a plan requires a high degree of understanding of the fundamental mechanisms involved in the phenomenon of air pollution.

The air resource management concept includes a clear definition of the air pollution problem and a technique for determining the optimum type and degree of control to achieve the desired air quality. The task of defining the problem can be facilitated by the use of the systems approach. The flow diagram shown in Figure 1 indicates the essential fac-

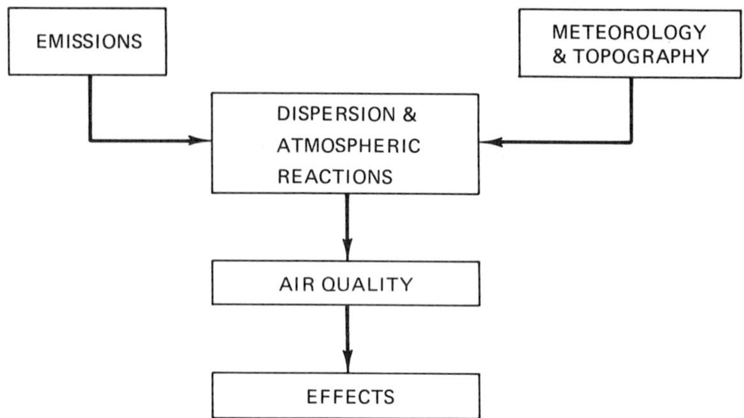

Figure 1. Systems analysis approach to the problem of air pollution.

tors in such a system and their interrelationships (*107*). In effect, the diagram illustrates the dynamics of air pollution.

It can be seen that the undesirable effects of air pollution result from a deterioration of air quality. In turn, the physical and chemical composition of air is affected by the interplay of the source factor and prevailing meteorological conditions and, to a lesser extent, topography. As an example, on a given day the sulfur dioxide emitted from a fuel-burning source may be efficiently diluted and transported from the source by favorable meteorological conditions with no observable ill effects. On the next day, a strong temperature inversion coupled with weak winds may cause the sulfur dioxide concentrations in the vicinity of the source to increase sharply, and adverse effects, such as vegetation damage, may result.

Thus, the systems approach provides a basis for understanding the role and significance of the principal factors involved in atmospheric pollution, and the manner in which they interact. This type of systems analysis provides a logical approach to the design of area-wide studies.

B. Objectives

From the above discussion, it follows that to arrive at a clear understanding of the status of pollution in a designated area it is necessary to obtain as complete as possible information on areal emission density of various pollutants, meteorolgical and topographical factors, air quality patterns, and adverse impacts on receptors. A community-wide study may cover an entire city, a metropolitan area consisting of several cities, or an entire region that may be conterminous with the airshed. The study may depend on an extensive network of air sampling and meteorological stations operated on a frequent, if not continuous, schedule supplemented by mobile air-monitoring facilities. In addition, the survey may include a determination of the principal pollutant sources, with a detailed estimate of their emission rates.

Other pertinent considerations include climatological and topographical factors, as well as an evaluation of effects of pollution on the human. Such surveys may extend over a period of one or more years and require the services of many professional disciplines, including physics, chemistry, mathematics, biology, and applied sciences, including engineering and the social sciences. Table II lists major community-wide air pollution surveys.

Stripped down to its essentials, a community air pollution survey must provide reasonable answers to the proverbial questions of where, what, why, which, and how. For example:

a. Where do the pollutants come from?
b. What are the pollution levels with respect to time and place?
c. Why do these levels vary?
d. Which undesirable environmental effects are produced?
e. How can the problem be solved?

The details of planning, design, and execution of surveys vary from study to study, since each must be tailored to the specific needs, requirements, and resources of the area involved. Nevertheless, there are some considerations that are common to all community air pollution surveys—planning and development, implementation, and evaluation of findings and report.

C. Planning and Development

This phase involves reconnaissance of the situation to size up the magnitude and nature of the problem, as well as the interests and resources of the community (*108*). The individual or organization selected to make the survey then reaches a general agreement with official representatives of the community or the sponsoring group on the purpose and scope of the proposed study, cost and duration, type of final report, and administrative or policy matters. Where several participating groups are involved it is advisable to clearly define in writing these terms as well as conditions of participation, responsibility, and authority. Since community studies are complex and expensive operations subject to public opinions and pressures, it is recommended that a clear written understanding

Table II Selected List of Major Air Pollution Surveys

Designation	*Date*	*Responsible organization*	*Scope*	*References*
Chicago, Illinois, smoke abatement survey	1911–1915	City of Chicago, Illinois	Comprehensive study of the causes and effects of air pollution in Chicago with emphasis on coal smoke, particularly from railroads. Smoke abatement recommendations were made.	*79*
Selby smelter study	1913–1914	U.S. Bureau of Mines	Fifteen-month investigation by a special commission of the dust and gaseous emissions from a large lead smelter in Solano County, California. Study included stack and atmospheric sampling; vegetation and soil analyses; effects on humans, livestock, and plants; and control recommendations.	*2*
Smoke abatement in Grafton, West Virginia	1924	U.S. Bureau of Mines	Smoke, airborne solids, sootfall, etc. in a coal-burning railroad center.	*80*
Smoke abatement in Salt Lake City, Utah	1926	U.S. Bureau of Mines	Study of coal consumption and air quality in terms of airborne solids; sootfall; oxides of sulfur, nitrogen, and carbon; chlorine, sulfuric acid, and ammonia. Data are related to meteorology, location, and time of year. Recommendations were offered.	*81*
Air Pollution in the Meuse Valley, Belgium	1930	Committee of experts appointed by the King's attorney at Liège, Belgium	An investigation following the Meuse Valley air pollution disaster cf December, 1930. A compilation of medical observations and opinions, meteorological information, data on chemical composition of dustfall, and estimates of sulfur dioxide and other emissions as potential causes.	*11, 12*
New York City, New York air pollution survey	1935–1937	Department of Health, New York City	Relation of fuel usage to air pollution, sootfall, atmospheric dust counts, and airborne bacteria measurements.	*82*

Leicester air pollution study	1937–1939	Atmospheric Research Committee of Department of Scientific and Industrial Research (Great Britain)	Study of the distribution, dispersal, and character of atmospheric pollution. Measurements of suspended and deposited dust, sulfur dioxide, ultraviolet light, and meteorological factors.	*83*
Trail smelter study (British Columbia)	1938–1940	Arbitral tribunal of the governments of Canada and the United States	Source study of smelter. Comprehensive program of air sampling for sulfur dioxide, meteorological investigation, and effects of sulfur dioxide on vegetation.	*84*
Air pollution in Donora, Pennsylvania	1948–1949	U.S. Public Health Service	A 5-month intensive investigation following the Donora crisis. Included medical interviews and examination; epidemiological studies; air sampling and analysis of gases, particulates, source evaluations; and meteorological investigations.	*13*
Air pollution in the Kanawha Valley West Virginia	1950–1951	West Virginia Department of Public Health assisted by the Kettering Laboratory (Cincinnati, Ohio) and the U.S. Public Health Service	A 19-month study of air pollution sources and airborne concentrations of oxides of nitrogen and sulfur, aldehydes, fluorides, ammonia, hydrogen sulfide, chlorine, particulates, as well as sootfall. Sampling done at a total of 28 locations by mobile units and fixed stations.	*85*
Air pollution in the Detroit, Michigan–Windsor, Ontario, area	1950–1960	Technical Advisory Board on Air Pollution of the International Joint Commission of Canada and the United States	A long-term study of the sources, concentrations, and effects of air pollution in the Detroit area, with recommendations for a legal control program on a local and international level.	*86*
Air pollution in Seattle, Washington	1952	University of Washington	Measurement of smoke and suspended particulates, dustfall, and gases. An extensive photographic documentation of emission sources and effects.	*87*

Table II (Continued)

Designation	Date	Responsible organization	Scope	References
Air pollution in the vicinity of Louisville, Kentucky	1952–1953	Battelle Memorial Institute (Columbus, Ohio) under the sponsorship of the Rubbertown Industrial Group	Measurement of particulate and gaseous pollutants by means of a mobile laboratory over a period of about $1\frac{1}{2}$ years. Estimates of industrial sources.	*88*
The Sarnia, Ontario, survey	1953–1954	Ontario Research Foundation with the support of six major industrial firms	Measurement of dustfall, sulfur dioxide, and airborne particulates at several fixed and one mobile sampling stations. Correlation with meteorological variables.	*89*
Los Angeles Basin, California, aerometric survey	1954	Air Pollution Foundation (San Marino, California) in cooperation with the Los Angeles County Air Pollution Control District	Continuous sampling and analysis for eight pollutants at ten stations, August through November. Estimation of degree and frequency of eye irritation and vegetation damage. Meteorological measurements. Dispersion studies aloft using a blimp.	*90*
Atmospheric pollution in Pretoria, South Africa	1955	National Physical Research Laboratory, Council for Scientific and Industrial Research, Union of South Africa	A 2-year study to determine degree of pollution in the city. Smoke, sulfur dioxide, and dustfall were measured at six fixed stations. Results are related to fuel usage and season and compared with data from Chattanooga, Tennessee.	*91*
The special air pollution study of Louisville and Jefferson County, Kentucky	1956–1958	U.S. Public Health Service in cooperation with several other federal agencies as well as state and local health and air pollution agencies	A 2-year comprehensive study, which included detailed source studies, emission inventory, and continuous air sampling at six network stations and mobile samplers. Measurements include airborne particulates, soiling, sulfur dioxide, hydrocar-	*92, 93*

			bons, auto exhaust, and several other gases; odor and human response; material deterioration; airborne bacteria; climatology; and meteorological variables at nine stations. Recommendations for abatement are offered.	
The Houston, Texas, air pollution surveys (1957 and 1966)	1956–1958 1964–1966	Southwest Research institute (San Antonio, Texas) under the sponsorship of the Houston Chamber of Commerce	Air sampling for sulfur compounds, oxidants, carbon monoxide, particulates, and other selected pollutants, and vegetation survey by means of mobile laboratory. Study of meteorological factors.	*94, 95*
New York–New Jersey interstate air pollution survey	1958	Interstate Sanitation Commission (New York, New Jersey, and Connecticut) in cooperation with the U.S. Public Health Service	Summary of the history as well as findings of earlier surveys of the area. Limited gas sampling. Tracer studies of interstate air movement. Appraisal of damage to vegetation, property, and health. Meteorological, climatological, and visibility studies. Review of the legislative aspects.	*96*
Nashville, Tennessee, air pollution study	1958–1959	Vanderbilt University for the medical phase and Robert A. Taft Sanitary Engineering Center, U.S. Public Health Service, for the engineering phase	A joint one-year medical-engineering study of air pollution in a typical coal-burning midwest community. Study of relationship between air pollution and public health, with emphasis on combustion products of coal. Detailed morbidity, mortality, clinical, and autopsy studies. Extensive aerometric, meteorological, and source investigation.	*6*

Table II (Continued)

Designation	*Date*	*Responsible organization*	*Scope*	*References*
Evaluation of air pollution in Spokane, Washington	1961	Washington State Department Health in cooperation with Spokane City Health Division and Spokane County Department of Health	Development and evaluation of data on population, topography, meteorology, past and current air sampling, and source factors.	*97*
In quest of clean air for Berlin, New Hampshire	1962	U.S. Public Health Service in cooperation with New Hampshire State Health Department and Berlin City Health Department	Measurement of sulfation, dustfall, weather factors, emissions and adverse effects.	*98*
Study of air pollution in the interstate region of Lewiston, Idaho, and Clarkston, Washington	1964	U.S. Public Health Service in cooperation with state and local air pollution authorities	Emission inventory, measurements of hydrogen sulfide, sulfur dioxide, suspended particulates, sulfate, and sodium, visibility, odor. Correlations with meteorological variables.	*99*
The Gothenburg air pollution study, Gothenburg, Sweden	1964	Gothenburg Air Pollution Committee (LUG)	Measurements of dustfall, sulfur dioxide, smoke, total suspended particulates, sulfate, chloride, and calcium. Correlation of air quality with meteorology. Pilot medical study.	*100*
St. Louis, Missouri, dispersion study	1963–1965	U.S. Public Health Service	A series of forty-three experiments of atmospheric transport and diffusion over an urban area, using airborne tracers.	*101*
Kanawha Valley, West Virginia, air pollution study	1964–1968	U.S. Department of Health, Education, and Welfare	Air pollution study including emission inventory, measurements of meteorologic parameters, air quality, odors, vegetation	*102*

			and materials damage, health effects, public attitudes, and an analysis of economics of pollution control measures.	
Air quality in Clark County, Washington	1965	Washington State Department of Health	Measurements of suspended particulates, dustfall, soiling, fluorides, and hydrogen sulfide. Meteorological observations and source inventory.	*103*
St. Louis, Missouri, and East St. Louis, Illinois, interstate air pollution study	1966	U.S. Public Health Service in cooperation with local air pollution authorities	Comprehensive study of emission sources meteorology, air quality, effects, and public opinion for the purpose of establishing an air resource management program.	*104*
Chattanooga, Tennessee–Rossville, Georgia, interstate air quality study	1967 and 1968	U.S. Department of Health, Education, and Welfare	Air pollution study in the vicinity of an explosives plant near Tyner, Tennessee. Included air quality and meteorological measurements, vegetative and materials effects, and emission inventory.	*105*
Los Angeles, California, smog project	1969	Collaborative effort involving three university and two state agencies	An intensive 4-week investigation of the physical and chemical characteristics of aerosols in Pasadena, California.	*21*
Joint Air Pollution Study of St. Clair, Ontario, and Detroit River, Michigan, Areas	1971	International Joint Commission of Canada and U.S.	Study of air pollution in the general vicinity of Port Huron–Sarnia and Detroit–Windsor. Pollutants included particulate matter, sulfur dioxide, hydrocarbons.	*106*
CHESS studies	1973	U.S. Environmental Protection Agency	Evaluation of existing environmental standards, collection of health data, and assessment of health benefits of air pollution control in six cities.	*19*
St. Louis, Missouri, regional air pollution study	1974	U.S. Environmental Protection Agency	Development of air quality diffusion models for use in air pollution control strategies.	*20*

among interested parties be achieved at the earliest possible time. Insofar as it is feasible, the general public should be kept informed initially and throughout the study.

The next step in this preliminary phase is the systematic and logical development of a detailed plan of study, including a budget, which specifies the broad objectives; the individual projects; the organization including the classification and grades of professional and subprofessional personnel; and the requirements for laboratory and office space, scientific and technical equipment and supplies, field facilities, and transportation.

D. Implementation

The actual execution of the various field and laboratory operations and projects that relate to the collection of information and the performance of the many technical measurements and observations represent the implementation phase. In general, the survey consists of the following eleven activities with appropriate modifications and emphasis to suit individual needs.

1. Emission Inventory

Knowledge of the types and rates of emissions is fundamental to any study of community air pollution (*107, 109*). There are several ways in which source information can be extremely useful. It can be of help in designing the air sampling and air analysis programs and interpreting the resulting air quality data. It provides an explanation for observable adverse effects and indicates the relative contribution of the various major pollution sources. In the establishment of a new abatement program, source data can serve as a basis for the adoption of ordinances and regulations, while in existing control programs, current emission data provide guidelines for modifications in engineering and legal control measures.

From the research standpoint, detailed emission data are necessary for constructing mathematical models of the atmosphere for purposes of predicting future concentrations of specific pollutants. The establishment of emission standards and planning and zoning regulations is greatly enhanced by the availability of current and predicted source information.

It must be stressed that, while knowledge of emission sources is fundamental to an understanding of the nature, causes, and effects of air pollution in a community, air pollution source information per se is definitely limited in its value. In other words, data on quantities of a pollutant emitted in one area are not necessarily directly applicable to the situation in another area. For instance, a given emission rate of a gas may be responsible for a serious problem in a specific location. The same emission rate applied to another area with differing meteorological, topographic, social, and economic conditions may produce only a barely perceptible effect.

A recent report of guidelines for the preparation of an air pollution source inventory describes two procedures for preparing a source or emission inventory. It includes a rapid survey method and a detailed source inventory (*110*). In the rapid survey method, primary information is collected in four general source categories—stationary combustion, transportation, refuse incineration, and industrial processes. The study area is divided into geographic subdivisions. Calculations of emission rates are performed through use of fuel combustion emission factors and collection efficiencies. The detailed source inventory includes determination of actual emissions from specific point sources and area-wide sources in the study area. This procedure may involve questionnaires to each establishment followed up by field visits (*111*).

While the inventory covers the entire region under study, attention must be also paid to sources and emissions that are not within the area but that may contribute significantly to the air quality in the study area. Though the principal effort is directed at manmade sources, the contribution of relevant natural sources must also be taken into account.

It is to be expected that the emission inventory will not produce highly precise information. Limitations in the state-of-the-art knowledge introduce uncertainties in a number of factors, such as reported fuel and solvent usage figures, fuel composition, process information, and emission factors. In spite of these limitations, the emission inventory can yield results with adequate accuracy for the purposes of the community air pollution survey.

2. Data Acquisition and Handling

A major consideration in an air pollution survey is the acquisition of adequate and reliable data. Air quality monitoring should be planned to acquire data at the frequency and in the form required by the applicable reviewing agency. In some cases, this will entail the use of SAROAD format, the Storage and Retrieval of Aerometric Data system developed and used by the United States Environmental Protection Agency. It is primarily a data bank that permits rapid retrieval of air quality information collected throughout the United States (*112*). A review of basic data acquisition, analysis, and utilization is in Chapter 12, Vol. III.

3. Meteorological and Topographic Studies

A thorough study of local climatology, meteorology, and topography is essential for defining the air flow patterns, which in large measure control the rate of intensification and attenuation of air pollution in the locality. The practicality of a new technique for on-site measurement of meteorological parameters is currently being investigated at the University of Washington. By use of acoustical radar, it is feasible to determine vertical temperature profile, inversion height, and wind shear (*113*). Climatological information is a necessary requirement for the proper design of air sampling and observation networks used in area-wide studies. Forecasts are needed for certain field activities such as intensive air sampling during specified weather conditions.

Meteorological parameters provide a rational basis upon which to select from a large number of collected air samples those to be given detailed analytical treatment. Perhaps the most important use of meteorology in an air pollution survey is in correlating weather factors with air sampling and allied data to permit reasonable interpretation of air quality measurements in terms of source and atmospheric variables. Tracer studies in which fluorescent, radioactive, or other type of tracer material is released into the air at air pollutant emission points provide excellent means for studying the behavior of pollutants under specified weather and source conditions (*92, 114*).

4. Air Sampling and Analysis

A thorough and accurate record of air quality is of fundamental importance in a survey. This should include information on the variations and fluctuations in pollutant concentrations with respect to time and space. Time considerations include diurnal, seasonal, and long-range trends and patterns, while spatial factors involve comparisons of pollutant concentrations on the basis of their horizontal and vertical distribution. In large studies, it is desirable to use an air sampling network in and around the area of interest. The design of such a network will depend on available resources; however, its configuration should take into account the sources, receptors, and local climatology.

A recent report of principles and procedures for the design and operation of air quality surveillance systems describes basic design considerations, an outline of a systems approach to air quality surveillance, and requirements for operation, performance, and training. In addition, methodologies for air quality monitoring and data acquisition, handling, and analysis are presented (*115*). The network of sampling stations employed in the Aerometric Survey of Los Angeles, California is shown in Figure 2 (*90*). It does not conform to any specific geometric pattern, the sites having been selected on the basis of typical surface wind trajectories for the season during which the study was conducted.

The Louisville, Kentucky, air pollution study networks are shown in Figure 3 (*92*). Unlike the Los Angeles study, the air sampling network was laid out according to a pattern consisting of two concentric circles around the presumed major sources. The selection of location of individual stations in this network was determined on the basis of topography and prevailing surface wind movements, as well as sources and receptors.

The aerometric network employed on the Nashville, Tennessee, study is shown in Figure 4 (*6*). This network possessed a high degree of symmetry and contained a much greater number of stations than the two previously discussed networks. Basically, it represents a triangulation pattern superimposed upon a circular grid system having its geometric center at the center of the city. In it, a total of 119 stations were situated at apexes of equilateral triangles having 0.87 miles (1.4 km) in length. Thus, no ground location was further than about a half mile (0.8 km) from the nearest observation station.

The large-scale study conducted by the U.S. Environmental Protection Agency in St. Louis, Missouri, known as the Regional Air Pollution Study, provided information on criteria for determining siting requirements for a

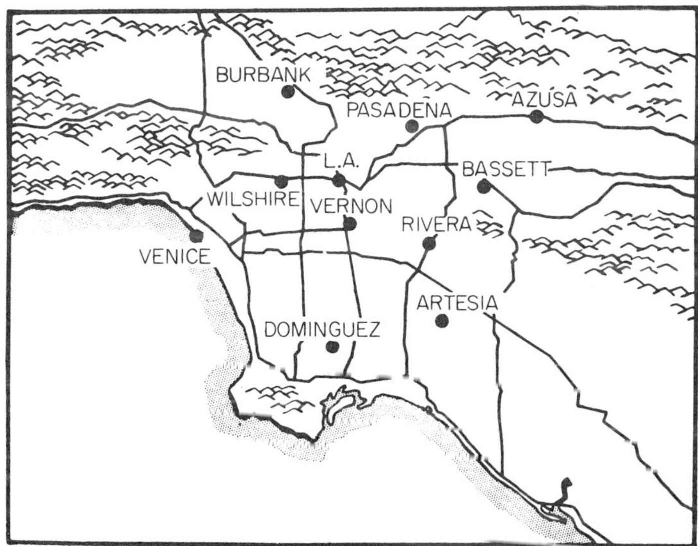

Figure 2. Network of sampling stations (*90*).

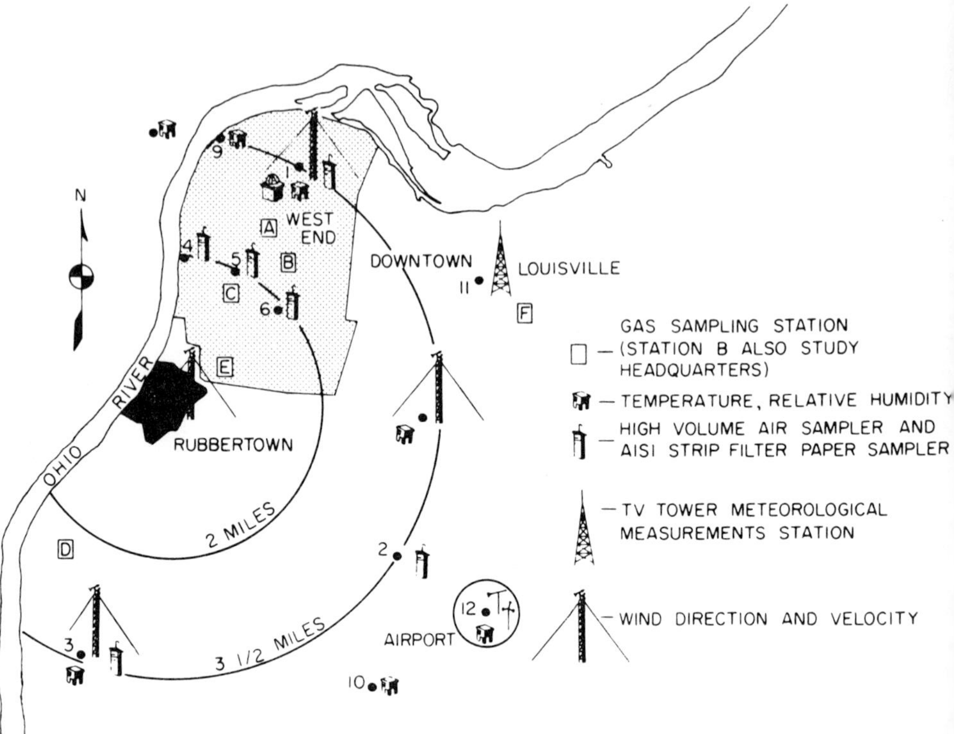

Figure 3. Air sampling network and meteorological observation network. Special Air Pollution Study of Louisville and Jefferson County, Kentucky (92).

monitoring network. The network employed is illustrated in Figure 5 (*20*). The network configurations utilized in other studies are described in several air pollution investigations and surveys listed in Tables I and II.

In some of these surveys, measurements were made on a full-time basis. The type and frequency of air sampling depends on local considerations. In general, the value of air sampling results will be maximum if the network stations are fixed, and operate simultaneously and continuously every day, including weekends and holidays, with the shortest practicable sampling period. Intensive physical and chemical analyses are performed only on samples selected on some rational basis. This system permits efficient use of limited analytic facilities, since more thorough examination of significant samples is possible.

A fixed air sampling network should be supplemented by mobile stations, built in suitable vehicles, to enable more intensive sampling of selected areas during periods of special interest. For obtaining samples

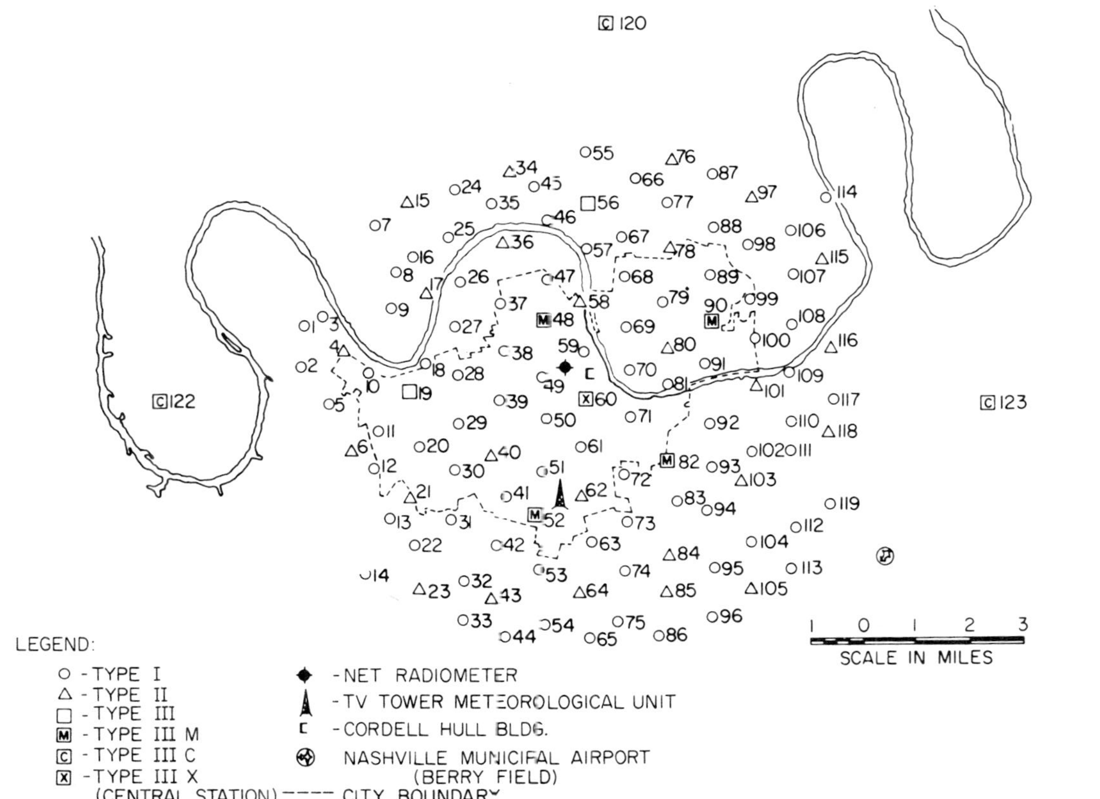

Figure 4. Aerometric station network. Nashville, Tennessee, Community Air Pollution Study (*6*).

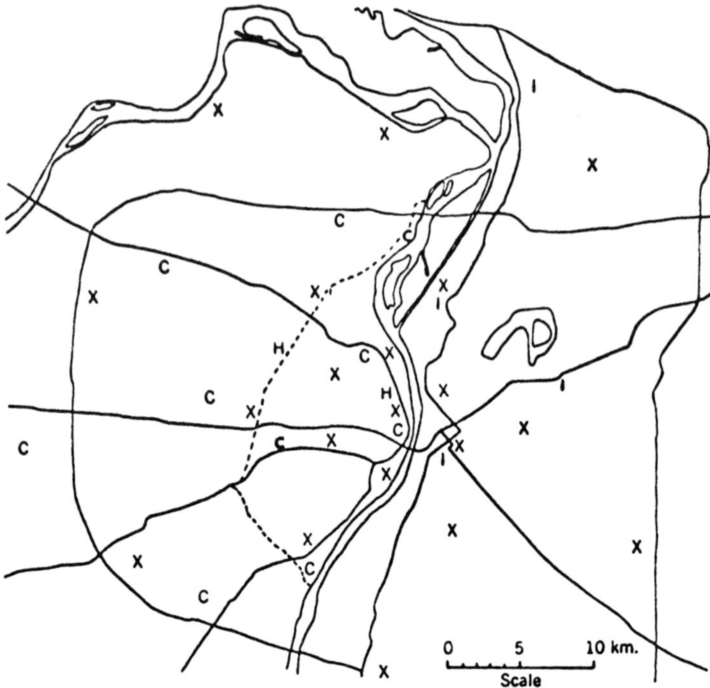

Figure 5. Siting plan for the St. Louis, Missouri Regional Air Measurement Network. Sites in existing networks indicated by letters: C, St. Louis City–County; H, CHESS; I, III, EPA. Crosses indicate the general locations of proposed sites. Four rural sites at distances of about 40 km from center of the network not shown (*20*).

vertically through the air mass, it may be necessary to employ towers, airplanes, helicopters, or captive balloons (*116*).

The type of shelter used at each station and the facilities and equipment for obtaining a sample should be such as to ensure the collection of representative samples. In the Louisville, Kentucky, study, the stations were standardized as to shape, size, construction, and height above ground. A typical station is shown in Figure 6. The Nashville, Tennessee, study made use of several standardized types of stations, as shown in Figure 7. An example of a movable monitoring station fully equipped with automatic air samplers and meteorological equipment, and telemetering capability is illustrated in Figure 8. In addition to routine network sampling, provision should be made for the collection of special samples, such as freeze-out, plastic bag, or evacuated container samples for gas analysis and slides or impactor samples for chemical, physical, or morphological identification of particulates. In some studies, viable organisms and airborne radioactivity can also be measured.

5. Measurement of Effects

Area-wide air pollution surveys are initiated because a problem is thought to exist, which implies the existence of an undesirable effect, such as reduced visibility, haze, irritation of the eyes or respiratory tract, obnoxious odors, damage to vegetation, livestock, property, or health, or other untoward effect. An air pollution survey, therefore, must seek to identify and evaluate the undesirable effects and relate them to air quality and source. In this manner, the cause-and-effect relationships are clarified and the way is opened for development of a rational basis for control and abatement of the problem.

Figure 6. Standardized type of air sampling station used in the Special Air Pollution Study of Louisville and Jefferson County, Kentucky.

Figure 7. Type II station showing the mounting of dustfall and lead peroxide instruments, as well as shelter for housing filter paper and sulfur dioxide sampling equipment and counter for the wind odometer shown at the top of the pole. Nashville, Tennessee, Community Air Pollution Study.

Most air pollution effects, such as visibility reduction or corrosion, lend themselves to physical or chemical mensuration (*20*). Others, such as the health effects, are not easily assayed, since they represent long-term and not easily demonstrated changes in biological systems. Still other effects, such as odors, are difficult to quantitate because of the subjective nature of the human response. Present state of knowledge makes measurement of these types of effects difficult. However, progress toward better identification and quantitation is encouraging, and it is hoped that research will soon provide more reliable measurement methods for these air pollution effects.

A survey may include measurements of at least the more important economic effects of air pollution. The corrosion of metals; deterioration of fabrics; damage to paint; and soiling of laundry, merchandise, and interior furnishings represent substantial economic losses to the individuals as well as commercial and industrial organizations of the typically urban society. Other forms of financial loss from pollution result from damage to or reduced yield of crops, injury to livestock, and general property devaluation. The relationship of air pollution to these damaging

Figure 8. A completely equipped movable air monitoring station. Courtesy of Westinghouse Environmental Systems Center, Raleigh, North Carolina.

effects can in itself be a subject of major study. Nevertheless, a reasonable effort to appraise the pertinent economic effects is a necessary part of community air pollution studies. A few relatively simple survey techniques already exist, and more will undoubtedly be developed as interest in this area grows (*117*).

6. Photographic Studies

The use of photographic techniques in the air pollution survey has not been sufficiently explored. Applications of photography include the already mentioned use of aerial maps in the emission inventory; the documentation of decreased visibility; the range of stack plumes under varying weather conditions; and the type and degree of damage to vegetation, materials of construction, and paints (*118, 119*). Time-lapse photography allows the diurnal study of sources and their influence on the development of smog.

7. Complaint Study

In an area afflicted by chronic air pollution, there generally is a high incidence of citizens' complaints. Many of these complaints are directed to such official agencies as air pollution control, public health, fire, and police, where records are kept. An insight into the nature and severity of the local problem may be gained by analyzing these records in terms of the number, type, time period and nature of complaints, and location of the complainant. A mapping of these data may assist in visualizing the character of the problem. Also of interest is the manner in which the complaints were handled and the extent of relief, if any, obtained.

8. Opinion Survey

Since air pollution is a matter of public concern, it is useful to know the degree of awareness as well as attitudes of the community. Such information may be obtained through a public opinion survey. Because opinions of individuals can be significantly influenced by such news media as radio, television, and press, it is advisable to conduct such a survey before the public is fully cognizant that a community-wide air pollution study is under way.

The collection of accurate information on attitudes toward the type and nature of local air pollution is most difficult because of inherent problems of personal bias. The design and implementation of public opinion surveys, therefore, should be entrusted to qualified investigators having experience in this field. Valuable information on methodologies may be obtained by a review of the studies conducted in Nashville, Tennessee, and the St. Louis, Missouri–East St. Louis, Illinois, area (*120, 121*).

9. Legal and Administrative Aspects

Since most area-wide air pollution studies ultimately lead to the adoption of new regulations or the establishment of control programs, the survey should include an evaluation of existing legal authority, ordinances, and regulations and in addition should explore the best methods for establishing an effective control program. Attention should be given to the type of administrative structure needed as well as the regulatory and enforcement powers required.

10. Special Investigations

In the majority of cases, satisfactory abatement of atmospheric pollution can be achieved by the application of existing technological knowledge. In certain important instances, however, the current state of the art neither provides a solution to the problem nor permits a sufficient level of control to meet the anticipated requirements of the future. This applies particularly to the pollution problem resulting from combustion products from such sources as burning of fuels for power and heating, incineration of refuse, and motor vehicles. In the development of an air resource management plant, it is essential to investigate these special problems to generate information and guidelines for long-range constructive action (*107*). The scope and content of special studies to evaluate the magnitude of these problems include:

Burning of fuels for power and heat

a. Types and quantities of solid, liquid, and gaseous fuels used in industrial, commercial, and residential applications.

b. Chemical composition of fuels, with particular reference to content of sulfur, volatiles, and ash.

c. Types and performance characteristics of fuel-burning equipment and installations.

d. Emissions of sulfur dioxide, sulfur trioxide, oxides of nitrogen, organic compounds, particulate matter, and odors.

e. Plume opacity and contribution to general visibility reduction.

f. Contribution of these sources to the overall air pollution problem.

g. Evaluation of methods for emission control currently in use.

h. Recommendations for the control of these sources.

Incineration of refuse

a. Types and quantities of solid refuse produced.

b. Relative quantities disposed of in municipal, industrial, commercial, and residential incinerators or by open burning.

c. Emissions of pollutants and odors.

d. Contribution of these sources to the local and area-wide air pollution problem.

e. Evaluation of incineration methods and equipment currently employed.

f. Recommendations for controlling the problem by (i) reduction in pollutant emissions or (ii) alternative methods of refuse disposal.

Motor vehicles

a. Number and geographical distribution of various types of motor vehicles.

b. Quantities of various fuels used.

c. Traffic density patterns.

d. Emissions of carbon monoxide, oxides of nitrogen, organic compounds, odors, and particulate matter, with particular reference to certain metals, including lead.

e. Relative contribution of motor vehicles to the overall pollution problem.

f. Evaluation of current methods of motor vehicle pollution control.

g. Recommendations for achieving satisfactory control of this problem.

11. Public Relations and Information

The success of an air pollution abatement program depends in a large measure upon the level of community understanding and cooperation that prevails. Thus, it is important that at the very outset of a survey and at frequent intervals during the survey the general public be kept closely informed on the purpose and progress of the study through the medium of press, radio, television, and addresses to local organizations. Photographic records of pertinent developments during the study not only provide a useful documentary but also furnish graphic material for press releases and for the final report.

In addition, the final report of the survey should include a recommendation pointing out the value of including a public relations and information activity in the abatement program. The role and value of this activity should be explained, and a program tailored to the local needs should be outlined.

E. Evaluation of Findings and Report

The evaluation of findings and preparation of the report is one of the most challenging aspects of the study, since it includes the difficult task of analyzing voluminous data, interpreting the findings, and formulating a logical conclusion. After this, appropriate recommendations must be developed, and finally, all this must be reported in a clear, understandable, and systematic manner. Perhaps the greatest difficulty in arriving at definite conclusions lies in the inherent complexity of air pollution, the incomplete state of knowledge in this field, and limitations in survey methods and techniques. These factors very frequently make firm and irrefutable recommendations most difficult.

One of the essential results of the survey is information on the frequency, concentration, duration, and geographical distribution of airborne contaminants in the area. Of prime importance is an understanding of how these air quality data relate to source and meteorological factors and observed effects. Thus, these measurements must be analyzed synoptically over some appropriate time factor. There is no standardized method of analyzing all the data. The techniques employed depend on the skill and resourcefulness of the investigator. An examination of methods utilized in past surveys such as Louisville, Kentucky; Nashville, Tennessee; and St. Louis, Missouri, suggests several useful approaches.

It is readily apparent that the routine handling and reduction of the voluminous data and information collected during a comprehensive study constitute a formidable task. Advice of and consultation with statistical analysts familiar with modern methods of data processing should be sought early in the planning and design stage to ensure the development of a realistic and efficient system for the collection and reduction of raw data. Decisions on selection of data-handling methods as well as on many of the specific types of correlative analyses should be made at this stage. Advantageous use should be made of electronic calculators and computer programming.

The importance of submitting a clear, concise, attractive, and easily understood report is not always apparent to technical people, who are accustomed to dealing mainly with other professional people and to reading the technical literature. It must be remembered that invariably the report will be put in the hands of intelligent but nevertheless nontechnical persons. These may include mayors, city managers, county judges, boards of aldermen, company presidents, and citizens who eventually will have to make decisions on a course of action. The content of the report must be made intelligible to them or much of the effort of the study will be wasted. In case of publicly financed surveys of sufficient magnitude, a formal presentation of the report to the leading local government officials in the presence of invited representatives from industry, civic organizations, press, and radio may have a beneficial effect on its general acceptance. Equally important is an appropriate announcement at the completion of a privately supported community survey.

IV. Air Resource Management Plan

An important end product of a community air pollution survey is basic data upon which to design a plan by which officials can decide the scope, character, and degree of restraints needed in the area. Ideally, the results should be presented in the form of an information system that can be used to predict air quality and adverse effects on a time and location basis. Such information forms the basis for an air resource management program (*107, 122*).

Required for this purpose are dynamic models that can relate source strength and configurations, meteorological variables, and air quality. A dynamic system embodying this approach to the development of an air resource management plan is shown diagrammatically in Figure 9. This diagram is similar to that shown earlier for illustrating the dynamics of air pollution. To the left has been added another subsystem for generat-

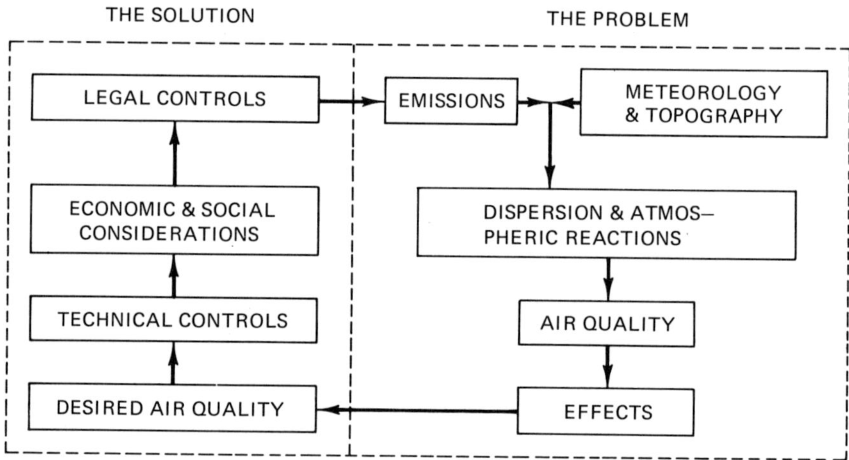

Figure 9. Systems approach to development of an air resource management plan.

ing optimum solutions to the problem of air pollution. The use of mathematical simulation models of such systems permits estimation of the degree of source control necessary to achieve desired air quality. In such models, a 1-hour interval is recommended for air quality and source strength estimates. Technical, socioeconomic, and legal aspects of alternative approaches are considered in selecting the optimum solution of the problem. The value of this approach is that it provides the type of objective information desired by air pollution control administrators, namely:

a. Present air quality levels throughout the area.

b. Air quality levels at some future date.

c. The type and degree of optimum source control to ensure the maintenance of desirable air quality now and in the future.

V. Summary

In summary, the air pollution survey in general concerns itself with a systematic study of a considerable segment of the environment to define the interrelationship of sources of pollution, atmospheric and geographic parameters, and measureable manifestations in order to evaluate the character and magnitude of an existing problem, to estimate the potential for future worsening, or to obtain the knowledge and understanding necessary for developing preventive and corrective measures.

The design, development, and execution of the survey demand the imaginative application of many skills and professional disciplines. The objectives and purposes must be made clear from the outset and strictly adhered to throughout the course of the survey, particularly in the final stages of interpretation and reporting of the findings. The results of a well-designed air pollution survey can provide valuable data with which air pollution control administrators can make objective decisions on the long-range management of the local air resource.

REFERENCES

1. G. England, *J. Inst. Fuel* **30**, 511 (1957).
2. J. A. Holmes, E. C. Franklin, and R. A. Gould, *U.S., Bur. Mines, Bull.* **98** (1915).
3. *Federal Register,* **38**, No. 117, Part II (1973).
4. "Environmental Impact Statement Guidelines." United States Environmental Protection Agency, Region X, Seattle, Washington, 1973.
5. State of Washington, Council on Environmental Policy, "Draft Guidelines to Implement the State Environmental Policy Act of 1971." Lacey, Washington, 1974.

6. L. D. Zeidberg, J. J. Schueneman, P. A. Humphrey, and R. A. Prindle, *J. Air Pollut. Contr. Ass.* **11,** 289 (1961).
7. "Airborne Particulate Emissions from Cotton Ginning Operations," Tech. Rept. A60-5. Robert A. Taft Sanit. Eng. Cent., Public Health Service, U.S. Dept. of Health, Education, and Welfare, Cincinnati, Ohio, 1960.
8. "Air Pollution Aspects of the Iron and Steel Industry." Robert A. Taft Sanit. Eng. Cent., Public Health Service, U.S. Dept. of Health, Education, and Welfare, Cincinnati, Ohio, 1961.
9. R. L. Chass, P. S. Tow, R. G. Lunche, and N. R. Shaffer, *J. Air Pollut. Contr. Ass.* **10,** 351 (1960).
10. C. V. Kanter, R. G. Lunche, F. Bonamassa, B. J. Steigerwald, and R. K. Palmer, "Emissions of Air Contaminants from Oil Refineries," Final Rept., Los Angeles County Air Pollution Control District, Los Angeles, California, 1958.
11. J. Firket, *Bull. Acad. Roy. Med. Belg.* **11,** 683 (1931).
12. J. Mage and G. Batta, *Chim. Ind. (Paris)* **27,** 961 (1932).
13. H. H. Schrenk, H. Heimann, G. D. Clayton, W. M. Gafafer, and H. Wexler, *U.S. Pub. Health Serv., Pub. Health Bull.* **306** (1949).
14. L. C. McCabe and G. D. Clayton, *Arch. Ind. Hyg. Occup. Med.* **6,** 199 (1952).
15. P. Drinker, *Arch. Ind. Hyg. Occup. Med.* **9,** 247 and 248 (1954).
16. Ministry of Health, "Mortality and Morbidity During the London Fog of December, 1952," Rept. No. 95. HM Stationery Office, London, 1954.
17. Bureau of Air Sanitation, "Fresno Air Pollution Study—A Joint State-Federal Project." California Dept. of Public Health, Berkeley, California, 1960.
18. J. McCarrol, E. J. Casell, D. W. Wolters, J. D. Mountain, J. R. Diamond, and I. M. Mountain, *Arch. Environ. Health* **14,** 178–180 (1967).
19. C. M. Shy and J. F. Finklea, *Environ. Sci. Technol.* **7,** 204–208 (1973).
20. R. J. Charlson, A H. VanderPol, D. S. Covert, A. P. Waggoner, and N. C. Ahlquist, *Science* **184,** 156–158 (1974).
21. G. M. Hidy, ed., "Aerosols and Atmospheric Chemistry." Academic Press, New York, 1972.
22. J. L. Sullivan, "A Report of a Survey of Air Pollution in New South Wales." Div. Ind. Hyg., New South Wales Dept. of Public Health, Sydney, Australia, 1953.
23. Stanford Research Institute, "Air Pollution in the Vicinity of Portland, Oregon," Final Rept. Menlo Park, California, 1956.
24. E. C. Tabor and J. E. Meeker, "Air Pollution in St. Bernard, Ohio," Tech. Rept. A 58-5. Robert A. Taft Sanit. Eng. Cent., Public Health Service, U.S. Dept. of Health, Education, and Welfare, Cincinnati, Ohio, 1958.
25. A. Giovanardi and E. Grosso, *Nuovi Ann. Ig. Microbiol.* **9** (1958).
26. Truesdail Laboratories, Inc., "An Aerometric Survey of the City of Honolulu, Territory of Hawaii, 1956–1957." Los Angeles, California, 1957.
27. E. R. Hendrickson, D. M. Keagy, and R. L. Stockman, "Evaluation of Air Pollution in the State of Washington." Report of a Cooperative Survey. Robert A. Taft Sanit. Eng. Cent., Public Health Service, U.S. Dept. of Health, Education, and Welfare, Cincinnati, Ohio, 1956.
28. Illinois Department of Public Health, "Something in the Air: A Report on Air Pollution in Illinois." Springfield, Illinois, 1956.
29. R. O. McCaldin and P. A. Kenline, "Air Pollution in Connecticut," Report of a Cooperative Survey. Robert A. Taft Sanit. Eng. Cent., Public Health

Service, U.S. Dept. of Health, Education, and Welfare, Cincinnati, Ohio, 1957.
30. P. A. Kenline, "Appraisal of Air Pollution in Tennessee," Report of a Cooperative Survey. Robert A. Taft Sanit. Eng. Cent., Public Health Service, U.S. Dept. of Health, Education and Welfare, Cincinnati, Ohio, 1957.
31. O. Paganini, M. D. High, and P. A. Kenline, "Appraisal of Air Pollution in Texas," Report of a Cooperative Survey. Texas Dept. of Health, Austin, Texas, 1958.
32. New York State Air Pollution Control Board, "A Review of Air Pollution in New York State." Albany, New York, 1958.
33. C. S. Maneri and W. H. Megonnell, *J. Air Pollut. Contr. Ass.* **10,** 374 (1960).
34. B. C Blakeney and M. D. High, "Cleaner Air for North Carolina," Report of a Cooperative Survey. North Carolina State Board of Health, Raleigh, North Carolina, 1959.
35. D. M. Anderson and V. H. Sussman, "Pure Air for Pennsylvania." Pennsylvania Dept. Of Health. Harrisburg, Pennsylvania, 1960.
36. Minnesota Dept. of Public Health, "An Appraisal of Air Pollution in Minnesota." Minneapolis, Minnesota, 1961.
37. C. I. Harding, "A Report of Florida Air Resources." Florida State Board of Health, Jacksonville, Florida, 1961.
38. R. P. Lewis, "Air Pollution in Georgia." Georgia Department of Public Health, Atlanta, Georgia, 1962.
39. C. E. Carl and G. L. Christensen, "Appraisal of Air Pollution in South Dakota." South Dakota State Dept. of Health, Pierre, South Dakota, 1962.
40. D. F. Metzler, "The Air Resources of Kansas." Kansas State Board of Health, Topeka, Kansas, 1962.
41. Montana State Board of Health and U.S. Public Health Service, "Air Pollution in Montana." Public Health Service, U.S. Dept. of Health, Education, and Welfare, Washington, D.C., 1968.
42. Utah Legislative Council, "Air Resources of Utah." Air Pollution Advisory Committee, Salt Lake City, Utah, 1962.
43. Maryland State Department of Health, "The Program for Appraisal of the Air Quality of Maryland." Bureau of Environmental Hygiene, Baltimore, Maryland, 1962.
44. J. J. Schueneman and S. M. Rogers, "The Air Pollution Problem in Steubenville, Ohio." Robert A. Taft Sanit. Eng. Cent., Public Health Service, U.S. Dept. of Health, Education, and Welfare, Cincinnati, Ohio, 1956.
45. J. J. Schueneman, "The Air Pollution Problem in Portland, Oregon." Robert A. Taft Sanit. Eng. Cent., Public Health Service, U.S. Dept. of Health, Education, and Welfare, Cincinnati, Ohio, 1956.
46. J. J. Schueneman, "The Denver Area Air Pollution Problem." Robert A. Taft Sanit. Eng. Cent., Public Health Service, U.S. Dept. of Health, Education, and Welfare, Cincinnati, Ohio, 1957.
47. D. M. Keagy and J. J. Schueneman, "Air Pollution in the Birmingham, Alabama, Area." Robert A. Taft Sanit. Eng. Cent., Public Health Service, U.S. Dept. of Health, Education, and Welfare, Cincinnati, Ohio, 1960
48. New York State Department of Health, "Air Pollution in Greater Elmira." Air Pollution Control Board, Albany, New York, 1958.
49. P. A. Kenline, "Air Pollution in Charleston, South Carolina," Tech. Rept.

A60-6. Robert A. Taft Sanit. Eng. Cent., Public Health Service, U.S. Dept. of Health, Education, and Welfare, Cincinnati, Ohio, 1960.
50. P. A. Kenline, "Air Pollution in Hamilton. Ohio," Tech. Rept. A60-8. Robert A. Taft Sanit. Eng. Cent., Public Health Service, U.S. Dept. of Health, Education, and Welfare, Cincinnati, Ohio, 1960.
51. G. Dyksterhouse, "Air Pollution in Lynchburg, Virginia." Robert A. Taft Sanit. Eng. Cent., Public Health Service, U.S. Dept. of Health, Education, and Welfare, Cincinnati, Ohio, 1961.
52. G. B. Welsh, "Publication 955." Robert A. Taft Sanit. Eng. Cent., Public Health Service, U.S. Dept. of Health, Education and Welfare, Washington, D.C., 1962.
53. G. B. Welsh, "Clean Air for Chattanooga" Robert A. Taft Sanit. Eng. Cent., Public Health Service, U.S. Dept. of Health, Education, and Welfare, Cincinnati, Ohio, 1964.
54. Oregon State Board of Health, "Air Pollution in the Portland Metropolitan Area." Oregon State Sanitation Authority, Portland, Oregon, 1963.
55. "Report on Bi-State Study of Air Pollution in the Chicago Metropolitan Area." Illinois Dept. of Public Health, Indiana State Board of Health, and Purdue University, Lafayette, Indiana, 1957–1959.
56. Bureau of Air Sanitation, "Fresno Air Pollution Study—A Joint State-Federal Project." California Dept. of Public Health, Berkeley, California, 1960.
57. J. Palomba, Jr., "Denver Metropolitan Area Air Sampling Survey." Colorado State Dept. of Public Health, Denver, Colorado, 1961.
58. S. Hochheiser, S. W. Horstman, and G. M. Tate, "A Pilot Study of Air Pollution in Birmingham, Alabama," Tech. Rept. A62-22. Robert A. Taft Sanit. Eng. Cent. Public Health Service, U.S. Dept. of Health, Education, and Welfare, Cincinnati, Ohio, 1962.
59. S. Hochheiser, M. Burchett, and H. J. Dunsmore, "Air Pollution Measurements in Pittsburgh." Robert A. Taft Sanit. Eng. Cent., Public Health Service, U.S. Dept. of Health, Education, and Welfare, Cincinnati, Ohio, 1963.
60. R. L. Mecham, J. S. Ameen, and R. W. Slater, Jr., "A Pilot Study of Air Quality in Winston-Salem, North Carolina." Robert A. Taft Sanit. Eng. Cent., Public Health Service, U.S. Dept. of Health, Education, and Welfare, Cincinnati, Ohio, 1963.
61. T. E. Kreichelt and F. W. Dahle, Jr., "Air Pollution Measurements in Baltimore, Maryland." Public Health Service, U.S. Dept. of Health, Education, and Welfare, Washington, D.C., 1964.
62. S. Hochheiser and R. H. Wetzel, "Air Pollution Measurements in Indianapolis." Robert A. Taft Sanit. Eng. Cent., Public Health Service, U.S. Dept. of Health, Education, and Welfare, Cincinnati, Ohio, 1964.
63. S. Hochheiser, "Air Pollution Measurements in Duquesne, Pennsylvania." Robert A. Taft Sanit. Eng. Cent., Public Health Service, U.S. Dept. of Health, Education, and Welfare, Cincinnati, Ohio, 1964.
64. J. P. Sheehy, J. J. Henderson, C. I. Harding, and A. L. Danis, "A Pilot Study of Air Pollution in Jacksonville, Florida," Rept. 999 AP3. Public Health Service, U.S. Dept. of Health, Education, and Welfare, Washington, D.C., 1963.
64a. W. J. Basbagill, "Air Contaminant Measurement at Roosevelt Field, Nassau County, New York." Robert A. Taft Sanit. Eng. Cent., U.S. Dept. of Health, Education, and Welfare, Cincinnati, Ohio, 1965.

65. H. Riehl and L. W. Crow, "A Study of Denver Air Pollution," Tech. Pap. No. 33. Dept. of Atmospheric Sciences, Colorado State University, Fort Collins, Colorado, 1962.
66. M. D. Hornedo and J. H. Tillman, "Air Pollution in El Paso, Texas Area." El Paso City-County Health Unit, El Paso, Texas, 1957–1959.
67. Arizona State Dept. of Health, "Air Pollution in Phoenix, Arizona." Rept. No. 1 on Air Pollution. Phoenix, Arizona, 1958.
68. H. Bravo, *J. Air Pollut. Contr. Ass.* **10**, 447 (1960).
69. Arizona State Dept. of Health, "Air Pollution in Tucson, Arizona," Rept. No. 2 on Air Pollution. Phoenix, Arizona, 1959.
70. Oregon State Sanitary Authority, "Report on Air Pollution Conditions in and Around Roseburg, Oregon." Portland, Oregon, 1960.
71. Oregon State Sanitary Authority, "The Air Pollution Problem in Medford, Oregon." Portland, Oregon, 1960.
72. N. Waggoner, "The Eureka-Arcata Ambient Air Quality Study." Bureau of Air Sanitation, California Dept. of Public Health, Berkeley, California, 1961.
73. R. L. Stockman and P. Hildebrandt, "The Air Pollution Aspects of Orchard Heating in Yakima, Washington, Area." Washington State Dept. of Health, Seattle, Washington, 1961.
74. Federal Abatement Activity Surveys: Air Pollution Technical Documents (APTD), U.S. Dept. of Health, Education, and Welfare, Public Health Service, Washington, D.C., and United States Environmental Protection Agency, Washington, D.C.: New York–New Jersey, 1967; Lewiston, Idaho and Clarkston, Washington, 1967 (APTD 1304); Kansas City, Kansas and Missouri, 1968 (APTD 68-1), Ironton, Ohio; Ashland, Kentucky; and Huntington, West Virginia, 1968 (APTD 68-2); New Cumberland, West Virginia and Knox Virginia, 1968 (APTD 68-2); New Cumberland, West Virginia and Know Township, Ohio, 1969 (APTD 69-13); Parkersburg, West Virginia and Marietta, Ohio, 1969 (APTD 69-50); Mt. Storm and Keyser, West Virginia, and Gormano and Luke, Maryland, 1971 (APTD #656); Helena Valley, Montana, 1972 (APTD 091).
75. Air Quality Control Region Consultation Surveys: Air Pollution Technical Documents (APTD), Public Health Service, United States Department of Health, Education, nad Welfare, Washington, D.C. Dated 1968: Denver, Colorado (1093); San Francisco, California (1098); New Jersey, New York, Connecticut (1195); Boston, Massachusetts (1209); St. Louis, Missouri (1210); Los Angeles, California (1211); Washington, D.C. (1216); Chicago, Illinois (1217); Philadelphia, Pennsylvania (1218). Dated 1969: San Antonio, Texas (1090); Baltimore, Maryland (1091); Birmingham, Alabama (1092); Louisville, Kentucky (1094); Providence, Rhode Island (1096); Steubenville, Ohio/Weirton and Wheeling, West Virginia (1097); Milwaukee, Wisconsin (1188); Indianapolis, Indiana (1190); Dayton, Ohio (1191); Phoenix and Tucson, Arizona (1205); Hartford and Springfield, Connecticut (1214); Buffalo, New York (1215); Cincinnati, Ohio (1219); Cleveland, Ohio (1220); Puget Sound, Washington (1221); Pittsburgh, Pennsylvania (1222); Kansas City, Kansas and Missouri (1223); Houston and Galveston, Texas (1224); Dallas and Fort Worth, Texas (1225). Dated 1970: Duluth-Superior, Minnesota (1095); Champlain Valley, New York, and Vermont (1099); Oklahoma City, Oklahoma (1186); Omaha, Nebraska (1187); Miami, Florida (1189); Atlanta, Georgia (1192); Virgin Islands (1193); Puerto Rico (1194); Sioux Falls, South

Dakota (1196); Portland, Maine (1197); Las Vegas, Nevada (1198); Fargo, North Dakota and Moorehead, Minnesota (1199); Cheyenne, Wyoming (1200); Charlotte, North Carolina (1201); Billings, Montana (1202); Hawaii (1203); Northwest Nevada (1204); Southern Louisiana and Southeast Texas (1206); Wasatch Front, Utah (1207); Merrimack Valley, Massachusetts and New Hampshire (1208); Albuquerque, New Mexico (1212); El Paso, Texas and Los Alamos and Alamagordo, New Mexico (1213).

76. Transportation Control Surveys, Air Pollution Technical Documents (APTD), United States Environmental Protection Agency, Research Triangle Park, North Carolina, Dated 1972: Los Angeles, California (1372); Houston, Texas (1373); Boston, Massachusetts (1442); Baltimore, Maryland (1443); Seattle, Washington (1444); Salt Lake City, Utah (1445); Pittsburgh, Pennsylvania (1446); Minneapolis and St. Paul, Minnesota (1447); Spokane, Washington (1448): Dated 1973: Dayton, Ohio (1367); Denver, Colorado (1368); Phoenix and Tucson, Arizona (1369); Philadelphia, Pennsylvania (1370); New York, New York (1371); Transportation Strategy Survey—Los Angeles, California (R4-73-012D); Vehicle Emissions Surveys—District of Columbia (1506). and New York, New York (1523); Undated Vehicle Emission Surveys: Chicago, Illinois, Denver, Colorado and Houston, Texas (1504); Los Angeles, California and St. Louis, Missouri (1505).

77. Federal Real Property Surveys: One for each state: Air Pollution Technical Documents (APTD) 1014–1064. United States Environmental Protection Agency, Research Triangle Park, North Carolina, 1971.

78. Federal Facility Surveys: One for each of thirty-three major metropolitan areas: Air Pollution Technical Documents (APTD) 983–1013; 1301–1302. United States Department of Health, Education, and Welfare, Public Health Service, and United States Environmental Protection Agency, Research Triangle Park, North Carolina, 1968–1971.

79. A Committee of the Chicago Association of Commerce, "Smoke Abatement and Electrification of Railway Terminals of Chicago." Rand McNally. Chicago, Illinois, 1915.

80. O. Monnett and L. R. Hughes, *U.S., Bur. Mines, Tech. Pap.* **338** (1924).

81. O. Monnett, G. St. J. Perrott, and H. W. Clark. *U.S., Bur. Mines, Bull.* **254** (1926).

82. A. C. Stern, L. Buchbinder, and J. Siegel, *Heat. Piping Air Cond.* **17,** 7–10 (1945).

83. Great Britain, Dept. of Science and Industrial Research, "Atmospheric Pollution in Leicester—A Scientific Survey," Atmos. Res. Tech. Pap. No. 1. HM Stationery Office, London, 1945.

84. R. S. Dean and R. E. Swain, *U.S., Bur. Mines, Bull.* **453** (1944).

85. West Virginia Department of Health, "Atmospheric Pollution of the Great Kanawha River Valley Industrial Area, West Virginia." Bur. Ind. Hyg., Charleston, West Virginia, 1952.

86. International Joint Commission (Canada and United States). "Air Pollution in the Detroit-Windsor Area," Report of the Technical Advisory Board on Air Pollution, 1960.

87. R. G. Tyler, "Report on an Air Pollution Study for City of Seattle." Environmental Research Laboratory, University of Washington, Seattle, Washington, 1952.

88. J. E. Yocum, I. M. Saslaw, S. Chapman, and R. L. Richardson, "Research In-

vestigations of Air Pollution in the Vicinity of Louisville, Kentucky," Summary Report to Rubbertown Industrial Group. Battelle Mem. Inst., Columbus, Ohio, 1954.

89. B. C. Newbury, *The Sarnia Survey—Action without Compulsion,* First International Congress on Air Pollution, New York, 1955. *In* "Problems and Control of Air Pollution: Proceedings," pp. 121–132. Reinhold, New York, 1955.
90. N. A. Renzetti, ed., "An Aerometric Survey of the Los Angeles Basin, August–November, 1954," Rept. No. 9. Air Pollut. Found., San Marino, California, 1955.
91. E. C. Halliday and S. Kemeny, *Air Pollut. Contr. Ass. News* **5,** 3–5 (1957).
92. A. T. Rossano, Jr., "Air over Louisville—Summary of a Joint Report." Robert A. Taft Sanit. Eng. Cent., Public Health Service, U.S. Dept. of Health, Education, and Welfare, Cincinnati, Ohio, 1958.
93. "The Louisville Air Pollution Study," Tech. Rept. A61-4. Robert A. Taft Sanit. Eng. Cent., Public Health Service, U.S. Dept. of Health, Education, and Welfare, Cincinnati, Ohio, 1961.
94. Southwest Research Institute, "Air Pollution Survey of the Houston Area," Tech. Rept. No. 4, Summary Rept., Phase I. San Antonio, Texas, 1957.
95. H. McKee, "Air Pollution Survey of the Houston Area, 1964–1966." Southwest Res. Inst., Houston, Texas, 1966.
96. Interstate Sanitation Commission, "Smoke and Air Pollution, New York–New Jersey." New York, 1958.
97. R. L. Stockman and P. W. Hildebrandt, "Evaluation of Air Pollution in the Greater Spokane Area." Washington Dept. of Health, Seattle, Washington, 1961.
98. P. A. Kenline, "In Quest of Clean Air for Berlin, New Hampshire," Tech. Rept. A62-9. Robert A. Taft Sanit. Eng. Cent., Public Health Service, U.S. Dept. of Health, Education, and Welfare, Cincinnati, Ohio, 1962.
99. "A Study of Air Pollution in the Interstate Region of Lewiston, Idaho, and Clarkston, Washington." Publ. No. 999-AP-8. Robert A. Taft Sanit. Eng. Cent., Public Health Service, U.S. Dept. of Health, Education, and Welfare, Cincinnati, Ohio, 1961.
100. The Gothenburg Air Pollution Committee (LUG), "The Gothenburg Air Pollution Study 1959–64." City of Gothenburg, Sweden.
101. National Air Pollution Control Administration, "St. Louis, Missouri Study," Vols. I and II, Publ. APTD-68-12. Public Health Service, U.S. Dept. of Health, Education, and Welfare, Durham, North Carolina, 1968.
102. National Air Pollution Control Administration, "Kanawha Valley Air Pollution Study," Publ. APTD-70-1. Public Health Service, U.S. Dept. of Health, Education, and Welfare, Durham, North Carolina, 1970.
103. P. W. Hildebrandt and R. L. Stockman, "Air Quality in Clark County, Washington." State of Washington Health Dept., Seattle, Washington, 1965.
104. National Center for Air Pollution Control, "Interstate Air Pollution Study of St. Louis–East St. Louis," Phase II Project-Report. Public Health Service, U.S. Dept. of Health, Education, and Welfare, Cincinnati, Ohio, 1966.
105. National Air Pollution Control Administration, "Chattanooga, Tennessee Rossville, Georgia Interstate Air Quality Study, 1967–1968." Publ. APTD-0583. Public Health Service, U.S. Dept. of Health, Education, and Welfare, Durham, North Carolina, 1970.
106. International Joint Commission, "Joint Study of St. Clair–Detroit Windsor

River Areas," Publ. APTD-1305. Public Health Service, U.S. Dept. of Health, Education, and Welfare, Durham, North Carolina, 1971.
107. A. T. Rossano, Jr. and TRC Corporation, "Management Plan and Survey Protocol for a Comprehensive Air Pollution Survey of the New York/New Jersey Metropolitan Area," Vols. I and II. TRC Service Corporation, Hartford, Connecticut, 1965.
108. A. T. Rossano, Jr., *J. Air Pollut. Contr. Ass.* **6**, 176 (1956).
109. A. T. Rossano, Jr., *Interdisciplinary Conf. Atmos. Pollut., Santa Barbara, 1959,* pp. 25–41, Amer. Meteorol. Soc., Boston, Mass. (1959).
110. A. T. Rossano and T. A. Rolander, "The Preparation of an Air Pollution Source Inventory." Manual on Urban Air Quality Management. World Health Organization, Regional Office for Europe, Copenhagen, Denmark, 1974.
111. B. H. B. Cooper and A. T. Rossano. "Source Testing for Pollution Control." McGraw-Hill, New York City, 1974.
112. D. H. Fair, J. B. Clements, and G. B. Morgan, "Storage and Retrieval of Aerometric Data, Parameter Coding Manual." United States Environmental Protection Agency, Research Triangle Park, North Carolina, 1971.
113. F. Casey, Dept. of Civil Engineering, University of Washington, Seattle, Washington (private communication).
114. W. H. Megonnell, "Interstate Atmospheric Transport of Tracer Particles in the New York–New Jersey Metropolitan Area," Report to the Interstate Sanitation Commission. Robert A. Taft Sanit. Eng. Cent., Public Health Service, U.S. Dept. of Health, Education, and Welfare, Cincinnati, Ohio, 1958.
115. A. T. Rossano, and J. F. Thielke, "The Design and Operation of Air Quality Surveillance Systems," Manual of Air Quality Management. World Health Organization, Regional Office for Europe Copenhagen, Denmark, 1974.
116. D. Ensor, "The Use of Light Aircraft to Measure Three-Dimensional Distribution of Air Pollutants," Annual Meeting of the Pacific Northwest—International Section. Air Pollution Control Association, Pittsburgh, Pennsylvania, 1972.
117. L. B. Barrett and T. E. Waddell, "Cost of Air Pollution Damage: A Status Report," AP-85. United States Environmental Protection Agency, Research Triangle Park, North Carolina, 1973.
118. S. A. Veress, *Photogramm. Eng.* **36**(8), 840–848 (1970).
119. S. A. Veress, *Photogramm. Eng.* **38**(2), 183–191 (1972).
120. W. S. Smith, J. J. Schueneman, and L. D. Zeidberg, *J. Air Pollut. Contr. Ass.* **14**, 418–423 (1964).
121. J. D. Williams and F. L. Bunyard, "Opinion Surveys and Air Quality Statistical Relationships," Vol. III, Interstate Air Pollution Study of St. Louis–East St. Louis, Phase II Project Report. National Center for Air Pollution Control, Cincinnati, Ohio, 1966.
122. J. D. Williams and N. G. Edmisten, "An Air Resource Management Plan for the Nashville Metropolitan Area," Publ. No. 999-AP-18. Robert A. Taft Sanit. Eng. Cent., Public Health Service, U.S. Dept. of Health, Education, and Welfare, Cincinnati, Ohio, 1965.

Part B

NATIONAL AND WORLDWIDE

7

Energy and Resources Planning—National and Worldwide

William A. Vogely

I. Introduction

At the three-quarter mark of the twentieth century, once again resource scarcity has become a major concern to the world. These concerns are expressed in their most virulent form as predictions of doomsday for mankind, such as the report prepared for the Club of Rome called "Limits to Growth" (*1*). At a less sensational level, several prestigious reports coming out of the United States National Academy of Sciences have sounded the same kinds of warning (*2, 3*). These approaches emphasize the depletion of liquid petroleum deposits of the world and the implications of this on the future of mankind. Other authors, convinced that we are not running out of mineral resources, see the problem as centering on the environmental impacts of high rates of exploitation of the fossil fuels and other mineral resources (*4*).

This chapter attempts to sort out these very complex issues, which relate resources to economic growth and the quality of life for the world at large. The first sections of the chapter are descriptive. In them resources of the world as they are known today are described, and the rate of use of those resources over the past decade is presented. The following sections examine the issue of future resource availability. In them concepts of reserves and resources and potential reserves are described and developed. The determinants of resource use are examined in sections after that. The final sections of the chapter address themselves to emerging resource policy issues on a worldwide basis. These involve the use of developed resources as an economic weapon to transfer wealth from the consumer countries to the producer countries and the use of interruptions in resource supply as a political weapon in the world.

II. Minerals and the World Economy

A. Overview

Society uses a wide variety of minerals in many forms as basic constituents of living. The mineral content of man's food is absolutely necessary for life. Whenever a habitation is constructed, a tool fashioned, a means of transport employed, a weapon designed, or an occupation persued, minerals are involved in a fundamentally essential way. Thus, the availability of minerals to mankind has been a continuing matter of concern.

The number of mineral commodities through their primary state as classified by the United States Bureau of Mines (*5*) is 95. Many of these, of course, are minor commodities of little immediate interest, such as yttrium and rubidium, while others are dominant in their quantity and importance. Table I presents world production of the major mineral commodities for 1973 and the percentage distribution of that production by major areas. It is impossible in the scope of this chapter to analyze all of the commodities listed in these tables. By using the criterion of importance in international trade, which will identify the commodities where the geographical distribution of supply differs markedly from the geographical distribution of use, the list can be reduced for further analysis. The result is the choice of the following commodities: the metals aluminum, copper, iron, lead, nickel, tin, and zinc; the nonmetals asbestos, phosphate rock, and sulfur; and the fuels coal and petroleum. For these commodities, Table II compares estimated consumption, production, and proved reserves for the major areas of the world.

Detailed information on the flow of these materials through the individual economies of the world are very difficult to come by and extremely voluminous to present. Therefore, as an illustration of the flow of these commodities into end uses, a series of charts (Figs. 1–16), each of which shows for 1973 world production, and flows through imports and production to supply and distribution of demand in the United States, are presented in the following pages. Finally, in order to achieve some feel for the time trend involved in these commodities, production broken down by the United States and the world for each of the commodities for the period 1964–1973 is shown in Tables III through XIV.

B. Aluminum

Aluminum is one of the most rapidly growing metals in the world economy. It is produced in metallic form from bauxite, which is a very abundant claylike material. As can be seen from Tables I and II, the world production of bauxite is approximately five times greater than the production of ingot metal. There is substantial flow of trade in bauxite in the world, as indicated by the differing percentage distribution at the three stages of major world areas.

In the decade ending in 1973, world production of aluminum increased

Table I World Production of Major Mineral Commodities and Its Percentage Distribution by Major Areas in 1973[a,b]

Commodity	World production[c] (metric tons except as noted)	Approximate percentage distribution by major areas[c] North and Central America	South America	Europe	Africa	Near East and Asia	Oceania
		Metals					
Aluminum							
Bauxite	70,694,000	24.5	16.9	21.6	5.7	6.1	25.2
Alumina	26,538,000	38.7	8.9	24.9	2.3	10.0	15.2
Unalloyed ingot metal	12,117,000	41.6	1.5	39.0	2.1	13.2	2.6
Antimony	70,000	6.8	22.2	17.1	24.3	27.4	2.2
Arsenic, white[d]	47,000	9.2	2.4	70.2	17.2	1.0	—
Beryl[d]	3,589	—[d]	41.2	40.5	16.3	NA	2.0
Bismuth[d]	3,923	16.0	35.6	7.2	0.1	30.9	10.2
Cadmium	17,013	28.0	1.2	44.5	2.3	20.0	4.0
Chromite	6,701,000	0.3	1.9	50.1	27.4	20.3	—
Cobalt							
Mine[d]	25,638	13.2	—	11.7	72.0	NA	3.1
Refined	22,849	4.6	—	21.1	74.3	NA	—
Columbium–tantalum concentrates[e]	24,039	11.2	81.3	NA	6.2	0.5	0.8
Copper							
Mine	7,136,000	34.4	13.6	17.8	20.9	7.7	5.6
Smelter	7,013,000	30.7	10.4	22.6	19.1	14.9	2.3
Gold	43,070,000 troy ounces	7.7	1.4	17.4	67.1	2.7	3.7
Iron and steel							
Iron ore	864,463,000	16.7	11.4	41.1	7.1	13.6	10.1
Pig iron[f]	504,412,000	20.7	1.6	48.6	1.0	26.6	1.5
Ferroalloys[f]	10,589,000	20.7	1.6	48.6	1.0	26.6	1.5
Crude steel	694,318,000	22.3	1.8	51.3	0.8	22.7	1.1
Lead							
Mine	3,532,000	32.1	8.0	32.4	6.1	9.9	11.5
Smelter	3,534,000	27.7	4.5	21.0	3.2	12.1	11.5
Magnesium	237,000	49.0	—	45.9	—	5.1	—
Manganese ore	22,153,000	1.6	10.0	37.3	31.2	12.9	7.0
Mercury	276,000 76 pound flasks	15.4	1.5	63.3	5.1	14.7	Neg
Molybdenum	82,191	79.0	8.1	10.8	—	2.1	0.1
Nickel	676,000	48.2	0.5	23.1	4.6	3.1	20.5
Platinum-group metals	4,314,000 troy ounces	7.2	0.6	56.8	34.9	0.5	Neg
Selenium[g]	1,115	53.3	0.7	13.6	NA	32.1	0.3
Silver	307,314,000 troy ounces	41.7	18.1	23.3	4.2	4.7	8.0

Table I ***(Continued)***

Commodity	World production[c] (metric tons except as noted)	Approximate percentage distribution by major areas[c]: North and Central America	South America	Europe	Africa	Near East and Asia	Oceania
Tellurium[e]	191	68.0	9.4	NA	NA	22.6	NA
Tin							
Mine[d]	233,000 long tons	0.2	14.5	15.1	7.2	58.5	4.5
Smelter	229,000 long tons	2.0	5.0	24.8	3.8	61.4	3.0
Titanium concentrates							
Ilmenite[e]	3,567,000	44.4	0.1	25.7	—	9.3	20.5
Rutile[d,e]	334,000	—	Neg	NA	—	1.7	98.3
Tungsten, mine output, metal content	38,365	14.9	10.1	25.7	3.3	42.7	3.3
Uranium oxide (U_3O_8)[e]	23,154	70.1	0.2	9.5	20.2	NA	NA
Vanadium[e]	19,223	20.7	5.0	28.2	46.1	NA	—
Zinc							
Mine	5,703,000	34.6	9.3	31.4	4.6	11.7	8.4
Smelter	5,231,000	20.9	2.4	46.7	3.4	20.9	5.7
	Nonmetals						
Asbestos	4,171,000	46.2	1.0	34.9	10.6	6.5	0.8
Barite	4,316,000	31.1	7.2	38.7	3.8	18.8	0.4
Cement, hydraulic	694,396,000	14.8	4.4	53.0	3.0	23.8	1.0
Diamond							
Gem	12,560,000 carats	—	3.4	15.1	81.3	0.2	—
Industrial	31,167,000 carats	—	2.3	24.4	73.3	NA	—
Diatomite	1,588,000	37.9	0.9	60.2	0.4	0.2	0.4
Feldspar	2,594,000	31.4	3.8	57.5	1.6	5.6	0.1
Fluorspar	4,495,000	32.3	2.8	41.6	6.3	17.0	Neg
Graphite[d]	370,000	17.7	1.1	34.6	4.1	42.5	—
Gypsum	60,575,000	36.3	1.8	48.2	2.3	9.7	1.7
Magnesite[d]	9,234,000	0.3	3.3	57.7	2.2	36.3	0.2
Mica[e]	222,000	72.7	1.8	5.1	3.2	17.2	—
Nitrogen fertilizers, contained nitrogen[h]	38,812,000	25.5	0.9	51.6	1.4	20.1	0.5
Phosphate rock	99,995,000	38.4	0.3	23.1	28.8	4.9	4.5
Potash (marketable), K_2O equivalent	21,564,000	29.5	0.1	65.3	1.2	3.7	—
Pumice[e]	14,332,000	27.3	1.2	71.0	0.1	NA	0.4
Pyrites, including cupreous, gross weight	22,110,000	3.2	—	71.1	4.8	20.0	0.9
Salt	150,749,000	33.8	3.3	38.0	1.4	20.7	2.8

Table I (*Continued*)

Commodity	*World production[c] (metric tons except as noted)*	*Approximate percentage distribution by major areas[c]* North and Central America	South America	Europe	Africa	Near East and Asia	Oceania
Strontium minerals[e]	93,559	82.6	1.2	14.0	1.9	0.3	—
Sulfur elemental							
Frash and from ores	15,897,000	58.3	1.0	37.3	—	3.4	—
By-product	16,670,000	59.9	Neg	27.0	0.3	10.8	2.0
Talc, soapstone, pyrophyllite	5,232,000	23.6	2.1	24.8	0.3	48.1	1.1
Vermiculite[e]	500,000	66.3	1.7	NA	31.5	0.5	—
Mineral fuels and related materials							
Coal[i]							
Anthracite	174,000,000	23.6	0.3	45.1	2.8	25.5	2.7
Bituminous	2,166,000	23.6	0.3	45.1	2.8	25.5	2.7
Lignite	819,000	2.0	—	93.7	—	1.3	3.0
Total	3,159,000						
Coke							
Metallurgical	366,850,000	18.4	0.9	54.8	1.2	23.4	1.3
Other types	17,841,000	—	0.3	52.0	0.8	46.4	0.5
Fuel briquets	86,761,000	—	—	84.9	Neg	13.8	1.3
Gas, natural, marketed	44,862 billion ft³	58.8	2.2	33.9	0.8	4.0	0.3
Peat	95,475,000	1.0	Neg	98.9	—	0.1	—
Petroleum, crude	20,361 million barrels	20.8	8.1	16.5	10.8	43.1	0.7

[a] Preliminary production figures.
[b] Neg = production negligible (less than 0.05% of world output); NA = production data not available and no basis available for reliable estimate of output level.
[c] Incorporates numerous revisions from world production tables and country production tables appearing in Volumes I and III, respectively, of the *Minerals Yearbook*. Percentages in this table have been calculated from the most reliable data available through May 15, 1975.
[d] United States data withheld to avoid disclosing individual company confidential data and not included in total upon which percentages have been calculated.
[e] Excludes production from Albania, Bulgaria, the People's Republic of China, Cuba, Czechoslovakia, East Germany, Hungary, Mongolia, North Korea, North Vietnam, Poland, Romania, the Soviet Union, and Yugoslavia, except in the case of vanadium, which includes a figure for the Soviet Union alone. The total upon which percentages are calculated includes production estimates for Yugoslavia.
[f] Data presented for pig iron includes relatively small quantities of ferroalloys (not duplicating quantities reported under ferroalloys) produced in a few countries that do not report ferroalloy production separately from pig iron production.
[g] Excludes production from countries listed in footnote *e* except for Yugoslavia.
[h] Year ending June 30.
[i] Production of coal by some countries is not reported divided into the three categories listed; such output has been distributed to the three listed grades according to best available information from supplementary sources relating to the quality of such coals.

by 131% (Table III). The United States production of bauxite remained virtually stable over this period and there were very substantial increases in bauxite imports to the United States.

The supply/demand flow chart for aluminum (Fig. 1) details the 1973 flow of aluminum to the United States economy. The major exporter to the United States is Jamaica, followed by Surinam and Guinea. However, the major producer in the world is Australia, which supplies the bulk of the alumina imports by the United States. On the demand side, aluminum is consumed in many sectors, the largest being the construction industries, transportation, the packaging industries, and the electrical manufacturing industries.

C. Copper

Mine production of copper increased over the 10-year period by 54%, indicating substantial replacement of copper by aluminum, primarily in electrical and construction uses (Fig. 2). Production in the United States increased somewhat less, 38%, than did that for the world (Table IV). There is a much closer correspondence between copper mine production and copper smelter production by area than is the case with aluminum. The major exception to this statement is the Near East and Asia, indicating the dominance of Japan in smelter production. The Japanese search worldwide for the concentrates to feed their copper smelters.

A glance at Table II, however, indicates the strength of the trade in copper metal on a worldwide basis. While North and Central Americas are in rough balance, Europe shows major imports of refined metal, matched by major exports from Africa and South America.

The largest producers of primary copper in the world are the United States, Canada, Chile, Zambia, the Soviet Union, and Zaire. The major uses of copper, as indicated by the relationships in the United States, are in the electrical, construction, and machinery industries.

D. Iron and Steel

As indicated by the relationships among iron ore, pig iron, and crude steel in Table I, there is very substantial shipments of the primary ore in world commerce. South America, Africa, and Oceania supply the bulk of this material (Fig. 3), which is shipped to the steel mills in North America, Europe, and the Near East and Asia, specifically Japan.

Raw steel production in the world has increased by 60% in the decade ending in 1973 (Table V, Fig. 4). Over this period, the United States showed a 19% increase, as contrasted to a 75% change for the rest of the world.

Table II Estimated Consumption, Production, and Proved Reserves of Selected Mineral Commodities for the Major Areas of the World in 1973[a]

	Element content (thousand short tons)[b]					
Commodity	*North and Central America*	*South America*	*Europe*[c]	*Africa*	*Near East and Asia*	*Oceania*
		Metals				
Aluminum (Ingot)						
Consumption	5,987	303	5,965	119	2,394	200
Production	5,555	200	5,208	280	1,763	347
Proved reserves	246,200	554,000	300,000	984,000	76,000	950,000
Copper (Refined)						
Consumption	2,745	234	4,594	92	1,790	150
Production	2,373	804	1,747	1,476	1,151	178
Proved reserves	137,000	81,000	64,000	53,000	7,000	24,000
Iron						
Consumption[d,e]	117,479	9,080	275,820	5,675	150,963	8,513
Production[d]	159,091	108,601	391,534	67,637	129,559	96,216
Proved reserves	14,110,000	18,330,000	39,530,000	3,380	11,160,000	10,210,000
Lead						
Consumption	1,418	130	2,567	56	582	80
Production	1,079	175	818	125	471	448
Proved reserves	58,955	6,175	46,625	5,200	9,495	18,500
Nickel						
Consumption	217	7	338	5	152	4
Production	359	4	172	34	23	153
Proved reserves	30,000	—	10,000	—	29,000	1,000
Tin						
Consumption	71	8	117	4	67	5
Production	5	13	64	10	157	8
Proved reserves	76	1,782	1,020	790	7,296	211
Zinc						
Consumption	1,727	189	3,060	91	1,350	155
Production	1,205	138	2,692	196	1,205	329
Proved reserves	92,000	18,000	64,000	15,000	46,000	23,000

Nonmetals						
Asbestos						
Consumption	1,043	117	3,514	155	756	77
Production	2,124	46	2,385	565	299	41
Proved reserves	154,000	—	—	23,000	—	—
Phosphate rock						
Consumption	—	—	—	—	—	—
Production	42,315	331	25,455	31,736	5,399	4,959
Proved reserves	1,500,000	98,000	—	4,000,000	530,000	206,000
Sulfur (native including Frasch)						
Consumption[f]	10,778	1,521	6,766	848	2,171	915
Production	10,213	175	6,534	—	596	—
Proved reserves	593,600	257,600	504,000	50,400	1,349,600	11,200
Mineral fuels						
Coal						
Consumption	536,068	22,811	1,997,320[i]	Not available	284,812	42,151
Production	626,619	7,736	2,008,663	72,203	669,296	96,700
Proved reserves	1,720,000,000[g]	20,000,000[g]	7,120,000,000	80,000,000	500,000,000	60,000,000
Petroleum (crude)						
Consumption	7,291,884[h]	538,755	8,661,861[i]	351,840	3,193,681	236,026
Production	4,235,088	1,649,241	3,359,565	2,198,988	8,775,591	142,527
Proved reserves	68,700,000	14,000,000	119,400,000	67,300,000	365,300,000	—[j]

[a] SOURCES: *Consumption:* "BP Statistical Review of the World Oil Industry 1973." British Petroleum Company Ltd., London, England, 1974. "Metal Statistics 1963–1973," 61st ed. Metallgesellschaft Aktiengesellschaft, Frankfurt am Main, Germany, 1974. "Mining—Annual Review 1974." Mining Journal, London, England, 1974. "World Trade Annual 1973." Statistical Office of the United Nations, New York, New York, 1974. *Production:* Mineral Industry Surveys, "World Mineral Production, Annual." United States Bureau of Mines, Washington, D.C., 1975. *Proved reserves:* "Commodity Data Summaries 1974." United States Bureau of Mines, Washington, D.C., 1974. "Mineral Facts and Problems." United States Bureau of Mines, Washington, D.C., 1970. "Professional Paper 820." United States Geological Survey, Washington, D.C., 1973.

[b] Except petroleum, which is in thousands of 42 gallon barrels.

[c] Includes Soviet Union.

[d] Iron ore.

[e] Assumed equal to pig iron + ferroalloys production.

[f] Brimstone.

[g] Central America figures included under South America.

[h] Includes figures for Columbia and Venezuela.

[i] Includes figures for People's Republic of China.

[j] Oceania's proved petroleum reserves are included under Near East and Asia.

Table III Aluminum Production, 1964–1973 (*5a*)

	Production (bauxite) (thousand short tons aluminum content)									
	1964	1965	1966	1967	1968	1969	1970	1971	1972	1973
United States	412	426	463	426	429	475	536	512	467	442
Rest of world	6,832	8,204	8,968	10,067	10,202	11,442	12,555	13,884	14,576	16,268
Total	7,244	8,630	9,431	10,493	10,631	11,917	13,091	14,396	15,043	16,710

Table IV Copper Production, 1967–1973 (*5a*)

	Mine production—primary (thousand short tons)									
	1964	1965	1966	1967	1968	1969	1970	1971	1972	1973
United States	1247	1352	1429	954	1205	1545	1720	1522	1665	1718
Rest of the world	3865	3967	4056	4270	4436	4679	4918	5131	5649	6139
Total	5112	5319	5485	5224	5641	6224	6638	6653	7314	7857

Table V Raw Steel Production, 1964–1973 (*5a*)

	Production (million short tons)									
	1964	1965	1966	1967	1968	1969	1970	1971	1972	1973
United States	127.0	131.5	134.1	127.2	131.5	141.3	131.5	120.4	133.2	150.8
Rest of world	352.0	371.6	385.0	420.4	451.0	490.7	522.7	519.5	561.3	616.4
Total	479.0	503.1	519.1	547.6	582.5	632.0	654.2	639.9	694.5	767.2

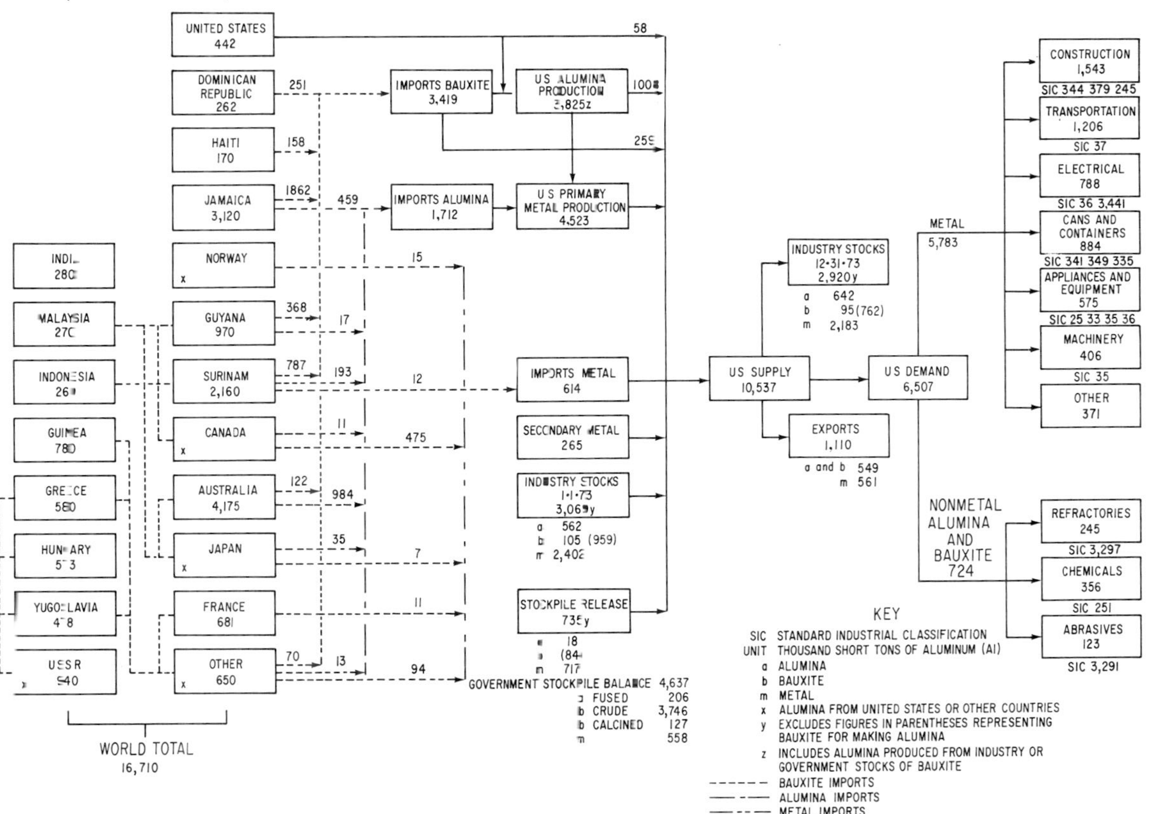

Figure 1. Aluminum supply–demand relationships—United States 1973.

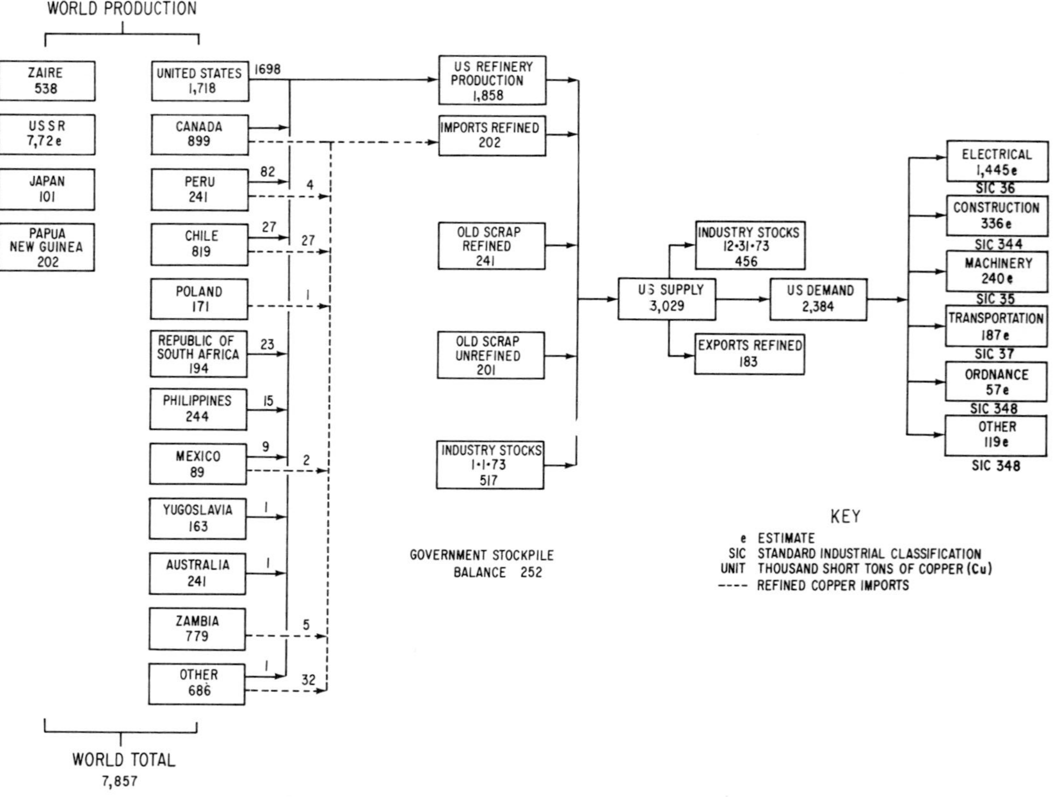

Figure 2. Copper supply–demand relationships—United States 1973.

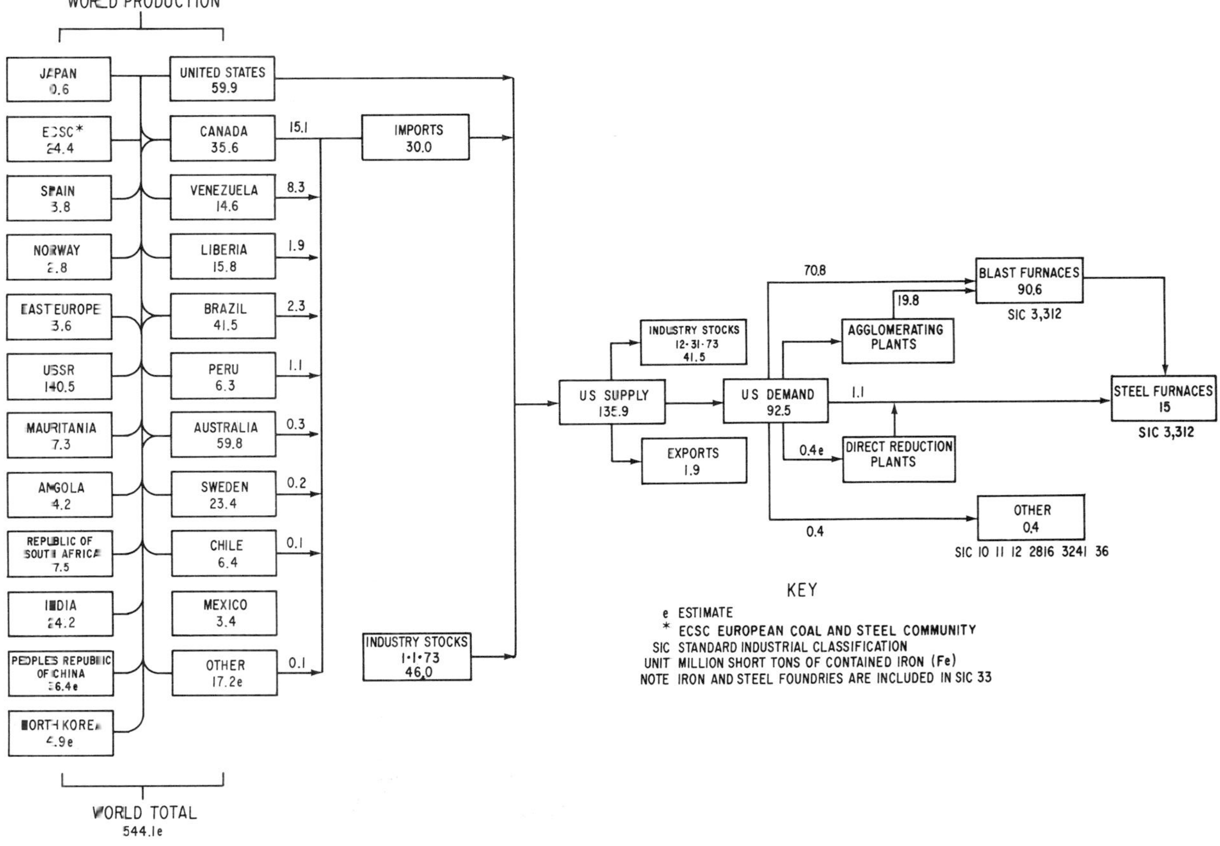

Figure 3. Iron ore supply–demand relationships—United States 1973.

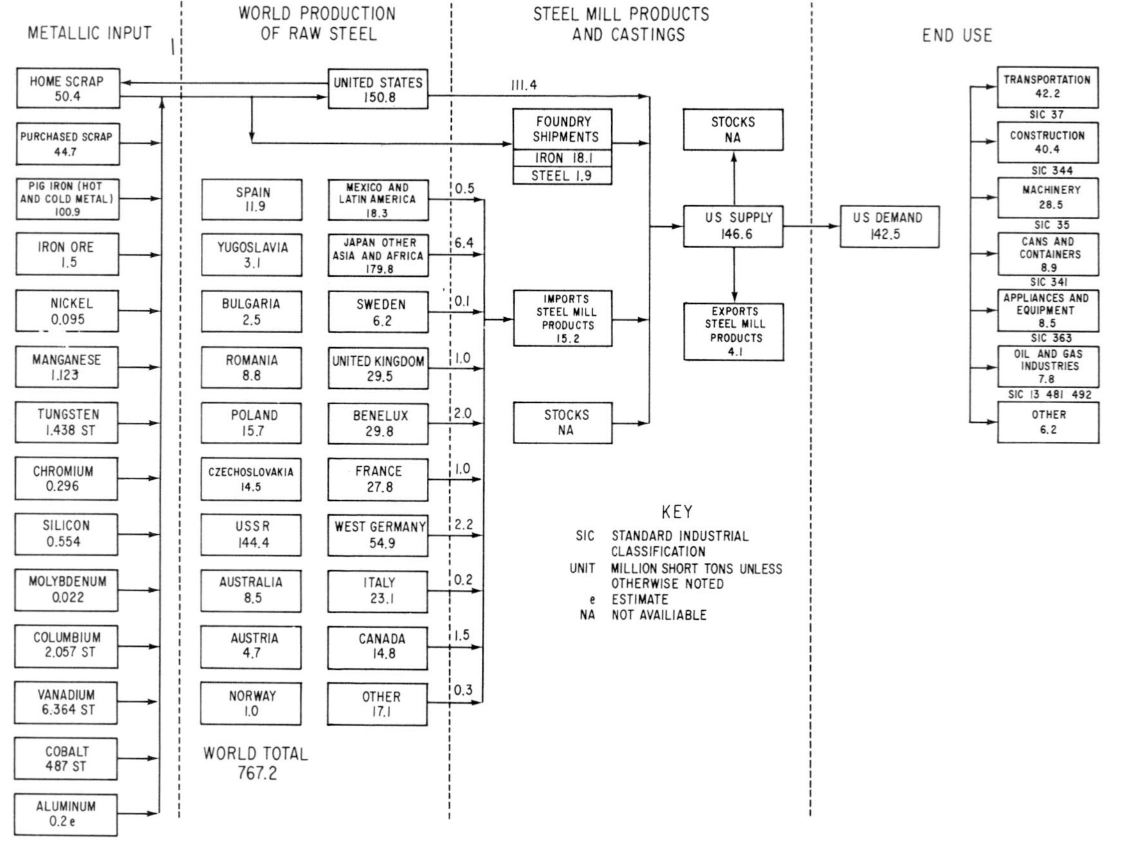

Figure 4. Iron and steel supply–demand relationships—United States 1973.

Iron ore is produced by many countries of the world; the leading producers are the Soviet Union, by a substantial margin; followed by the United States, Australia, Peoples Republic of China, Brazil, and Canada. The iron ore, of course, flows into blast furnaces and steel furnaces and then gets transformed into that most used of all metals—steel. As can be seen by the chart on iron and steel (Fig. 4), the major industrial consumers are transportation, construction, and machinery.

E. Lead

This metal presents quite a contrasting picture to the others analyzed. United States mine production has increased by 111% over the decade, as contrasted to a world increase of only 39% (Table VI). This results from the discovery of a major lead belt in Missouri in the late 1950s. The significant finding flowing from the comparison between mine and smelter production contained in Table I lies in the rather large flow to Japan of lead ores. The major world producers of primary lead are the United States, the Soviet Union, Australia, and Canada (Fig. 5). On the consumption side, the bulk of the lead flows into the transportation sector in the form of storage batteries, gasoline additives, and the electrical industries.

F. Nickel

World production of nickel increased by 75% over the decade (Table VII). Production is concentrated in North America (Canada), the Soviet Union, and Oceania. Nickel is a very important ferro-alloy metal used in the manufacture of stainless steel. As can be seen from the supply/demand chart (Fig. 6), it is used in significant amounts in a number of industrial sectors.

G. Tin

Although the world production figures on tin are withheld to avoid disclosure for the United States producer, the rest of the world figure shown in Table VIII can be taken as an excellent proxy of world production. Tin is one of the slower-growing metals in world production, increasing by 20% over the decade covered. Its major use is in cans and containers and the electrical industries (Fig. 7). In both of these uses, it is facing severe competition from other materials, which accounts for its slow rate of growth in production.

Table VI Lead Production, 1964–1973 (*5a*)

	Mine production (thousand short tons)									
	1964	1965	1966	1967	1968	1969	1970	1971	1972	1973
United States	286	301	327	317	359	509	572	578	619	603
Rest of world	2449	2666	2812	2842	2940	3014	3168	3165	3145	3203
Total	2735	2967	3139	3159	3299	3523	3740	3743	3764	3806

Table VII Nickel Production, 1964–1973 (*5a*)

	Mine production (thousand tons)									
	1964	1965	1966	1967	1968	1969	1970	1971	1972	1973
United States	12.2	13.5	13.2	14.6	15.2	15.8	15.6	15.6	15.7	13.9
Rest of world	396.8	454.8	426.9	480.2	532.9	516.9	676.8	685.0	666.2	707.7
Total	409.0	468.3	440.1	494.8	548.1	532.7	692.4	700.6	681.19	721.6

Table VIII Tin Production, 1964–1973 (*5a*)

	Production (long tons)									
	1964	1965	1966	1967	1968	1969	1970	1971	1972	1973
United States	65	47	97	—[a]	—[a]	—[a]	—[a]	—[a]	—[a]	—[a]
Rest of world	193,392	201,068	207,974	214,233	228,332	225,725	228,500	232,232	239,602	232,404
Total	193,457	201,115	208,071	—	—	—	—	—	—	—

[a] Withheld to avoid disclosing company confidential data.

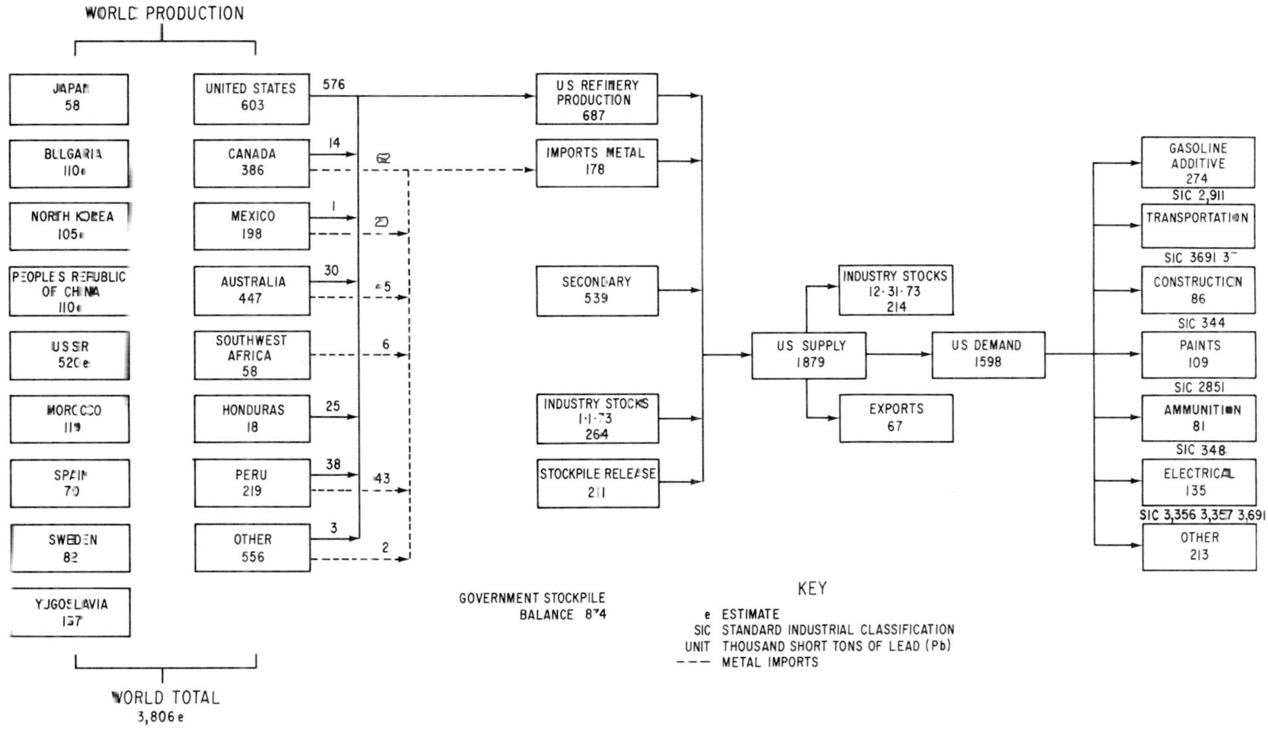

Figure 5. Lead supply–demand relationships—United States 1973.

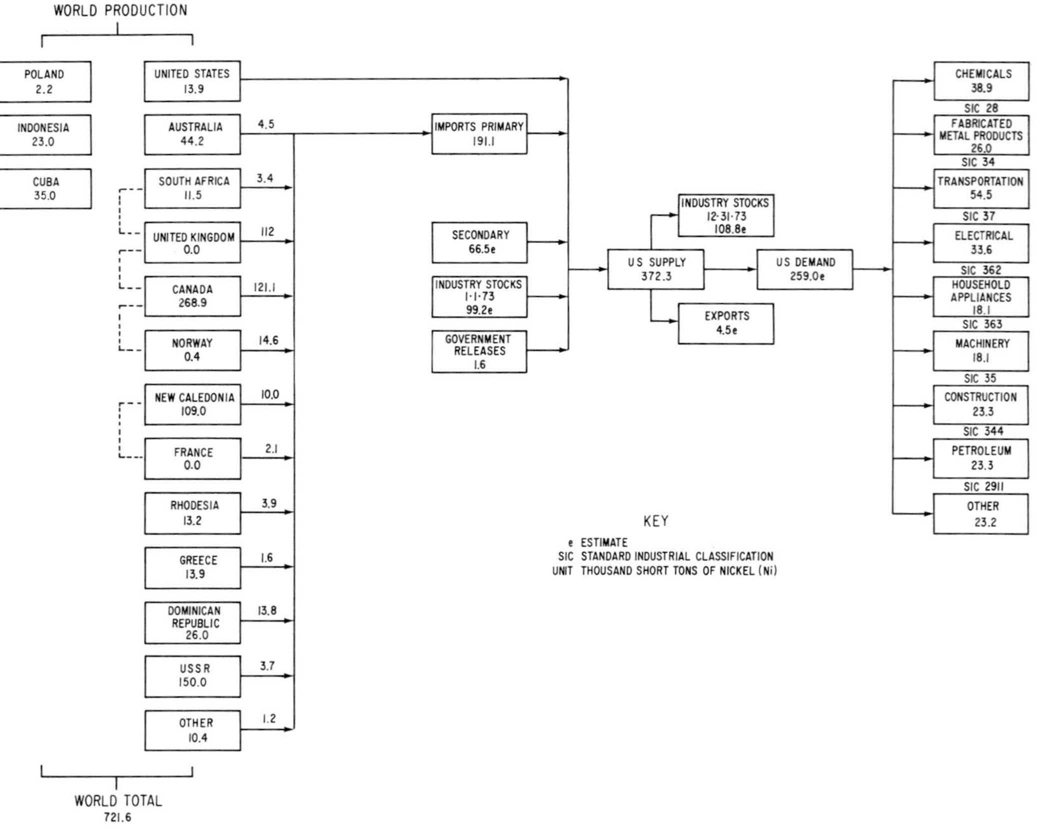

Figure 6. Nickel supply–demand relationships—United States 1973.

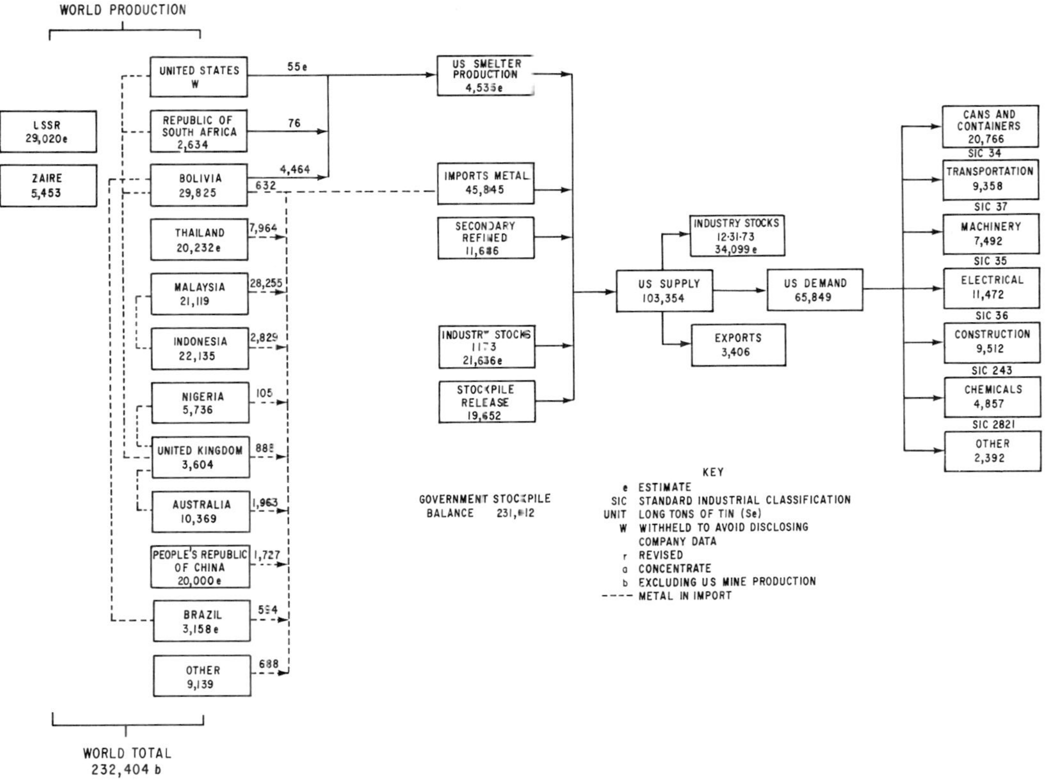

Figure 7. Tin supply–demand relationships—United States 1973.

H. Zinc

World production of zinc increased by 44%, while United States production actually declined over the decade ending in 1973 (Table IX). The major producers of zinc are Canada, the Soviet Union, Australia, and the United States (Fig. 8). Zinc is used both as a metal and as a compound in such things as paints and chemicals. In its metal use, the major consuming sectors are construction and transportation.

I. Asbestos

Asbestos production in the world increased by 51% over the 10-year period (Table X). United States production is very small; the major producers in the world are Canada, the Soviet Union, and South Africa (Fig. 9). Asbestos is used to fabricate many products used in construction, such as flooring, cement pipe, roofing, and insulation. It is also used in friction products, such as brakeshoes for automobiles.

J. Phosphate Rock

World phosphate rock production increased by 72%, while the United States production rose by 64% in the decade covered by Table XI. The bulk of the world's supply of this necessary fertilizer mineral comes from the United States and Morocco (Fig. 10). The Soviet Union is a major producer, but mainly for internal consumption. Except for a relatively small part of phosphate that goes into detergents, virtually all of the material flows into the food industry, either as fertilizer or as ingredients in animal feed and other food products. Unlike the other materials covered in this brief survey, phosphate rock demand is primarily related to the level of agricultural production in the world, rather than to the industrial sector of the economy.

K. Sulfur

World sulfur production increased by 65%, while that of the United States rose by 54% in the period covered in Table XII. Sulfur is a ubiquitous chemical used in fertilizer and virtually all of the industrial sectors in the form of sulfuric acid. The major producers of the world are the United States, Canada, the Soviet Union, and Poland (Fig. 11). This commodity will be heavily affected by future recovery of elemental sulfur removed from the products of combustion of high-sulfur fuels.

Table IX Zinc Production, 1964–1973 (*5a*)

	Production (thousand short tons zinc content)									
	1964	1965	1966	1967	1968	1969	1970	1971	1972	1973
United States	575	611	573	549	529	553	534	503	478	473
Rest of world	3865	4131	4369	4781	4955	5335	5489	5576	5743	5893
Total	4440	4742	4942	5330	5484	5888	6023	6079	6221	6377

Table X Asbestos Production, 1964–1973 (*5a*)

	Mine production (thousand short tons)									
	1964	1965	1966	1967	1968	1969	1970	1971	1972	1973
United States	101	118	126	123	121	126	125	131	132	150
Rest of world	2950	2984	3149	3084	3170	4042	3672	3816	4050	4445
Total	3051	3102	3275	3207	3291	4168	3797	3947	4182	4595

Table X Phosphate Rock Production, 1964–1973 (*5a*)

	Production (thousand short tons)									
	1964	1965	1966	1967	1968	1969	1970	1971	1972	1973
United States	25,715	29,482	39,044	39,770	41,251	37,725	38,739	38,886	40,831	42,137
Rest of world	37,004	40,816	44,150	46,144	51,326	52,873	54,896	53,622	58,150	65,923
Total	62,719	70,298	83,194	85,914	92,577	90,598	93,635	92,508	98,981	108,060

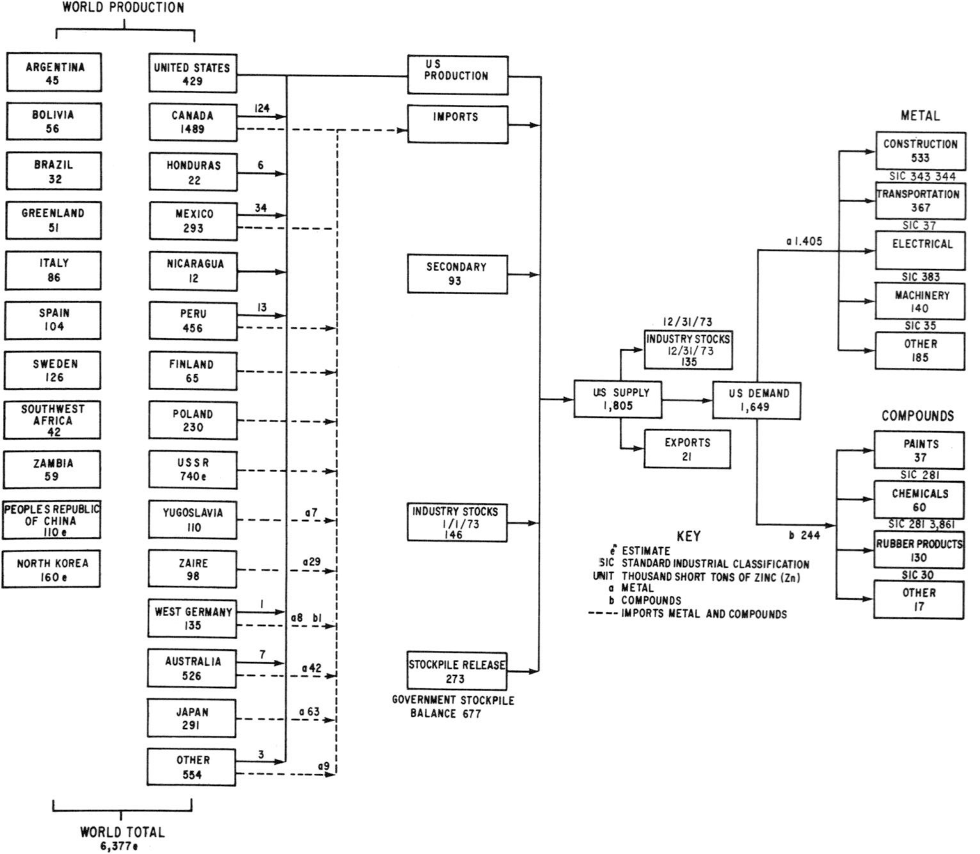

Figure 8. Zinc supply–demand relationships—United States 1973.

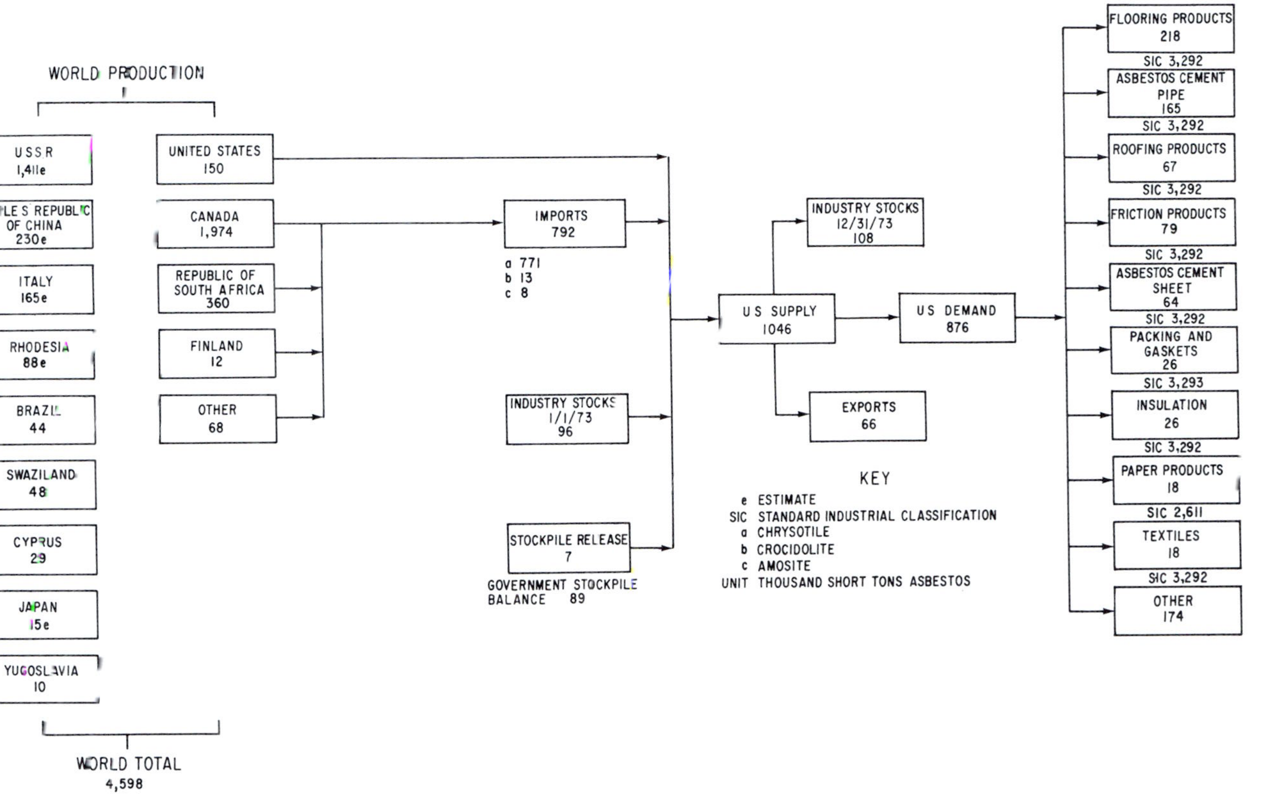

Figure 9. Asbestos supply–demand relationships—United States 1973.

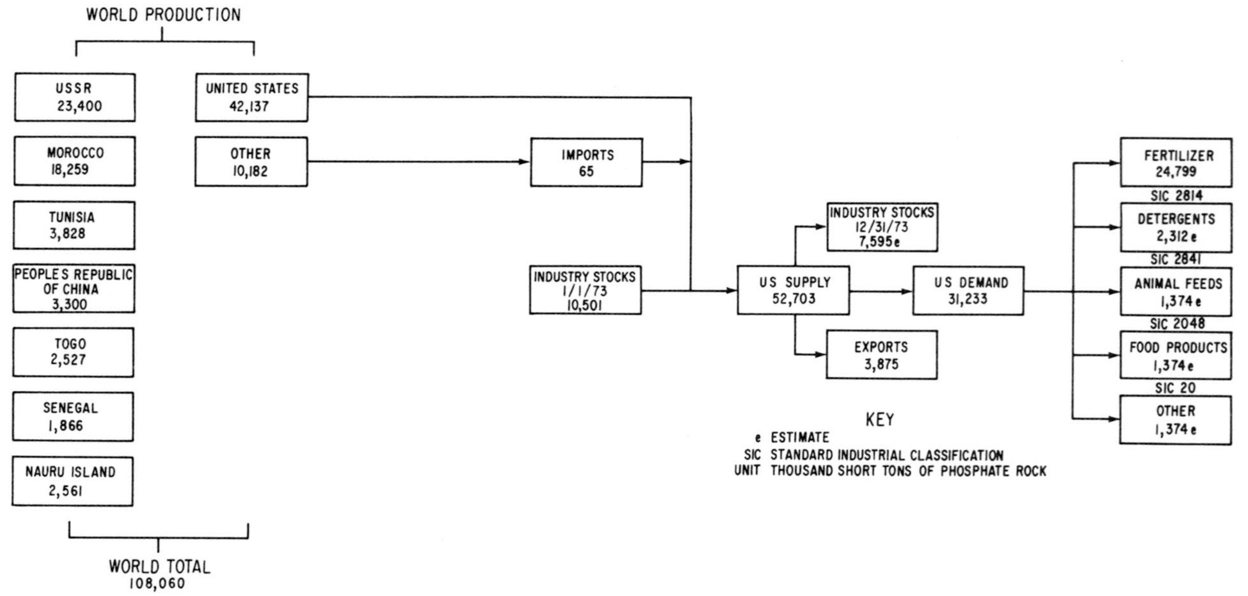

Figure 10. Phosphate rock supply–demand relationships—United States 1973.

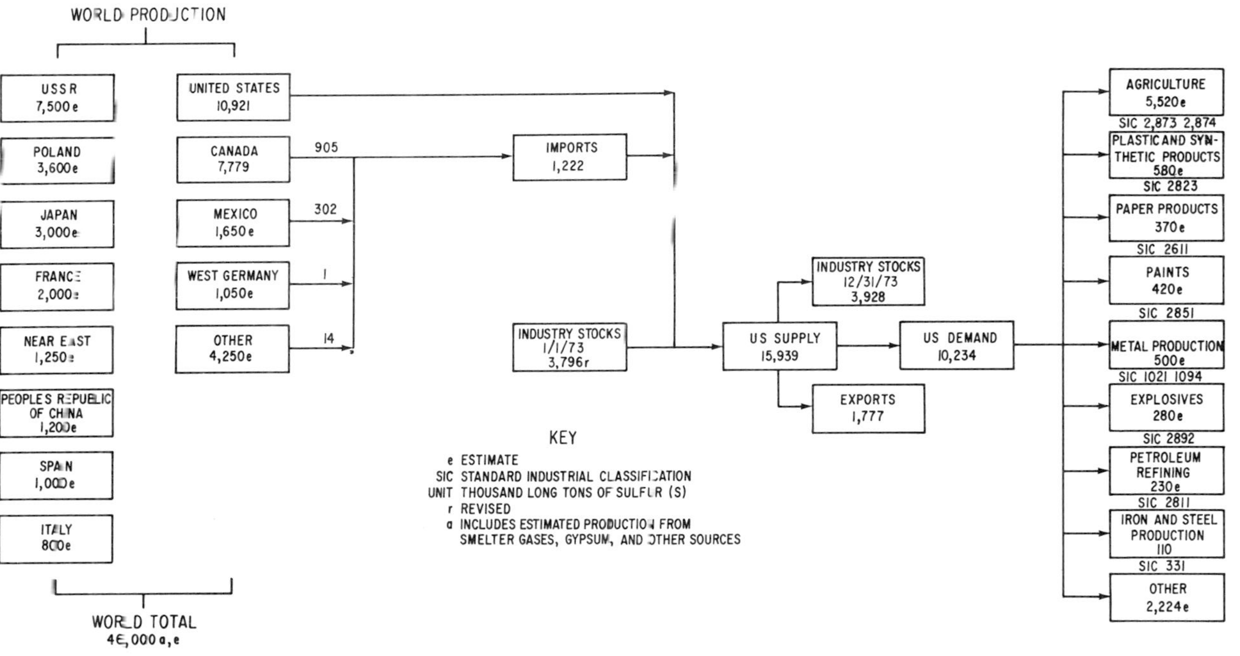

Figure 11. Sulfur supply–demand relationships—United States 1973.

Table XII Sulfur Production, 1964–1973 (*5a*)

	Production (thousand long tons)									
	1964	1965	1966	1967	1968	1969	1970	1971	1972	1973
United States	7,087	8,196	9,141	9,121	9,735	9,545	9,557	9,580	10,218	10,921
Rest of world	20,813	21,804	22,559	24,579	25,565	27,655	29,643	31,320	32,782	35,079
Total	27,900	30,000	31,700	33,700	35,300	37,200	39,200	40,900	43,000	46,000

Table XIII Coal Production, 1964–1973 (*5a*)

	Mine production (million short tons)									
	1964	1965	1966	1967	1968	1969	1970	1971	1972	1973
Anthracite coal										
United States	17.2	14.9	12.9	12.3	11.5	10.5	9.7	8.7	7.1	6.8
Rest of world	192.5	196.5	201.8	187.5	189.5	188.5	191.9	190.0	185.5	185.1
Total	209.7	211.4	214.8	199.8	200.9	199.0	201.6	198.7	192.6	191.9
Bituminous coal and Lignite										
United States	487.0	512.1	533.9	552.6	545.2	560.5	602.9	552.2	595.4	591.7
Rest of world	2334.4	2355.0	2365.8	2244.3	2344.4	2410.4	2482.4	2558.2	2564.6	2696.8
Total	2821.4	2867.1	2899.7	2796.9	2889.6	2970.9	3085.3	3110.4	3160.0	3288.5

L. Coal

Table XIII presents figures on anthracite and bituminous coal. Anthracite is relatively unimportant in the total coal picture (Fig. 12). It is a low-volatile, low-sulfur coal used in high-priority uses, such as domestic heating. Its use in the United States and in the rest of the world has been declining for many years. Bituminous coal (Fig. 13) is, by contrast, a major energy resource. United States production over the 10-year period has shown a modest increase of 21%, while the rest of the world has increased only by 16%. During this period, of course, coal has been substituted against by other energy forms that are less environmentally damaging, especially in the areas of air pollution. The major users of coal in the United States are the electrical utilities, followed by the primary metals industries, where it is used as coke in steel making. To prevent source disclosure, China is included as part of Europe in the tables.

M. Petroleum

World petroleum production has almost doubled in 10 years (Table XIV). The United States shows a much less rapid increase, rising by only 21% over the 10-year period and actually declining since 1970. The rest of the world, except the United States, shows an increase of 129% over the 10-year period. The bulk of the petroleum used goes to the transportation sector, followed by the household and commercial sector for heating, and into electric utilities for power generation (Fig. 14). The major world producers are the Middle East countries, the United States, the Soviet Union, and the South American countries. Petroleum is the major commodity moved in international commerce with all developed areas of the world showing major deficits between production and consumption, made up by the exports from South America, Africa, and the Middle East.

N. Uranium

Uranium production in the world and the United States fluctuated substantially during the 10-year period and in 1973 was slightly below the level reached in 1964 (Table XV) (*5a*) This production history masks the fact that during this period uranium moved from a material primarily used for military purposes to a significant energy material in nuclear reactors. Of significance in the supply–demand relationship indicated in Figure 15 is the very large industry stocks of uranium, equiva-

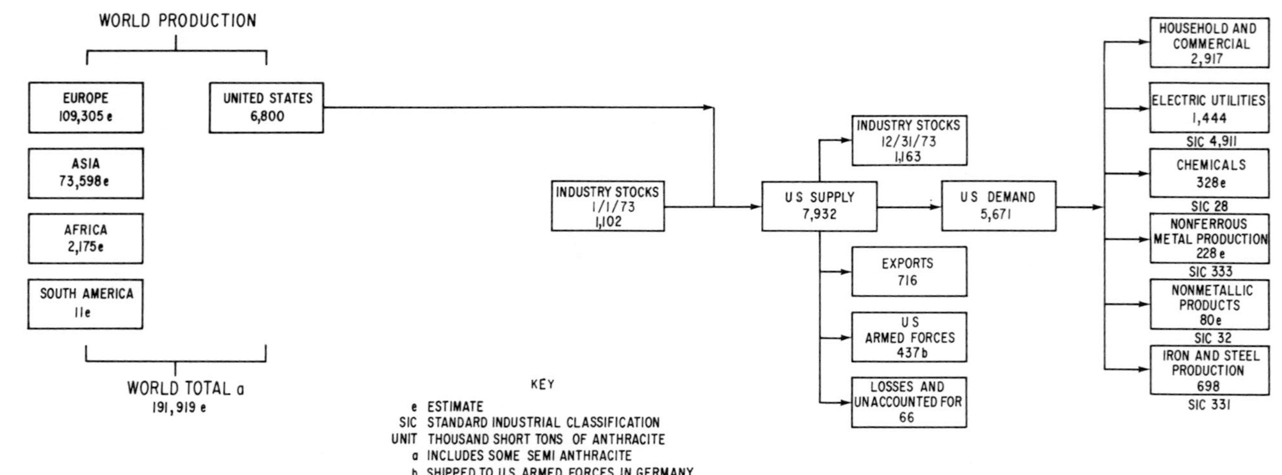

Figure 12. Anthracite supply–demand relationships—United States 1973.

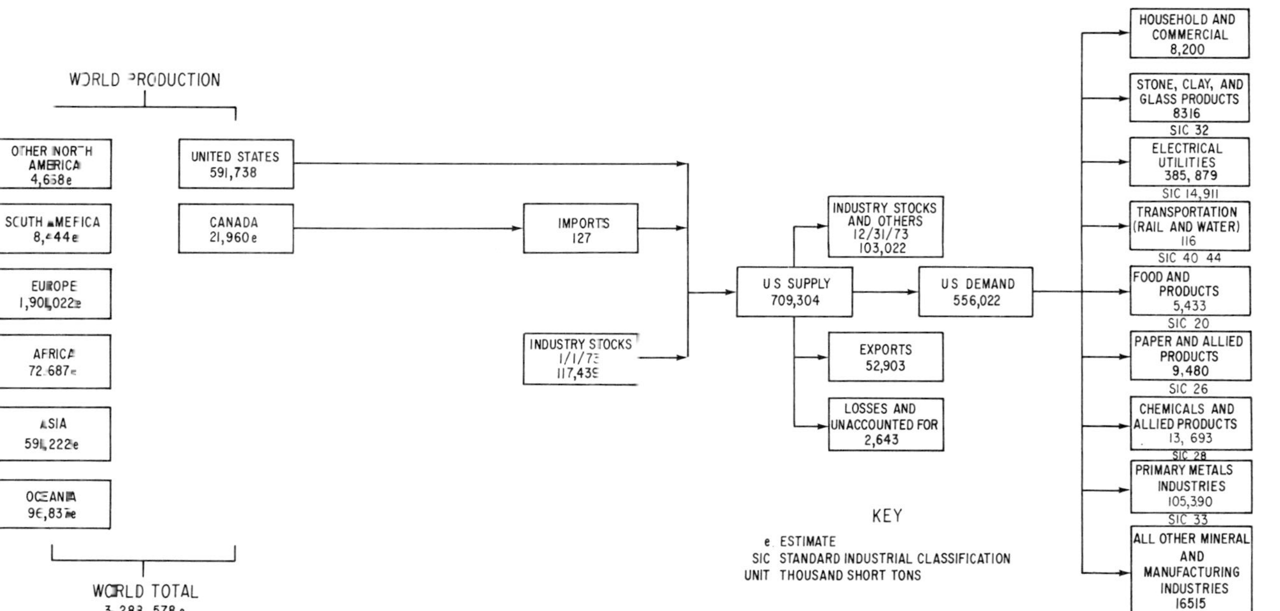

Figure 13. Bituminous coal and lignite supply–demand relationships—United States 1973.

Table XIV Petroleum Production, 1964–1973 (*5a*)

	Production (million 42-gallon barrels)									
	1964	1965	1966	1967	1968	1969	1970	1971	1972	1973
United States	2,786.8	2,848.5	3,027.8	3,215.7	3,329.0	3,371.8	3,517.5	3,453.9	3,455.4	3,360.9
Rest of world	7,524.3	8,210.0	8,992.0	9,673.6	10,775.3	11,843.1	13,193.3	14,208.9	15,142.4	17,200.0
Total	10,311.1	11,058.5	12,020.0	12,889.3	14,104.3	15,214.9	16,710.8	17,662.8	18,597.8	20,560.9

Table XV Uranium Production, 1964–1973 (*5a*)

	Production[a] (short tons uranium)									
	1964	1965	1966	1967	1968	1969	1970	1971	1972	1973
United States	10,625	8,769	8,979	9,543	10,488	9,844	10,943	10,408	10,939	11,223
Rest of world[b]	11,596	8,690	7,574	6,652	9,020	9,730	9,546	9,928	10,868	10,458
Total[b]	22,221	17,459	16,553	16,195	19,508	19,574	20,489	20,336	21,807	21,681

[a] Does not include centrally controlled economies.
[b] Estimate.

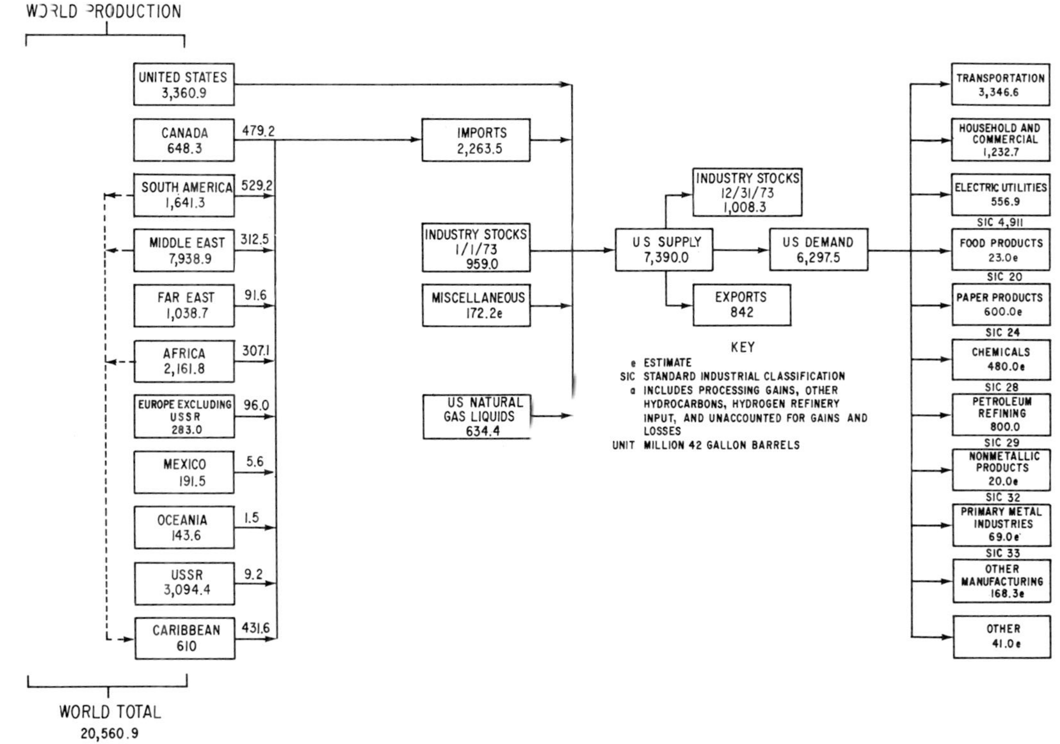

Figure 14. Petroleum supply–demand relationships—United States 1973.

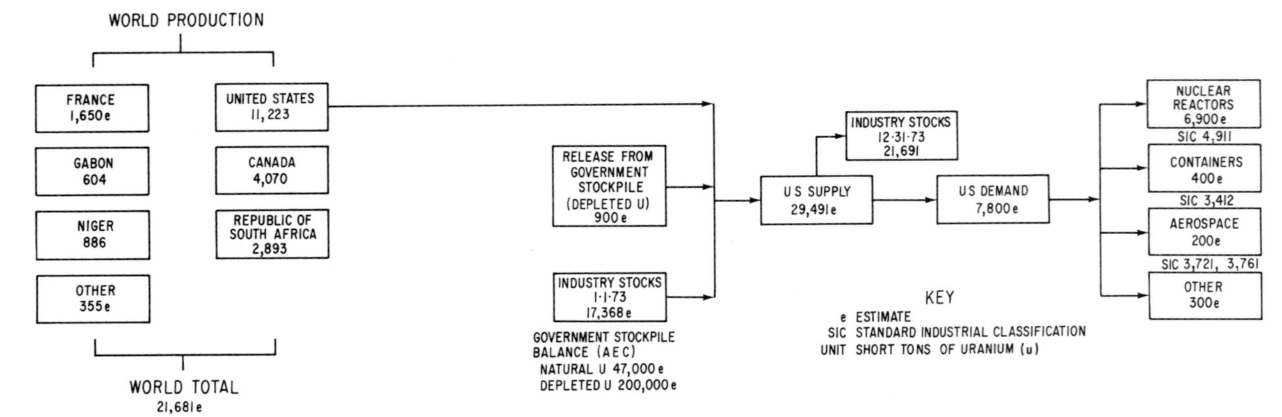

Figure 15. Uranium supply–demand relationships—United States 1973.

lent to approximately 3 years' supply for electricity generation. This does not include the very large stocks in the government stockpile that are also indicated in the figure. There is a worldwide feeling that uranium will be a major energy material with a very rapid growth rate in demand, at least through the year 2000.

O. Thorium

Unfortunately, because of disclosure regulations, thorium production figures are not available for the United States and the rest of the world. However, total world production of thorium for the years 1964–1973 increased from 570 short tons to a little over 1000 tons from the beginning to the end of the period. Figure 16 does show the 1973 supply–demand relationships in the United States. Notice that United States supply was much larger than United States demand, which indicates very substantial industry stocks. The bulk of thorium used, over 55%, goes to nuclear reactors, although there are significant uses in lamps and lighting, where it is consumed in gas mantles.

III. Resource Availability

The question of the availability of mineral resources in the future is one that has been addressed by each generation since the advent of the industrial revolution. In each generation, there are scholars who raise the spector of resource exhaustion and consequent decline in the quality of human life. Also, in each generation, there are scholars who are optimistic and argue the alternative position. In this section, the attempt will be to report this age-old debate in the context of current supply and to provide a frame of reference for evaluation of the doomsday versus cornucopian positions.

A. Mineral Availability Concepts

1. General Principles

Mineral resources can be distinguished from other economic goods by the fact that such resources are obtained by a complex process of exploration and development from the earth's crust or the waters of the sphere. The entire earth's crust is made up of minerals in varying abundance. Geologists have estimated abundance in terms of grams per metric ton of the total earth's crust. As can be seen from Table XVI,

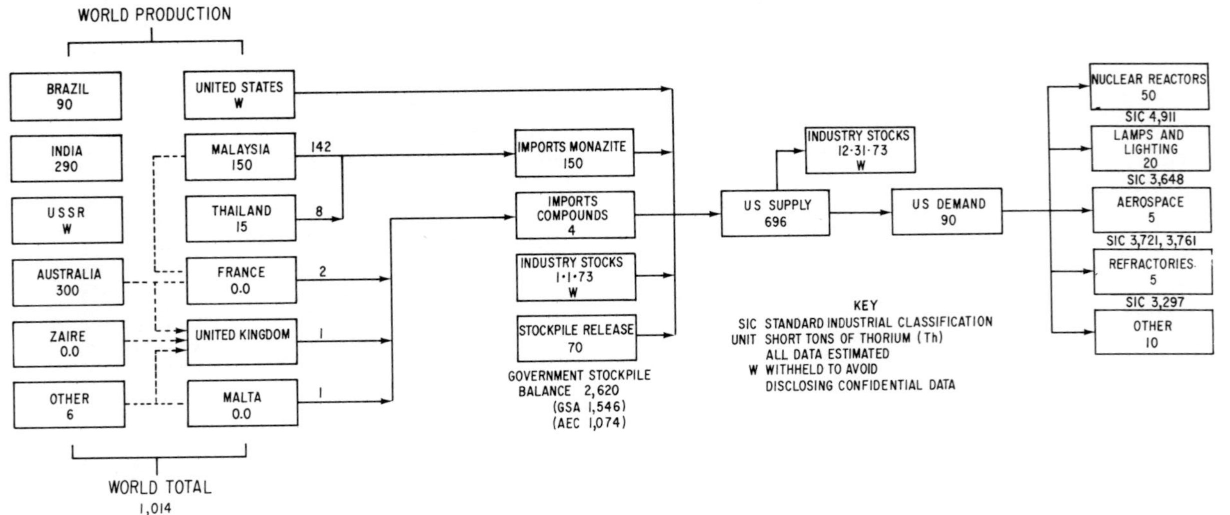

Figure 16. Thorium supply–demand realtionships—United States 1973.

Table XVI Crustal Abundance of Selected Mineral Elements *(7,20)*

Element	*Abundance—total earth's crust (gm/metric ton)*	*Element*	*Abundance—total earth's crust (gm/metric ton)*
Aluminum	83,000	Antimony	0.62
Barium	390	Beryllium	1.3
Chromium	110	Bismuth	0.0043
Fluorine	450	Cobalt	25
Iron	58,000	Copper	63
Manganese	1,300	Gold	0.0035
Phosphorus	1,200	Lead	12
Titanium	6,400	Lithium	21
Vanadium	140	Mercury	0.089
		Molybdenum	1.3
		Nickel	89
		Niobium	19
		Platinum	0.046
		Selenium	0.075
		Silver	0.075
		Tantalum	1.6
		Tellurium	0.00055
		Thorium	5.8
		Tin	1.7
		Tungsten	1.1
		Uranium	1.7
		Zinc	94

this crustal abundance varies greatly between the metals, with aluminum being by far the most abundant, followed by iron, and then, orders of magnitude less, the other elements.

Some writers have developed the concept of "resource base," which defines the crustal abundance (and similar kinds of data for the non-elemental minerals such as petroleum and coal) multiplied by the total volume of the earth's crust to a depth of 1 km or 1 mile as the resource base for any material (see Table XVI). Obviously, this measurement of resource base gives extremely large numbers; and when combined with the fact that many materials are not consumed in use, but are available in recycled form, such computations indicate that the idea of physical exhaustion so long as the earth remains is absurd. But, of course, physical exhaustion is not the issue.

The critical issue is the availability of resources to society through a production function and the terms of that availability in relative costs. Thus, to examine the impact of resources on the quality of life, the costs of resources in terms of capital, labor, and externalities must be considered, rather than any measures of physical abundance.

2. Proved Reserves

To indicate these various dimensions of resource availability, Dr. Vincent E. McKelvey of the U.S. Geological Survey (*6*) has developed a resource classification, reproduced in Figure 17. This figures covers the entire resource base as defined above. As one moves toward the left side of the figure, the degree of certainty of existence of an ore body of any given economic level increases. Movement up the diagram indicates the economic feasibility of any ore body known with a given degree of certainty. Thus, the extreme upper left-hand quandrant of the diagram contains those resources that are known with a high degree of certainty and that are recoverable under current economic conditions. That quandrant, then, contains information on the "proved reserves" of any given commodity. This proved reserve figure is the only one that has a degree of numerical exactitude, and it is the figure that is normally quoted in the literature concerning resource availability. The common statement that there is only enough oil for 8 years in the United States is based upon a division of this proved reserve figure by an annual consumption figure.

However, proved reserves at any point in time are a function of the exploration effort that has been devoted to search for the material in the previous few years and the investments in development. It takes considerable expenditures to find and develop a deposit of minerals. Thus, an industry will trade off such expenditures against alternatives available for the expenditure of that amount of its fiscal resources. It will not

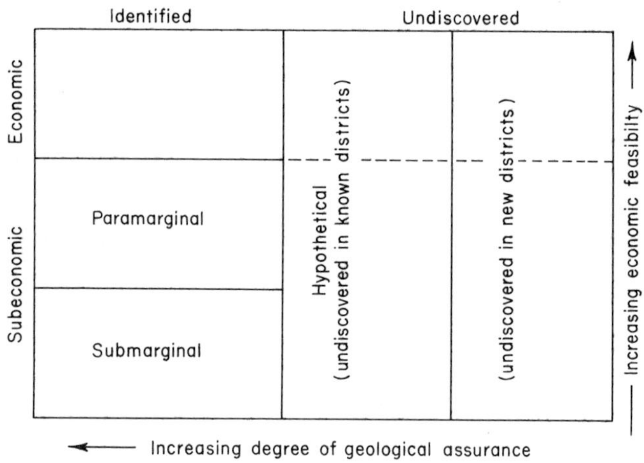

Figure 17. Resource classification.

pay an industry to develop proved reserves far in excess of current industry demand, for like any inventory, proved reserves carry cost. Thus, it is to be expected at any point in time the proved reserve figure will bear some relationship to industry capacity and demand, but will not measure, in any sense at all, the total potential availability of the mineral in question. Table XVII indicates the reserves and production for petroleum in the United States and the world for the past 50 years. It can be noted from this table that the ratio between these volumes fluctuates in response to major discoveries. At any point in time, the reserve production ratio indicates exhaustion within a short time period. However, cumulative production has greatly exceeded proved reserves at any point in history.

The critical question, then, for resource availability in the future is not the size of the proved reserve box, but the possibility of movement to the right and up in the diagram by exploration to discover new deposits and by technological advance to improve the feasibility of economic recovery of known or to-be-discovered poor-quality resources. As indicated above, the size of the total box is extremely large and poses no constraint upon mankind

3. Minerals

The future is unknowable. Nevertheless, in order to plan, mankind must make judgments about it. For those minerals for which crustal abundance data are available, a method of determining the size of the block taking the recoverable quadrant across the diagram to the end has

Table XVII Production and Proved Reserves of Crude Oil (Thousand 42-gallon barrels)[a]

	United States		*World*	
Year	*Production*	*Proved reserves*[b]	*Production*	*Proved reserves*[b]
1930	898,011	13,600,000	1,411,904	25,790,400
1940	1,353,214	19,024,515	2,149,821	41,465,515
1950	1,973,574	25,268,398	3,803,027	90,223,798
1960	2,574,033	31,613,211	7,674,493	299,065,011
1970	3,508,138	44,713,000	17,211,573	616,453,000

[a] Sources: "Petroleum Facts & Figures." American Petroleum Institute, Washington, D.C., 1971. "BP Statistical Review of the World Oil Industry, 1970." British Petroleum Company, Ltd., London, England, 1971. "The Petroleum Almanac." National Industrial Conference Board, New York, New York, 1946.
[b] Through December 31.

been developed by scientists at the U.S. Geological Survey (7). This concept measures the "recoverable resource potential" under current economic conditions, on the assumption that exploration efforts will be undertaken to discover all currently economic undiscovered ore bodies in the world.

The methodology is relatively simple. The beginning point is an assumption that the elements in the total earth's crust will be distributed with respect to concentration in a log-normal pattern. If this is so, then the past findings of an element in a fully explored area, assuming that area is large enough to be representative of the entire earth's crust, can serve as a yardstick for the predicted findings of similar concentrations in the entire earth's crust. An estimate of the potential economically recoverable resources can be made by developing the relationship between crustal abundance and the findings of a single resource in a representative area of the crust.

Lead has been chosen by the U.S. Geological Survey as the base of estimates, because in the United States as of 1970 the known reserves of mineral lead ore already exceed the ratio of abundance times 10^{10}, which is the ultimate relationship. In fact, in the United States, the factor by which lead reserves exceed abundance times 10^{10} is greater than for any other mineral that has been sought for many years. Thus, to develop a minimum estimate of the potential recoverable resources for all elements based upon their crustal abundance, the following formula has been applied:

$$R = 2.45A \times 10^6 \tag{1}$$

where R = potential recoverable resources and A is crustal abundance expressed in parts per million. Using this calculation, Table XVIII shows the ratio of potential resources to known reserves for the United States and the world for selected minerals. It is interesting to note that the metals that have the smallest ratio are those that have been looked for for the longest time—lead, aluminum, copper, silver, gold, and zinc. Uranium also shows a relatively low ratio, reflecting the intensity with which it was sought during the last two decades. These relationships, in a sense, validate the approach. They seem to indicate that if the search is intense enough, the findings of economically recoverable resources will approach as a limit the estimates developed by the above formula.

4. Fossil Fuels

Unfortunately, the above scheme of measuring recoverable resource potential in the unknown area of the resources diagram cannot be used

Table XVIII Ratio of Potential Resources to Known Reserves[a]

United States		*World*	
Lead	1	Antimony	5
Molybdenum	1	Copper	10
Copper	1.6	Lead	10
Silver	3.2	Tin	12
Gold	4.1	Gold	14
Zinc	6.3	Silver	18
Antimony	11	Molybdenum	23
Mercury	15	Mercury	30
Uranium	20	Nickel	38
Thorium	31	Zinc	42
Tungsten	37	Tungsten	42
Nickel	830	Uranium	112
Tin	Very high	Thorium	288

[a] Source: "Professional Paper 820." United States Geological Survey, Washington, D.C., 1973.

for the fossil fuels, and it is precisely these fuels about which the most questions have been raised concerning future availability.

In the case of petroleum, there are two fundamental methods used to estimate undiscovered economic resources. The first of these is to make estimates based upon the area and volume of sedimentary rocks that are possible hosts for petroleum accumulations in the major basins of the world multiplied by the amount of oil discovered in similar rocks in fully developed petroleum provinces. The second method is to use statistical projections of past oil and gas production and drilling experience for the continental United States as a test case.

A recent study by the National Academy of Sciences (*3*) looked at these two methods and found that the result were orders of magnitude apart. Table XIX presents the results reported in the National Academy study plus an additional recent estimate by Vincent McKelvey of the Geological Survey (*8*). Comparative estimates for the world are not available from all of the sources, although there are three estimates—two from ore companies (*9, 10*) and one from Weeks (*11*). The range in these estimates is very great, ranging from 68 billion tons to 206 billion tons.

Given the very wide estimates of ultimate petroleum recovery, there is some significant doubt as to usefulness of these estimates for public policy purposes. An in-depth survey of this problem in 1972 (*12*) lead to the conclusion that all such estimates are based on past exploratory

Table XIX Estimates of Undiscovered Recoverable Oil and Gas Resources of the United States[a]

	Oil and natural gas liquids (10^9 *bbl*)	*Gas* (10^{12} ft^3)
Oil companies		
1. Company A (Weeks, 1960)	168	—
2. Company C (1973)	55	—
3. Company D (1974)	89	450
4. Company E	90	—
U.S. Geological Survey		
5. Hendricks (1965)	346	1300
6. Hubbert (1967)	24–64[b]	180–500[b]
7. Theobald *et al.* (1972)	458	1980
8. McKelvey (1974)	200–400	990–2000
9. Hubbert (1974)	72	540
10. McKelvey (1975)	61–149	322–655

[a] Sources: Table as a whole: "Mineral Resources and the Environment," p. 89. National Academy of Sciences, Washington, D.C., 1975. Individual Entries: 1. L. G. Weeks, *Geotimes* **5,** No. 1, 18–21 and 51–55 (1960). 2–4. Not identified. 5. T. A. Hendricks, *U.S., Geol. Surv., Circ.* **522** (1965). 6. M. K. Hubbert, *Amer. Ass. Petrol. Geol., Bull.* **51,** 2207–2227 (1967). 7. P. K. Theobald, S. P. Schweinfurth, and D. C. Duncan, *U.S., Geol. Surv. Circ.* **650** (1972). 8. V. E. McKelvey, "Revised U.S. Oil and Gas Resources Estimates," News Release, March 26. U.S. Geological Survey, Washington, D.C., 1974. 9. M. K. Hubbert, Statement for Joint Hearings on May 10, 1974 of the Senate Commerce and Government Operations Committees to Consider Legislation Designed to Monitor and Alleviate Product and Material Shortage: Multilithed 1974. 10. V. E. McKelvey, News Release, April 13. U.S. Department of the Interior, Washington, D.C., 1975.
[b] Exclusive of Alaska.

records and therefore are unreliable. A prominent statistician, when asked to review the approach of these analysts and looking at the methodology, came to the conclusion that "projections of oil resources have never been sufficiently precise to provide any useful guidance for determining long-run policies with respect to energy—such remains the case today" (*13*).

Coal presents a very different picture. The identified coal resources of the world, see Table XX, are very large. This is true because different criteria are applied by the geologist in estimating coal reserves. Coal occurs in bedded deposits that tend to cover a very wide area. Geologists do not hesitate to interpret between very widely spaced drill holes and thus count as "proved reserves" very large amounts of coal that require

Table XX World Coal Production and Resources in 1974[a]

Coal type	*Production (thousand short tons)*	*Identified resources (thousand short tons)*
Anthracite	195,100	11,759,000,000[b]
Bituminous and lignite	3,355,800	

[a] Source: "Commodity Data Summaries 1974." United States Bureau of Mines, Washington, D.C., 1975.
[b] This estimate encompasses anthracite, bituminous, and lignite in accordance with non-U.S. countries' combination of data. Included are U.S. figures of 21 billion short tons for anthracite and semianthracite and 1,560 billion short tons for bituminous and lignite.

extensive exploration and development before they can be produced. Thus, the reserves figures for coal cannot be taken as comparable to those for the metals in the proved reserve quadrant of our resources diagram. The figures actually quoted for coal are probably much closer to the potential economically recoverable resources estimated by the McKelvey equation above for the metals and lie within the range of the ultimate petroleum recoverable resources estimates presented in Table XIX.

B. The Production Process

As the above discussion indicates, the ultimate availability of any single mineral commodity is subject to great uncertainty. While ultimate availability questions are of interest in very long-range planning, of much more immediate concern is the existing productive capacity for a mineral commodity relative to the demands for that commodity on a worldwide basis. For most mineral commodities, the relationship between demand and capacity at current market prices is one of alternate periods of glut and shortage. The explanation for these alternate periods of glut and shortage lie in the way that mineral markets operate. The demand for mineral commodities will be discussed in the next section of this chapter. Here the process that generates current mineral capacity and supply will be described.

Minerals are produced from the earth's crust in crude or raw form. They must go through a series of processing steps, which varies with each commodity, to bring them to their first usable form, and then they are incorporated through technologies into final goods and services delivered to the economies of the world. The production process flowing from the crude material to its ultimate form can be described in five stages—exploration and development, production, transportation, primary production, and manufacture.

1. A Typical Metal

The above process for a typical metal will involve exploration for an ore body containing the metal in sufficient concentration and in special relationship to the earth's surface so as to allow the development of a paying mine. It is impossible to put a time dimension on these exploration activities; however, generally there is a very significant passage of time between exploration in a prospective mining district and the initial discovery of an ore body that warrants development (*10*).

After discovery of such an ore body, the development phase begins, which involves the building of the primary structures to facilitate the mining of the ore body and the construction of the concentrating plant to remove the metal from the host rock. Such an operation will typically take at least five, and often as much as ten, years to complete. This is the period, of course, of major capital investment in the specific mine or ore body.

Once the concentrator has been built and the mine opened, the production phase begins. Production from a given ore body can proceed for many years or may be limited by the confines of economic concentration of the metal.

The concentrated ore will move to a smelting/refining operation, which may or may not be continguous to the mine. The result of the smelting and refining operation is primary metal in various shapes, such as ingots, bars, or slabs. This material is then transported to a manufacturing plant where it is incorporated into a wide variety of investment or consumer goods.

The above process inherently generates very long lead time for capacity expansion of primary metals. Potential mineral producers are constantly faced with the fact that major investments in exploration for and development of mining properties will not bring new production onto the market for up to a decade. Thus, they are making investment decisions based upon estimates of the future market for the mineral, rather than the current market. This short-run inelasticity in mineral capacity gives rise to instability in mineral prices over time and to recurrent periods of capacity glut and shortage.

2. Petroleum

The situation with respect to the fossil fuels, e.g., petroleum is different only in kind with respect to the above description for a typical metal.

Exploration activity must be undertaken through geophysical, geochemical, and geological means to discover formations or structures in the earth's crust that are necessary for the accumulation of economic amounts of petroleum. Then these structures must be tested by drilling to determine whether or not petroleum is contained in paying quantities. Upon discovery of a pool of petroleum, either oil or gas, then there is a period of development drilling to put in place producing wells and investment in transportation facilities to transport the petroleum and gas to a refinery or point of consumption. The crude petroleum moves then to a refinery where it is transformed into petroleum products, which in turn are either consumed directly by consumers or incorporated as fuel for manufacture. The refineries also produce feed stocks for petrochemical operations to produce a wide variety of organic chemicals and plastics.

Once again, it is impossible to put a time dimension on the exploration process. Experience under forced draft development, for example, in the North Sea area of Europe and in the outer continental shelf areas of the United States, indicate that from the time of acquisition of the right to drill, which follows a long period of geophysical and geological work, there is a minimum period of five years before petroleum begins to flow in significant volume from a discovery. Thus, here, too, there is a long period of time between foreseen demands and the building of capacity to meet them.

3. Summary

This excursion into the production process has the purpose of indicating the uncertainties that surround mineral resources. The long lead times involved, the uncertainty of the exploration process, and the major front-end investment involved in bringing the crude material into the economic system result in periodic periods of glut and shortage. These periods are exacerbated by the characteristics of mineral demand discussed in the next section.

Many scholars, who have insufficient appreciation of this cyclical phenomenon, tend to overreact greatly when the periods of shortage arise. The first five years of the decade of the 1970s were a period of such shortages in many major metals, as well as in petroleum and natural gas. This period has lead many scholars of the doomsday persuasion to argue that the abundance of mineral resources that characterized the industrial societies and the world as a whole for the last century has come to an end, and society must now adjust to a period of resource scarcity in absolute terms.

It is a thesis of this writer that such an analysis is incorrect. As indicated above, physical scarcity, at least of the elements in the earth's crust, simply is unthinkable. The question is the relative costs to society of obtaining minerals, not whether minerals can be obtained at all. In the case of the fossil fuels, it is equally true that physical constraints are not present for many, many years. Such fuels are not recycleable—once burned, the heat is dissipated into the atmosphere. Therefore, there is a loss of entropy in transferring the stored heat that is available into released heat, which is not. But the increase in prices of these materials in the past five years has been a function of a political and social phenomenon, not increased resource scarcity. High prices of these fuels result from the extraction of monopoly rents through cartel action, primarily in the area of crude petroleum operating through the Organization of Petroleum Exporting Countries. There is no shortage of petroleum capacity in the world. In fact, petroleum production on a worldwide basis is operating only in the range of three-quarters capacity. Major portions of the earth's crust have not been touched by the exploratory drill for petroleum. As indicated above, there are substantial differences of opinion as to the ultimate amount of petroleum that will be available to mankind in the earth's crust; however, all the estimates indicate that this amount still remains as a very large figure. Thus, the problem of resource scarcity is one of investments in exploration and development of resource sources, not in the exhaustion of the occurrence of these resources in the earth's crust.

IV. Resource Demand

The major determinants of resource use by the world society can usefully be grouped under seven headings (*11*):

1. Demographic variables. This involves the size of population, the rate of growth, the age and sex distribution, the number of households, and the labor force participation rates. These variables are important, because they influence the amount and composition of consumption goods, as well as the level of industrial activity.

2. Standard of living. Although there are no unambiguous measurements of standard of living, this variable can usually be represented by per capita gross national product (GNP) plus some information on the distribution of income among various groups of the population.

3. Style of living. This factor refers to the pattern of preferences for various kinds of living styles. A style of living that includes heavy emphasis on throw-away bottles, suburban living, and a high use of the

automobile as a transportation means will involve greater resource use than alternative life styles.

4. Geographic distribution. Where the population lives relative to each other and where economic activities take place will have a significant impact on resource use. The more dispersed the population and the economy, the more resources for transportation will be involved.

5. Technological structure. There are many available technologies for producing any given set of final goods demands. If these technologies are resource-intensive in terms of energy and raw materials as contrasted to alternative methods, total resource demands will be subsequently greater.

6. International relations. This factor is becoming of increasing importance with the development of cartel-like organizations among the raw-material producing countries of the world. Its impact on resource demands will be felt if the industrial nations pursue autarchic policies, which will, in the end, tend to reduce their resource use.

7. Institutions and policies. Resource demands are going to be effected by the economic institutions surrounding their production and use, as well as important policy decisions with respect to pollution, international trade, public investments, research and development, etc.

A. Demographic Variables

Although it is axiomatic that the overall size of the population must play a determining role in the consumption level for mineral resources, it must be remembered that changes in population growth play a relatively minor role in changes in resource use. For example, the world population was growing at a rate of something under 2% a year in the decade 1960–1970. A glance at the tables of specific commodities presented earlier in this chapter indicate that the world rate of growth in these commodities is generally greatly in excess of these figures. For example, petroleum over the 10-year period grew at 7.9% a year, sulfur at 5.7%, phosphate rock at 6.2%, asbestos at 4.7%, zinc at 4.0%, aluminum at 9.7%, and raw steel at 5.4%. It is clear from these figures that the change in the level of raw material demands must be explained by factors other than the numbers of people on the earth.

B. Standard of Living

The factors that relate to standard of living, as measured by per capita GNP, are the most important in determining resource demands. Economic theory relates the demands for any particular resource to two major variables, given the preferences and the structure of production in the society in question. These variables are the price of the resource relative to its substitutes and the level of income in the economy. Virtually all studies that have been done on mineral resources indicate that at least in the short run, price plays a relatively small role in level of consumption. That is, to use the economic jargon, the demands for these resources are price-inelastic. Although there are substantial price elasticities in the long run, for many years the prices of these commodities relative to all other prices have been declining. On the other hand, econometric studies indicate there are very high elasticities with respect to income. Stated another way, changes in the level of income are reflected more than proportionally in the increase in demand for these commodities. Thus, it is growth in income in the aggregate and per capita that is the major determinant of the level of raw material demand in the world.

C. The Other Factors

The other five factors mentioned above account for the differences in resource consumption between societies that are not explained by the major variable of per capita income and population size. They play a relatively small role in determining the overall level of aggregate resource demand; but they are nevertheless significant in any given country.

D. Commodity-Demand History

Recent studies on the demand for mineral resources have emphasized the relationship between the level of demand (consumption) and the size of the income of the economy. The basic hypothesis developed is that the level of consumption is a function of the size of the gross national product (GNP) of an economy, and that the use of a mineral per unit of GNP declines as GNP grows larger. Stated another way, the "intensity of use" of a material declines as the economy grows (*12*).

The only country in the world that can be used as a model to test this hypothesis is the United States, which, as a fully developed industrial economy, should exhibit such tendencies if they, in fact, occur. For the other countries of the world, the statistical data are simply not adequate to test the hypothesis, without very considerable basic research.

The major commodities analyzed in Section II of this chapter have been chosen for analysis here. Figure 18 shows time series for eleven commodities of short tons per billion dollars of real GNP from 1929 to 1974 (except for crude petroleum, which is in thousand barrels per billion dollars GNP). Figure 19 shows these relationships between use and GNP.

Aluminum. Aluminum illustrates a commodity in its relatively early stages of development when consumption per unit of real GNP is rapidly increasing. This increase indicates aluminum is replacing other materials in many uses, primarily copper and steel and forest products in the container industry. Aluminum is in the position of increasing use relative to GNP even at today's GNP level.

Copper. Copper demonstrates clearly the intensity of use hypothesis; copper shows declining use relative to GNP.

Iron ore. The analysis is conducted here on the basis of total iron input, rather than steel production, as was the case in Section II. Once again, the declining relationship between intensity of use and GNP is obvious.

Lead. Lead also clearly demonstrates the hypothesis.

Nickel. The statistical picture for nickel is not nearly as clear as that for the previously described metals. The relationship seems to be relatively constant over the period, but in any case, there is no clear correlation between the two variables.

Tin. Tin shows a sharp decline in use over time and also per unit of GNP.

Zinc. Zinc clearly demonstrates the hypothesis.

Asbestos. Asbestos demonstrates a clearly developed pattern of behavior. This, even though it is consumed mainly in the construction industry does not necessarily parallel gross national product.

Phosphate rock. This is a fertilizer material and is more responsive to the growing needs for food and changing technologies in that sector than to gross national product. The data do not demonstrate the intensity of use hypothesis.

Sulfur. This ubiquitous chemical is an excellent example of the lack of decline in per unit consumption as GNP increases.

Coal. This fuel, which has been substituted against by other competing fuels, clearly shows a decline both in the time series analysis and in the regression approach.

Crude petroleum. Even though crude petroleum did substitute for coal in electric power generation, it has in recent decades demonstrated the decline in use hypothesized.

Uranium. Unfortunately, uranium does not have a significantly long history of consumption, since it only within the last decade became a commercial material for power generation. Therefore, no data are presented.

E. Summary of Demand Analysis

The life cycle of a mineral is one of rapid increase in intensity of use during the early years of its movement into the industrial sector, fol-

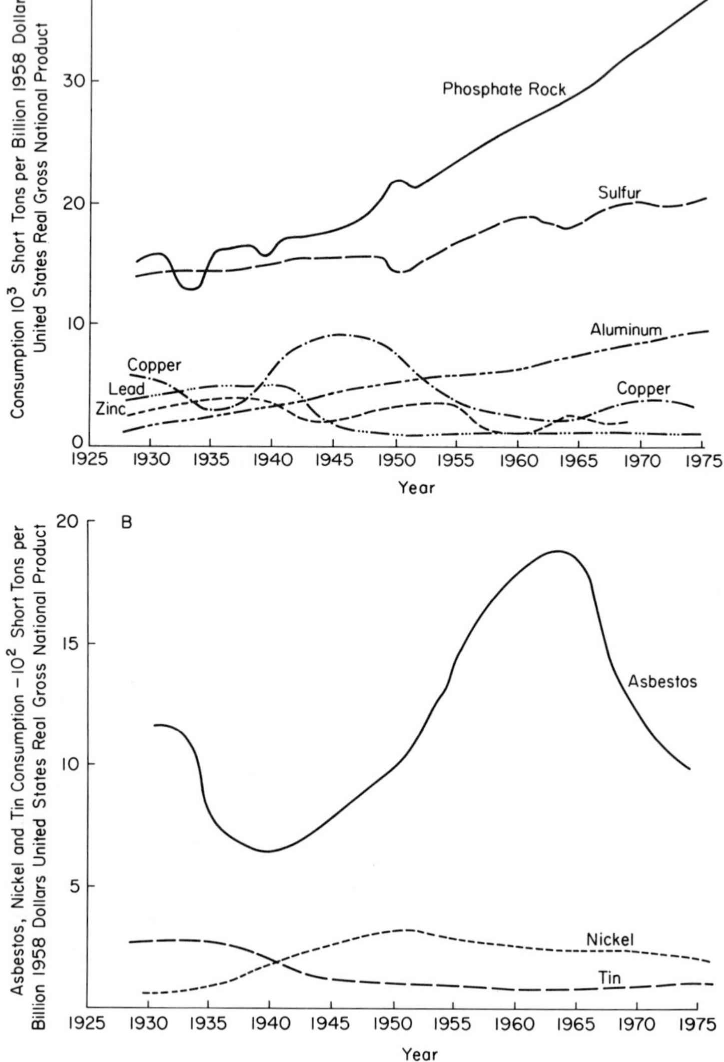

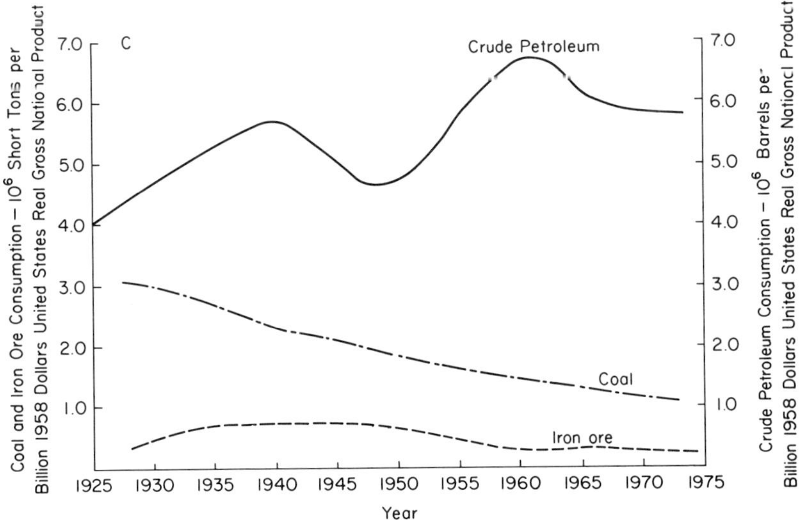

Figure 18. A, consumption, United States 1929–1974—aluminum, copper, lead, phosphate rock, sulfur, and zinc. B, consumption, United States 1929–1974—asbestos, nickel, and tin. C, consumption, United States 1929–1974—coal, iron ore, and crude petroleum.

lowed by a relative decline as the economy continues to grow. Such behavior is best explained by the substitution phenomenon. When a new material comes in, such as aluminum, it finds a ready market as a substitute for older and less suitable materials in certain specific uses. The material then grows more rapidly than the total economy, because it is substituting for a material that was previously used. Once this market spread or penetration is accomplished, however, the use of material becomes related to the total demand for the products in which it is used, which is in turn related to the size of the income in the economy. As the economy's income grows, there is a shift, at least in the case of the highly industrialized, wealthy economies, away from goods and into services. Thus, the goods component of income becomes a declining proportion of total income. This is reflected in the declining proportion of income represented by the basic raw materials.

This description of the life history of consumption of a material is what lead to the intensity of use hypothesis described above. Our analysis of the individual commodities indicated clearly that in case of the well-established, well-developed commodities the hypothesis held. Certain commodities, which are still penetrating new markets through substitution are still on the upswing relative to income levels.

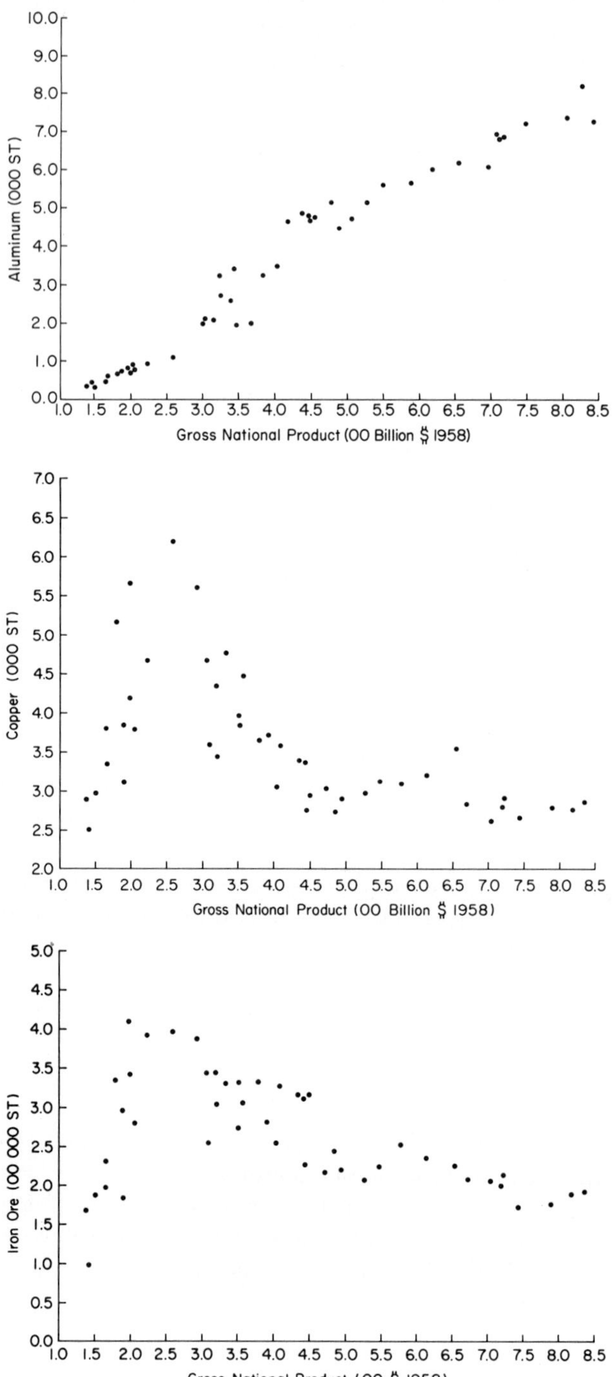

Figure 19. Abscissa: Real gross national product (GNP); ordinate: Consumption per unit GNP (ST = short tons). United States, 1929–1974.

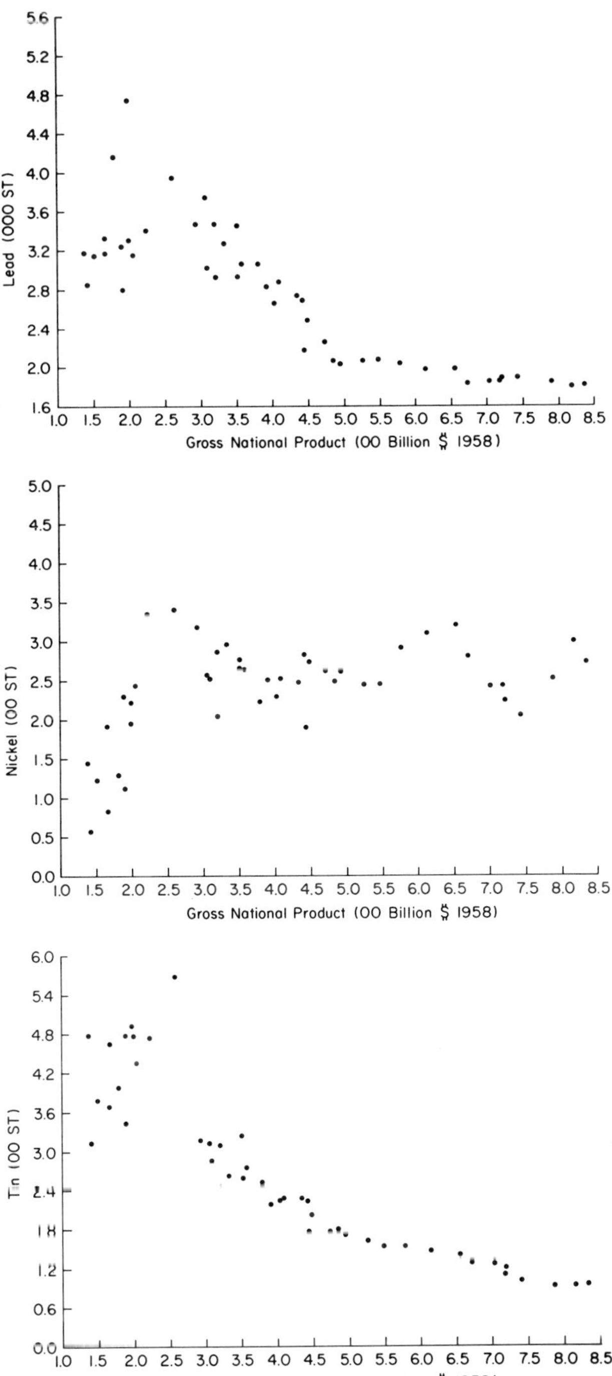
Lead (000 ST)
5.6
5.2
4.8
4.4
4.0
3.6
3.2
2.8
2.4
2.0
1.6
1.0 1.5 2.0 2.5 3.0 3.5 4.0 4.5 5.0 5.5 6.0 6.5 7.0 7.5 8.0 8.5
Gross National Product (00 Billion $ 1958)
Nickel (00 ST)
5.0
4.5
4.0
3.5
3.0
2.5
2.0
1.5
1.0
0.5
0.0
1.0 1.5 2.0 2.5 3.0 3.5 4.0 4.5 5.0 5.5 6.0 6.5 7.0 7.5 8.0 8.5
Gross National Product (00 Billion $ 1958)
Tin (00 ST)
6.0
5.4
4.8
4.2
3.6
3.0
2.4
1.8
1.2
0.6
0.0
1.0 1.5 2.0 2.5 3.0 3.5 4.0 4.5 5.0 5.5 6.0 6.5 7.0 7.5 8.0 8.5
Gross National Product (00 Billion $ 1958)

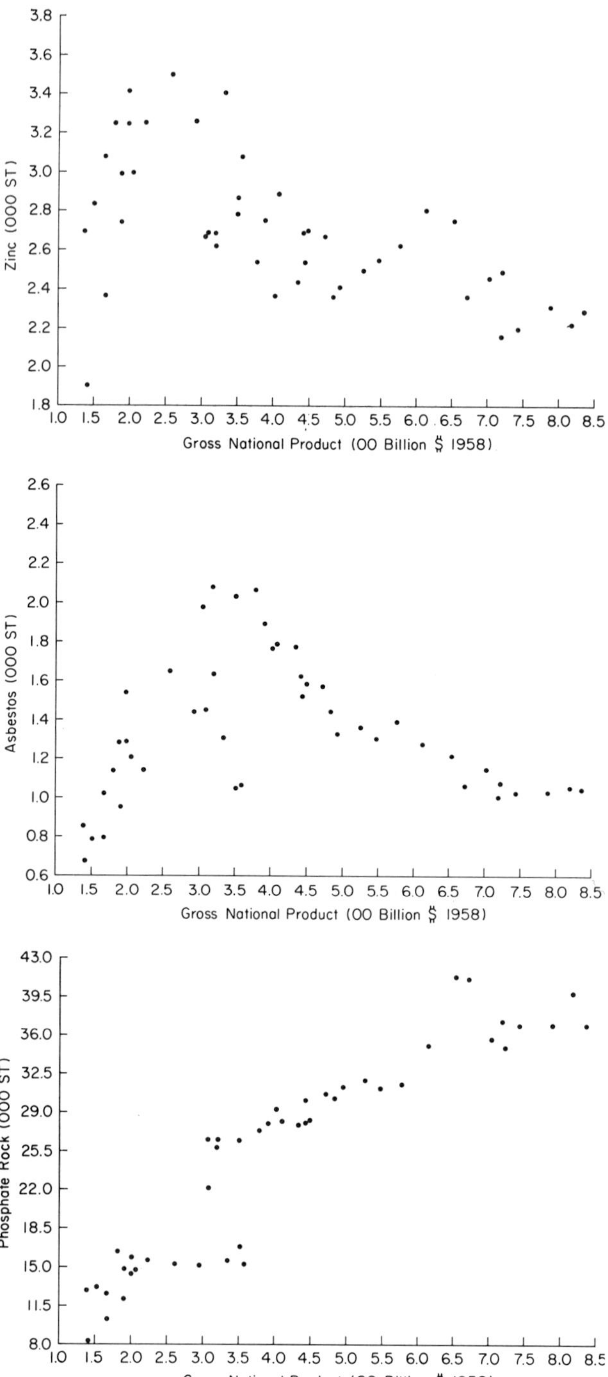

Figure 19 (*Continued*)

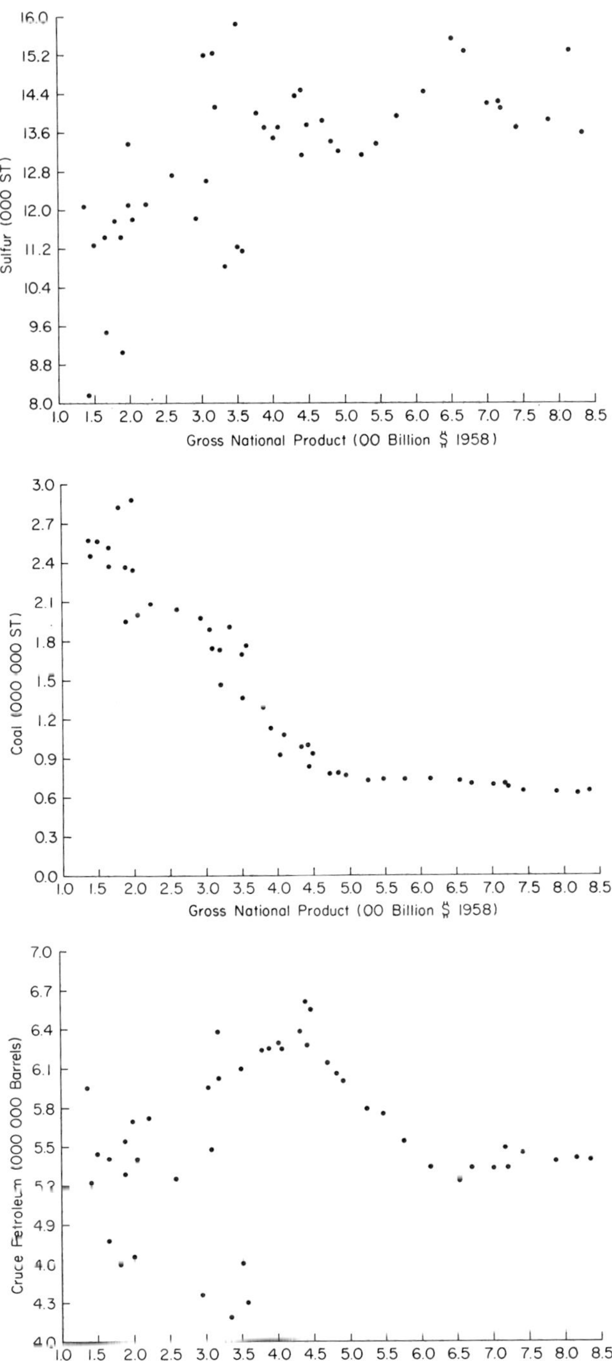
Sulfur (000 ST)
16.0
15.2
14.4
13.6
12.8
12.0
11.2
10.4
9.6
8.8
8.0
1.0 1.5 2.0 2.5 3.0 3.5 4.0 4.5 5.0 5.5 6.0 6.5 7.0 7.5 8.0 8.5
Gross National Product (00 Billion $ 1958)
Coal (000 000 ST)
3.0
2.7
2.4
2.1
1.8
1.5
1.2
0.9
0.6
0.3
0.0
1.0 1.5 2.0 2.5 3.0 3.5 4.0 4.5 5.0 5.5 6.0 6.5 7.0 7.5 8.0 8.5
Gross National Product (00 Billion $ 1958)
7.0
6.7
6.4
6.1
5.8
5.5
4.9
4.3
1.0 1.5 2.0 2.5 3.0 3.5 4.0 4.5 5.0 5.5 6.0 6.5 7.0 7.5 8.0 8.5
Gross National Product (00 Billion $ 1958)

V. Resource Adequacy

The classic economists were given the name of "the dismal scientists" because of their analysis of the constraints of resources on economic growth. Most clearly enunciated by Thomas Malthus (*13*), the theories of these economists stated that the economy would eventually reach a steady state, with the bulk of society at subsistence levels because of resource constraints. The century and a half that followed their analysis exhibited an absolutely contrary behavior. In fact, resources were available to sustain what, for them, would be unbelieveable levels of real income, and to do it with a decreasing cost of resources in real terms over the period. This, over a century of experience, led scholars to the conclusion that technology was the key that could overcome any presumed resource scarcity.

Nevertheless, the spectre of running out is still with us. The Club of Rome in its first report (*1*) modeled a collapse based upon resource scarcity. Its second report (*14*), while not as cataclysmic, foresees the necessity of living in a society without growth. The American Academy of Arts and Sciences devoted a major volume (*15*) to the problems of a no-growth society. Thus, once again the issue of resource adequacy to sustain economic growth is very much in the forefront.

There are three other analyses of this problem which throw some light on the issue. They are (a) an across-the-board comparison by the United States Bureau of Mines of cumulative primary mineral demand for the United States and the rest of the world to the year 2000 contrasted to proved reserves as of 1973; (b) a United States Geological Survey report "Mineral Resource Perspectives, 1975" (*16*); and (c) a series of analyses for the Commission on Population Growth in the American Future (headed by John D. Rockefeller, III), which look at the adequacy of nonfuel and fuel resources (*17, 18*). These are discussed below.

A. Cumulative Demand versus Reserve Analysis

Table XXI (*19*) presents the results of the United States Bureau of Mines analysis in terms of the ratio of mineral reserves recoverable at 1973 prices to the cumulative primary metal demand for the United States, the rest of the world, and the world as a whole. If this ratio is well above 1, it indicates that the reserves already discovered in 1973 are sufficient to supply the demands of the world through the year 2000. If it is smaller than 1, it indicates that these reserves are inadequate to meet these demands. An asterisk in the table indicates a ratio of ten or more.

Table XXI Ratio of Mineral Reserves Recoverable at 1973 United States Prices to Cumulative Primary Mineral Demand, 1973–2000[a] (19)

Commodity	*United States*	*Rest of the World*	*Total*
Aluminum	—	5.5	3.6
Antimony	0.1	2.3	1.6
Arsenic	1.2	4.0	2.7
Barium	1.0	0.7	0.8
Beryllium	1.4	*	*
Bismuth	0.1	0.5	0.3
Boron	6.7	7.0	6.9
Bromine	0.9	0.5	0.7
Cadmium	0.7	1.2	1.1
Cesium	—	*	*
Chlorine	*	*	*
Chromium	—	6.4	5.1
Cobalt	—	3.2	2.2
Columbium	—	*	9.4
Copper	1.1	1.1	1.1
Fluorine	0.1	0.4	0.3
Gallium	2.7	*	*
Germanium	0.5	0.8	0.7
Gold	0.4	1.3	1.1
Hafnium	*	*	*
Indium	0.3	1.1	0.7
Iodine	0.9	3.2	2.5
Iron ore	0.7	5.0	4.4
Lead	1.7	0.9	1.1
Lithium	1.1	3.1	1.9
Magnesium	*	*	*
Manganese	—	5.2	4.7
Mercury	0.1	1.3	1.1
Molybdenum	2.0	1.2	1.4
Nickel	—	2.2	1.6
Nitrogen	*	*	*
Palladium	—	2.9	2.0
Platinum	—	4.2	3.3
Rare earths and yttrium	7.6	4.3	6.0
Rhenium	*	*	*
Rhodium	—	5.7	3.4
Rubidium	—	*	*
Scandium	*	*	*
Selenium	1.5	3.7	2.8
Silicon	*	*	*
Silver	0.3	0.5	0.4
Strontium	—	2.1	1.4
Sulfur	0.5	1.2	1.0
Tantalum	—	2.9	1.7
Tellurium	1.6	*	5.1
Thallium	1.9	0.9	1.0
Thorium	*	*	*

Table XXI (*Continued*)

Commodity	*United States*	*Rest of the World*	*Total*
Tin	—	1.6	1.3
Titanium	1.1	4.2	3.3
Tungsten	0.3	1.4	1.2
Vanadium	0.2	*	7.1
Zinc	0.5	0.6	0.6
Zirconium	1.5	2.0	1.8
Asbestos	0.4	0.9	0.8
Clays	*	*	*
Corundum	—	1.2	1.2
Diatomite	1.5	0.5	0.7
Feldspar	*	4.1	7.6
Garnet	0.8	3.5	1.7
Graphite	—	0.5	0.5
Gypsum	0.5	0.9	0.8
Kyanite	3.8	4.4	4.2
Lime	*	*	*
Mica			
Scrap and flake	*	*	*
Sheet	—	0.1	0.1
Perlite	8.0	*	*
Phosphate Rock	1.9	9.1	7.7
Potash	0.8	*	9.6
Pumice	6.8	1.4	2.7
Salt	*	*	*
Sand and gravel	*	*	*
Soda ash	*	*	*
Stone			
Crushed	*	*	*
Dimension	*	*	*
Talc	3.7	0.9	1.4
Vermiculite	5.9	6.0	5.9
Anthracite coal	7.6	1.0	1.2
Bituminous coal and lignite	*	*	*
Natural gas	0.3	1.2	0.9
Peat	*	3.4	3.7
Petroleum	0.1	1.1	0.8
Shale oil	—	*	3.8
Uranium	0.3	0.4	0.4
Argon	*	*	*
Helium	4.1	0.1	3.2
Hydrogen	*	*	*
Oxygen	*	*	*

[a] Asterisk indicates ratio of 10 or more.

This table should be interpreted with care, because it is based upon proved reserves. The reader should reread Section III,A of this chapter, which indicates the relationship of proved reserves to ultimately available resources. Even on the basis of proved reserves, only a relatively few minerals pose substantial problems, i.e., bismuth, bromine, fluorine, germanium, indium, silver, zinc, asbestos, graphite, gypsum, sheet mica, natural gas, petroleum, and uranium.

Although only a few minerals are listed above, they include three of the major fuels. The United States National Academy of Sciences study on petroleum (Section III,A,4) and the fact that uranium has a very low potential reserve to proved reserve relationship (Table XVIII) show the necessity of developing new energy sources. The major worldwide effort to develop a breeder reactor to stretch out uranium resources, to eventually develop a fusion power reactor, and to develop solar and geothermal energy are illustrative of this point. The Bureau of Mines analysis gives reason for complacency for most of the major materials but identifies emerging shortages in energy and a few major metals.

B. *United States Geological Survey Analysis*

The United States Geological Survey (USGS) (Table XXII) (*16*) has looked at the United States position through the year 2000 A.D. on a somewhat more general basis than that of the Bureau of Mines reported above. This analysis identifies six commodities—tin, asbestos, chromium, antimony, mercury, and tantalum—that the Geological Survey feels may be in such short supply as to require new exploration methods or the development of substitutes. (Their analysis does not cover petroleum or natural gas.)

The major difference between the U.S.G.S. and Bureau of Mines analyses is that the former broadens the resource base to include a larger area of the resource diagram shown in Figure 14. The survey emphasizes the imprecision of this approach.

C. *The Population Commission Reports*

On behalf of the Commission on Population Growth in the American Future, Resources for the Future identified the principle resource and environmental consequences of future population growth in the United States (*17, 18*). The approach taken for nonfuel resources was to project demand-and-supply balances for the United States and the world and to look at possible substitution or resource availability extension through

Table XXII General Outlook for Domestic Reserves and Resources through 2000 A.D.—United States[a] (*16*)

Group 1: **Reserves** in quantities adequate to fulfill projected needs well beyond 25 years.

Coal	Clays	Boron
Construction stone	Potash	Argon
Sand and gravel	Magnesium	Diatomite
Nitrogen	Oxygen	*Barite
Chlorine	Phosphorus	Lightweight aggregates
Hydrogen	Silicon	Helium
Titanium (except rutile)	Molybdenum	Peat
Soda	Gypsum	*Rare earths
Calcium	Bromine	*Lithium

Group 2: **Identified subeconomic resources** in quantities adequate to fulfill projected needs beyond 25 years and in quantities significantly or slightly greater than estimated undiscovered resources.

Aluminum	Vanadium
*Nickel	*Zircon
Uranium	Thorium
Manganese	

Group 3: **Estimated undiscovered resources** (hypothetical and speculative) in quantities adequate to fulfill projected needs beyond 25 years and in quantities significantly greater than identified subeconomic resources; research efforts for these commodities should concentrate on geologic theory and exploration methods aimed at discovering new resources.

Iron	Sulfur	*Cobalt
*Copper	*Silver	*Cadmium
*Zinc	*Fluorine	*Bismuth
Gold	Platinum	Selenium
*Lead	Tungsten	*Niobium
	*Beryllium	

Group 4: **Identified subeconomic and undiscovered resources** together in quantities probably not adequate to fulfill projected needs beyond the end of the century; research on possible new exploration targets, new types of deposits, and substitutes is necessary to relieve ultimate dependence on imports.

Tin	*Antimony
Asbestos	*Mercury
Chromium	*Tantalum

[a] Within each group, commodities are listed in order of relative importance as determined by dollor value of United States primary demand in 1971. An asterisk marks those commodities that may be in much greater demand than is now projected because of known or potential new applications in the production of energy.

recycling to account for perceived shortfalls. The conclusion of the non-fuel study was, "Having sorted out the problems by degree of concern, we have found, by and large, that in the near perspective of three decades none loom threateningly" (*17*). The report does go on to say, however, that as the time horizon is pushed back, the authors are left with misgivings for several reasons. The first has to do with the inability to forecast either demand or supply over such a long period of time. The second (and more serious) has to do with whether or not the market can adequately handle emerging shortages. Can the process of substitution foreseen by the classical economist in fact take place? Third, while for any one material substitutes can be developed, would this be true for all materials taken together? Finally, will the very long lead times for supply augmentation identified above (Section III,B,1), prevent society from promptly adjusting to emerging shortages without substantial disruption.

The Resources for the Future report on energy (*18*) also projected demand and supply and looked for shortages, emphasizing the impact of environmental degradation on the pattern of energy use, although finding that resources are adequate to meet demand through substitution of abundant for scarce resources. The study concludes that "it could turn out that environmental repercussions will constitute a much earlier constraint on energy growth than demographic factors."

D. Summary

Reviewing the work on resource adequacy, there are nagging problems that appear on the surface to be unsolvable. The professional economist's view of resources appears to have been born out in the past centuries of human development. Resources became available to a society as they were needed. Availability has been at declining real costs in spite of rapidly burgeoning resource demands. Nevertheless, these resources are finite so that one is lead to wonder whether technological advance, especially when faced with the environmental challenges of exponentially increasing use of the resources can, in fact, continue to make these resources available in a timely fashion so that they do not provide a constraint upon economic growth.

Scholarly studies of a detailed nature indicate a problem of the magnitude to which society can adjust through time. However, as the Club of Rome warns, this can be true only if the adjustment mechanisms of society are working.

Thus, resource adequacy for the world becomes a matter of resource policy for the world. If the world will develop its resources in the orderly, optimum fashion as prescribed by the economists' analysis of costs, prices, supply, and demand, then there is no reason to believe that society will suffer catastrophic shortage. However, if large portions of the earth's crust are not opened for resource development because of environmental constraints, if market organizations such as cartels and nationalist companies prevent orderly investment and development of these resources, and if national policies of autarchy and "independence" prevent the free flow of materials throughout the world, then individual countries, and the world as a whole, can suffer the sharp economic costs of nonavailability of critical resources.

REFERENCES

1. D. H. Meadows, D. L. Meadows, J. Randers, and W. W. Behrens, III, "The Limits to Growth." Universe Books, New York, New York, 1972.
2. The Committee on Resources and Man, "Resources and Man." Freeman, San Francisco, California, 1969.
3. "Mineral Resources and the Environment." National Academy of Sciences, Washington, D.C., 1975.
4. D. B. Brooks and P. W. Andrews, *Science* **185**, 13–19 (1974).
5. "Commodity Data Summaries 1975." United States Bureau of Mines, Washington, D.C., 1975.

5a. "Minerals in the U.S. Economy." United States Bureau of Mines, Washington, D.C., 1975.

6. V. E. McKelvey, *Amer. Sci.* **60**, 32–40 (1972).
7. R. L. Erickson, *U.S., Geol. Surv., Prof. Pap.* **820**, 21–25 (1973).
8. News Release, April 13, U.S. Department of the Interior, Washington, D.C., 1975.
9. "Initial Report on Oil and Gas Resources, Reserves, and Productive Capacities." Federal Energy Administration, Washington, D.C., 1975.
10. J. F. McDivitt and F. Manners, "Minerals and Men." Johns Hopkins Press, Baltimore, Maryland, 1974.
11. R. G. Ridker, *In* "Economic Factors in Population Growth" (A. J. Coale, ed.) pp. 324–351. MacMillion, London, 1976.
12. W. Malenbaum, *In* "Proceedings of the Council of Economics," pp. 147–156. American Institute of Mining and Metallurgical Engineers, New York, New York, 1975.
13. H. J. Barnett and C. Morse, "Scarcity and Growth." Johns Hopkins Press, Baltimore, Maryland, 1975.
14. M. Mesarovic and E. Pestel, "Mankind at the Turning Point." Dutton, New York, New York, 1974.
15. The No-Growth Society, *Daedalus* (*Boston*) **102,** 1-253 (1973).
16. "Mineral Resource Perspectives 1975," U.S., Geol. Surv., Prof. Pap. No. 940. U.S. Department of the Interior, Washington, D.C., 1975.

17. L. L. Fischman and H. H. Landsberg, *in* "Population, Resources, and the Environment" (R. G. Ridker, ed.), Vol. 3, pp. 77–101. Resources for the Future, Washington, D.C., 1972.
18. J. Darmstadter, *in* "Population, Resources, and the Environment" (R. G. Ridker, ed.), Vol. 3, pp. 103–149. Resources for the Future, Washington, D.C., 1972.
19. News Release, March 24. United States Bureau of Mines, Washington, D.C., 1975.
20. T. Lee and C. L. Yao. *Eighth Int. Geol. Rev.* **12**(7) 778–786 (1970).

Air Pollution Legislation and Regulations

William A. Campbell and Milton S. Heath, Jr.

I. Common-Law Remedies *(1)*

Before considering legislative approaches to air quality control, it is helpful to examine the common-law remedies available to private citizens to bring about pollution abatement, because such legislation often arises, in part, from dissatisfaction with these common-law remedies and from recognition of their shortcomings. Yet, in almost all jurisdictions, common-law remedies continue to be available to supplement regulatory programs. Suits by private parties to abate air pollution or to recover damages caused by air pollution fall under the general heading of tort law, the branch of law dealing with harm to persons and property and the legal remedies available to rectify that harm.

A. Nuisance (2)

Actions in nuisance to abate air pollution are historically the oldest of the existing tort remedies. At the same time, the law of nuisance is a body of law with extremely amorphous boundaries, flowing out and drawing in like a jellyfish buffeted by waves on the sand. The gravamen of a nuisance action is unreasonable interference with the use and enjoyment of one's land. This unreasonable interference may be caused through either intentional or negligent action; lack of due care is not controlling. In this regard, it is well to consider the words of the late William B. Prosser, a leading torts scholar: "Nuisance, in short, is not conduct, nor is it even a condition. It is the invasion of an interest, a type of harm or damage, through any conduct which falls within the three traditional categories of liability (intent, negligence, or strict liability)" (*3*). It should be emphasized that a landowner must show more than just interference; he must show unreasonable interference, which is usually considered to be relatively high levels of pollution at fairly frequent intervals, or substantial levels of pollution over extended periods of time. Whether or not the interference caused by the pollution is unreasonable is ultimately a question of fact.

If the injured landowner is able to convince a jury or judge that the interference is unreasonable, then he is entitled to recover damages. In most cases, however, the landowner will also seek an injunction to stop or at least reduce the air pollution. It is at this point that the most serious weakness in a nuisance action becomes apparent; an application to a court for an injunction is a request for equitable relief, and before a court will issue an injunction, it must go through the process of "balancing the equities." This means that the court must take into account such matters as the extent of the damage to the affected landowner, whether the polluter was on the scene first, the cost to the polluter of abatement, and the economic position of the polluter in the community. When the case involves one householder or small business firm creating a nuisance condition for another householder or small business firm, the equities will usually come out about even or slightly in favor of the injured landowner and the injunction will issue. When the dispute has been between a major industrial concern and a landowner, or even a group of landowners, the courts have in almost every case given overriding importance to the economic position of the polluter in the community and have denied the application for injunction if the injunction would force the polluter out of business (*4*). When an injunction is denied but damages are allowed, most courts hold that permanent damages must be awarded, which is in effect a sort of private inverse condemnation award; successive suits for damages are not permitted (*5*). As Harper and James point out (*6*), there are really two balancing processes that take place when a suit is brought to enjoin air pollution as creating a nuisance; the first occurs when the jury or judge determines whether the interference with the plaintiff's enjoyment of his land is unreasonable; the second occurs when the court determines whether to enjoin the polluting activity. Apart from the environmental harm caused by the balancing of the equities doctrine, the policy decisions involved in many of these cases—location of industrial plants, cessation of agricultural and forestry activities, and economic trade-offs—are political questions that lie beyond the scope of the courts in most governmental structures, do not easily lend themselves to decision in a judicial forum, and in many cases are beyond the technical competence of the courts.

A second impediment to a successful nuisance action in some circumstances is the public–private nuisance distinction. As discussed above, a private nuisance action is founded on an unreasonable interference with the use and enjoyment of land. A public nuisance is a variety of low-grade crime that can only be abated by a criminal prosecution in the name of the state (*7*). If one person is harmed by air pollution that is generally affecting everyone in the community, he cannot recover damages for his injuries unless he can show that he is harmed in some special way different from that suffered by the general public (*8*). If the person harmed is a landowner and the damage is to the use of his land, it appears that he is always considered harmed in a special way, so that he can maintain an action for nuisance (*9*).

B. Trespass

The essential element in a trespass action is the physical invasion of a landowner's right to exclusive possession of his property (*10*). Traditionally, in imposing liability, courts have not inquired whether the invasion was unreasonable or whether any damages occurred. The modern trend has been to narrow the scope of liability for trespass by requiring that harm be shown where the invasion was not intentional (*11*). There are at least three advantages in bringing an action on a trespass theory rather than on a nuisance theory: (a) the period for the running of the statute of limitations is often longer for trespass (*12*); (b) the court will not inquire into the reasonableness of the invasion; and (c) when the trespass is intentional, the trespasser may be liable for all of the injurious consequences flowing from the trespass that are the natural and proximate result of his conduct, even though he did not foresee them (*13*). If the plaintiff in a trespass action seeks an injunction, he must still face the problems of balancing the equities.

Two fairly recent cases involving air pollution from an aluminum reduction plant have permitted recovery of damages based on a trespass theory (*14*). The following quotation from the *Martin* case illustrates the view that air contaminants cause a trespass: "[W]e may define trespass as any intrusion which invades the possessor's protected interest in exclusive possession, whether that intrusion is by visible or invisible pieces of matter or by energy which can be measured only by the mathematical language of the physicist" (*15*). The *Fairview Farms* case pointed out that, in air pollution cases, the torts of nuisance and trespass are not mutually exclusive and that in appropriate situations an action may be brought under either theory (*16*).

C. Negligence

Private actions based on negligence theories have become increasingly important in the air pollution field, especially where the pollution results in personal injuries, as distinguished from property damage. The necessary elements of a negligence action are: (a) a duty owed by the defendant to conform to a certain standard of conduct; (b) a failure on the defendant's part to meet this duty; (c) a reasonably close causal relationship between the defendant's conduct and the plaintiff's injury, known as proximate cause; and (d) actual injury to the plaintiff (*17*). The standard of conduct to which the defendant has a duty to conform is typically stated as that expected of a reasonable and prudent person under the same or similar circumstances (*18*). One of the leading air pollution cases in which suit was based on negligence is *Reynolds Metals Co.* v. *Yturbide* (*19*). This case is important because it shows the quality and type of evidence that is usually necessary in an air pollution negligence case and because the doctrine of *res ipsa loquitur* was used as a means of imposing liability on the defendant. *Res ipsa loquitur* is a procedural rule that governs the effect of certain circumstantial evidence of negligence. When an injury is caused by an intrumentality that is in the defendant's exclusive control and the event causing the injury is one that does not happen in the usual course of events unless someone was negligent, a court may employ the doctrine of *res ipsa loquitur* to create an inference that the defendant was negligent (*20*). The doctrine permits the plaintiff to get his case to the jury and then permits, but does not require, the jury to make a finding of negligence.

D. Strict Liability

Strict liability, or liability without fault, is imposed in certain circumstances where the defendant's activities harmed the plaintiff even though the defendant was not negligent (*21*). Of course, in one sense trespass and nuisance involve liability without fault, but cases applying strict liability principles are put in a separate category from traditional nuisance and trespass cases. Typically, this rule of liability is applied when the defendant's activities can be classified as "abnormal" or "ultrahazardous," and it has been applied in some air pollution cases (*22*). There is some disagreement about the necessary elements of strict liability (*23*), and different jurisdictions have applied the principle in different situations (*24*), but the concept is felt by some scholars to have potential value in the area of private suits for air quality control (*25*).

Although the common-law remedies briefly summarized above predate contemporary statutory approaches to air quality control and continue to supplement those approaches, they are clearly no substitute for comprehensive regulatory programs. They are haphazard, usually occurring only when persons subject to air pollution are harmed seriously enough to impel them to bring legal action. Problems of proving causation are often insurmountable for plaintiffs, especially in urban areas where several sources may have caused the harm. The legal theories available are often inadequate in many circumstances, as demonstrated by the balancing of the equities doctrine. Also, private suits can be expensive and involve long delays.

On the positive side, private suits provide an avenue by which persons harmed by air pollution may recover compensation for the harm and in that way provide a means for direct citizen participation in air quality control. They can also serve as a vehicle for pointing up the weaknesses in existing legal theories and for devising new ones.

This discussion of common-law remedies deals with remedies developed and refined in the state and federal courts of the United States. These remedies may not be available in the forms discussed, or at all, in countries where there is no private ownership of land or in countries with different legal traditions.

II. Legislation

A. Constitutional Basis

In countries with a constitutional form of government, before any set of activities may be subjected to the power of the government through regulation and control, some basis for this control must be found in the constitution. The constitutional authority for air quality control is different for each level of government—national, state or provincial, and local.

At the national level in the United States, the primary authority for enacting legislation is the power of the Congress to "regulate commerce . . . among the several states" (*26*)—the Interstate Commerce clause—and to tax and spend for the general welfare (*27*). There are today practically no judicial restraints on the power of Congress to legislate on almost any aspect of interstate commerce, so long as the legislation does not contravene other specific provisions of the Constitution, such as one of the provisions of the Bill of Rights or the Fourteenth Amendment. With the cases of the 1930s upholding various New Deal legislation (*28*) and of the 1960s upholding the Civil Rights Acts (*29*), the issues have resolved in favor of Congress's broad legislative authority.

At the state or provincial level, the constitutional authority is more general. Every state or province has residual authority to enact laws necessary and useful to protect and enhance the "public health, safety, and welfare." This is usually referred to as their general police power. Under this power, states and provinces may legislate to control all manner of environmental harms and undesirable activities (*30*). Again, legislation under the general police power must not contravene other specific provisions of the constitution, especially those guaranteeing equal protection of the law and due process of law.

The constitutional basis for air pollution control programs at the local level of government is also derived from the general police power. In general, a local governmental unit may legislate concerning any subjects falling within the scope of the police power. The authority of local governmental units to legislate is, however, usually delegated by the state or provincial legislature or constitution and subject to whatever restraints and conditions the legislature or constitution imposes. Local authority is often more narrowly construed than the state's or province's authority.

B. General Principles (31)

The development of air pollution control legislation is largely a problem of the specialized application of accepted principles of administrative law. As in most cases where the administrative law approach has been applied, the efficacy of a particular program depends on many extralegal elements, such as administrative personnel, size of appropriation, and public understanding. The goal of air pollution legislation should be to maintain a reasonable degree of purity of our air resources consistent with: (a) the public health and welfare; (b) the protection of plant and animal life; (c) the protection of physical property and other resources; and (d) the maintenance of adequate visibility.

The solution of local problems by local authorities should be encouraged. Municipal and other local agencies should be authorized to establish control programs, and in cases where a need is shown, regional authorities should be authorized. The state or provincial agency should, however, be authorized to take action in the absence of a local or regional authority or its failure to act.

In the light of the above considerations, the following criteria designed to establish and carry out an air pollution control program are suggested for legislation. While these criteria are not all-inclusive, they are intended to provide a convenient checklist of considerations involved in the development of such legislation.

1. Statement of Policy

A declaration of policy in the statute is desirable. Such a declaration indicates the basic objectives that are to guide the agency in the exercise of authority delegated to it by the legislature and is an aid to the administrative agency and to the courts in the interpretation of the statute.

2. Definitions

A definition section is an integral part of a comprehensive air pollution control statute. Such a section is important to the administration and interpretation of the statute, and unduly restrictive or vague definitions should be avoided. The definition of the term "air pollution" will control the scope of enforcement activity and should be broad enough to permit comprehensive control measures. The jurisdiction of the agency should be made to extend to those emissions into the atmosphere causing or contributing to air pollution that may create a nuisance or be actually or potentially harmful, detrimental, or dangerous to human, animal or plant life, or to property, or the enjoyment of property, or that adversely affects visibility. Similarly, the definition of "person" should not be restrictive but should include any governmental jurisdicton or agency; public or private corporation, partnership, firm, or other entity; or individual.

3. Administrative Agency

Administrative authority and responsibility should be vested in a single agency with comprehensive powers. The agency should be so constituted as to take into account the interests and views of affected groups in addition to public health considerations. An air pollution control agency should either be independent or be placed in an overall environmental quality department, ministry, or agency. The agency may have (a) a single administrator or (b) a board, commission, or authority appointed in a prescribed manner, or both. If the agency has a single administrator, an advisory council is generally considered desirable and useful.

4. Comprehensive Planning

The agency should be authorized to undertake comprehensive planning for the control and prevention of air pollution both as regards the maintenance and improvement of air quality and the potential needs of various areas of the jurisdiction. The relationship of air quality planning and land use planning is a crucial factor. (see also Chapter 2, Volume V). The agency should be authorized to engage with agencies of other jurisdictions in joint planning, either directly or through a jointly established agency, for the prevention and control of air pollution.

5. Enforcement

Discharges into the atmosphere from any source constructed or operated without a permit (if one is required) or in violation of the terms of a permit or of rules, regulations, or orders of the agency should be prohibited. Civil and criminal penalties, abatement authority, and judicial review of agency orders should be provided for. Provisions establishing the basis for citizen suits to enforce the provisions of the law are also desirable.

6. Administration

The agency should have sufficient power to carry out its responsibilities. Necessary authority should include power to hold hearings, to subpoena witnesses, to enforce subpoenas, to administer oaths, to examine plans and specifications, to require maintenance of records and making of reports, and to enter on property at reasonable times for purposes of inspection and investigation. The agency should, of course, be empowered to adopt rules and regulations, to issue orders, and to take action necessary and appropriate for the administration and enforcement of the statute.

7. Citizen Participation

Opportunities for active citizen participation in air pollution control programs should be maximized through formal procedures (such as public hearings).

8. Severability

A severability clause may be included to provide that if any part of the statute is declared unconstitutional, the other parts shall continue in force and effect.

9. Specific Criteria

In addition to the general guides set out above, the legislation should:

a. Provide for the appointment powers and duties of officers and employees of the agency, and authorize the agency to carry out the purposes of the act.

b. Authorize appropriations to the agency.

c. Provide for injunctive relief against violations and for criminal and civil penalities.

d. Authorize the agency to conduct a program of research in air pollution and to collect and disseminate information relating to air pollution and its control and prevention.

e. Authorize the agency to establish air quality and emission standards, which may be applicable jurisdiction-wide or in designated air quality control regions where stricter controls may be necessary, taking into account the purposes of the statute, current scientific knowledge, available control techniques, and the need for control after appropriate investigations and a hearing and to revise such standards from time to time.

f. Authorize the agency to issue rules and regulations to implement the statute, and hold public hearings on proposed rules and regulations.

g. Authorize the agency to conduct surveys and investigations, either on its own motion or on complaints, into alleged violations and issue orders, after hearing, to abate discharges or to require remedial measures.

h. Authorize the agency to advise, consult, and cooperate with other agencies and authorities, with other jurisdictions, interjurisdictional or international agencies, and industries in the formulation and carrying out of its program; to accept and administer funds, loans and grants from appropriate sources for carrying out any of its functions; and to provide technical assistance and advice to other governmental organizations, industry, and other affected groups.

i. Authorize the agency to employ personnel, including consultants and hearing officers, and utilize the personnel and facilities of other agencies and make reimbursement therefor.

j. Authorize the agency to take emergency action to abate air pollution that presents an imminent and substantial hazard to the public health.

C. Citizen Suits

Several obstacles have traditionally confronted the citizen or group seeking legal redress against air pollution and other environmental problems. First, a showing of "standing to sue" has been required before the courts would be open to such complaints, and the usual requirement for "standing" has been an injury to a property or contract interest of the plaintiff. In other words, the plaintiff has been required to show an injury to a recognized legal interest. Second, questions are sometimes raised, especially in the federal courts, as to whether the lawsuit involves the kind of controversy that may properly be resolved in the courts. Under the United States Constitution, the kinds of suits that federal courts will entertain are limited to "cases or controversies." This language has historically been construed in a rather narrow way by the United States Supreme Court to "limit the business of the federal courts to questions presented in an adversary context and in a form historically viewed as capable of resolution through the judicial process" (*32*). State courts are not necessarily subject to similar restraints. In addition, if the defendant happens to be a governmental entity, the defense of sovereign immunity has often been interposed. In recent years, all of these barriers have been lowered in some measure by a combination of legislative and judicial developments.

To some extent, the courts independently have narrowed and eroded the defense of sovereign immunity. National and state legislation have complemented and aided the trend toward elimination of the ancient defense that "the King can do no wrong." Some vestiges or variants (such as the "discretionary function" defense) still remain, but the defense of sovereign immunity will not bar as many citizen suits against government agencies today as it would have in earlier times. The courts have also independently taken some steps toward liberalizing standing-to-sue doctrines. Legislative developments have greatly accelerated this evolution.

On the federal level, Congress has made the federal courts more accessible to citizen suits both against polluters and against governmental regulatory agencies. The pattern has been developed in a series of environmental protection statutes such as the Clean Air Act and the Federal Water Pollution Control Act of authorizing suits to be brought either against the polluter who violates standards developed under the legislation or against the agency that fails to carry out its statutory mandate for environmental protection. For example, under Section 304 of the Clean Air Act, any person may bring suit in federal court against a violator of an emission standard or may sue the Administrator of the United States Environmental Protection Agency to require him to perform his nondiscretionary duties under the Act. These provisions have served as the vehicle for a number of important test cases brought by environmental groups concerning the administration of key provisions of the Clean Air Act, e.g., state abatement plans, transportation plans, nondeterioration, etc.

Similar developments have made the courts of some (though not all) states more accessible to citizen suits. For example, a Michigan statute recognizes a legal interest in the protection of the environment that may be enforced by state or local governments or by private citizens. It also establishes procedures for enforcement of this legal interest, through lawsuits brought "for the protection of the air, water and other natural resources and the public trust therein" (*33*).

Despite these liberalizing trends, the road to successful citizen suits may still be a rocky one. The plaintiff may encounter problems in proving the requisite interest or connection with the matter in controversy (*34*). Jurisdictional limitations, such as the minimum amount in controversy required for access to the federal courts, may effectively preclude class actions (*35*). And—subject to possible correction on appeal—resourceful trial judges may make it difficult for citizen groups to pursue lawsuits by, e.g., setting substantial bond requirements as a condition of granting plaintiff's requested relief. Even where such requirements are later reversed or modified on appeal, the burden of contesting them may be onerous (*36*).

D. Environmental Impact Evaluation (see also Chapter 6, Volume V)

The United States National Environmental Policy Act of 1969 (NEPA) lays down broad statutory principles for environmental impact evaluation but has left many unanswered questions for later interpretation. One important unsettled issue is the relationship between environmental impact statements and environmental pollution control programs. For several years, administrative agencies and courts have grappled with a series of issues concerning the applicability of NEPA's requirements for impact statements to various aspects of pollution control. The water quality issues were largely resolved by Congress in the Federal Water Pollution Control Act Amendments of 1972 (*37*) in favor of nonapplicability of NEPA to water pollution programs. The air pollution issues thus far have been left for the agencies and the courts to resolve.

NEPA itself, read literally, offers no obvious basis for exempting actions taken in furtherance of the federal air pollution control program from the general requirement for environmental impact statements concerning federal "actions" significantly affecting environmental quality. Nor did the Clean Air Amendments of 1970 and 1977 speak expressly to the issue. However, some courts have read into this complex of legislation an implied exemption of actions pursuant to environmental protection programs, based on the redundancy of requiring an environmental watchdog agency to file an environmental impact statement (*38*).

Having successfully fought off efforts to compel compliance with NEPA as a matter of statutory interpretation, the Environmental Protection Agency (EPA) has yielded back some of this ground in the exercise of its administrative discretion. In a policy statement adopted in 1974, EPA agreed to voluntarily prepare impact statements for certain major regulatory actions involving air pollution, noise control, atomic energy, marine sanctuary, and pesticide programs. The subjects on which EPA agreed to prepare impact statements under the Clean Air Act were national ambient air quality standards, substantive criteria for state implementation plans, new source performance standards, hazardous pollutant emission standards, motor vehicle emission standards, and fuel or fuel additive regulations (*39*).

Section 309 of the Clean Air Act gives EPA an additional oversight responsibility for environmental evaluation (*40*). The agency is instructed to review and comment on the environmental impact of federal actions affecting EPA's functions under the Clean Air Act. The scope of this oversight responsibility extends to legislative proposals of federal agencies, proposed federal regulations, and newly authorized federal construction projects.

Almost half of the states have followed the federal lead and adopted environmental impact statement requirements for state government activities similar to those provided by NEPA for federal activities (*41*).

Depending on the wording of the state legislation, some of these statutes may require impact statements for some aspects of state air pollution control programs.

III. The United States Clean Air Act

The first national air pollution legislation in the United States was enacted in 1955. The original responsible agency was the Public Health Service of the Department of Health, Education, and Welfare. This legislation was amended in 1963, 1965, 1966, 1967, 1970 (*42*), and 1974 Amendments to the Clean Air Act, passed by the House and Senate in 1976, failed on compromise before the Congress ended. It was, however, extensively amended in 1977. The responsible federal agency since 1970, the United States Environmental Protection Agency (USEPA), establishes national air quality standards, and then the states are required to submit plans that demonstrate how each state is going to achieve and maintain those standards. In addition to setting the national ambient air quality standards, the federal government is charged with establishing performance standards for new stationary sources, for hazardous pollutants, and emission standards for motor vehicles. In their implementation plans that are submitted to the USEPA for approval, the states must establish performance standards and, where necessary, compliance and abatement schedules for sources of pollution. A state may delegate a portion of its authority for air quality control to cities, counties, and regional governments, but the delegation must be made pursuant to the Federal Clean Air Act and its implementing regulations, and any local governmental unit to which authority is delegated must see to it that the national air quality standards are enforced within its jurisdiction. The Clean Air Act has thus significantly altered the relationships among the federal, state, and local governments in air pollution control. Air pollution is viewed as a matter of national concern requiring federal leadership, and the federal government is seen as the only level of government with the financial resources and legal authority necessary to deal adequately with the many aspects of air quality control.

Historically, the states have not usually had strong air quality control programs. Typically, there was a division somewhere in the state department of public health with some authority over air pollution, often of a technical or advisory nature. The Clean Air Act has, of course, changed all of that. The states are now seen as major partners with the federal government in the control of air pollution. This up-grading of the states' role has required substantial increases of funds and personnel in their air quality divisions and has frequently been accompanied by a transfer of air quality control functions from the health department to a state agency concerned with environmental quality or natural resources.

Before the federal government greatly expanded its role in air pollution control and at the same time promoted an increase in the activities of state governments, the level of government with the most concern about, and the largest role in, air quality control was the city. Concern of municipal governments with the air quality within their jurisdictions has ancient historical antecedents (*43*). The earliest and simplest forms of legislative controls were ordinances prohibiting the burning of certain substances (*44*), smoke or visible emission ordinances (*45*), and nuisance ordinances. Municipal air quality programs varied in effectiveness, but in virtually every large urban center there existed machinery for air pollution control and experience in attempts to control air pollution at the time the federal government enlarged its role and that of the states. The Clean Air Act, with its relatively strict standards and deadlines, gave these local programs new strength and, in many states, they continue to be the leading edge for air quality control.

A. Ambient Air Quality Standards

No matter what level of government is involved, there are essentially two legislative approaches to the direct regulation of air quality—ambient air quality standards and performance standards. (Economic techniques for controlling air pollution, such as effluent charges, would of course have to be imposed through the legislative process, but such measures are outside the scope of this chapter.) Ambient air quality standards prescribe the permissible levels of contaminants for the region covered by the standards. One of the first questions to be answered concerning ambient air quality standards is what level or levels of government should be responsible for setting the standards and how extensive a region they should cover. Under the Clean Air Act, the federal government has been assigned the task of establishing both national primary ambient air quality standards—those designed to protect the public health—and national secondary ambient air quality standards—those designed to protect the public welfare. The states, and to some extent local governmental units, are free to establish standards that are stricter than the secondary standards, but in no event are they permitted to establish less strict standards. Strong arguments can be made that air quality standards adequate to protect the public health ought to exist in every state and locality. That is, it appears reasonable to begin with a minimum base for protecting health, such as the national primary standards. The case for national secondary standards, however, appears to be somewhat weaker. States and local governmental units may be able to muster cogent arguments that so long as the public health is not endangered, they ought to be able to make any tradeoffs against air quality that they deem appropriate. This approach would leave to the states the responsibility for establishing secondary standards, if any are to be established at all. The arguments favoring a larger state role in setting the standards are that states should have a larger scope of action in furthering their economic goals, in land-use planning, and in self-determination generally. Arguments favoring the federally established national standards emphasize the realities of a national economy, the mobility of large numbers of citizens, the impropriety of permitting industries with relatively high levels of emissions to shop around for the cleanest air or the most permissive jurisdictions, and the likely inability of many states to adopt and enforce adequate standards.

Where local units of government are concerned, some states may wish to involve them only to the extent of encouraging them to assist in enforcing state-adopted federal standards or more restrictive state standards. Others may wish to encourage them to adopt their own standards where that is practicable. Many units of local government are not large enough geographically to be able to adopt and enforce air quality standards different from those in the rest of the state.

A second group of questions about air quality standards concerns the methods by which the standards are set. By this is meant the legal and administrative techniques by which the standards are arrived at and promulgated. One possibility is for the legislative body, Congress under the Federal Clean Air Act, to set the actual standards in the statutes. The weaknesses in following this approach are obvious—the scientific information upon which the standards are based is almost certain to change, and economic and social conditions may change; if the standards are locked into a statute, the process of conforming them to these changes may be difficult and time-consuming. On the other hand, the one great strength of such a method is that it eliminates the possibility of pressure by the controlled sources being brought to bear on an administrative agency to change the standards or to delay their implementation.

The alternative approach, and the one adopted by Congress in the Federal Clean Air Act (*40*), is to establish general air quality goals or objectives in the statute and then leave to the appropriate administrative agency the task of working out the detailed standards necessary to achieve the goals. When this approach is taken, great care must be exercised in stating the goals, because the standards, however necessary they may be from the standpoint of the public health and welfare, will be subject to legal challenge if they cannot be shown to be directly related to the achievement of the air quality goals. Also, the detailed standards must not be subject to the unfettered discretion of an administrator, but must be based on sound scientific data. If economic and technological considerations are to be given weight in achieving and maintaining the standards, they must be carefully articulated in the statutes and implemented in the regulations.

A third issue to be considered is that of how the air quality standards are to be enforced. Clearly, monitoring and sampling systems and techniques, including diffusion models, that are adequate to provide information upon which realistic assessments of what regions are meeting the standards and what regions are not must exist before the standards can be enforced. In addition, however, there is the problem of potential inequities when emissions from several different sources must be reduced in order to achieve the standards. One method of reduction that has a certain appeal is that of requiring each source to make a proportionate reduction of its emissions. The problem with this method is that it may place such a heavy burden on sources that are using old or outmoded equipment or that are in a highly competitive market situation that they have to close. On the other hand, using a method other than proportionate reduction, taking into account such factors as age of equipment and competitive position, may very likely place an unfair burden on newer plants and on those that are well-managed. Lurking behind all of these problems of equity and fairness are possible legal attacks on enforcement procedures on the ground that the sources have been denied equal protection of the laws.

Another enforcement problem is that regarding the capabilities and limitations of state and local enforcement agencies. It would seem to be sound public administration theory that the level of government that is responsible for enforcing environmental quality standards, including those dealing with air quality, should also be the level of government responsible for setting those standards (*47*). The practical realities underlying this theory are that generally government policy makers will not attempt to establish legislative standards that they know they will be unable to enforce. Clearly, it is poor public policy to have laws that the state is incapable of enforcing and knows it is incapable of enforcing. Thus, if the standard-setting and enforcing levels are the same, the standards are not likely to exceed the enforcement capabilities. The United States 1970 and 1977 Acts disregards this theory. The federal government is the standard-setting level, but the states are the primary enforcers. In this state of affairs, it is incumbent upon the federal government to pay close attention to the enforcement capabilities of the states when the air quality standards are set. Personnel, training, equipment, and funding levels must be adequate to do the job.

The last issue to be discussed concerning air quality standards is that of nondegradation of air quality. Any legislation imposing air quality standards should deal with this question in a clear, straightforward manner. The arguments for and against some type of nondegradation regulation are many (*48*). Generally, it can be said that strong arguments can be made for not permitting air that is already purer than that required by the standards to be degraded down to the level of the standards in regions of natural beauty, wilderness areas, and areas set aside for outdoor recreation. For the remaining areas, the issues are very complex. If the air quality standards call for air of a high degree of purity, then the need for a nondegradation policy is less urgent. If, however, the standards do not call for a relatively pure quality of air, then some type of nondegradation policy is essential. Some of the questions to be faced in including nondegradation provisions in air quality standards legislation are how the concept is to be defined in the statute, the degree of state and local control over nondegradation questions, blending nondegradation policies with other land-use planning decisions, and how economic considerations are to be dealt with. In the United States, the principle of nondegradation was initially established by judicial interpretation of the Clean Air Act of 1970; after extensive litigation, Congress dealt with the matter in detail in the 1977 Amendments to the Act (*49*).

B. Performance or Emission Standards

The second major legislative approach for controlling air quality is the use of performance or emission standards. This approach focuses the control mechanism on what comes out of the stack rather than on the general level of contaminants in the ambient air. Most legislative control systems employ a combination of ambient air quality standards and performance or emission standards, as do both the 1970 and 1977 Acts. While it is possible to have a system based solely on emission standards, it is not possible to have an enforceable system based solely on ambient air quality standards. In order to translate ambient air quality standards into a practical control scheme, some sort of performance or emission standards must be used. Also, where mobile sources of pollution are involved, performance standards are essential (*50*).

There are several types of emission standards. Perhaps the simplest is a visible emissions standard imposed by statute or ordinance. This standard is usually written in terms of the Ringlemann chart, typically making it a violation to emit smoke of an opacity darker than number 2 on the chart. This type of performance standard has been upheld by the United States Supreme Court in the face of an argument that such a standard violated due process of law (*51*). Other performance standard formulations are more sophisticated. There is, for example, the "best available technology" approach that is used in the Clean Air Act's performance standards for new stationary sources (*52*). Yet another performance standard approach is to restrict the fuels or other materials that undergo combustion and control emissions in that manner. Regulations requiring the burning of low-sulfur coals and oil in utility plants are examples of this approach.

With any type of emission standard legislation, several questions arise. One of these questions is to what extent and in what way are economic considerations to be taken into account. For example, if a fuels policy is to be used, to what extent is the cost of this policy to be taken into account? Usually, where a "best technology" approach is employed, a cost qualification will be added, such as "affordable" or —in the language of the Clean Air Act—"taking into account the cost of achieving such reduction" (*53*). Another question is the coverage of the emissions standards and whether the same standards are to be applied to all sources. By "coverage" is meant whether the performance standards are to apply to all existing sources, only to new sources, or to new and modified sources. Also, consideration must be given to the time within which compliance with the standards will be required.

Another major problem area is that of monitoring emissions and gathering sufficient information to enforce adequately the performance standards. For there to be effective enforcement of performance standards, the responsible air quality agency must be able to determine with precision when a source has violated the standards. This requires an extensive monitoring program. Sufficient numbers of adequately trained personnel are required, and they must have reliable equipment. Nothing will sabotage an abatement action faster than an inexperienced or ill-trained enforcement officer using unreliable monitoring equipment. In many instances, the monitoring and reporting may be performed by the source itself, rather than by the enforcement agency. This raises other issues, such as the integrity of a self-policing procedure and the usefulness of the information obtained in an abatement action against the source.

C. Interstate Pollution Abatement

As with many problems sought to be regulated by government action, air pollution is no respecter of jurisdictional boundaries. In many localities, the sources of pollution are in a jurisdiction different from the recipients; in other cases, sources from more than one jurisdiction create the problem, which is then exacerbated by the synergistic effects of the air contaminants. Thus, air pollution problems may in many cases extend beyond the jurisdiction of any state, province, or locality to deal adequately with the matter. This situation requires a device or devices for dealing with interjurisdictional pollution. There are essentially three methods of dealing with this matter. The first is to let the state, provincial, and local governments involved deal with the problem as best they can, but without national government restraint or encouragement. The second is for the national government to exert its unquestioned authority over interjurisdictional air pollution and deal with the problem at the national level. The third is for the jurisdictions involved to form an interjurisdictional compact to deal with the problem. Prior to 1967 in the United States, the states and localities were left to their own devices to deal with the problem as best they could. The result in many cases was informal arrangements between neighboring jurisdictions for the control of pollution under certain conditions. These informal arrangements may have worked satisfactorily in many instances, but they all suffered from the fatal weakness that there were no legal sanctions to back them up. If one of the participating jurisdictions did not take the necessary measures to control pollution, there was nothing the other jurisdictions could do about it.

The Clean Air Acts of 1970 and 1977 illustrates two ways in which the federal government has exercised its authority to deal with interstate air pollution. In the first instance, a requirement is imposed upon the states to establish governmental machinery adequate to deal with interstate pollution. In the section of the act that sets forth the requirements for state implementation plans that must be submitted to the United States Environmental Protection Agency for approval is the following requirement:

> [It must contain] adequate provisions for intergovernmental cooperation, including measures necessary to insure that emissions of air pollutants from sources located in any air quality control region will not interfere with the attainment or maintenance of such primary or secondary standard in any portion of such control region outside such state or in any other air quality control region (*54*).

The effect of this requirement is to put a burden on each state to ensure that the control measures it proposes are sufficient to control interstate pollution, at least to within the limits of the national primary and secondary ambient air standards.

In the second instance, the federal government has set up three mechanisms by which it can exercise direct control over cases of interstate air pollution. The first of these mechanisms is the conference procedure. Whenever the administrator of EPA has reason to believe that air contaminants for which national ambient air quality standards have not been placed in effect are endangering the health and welfare of persons in a state other than the state in which the discharges originate, he may call an abatement conference *(55)*. The purpose of the abatement conference is to bring together the sources of the air pollution with the appropriate control agencies and the EPA in an attempt to work out an abatement schedule satisfactory to all concerned. It is a cumbersome procedure and generally regarded as an ineffective enforcement tool *(56)*.

The second mechanism for direct federal involvement in abating interstate pollution is the provision for federal assumption of the enforcement of state implementation plans. If a state's implementation plan contains adequate procedures for controlling interstate pollution, but the state has failed or refused to implement them, the EPA may assume enforcement of the implementation plan and take direct action against the sources of pollution *(57)*. This not only has the advantage of avoiding the delays of the abatement conference procedure, but it requires imposition of the relatively stringent federal civil remedies and criminal penalties as well *(58)*.

The third possible means for controlling interstate pollution is with the use of an interstate compact among the states involved. Interstate compacts are provided for in Article I, Section 10, of the United States Constitution *(59)*. They are legal instruments in the nature of a contract, created by legislation enacted by the states that are members of the compact. They typically set forth the purposes of the compact, the machinery—usually a commission—by which the compact is to be operated, the details through which the objectives of the compact are to be accomplished, and the means for dissolution of the compact. Generally, interstate compacts must be approved by Congress before they can be implemented. Congress has given its general consent in the Clean Air Act to states to enter into compacts not in conflict with federal law for the control of air pollution *(60)*. Consent of Congress is still required before any compact entered into is binding on the party states and therefore effective *(61)*. At this writing, no interstate compacts for the control of air pollution have been submitted to and approved by Congress. This indicates either that the party states cannot agree on the provisions of the compact or that Congress has considered that the compacts submitted are not in consonance with the Clean Air Act.

IV. Air Pollution Legislation Worldwide (See also Chapter 7, Volume V)

There are two major kinds of air pollution legislation worldwide, those that depend primarily on air quality standards (i.e., ambient air, deposited matter, soiling index, emergency procedure, and point-of-impingement standards) and those that depend primarily on emission standards (i.e., specific pollutant, particulate matter, soot, visible emission, fuel, and mobile source standards) (Table I) (*62*). From among the European countries, it will be seen from Table I that there is a group of eastern European countries (Soviet Union, Democratic Republic of Germany, Bulgaria, Yugoslavia, Hungary, Romania, Poland, Finland, and Turkey) essentially all of whose published air quality management standards are air quality standards, as defined above, and

Table I Classification of Countries Based on the Number of Their National Air Pollution Regulatory Standards in 1974 *(62)*

Country	*Air quality standards*	*Emission standards*	*Country*	*Air quality standards*	*Emission standards*
Predominantly air quality standards[a]			Predominantly emission standards[b]		
Soviet Union	117	—	Fed. Rep. of Germany[c,e]	75	139
Dem. Rep. of Germany	113	2	Japan	7	125
Bulgaria	99	—	Sweden[d]	3	62
Yugoslavia[c,d]	96	—	USA[c,d]	11	62
Hungary	62	—	France[d]	5	46
Israel	46	5	Great Britain	—	44
Romania	35	—	Australia[c]	—	31
Italy	32	17	Switzerland	2	23
Poland	31	—	Singapore	—	22
Philippines	23	4	Mexico	—	14
Spain[d]	13	2	New Zealand	—	8
Canada[c,d]	13	7	Denmark[d]	—	7
Finland	10	2	Belgium	1	7
Argentina	8	—	Greece	—	6
Netherlands	7	—	Austria[d]	—	4
Turkey	3	—	Ireland	—	4
Colombia	3	1	Brazil[c,d]	—	1
Balanced			Hong Kong	—	1
Czechoslovakia	25	25	Malta	—	1
			Norway[d]	—	1
			Portugal	—	1

[a] Air quality standards comprise ambient air, deposited matter, soiling index, emergency procedure, and point of impingement standards.
[b] Emission standards comprise specific pollutant, particulate matter, soot, visible emission, fuel, and mobile source standards.
[c] Also significant state and provincial standards.
[d] Also significant municipality standards.
[e] Includes quasi-governmental Verein Deutcher Ingenieure Standards.

that there are a group of western European countries (Federal Republic of Germany, Sweden, France, Switzerland, Denmark, Belgium, Greece, Austria, Ireland, Malta, Norway, and Portugal) essentially all of whose published air quality management standards are emission standards as defined above. There are no eastern European countries in the list of those with predominantly emission standards, but Italy, Spain, Finland, and the Netherlands are among the group with predominantly air quality standards. The remaining European country on the table, Czechoslovakia, has the same number of air quality and emission standards and therefore fits into neither category.

There are two South and Central American countries in each group—Argentina and Colombia in the air quality standards group, Mexico and Brazil in the emission standards group. Of the reporting jurisdictions in Asia, Asia Minor, and the Antipodes, two (Israel and the Phillipines) are in the air quality standards group and five (Australia, Japan, Singapore, New Zealand, and Hong Kong) are in the emission standards group.

The two North American countries are more difficult to categorize because in both of them the national air pollution regulatory standards listed tell only a part of the story. In Canada there are 226 provincial standards, 164 of which are in the air quality standard group, 62 of which are in the emission standards group. This makes the mix of the total of national and provincial standards 177 air quality and 69 emission standards, a predominance fo air quality standards for the country as a whole. In the United States, there are 278 state air quality standards and over 1250 state emission standards. Thus, the mix of the total of national and state standards, 289 air quality and over 1300 emission standards, puts the United States predominantly in the emission standards category, despite the strong air quality management slant of the United States Clean Air Act.

Much, but not all, of the above data were derived from a 1974 World Health Organization inquiry to its member nations around the world. Of them, 48 countries responded that they had no numerical national air quality management standards (Table II) (*62*). The absence of numerical standards may be interpreted to indicate that the national air pollution control legislation of these countries, if such has been enacted, does not strongly commit the country to one or the other of these two air pollution control approaches. Those countries that did not respond to the WHO inquiry and are listed on neither Table I nor Table II may be considered as having no national air quality management standards and either no national air pollution control legislation or, if they have such legislation, no strong commitment to either air pollution control approach.

Table II Countries Reporting No National Air Quality Management Standards in Their Response to the WHO Inquiry *(62)*

Afghanistan	Gambia	Liberia	Panama
Bangladesh	Ghana	Libya	Peru
Bolivia	Honduras	Luxembourg	Portugal
Burma	India	Madagascar	South Africa
Chili	Indonesia	Malawi	Tanzania
Comoro Islands	Iran	Malaysia	Thailand
Costa Rica	Iraq	Mauritius	Togo
Dahomey	Ivory Coast	Morocco	Tunisia
Ecuador	Kenya	Nepal	Uganda
El Salvador	Korea	New Guinea	Venezuela
Ethiopia	Laos	Nigeria	Zaire
Fiji	Lebanon	Pakistan	Zambia

REFERENCES

1. See Jurgensmeyer, *Control of Air Pollution Through the Assertion of Private Rights,* DUKE L. J. 1126 (1067), for a more detailed discussion of private suits.
2. See 1 F. Harper & F. James, TORTS 64–93 (1956) for a perspicuous discussion of the law of nuisance.
3. Prosser, *Private Action for Public Nuisance,* 52 VA. L. REV. 997, 1004 (1966).
4. For one of the most notorious of the "balancing" cases, see Madison v. Ducktown Sulphur, Copper & Iron Co., 113 Tenn. 331, 23 S.W. 658 (1904). Compare Hulbert v. California Portland Cement Co., 161 Cal. 239, 118 P. 928 (1911).
5. See 1 F. Harper & F. James, *supra* note 2, at 91.
6. *Id.* at 90.
7. Many early smoke ordinances were based on a public nuisance theory; see H. Kennedy and A. Porter, *Air Pollution: Its Control and Abatement,* 8 VAND. L. REV. 854 (1955).
8. See Prosser, *supra* note 3, at 1009.
9. *Id.* at 1018. Prosser, in his article cited in note 3, appears to take a more sanguine view of the chances of bringing a successful private action for recovery of damages where a public nuisance may be involved than does Jurgensmeyer in his article cited in note 1.
10. See W. Prosser, TORTS 63, 3d ed., St. Paul, Minnesota (1964).
11. Keeton, *Trespass, Nuisance, and Strict Liability,* 59 COLUM. L. REV. 457, 464, 465 (1959).
12. See Jurgensmeyer, *supra* note 1, at 1138.
13. See Fairview Farms, Inc. v. Reynolds Metals Co., 176 F. Supp. 178 (D.C. Ore. 1959) and Keeton, *supra* note 11, at 465.
14. Martin v. Reynolds Metals Co., 221 Ore. 86, 342 P. 2d 790 (1959), cert. denied 362 U.S. 918 (1960) and Fairview Farms, Inc. v. Reynolds Metals Co., 176 F. Supp. 178 (D.C. Ore. 1959).
15. 221 Ore. 86, 94, 342 P. 2d 790, 794 (1959).
16. 176 F. Supp. 178.

17. W. Prosser, Torts 146, 3d ed. St. Paul, Minnesota (1964).
18. *Id.* at 153.
19. 258 F. 2d 321 (9th Cir. 1958), cert. denied, 358 U.S. 840 (1958).
20. See W. Prosser, Torts 218, 3d ed., St. Paul, Minnesota (1964); see also Reynolds Metals Co. v. Yturbide, 258 F. 2d 321 (9th Cir. 1958), cert. denied, 358 U.S. 840 (1958).
21. See W. Prosser, Torts 506, 3d ed., St. Paul, Minnesota (1964).
22. See *id.* at 526 and cases collected in footnote 45. See also *id.* at 528 for acceptance of the principle under the label of "absolute nuisance."
23. See Jurgensmeyer, *supra* note, 1149–51, and J. Krier, Environmental Law and Policy 174–175, Indianapolis, Indiana (1971).
24. See W. Prosser, Torts 524–532, 3d ed., St. Paul (1964).
25. See Krier, *The Pollution Problem and Legal Institutions: A Conceptual Overview,* 18 U.C.L.A. L. Rev. 429, 455 (1971).
26. Art. I, § 8[3].
27. Art. I, § 8[1].
28. See, e.g., Wickard v. Filburn, 317 U.S. 111 (1942).
29. See Heart of Atlanta Motel, Inc. v. United States, 379 U.S. 241 (1964).
30. See, e.g., Northwestern Laundry v. Des Moines, 239 U.S. 486 (1916) and Hutchinson v. City of Valdosta, 227 U.S. 303 (1913).
31. This material is drawn from Edelman, Air Pollution (A. C. Stern, ed.) (New York, Academic Press, 1968), 2d ed., Vol. III, Ch. 50.
32. Flast v. Cohen, 392 U.S. 83, 95 (1968).
33. Michigan Stat. Ann., § 14.528 (202) (1).
34. Sierra Club v. Morton, 405 U.S. 727 (1972)—the *Mineral King Case.*
35. Zahn vs. International Paper Co., 414 U.S. 291 (1973).
36. A case in point, the *Chicod Creek* case involved a $75,000 bond requirement imposed by the trial judge, but later set aside on appeal in favor of a nominal bond because plaintiffs were acting as "private attorneys-general." Natural Resources Defense Council v. Grant, 341 F. Supp. 356 (E.D.N.C. 1972), remanded 2 ELR 20555 (4th Civ. 1973).
37. 33 U.S.C. 1371 et seq.
38. Portland Cement Assn. v. Ruckelshaus, 486 F. 2d 375 (D.C. Civ. 1973); Anaconda Co. v. Ruckelshaus, 482 F. 2d 1301 (10th Cir. D.C. Civ. 1973); Buckeye Power Co. vs. EPA, 481 F. 2d 162 (6th Cir. 1973); Duquesne Light Co. v. EPA, 481 F. 2d 1 (3rd Cir. 1973); Appalachian Power Co. v. EPA, 477 F. 2d 495 (4th Cir. 1973); Getty Oil Co. v. Ruckelshaus, 467 F. 2d 349 (3rd Cir. 1972), cert. denied 409 U.S. 1127 (1973). For a criticism of these decisions see *Comment: Coordinating the EPA, NEPA and the Clean Air Act,* 52 Texas L. Rev. 527 (1974). Buckeye, Duquesne, and Getty were modified on other grounds in *Union Electric Co. v. EPA,* 49L. Ed. 2d 474 (1976).
39. 39 Fed. Reg. 16187, May 7, 1974.
40. 42 U.S.C. 1857h-7.
41. *Council on Environmental Quality* (*United States*), 5th Annual Report (Washington: Government Printing Office, 1974), pp. 401–409.
42. See 42 U.S.C. § 1857 et seq. For a more detailed discussion of the Clean Air Act of 1970 see Dolgin and Guilbert, *Federal Environmental Law,* West Pub. Co., St. Paul, Minn., 1974. Useful articles on various legal issues arising from the Clean Air Act include Krier, *The Irrational National Air Quality Standards: Macro- and Micro-Mistakes,* 22 UCLA L. Rev. 328 (1974); Ayres, *Enforcement*

of Air Pollution Controls on Stationary Sources Under the Clean Air Amendments of 1970, 4 ECOLOGY L.Q. 441 (1975); and Note, *Review of EPA's Significant Deterioration Regulations: An Example of the Difficulties of the Agency Court Partnership in Environmental Law,* 61 VA. L. REV. 1115 (1975).

43. Kennedy and Porter, *Air Pollution: Its Control and Abatement,* 8 VAND. L. REV. 854 (1955).
44. *Id.*
45. *Id.*
46. See 42 U.S.C. § 1857c-4.
47. See Heath and Hufschmidt, *State, Federal, and Local Responsibilities Under the Clean Air Act, Proceedings, The National Conference on the Clean Air Act,* 31 at 34 (Air Pollution Control Association, Pittsburgh, Pennsylvania, 1974).
48. See Campbell, *Air Quality Standards, Proceedings, The National Conference on the Clean Air Act,* 122 at 127 (Air Pollution Control Association, Pittsburgh, Pennsylvania, 1974), and A. C. Stern, Critical Review of Significant Deterioration of Air Quality, J. Air Pollut. Cont. Ass. **27** (5) (May 1977) and 27(9) (September 1977) for a discussion of non-degradation issues under the Clean Air Act of 1970.
49. See P. L. 95–96. The 1977 Amendments establish a statutory procedure for administering the nondegradation policy and set forth maximum allowable concentrations of sulfur dioxide and particulates for three classifications of land areas. National Parks and Wilderness Areas are selected for special protection.
50. See 42 U.S.C. § 1857f-1 for the Clean Air Act's approach to motor vehicle emissions.
51. Northwestern Laundry v. Des Moines, 239 U.S. 486 (1916).
52. 42 U.S.C. § 1857c-6(a) (1).
53. *Id.*
54. 42 U.S.C. § 1857c-5(a) (2) (E).
55. 42 U.S.C. § 1857 (e).
56. See Esposito, *Vanishing Air,* p. 114 (1970) (the Bishop Co. case).
57. 42 U.S.C. § 1857c-8(a).
58. See 42 U.S.C. § 1857c-8.
59. See Frankfurter and Landis, *The Compact Clause of the Constitution— A Study in Interstate Adjustments,* 34 YALE L. J. 685 (1925) for the seminal work in this field.
60. 42 U.S.C. § 1857a(c).
61. *Id.*
62. W. Martin and A. C. Stern, *The World's Air Quality Management Standards,* "The Air Quality Management Standards of the World, Including United States Federal Standards." (Washington, D.C.: Office of Research and Development, United States Environmental Protection Agency, 1974). Vol. 1, EPA Misc. Ser. 65019-75-001a.

Organization and Operation of National Air Pollution Control Programs

Goran Persson

I. Introduction

Air pollution control policy is a part of overall environmental policy, which is based on political decisions. Scientific and technical information, the state of the national economy, policy priorities in other areas, and public opinion all influence these decisions. Flexibility is needed in establishing and implementing national air pollution control programs. Different regions have different types of air pollution problems. Conditions change, new problems emerge and new abatement techniques and industrial processes are developed, but at the same time, long-term planning is necessary for solving the problem.

II. Approaches to Legislation

Three different approaches to environmental protection can be recognized. The first is characterized by legislation with limited scope. Each law is concerned with a single aspect of environmental protection or pollution. With this approach, air pollution alone is the subject of a set of laws. This is the approach of the United States Clean Air Act. The second approach is more comprehensive. A single law regulates air and water pollution, noise, and other nuisances. The 1969 Swedish Environment Protection Act is an example of this approach. The third approach consists of integration of environmental legislation with land-use planning and with national planning in general. This legislative approach, chosen by Yugoslavia and under consideration or development elsewhere, ensures an integrated treatment of both rural and urban environments. No final comparative evaluation of these three approaches can be attempted at this stage since the various pieces of legislation concerned have not been applied for a sufficiently long period of time.

The single-medium legislative approach may be used in the stage-by-stage implementation of a comprehensive policy, in which case it is entirely reasonable. It may, however, represent a fragmented policy, which makes it difficult to fight pollution in all its complex aspects. While, in such a case, it is possible to multiply and improve existing laws, the interaction between them can create new problems, each of which will have to be dealt with by still another set of measures.

The comprehensive legislation approach expresses intention to deal with the environmental problem in all its complexity. These diverse problems show a greater similarity than would be expected. It is obvious, however, that this policy requires a clear definition of its scope—urbanized as well as rural; stationary as well as mobile sources; and not only water, air, and soil pollution, but other environmental problems as well. However, this combination may require such massive legislation as to deter its assembly and passage.

The approach that combines environmental protection and land-use planning has the advantage of offering joint or complementary solutions and synchronizing pollution control and planning intervention. There is a risk, however, that, with such a comprehensive policy, good solutions may be unduly delayed while waiting in vain for the best solution. The experience of Yugoslavia seems to confirm this. Even with this approach, specific pollution laws seem to remain indispensable.

The first thing to consider with new legislation is whether the law should deal with air pollution only, with all types of pollution, or with the whole environment. The second is whether and to what extent environmental legislation should be complemented by economic measures. Regulations tend to be rather inflexible and not always well suited to reducing pollution damage to an optimum level from the point of view of the best allocation of available resources. This is a serious disadvantage, because resources are too scarce to satisfy fully all the needs of society. It is the political task to weigh the urgency or utility of environmental improvements against all other needs and goals of society. Economic considerations help to determine how these needs can be most efficiently satisfied.

One approach involves charging a fee for environmental damages. The optimum fee for a particular activity is equal to the social damage caused by that activity. In practice, it is impossible to arrive at an optimum fee, because reasonable cost estimates of damage in terms of money usually cannot be determined. There are, however, alternative possibilities to use fees that have been set as the result of political considerations as tools to reach environmental goals, e.g., ambient air quality standards or a specific reduction of total emissions. This enables the targets to be attained at minimum cost, because it is generally agreed that economic incentives provide greater flexibility to the source and therefore result in least-cost solutions. Another merit appears to be the potentialities of such fees to give producers and, indirectly, technicians, product designers and researchers, incentives to find new techniques and new products with less damaging effects.

Even where air pollution control policy is the same, legislation varies from country to country due to differences in administrative and legislative traditions. In general, basic acts are supplemented by regulations and ordinances that provide flexibility yet are specific enough to meet the needs.

III. Elements of National Air Pollution Control Programs

The preparation of an air pollution control program includes (a) the establishment of goals, (b) the formulation of strategies for attaining these goals, and (c) the introduction of a system of supervision.

A. Formulation of Goals

Long-term goals can be defined only in a general way. Examples of such declarations of intent are found in the national reports prepared for the United Nations Conference on Human Environment in Stockholm, Sweden, in 1972:

> The goal of the United States is to adopt an environmental program that will not force us to abandon growth but to redirect it. To do so, we must balance economic development with conservation of land, water, air and other resources in such a way as to maintain a life of dignity and social justice. (*1*)
>
> The objective of Swedish environmental protection policy is to stop current damage to the environment and, insofar as is possible, to restore environments which have already deteriorated. (*2*)

The 1977 United States Clean Air Act defines one of the purposes of air pollution prevention and control as "to protect and enhance the quality of the Nation's air resources so as to promote the public health and welfare and the productive capacity of its population" (*3*).

In 1973, the Government of the Federal Republic of Germany declared that a decent environment is a fundamental right of every citizen. By law, the environment will be placed under the special protection of the state, thus providing legislators and administrators with a standard that gives ecological requirements the rank they deserve within the value system of society.

In the German Democratic Republic, Article 15 of the constitution provides for environmental protection.

In the Soviet Union, national air quality standards for more than one hundred air pollutants and combinations of substances have been established. A concentration is considered permissible if it does not have any direct or indirect harmful or unpleasant effect upon man and if it does not reduce the working capacity or worsen the mood of man. These Soviet Union national standards are declarations of intent equivalent to long-term goals.

An expert committee of the World Health Organization has recommended long-term goals for maximum levels of a number of air pollutants (*4*). These levels, with some modifications, will probably be used in most countries as goals in the planning for cleaner air in the urban areas.

The 1977 United States Clean Air Act provides for promulgation of primary and secondary national air quality standards. The primary standards are those requisite to protect public health with an adequate margin of safety and are examples of short-term goals. The United

States secondary standards, to protect welfare as well as health, are examples of long-term goals.

It is not easy to detect significant differences between countries in formulating declarations of intent and long- and short-term goals. There is tendency in countries with infrequent elections to put more emphasis on long-term goals, whereas countries with frequent elections tend to emphasize short-term goals. In theory, the long- and short-term air pollution goals of a nation could be formulated on the basis of the total allowable national emissions of a substance. Such goals have not yet been formulated but will be necessary in the future when continental and global air pollution problems have to be tackled.

B. Strategy Development

Strategies for attaining the established goals will vary, depending on whether they are long or short term. Long-term goal strategies emphasize systems planning in sectors like transportation, energy supply, and solid waste treatment, whereas short-term goal strategies emphasize control of individual sources through a system of permits, registration, emission limitation, and control of the use of fuels.

The two generally recognized approaches to maintaining air quality are the air resource management approach and the best practicable means approach. The air resource management approach uses air quality guides and standards as their basic elements, whereas the best practicable means approach stresses the principle that air pollution must be prevented as far as is practicable without reference to air quality guides or standards. Those supporting the latter approach say, "Why discharge any pollutant into the air when we have an adequate economically feasible technology to avoid so doing?" Supporters of the former approach call this "control for control's sake," and would allow pollutants to be emitted into the air up to its so-called "assimilative capacity," as measured by air quality guides or standards. It is argued that the best practicable means approach does not guarantee safe air quality and that it does not provide a useful basis for controlling pollution from numerous small sources, such as motor vehicles, home heating, etc. The best practicable means supporters counter by saying that air resource management can entail a degradation of air quality where it is now good and may result in widely varying control requirements for competing companies.

It is clear that the air resource management approach is the most logical one. The best practicable means approach, however, is much simpler and can easily be applied to single sources, especially large stationary sources, which often are in focus when a national air pollution control program is initiated. It is now generally agreed that full consideration must be given to both approaches to protect and maintain air quality adequately (Fig. 1). For single sources and small industrial areas with good ventilation (A–B in Fig. 1), emission standards and requirements on stack heights based on best practicable means will guarantee an air quality better than the air quality standard. A special nondegradation policy is not needed. In large industrial and urban areas (B–C in Fig. 1), the air quality standard should be the determining factor for regulations and systems planning concerning industrial development, energy supply, and transportation.

Air pollution control standards, established through legislative, regulatory or administrative action, mutual agreement or voluntary acceptance, include air quality, product, emission, process, and land-use standards. In short-term programs, implementation time 1–5 years, emission and process standards are favored by control agencies. There is a trend toward the establishment of uniform standards throughout a country. A typical example is the United States, which has established national air quality standards, national standards of performance for certain new stationary sources, and national emission standards for hazardous pollutants.

There are differences among countries on the relative importance of air quality standards and other types of standards. The Soviet Union, with more than one hundred national air quality standards, has no national emission standards. On the other hand, the United Kingdom has no national air quality standards but has set national point-of-emission

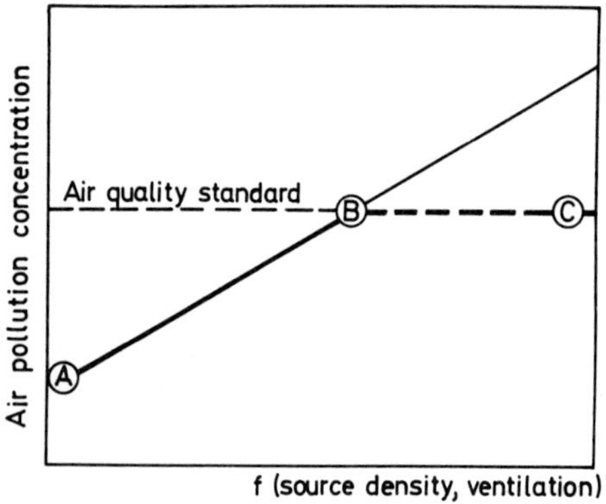

Figure 1. Air quality in a national air pollution control program using the combined approach of emisson standards based on best practicable means and air quality standards.

standards, wherever possible, to help industrialists and inspectors to interpret the "best practicable means" requirements for design and control purposes. However, differences in practice are not so great. In the Soviet Union, maximum permissible emissions are calculated from air quality standards by a simple formula and used as a basis for the design of control equipment. In the United Kingdom, in order to disperse residual pollutants and make them harmless, chimney heights are based on the known and expected effects of the pollutants on man, animals, and vegetation and on the quality of the surrounding environment. In Figure 1, it was shown how air quality standards and other types of standards can be combined in a strategy to utilize the advantages of different approaches. Most countries are moving in this direction, but in both the United States and the Soviet Union, the concept of air quality standards dominates.

Long-term strategies for air pollution control have to take account of the potential for improving air quality inherent in changes in population, transportation, manufacturing, energy requirements, solid waste disposal, and similar matters that have to do with the proper design and operation of communities. Planning has long been used as a decision-making instrument in land-use management and the design and development of communities and their contiguous rural areas. While such questions as water supply, sewage treatment, transportation, energy availability, facilities for recreation, and provision of open spaces have always been taken into account in community planning, there has been a growing awareness that consideration ought also to be given to air quality. Notable reductions in pollution can be expected through the development of community plans for energy supply, solid waste treatment, and transportation in which air quality has been one of the determining factors. A plan for energy supply could include proper siting of significant energy sources and efficient use of their waste heat in the community. With planned disposal of solid waste, domestic and commercial incineration could be eliminated. Restriction of the use of private motor vehicles in many of our large cities must be carefully considered in relation to the use of cleaner, cheaper, and possibly more convenient, systems of mass transportation.

Air pollution is created and experienced principally in the large metropolitan regions. Stricter measures than those adopted for the whole country are therefore justified for these regions. Unfavorable topographic and meteorological conditions also justify special measures. To provide a comprehensive plan for the air pollution control problems of an entire metropolitan area and that part of its surrounding territory that should be considered with it from an air-pollution point of view, it may be administratively advantageous to designate the defined geographical area as an air quality control region. In many countries, the designation of air quality control regions may be in conflict with the autonomy of local administration. The legislation creating such regions has, therefore, to be clear as to the responsibilities of the various authorities within the region.

Positive planning to ensure wise use of the environment is indispensable. This is especially true where there are great and competing pressures for limited natural resources—and no country in the world can avoid such a situation indefinitely. Intelligent land-use planning and good urban design are themselves valuable safeguards against pollution and squalor. The rational siting of industries within a country is of great significance in solving the problems of air pollution. In the Soviet Union, considerable attention has been devoted to uniformly siting industries throughout the entire territory of the country, with priority given to building new industries in medium-sized and small towns. This type of national planning is easier to execute, at least in theory, in countries where the central government controls the economy.

C. Supervision of Goal Achievement

Long- and short-term goals are normally expressed as air pollution control standards. The supervision of goal achievement, therefore, requires monitoring and surveillance programs. Monitoring programs must be related to the needs of governments to manage the environment and prevent damage to health, amenities, and resources; and surveillance programs must be related to their need for information about potential threats. There are many parameters that can be measured; what is needed is to choose those that can be measured accurately and economically and are valuable indicators of the state of an environmental system or that of themselves are so likely to cause hazard that they cannot be ignored.

Some plant and animal species accumulate materials in their tissues, and thus can be useful integrators of substances present only in great dilution in the environment. Systems and techniques of sampling and measurement should, if not completely standardized, be intercalibrated. It is even more important to ensure proper standards of procedure and craftsmanship, including proper statistical design and validation of results.

The problems of emission measurements are different. The pollutant concentrations measured are much greater than those that must be measured in the ambient air. Legislation should provide the administrator with the authority to require the owner or operator of any source of emission to install, use, and maintain monitoring equipment and to sample emissions in accordance with defined methods at such locations, at such intervals, and in such a manner as may be prescribed. The owner or operator should also have the duty to establish and maintain such records and keep such reports as reasonably may be required. Representatives of the administrator should have the right to enter any premises, to have access to and to copy any records, and to inspect any monitoring equipment and sample any emission. Reports, records, or information, and especially emission data, should be available to the public, except when it would divulge methods or processes entitled to protection as trade secrets.

The supervisors of stationary pollution sources are engineers. Therefore, it is important that there be engineers in local and regional level air pollution control organizations who are aware of the problems of industry. Each industrial firm or installation should have an officer designated as responsible for air pollution control or environmental protection. He must be at a high level in the organization; normally, he should report directly to the managing director.

IV. Organizational and Institutional Arrangements

Central environmental control agencies have been created in several national governments. Along with this trend, there is a corresponding trend toward establishing uniform regulations throughout a given country. Several of these agencies are independent ministries, whose principals are officers of cabinet rank. This is a trend likely to be pursued more strongly in the future. It is probable that environmental ministries will not attain the political power of the older, established ministries. However, environmental ministers can be expected to play increasingly influential roles in national political affairs, and effective political power in environmental affairs will gravitate increasingly toward central governments. Some highlights of national air pollution control programs in selected industrial nations are given in Table I.

At the 1971 Economic Commission for Europe Conference on the Environment, governmental institutional arrangements for dealing with environmental problems were critically examined (*5*). A pilot survey was made of five countries with considerable differences in their institutional arrangements for environmental management: the Federal Republic of Germany, Poland, Sweden, the United Kingdom, and Yugoslavia.

Table I Highlights of National Air Pollution Control Programs in Selected Industrial Nations

Nation	*Comprehensive pollution control ministry*	*Clean air law*	*Status of problems and programs*
Belgium	None.	The Clean Air Act (1964) is a basic law, or enabling act, that confers broad powers for issuance of executive decrees regulating air pollution. However, no decrees establishing industrial air pollution standards have been issued. Regulations limiting domestic trash burning and the sulfur content of fuels used for domestic heating have been adopted.	Serious air quality problems result from autos, domestic heating, and powerplants. Belgium is one of the world's most densely populated nations.
Canada	Department of the Environment (1970). Until the 1970s, most authority for pollution control was left to provinces and municipalities.	Clean Air Act of 1971 authorizes federal government to set ambient air objectives and to set emission standards necessary to protect health. Other emission standards are set by provincial and local governments under federal guidelines. Auto emissions and fuel composition are federally regulated.	Federal government has become significantly involved in pollution control since 1970. It is still too early to judge implementation of the new air laws.
France	Ministry for the Environment and Protection of Nature (1971).	Law of 1961 authorizes controls to protect health and prevent public nuisance; subsequent decrees regulate auto emissions and electric utilities and institute general controls in specified urban areas. A decree of 1973 coordinates in-	Broad legal authorities have not been fully implemented. Control of air pollution in Paris has progressed substantially since the early 1960s.

		terministerial efforts against air pollution.	
Federal Republic of Germany	None.	Under the 1971 Clean Air Law, the federal government has authority to regulate industrial emissions, but enforcement powers reside with the states and often are not used. Federal government regulates auto emissions and lead content in gasoline.	Substantial air and noise problems persist. Federal authority is very limited.
Italy	Ministry for the Environment (1973)	Air Pollution Law 615 (1966) authorizes emission limitations for some industries and for autos, with variations by zones.	Intense efforts to increase industrialization outweigh environmental concerns, and federal jurisdiction for protective action is limited. Air pollution is widespread.
Japan	Environmental Agency (1971).	Basic Law for Environmental Pollution Control, as amended in 1970 (1967), is designed to protect health and preserve the environment. Subsequent laws impose criminal sanctions (1970) and provide no-fault relief for damage to health (1972) and control odors (1971). Air Pollution Control Law (1968) authorizes auto emission standards, controls on lead in gasoline, and minimum national controls on stationary sources. Japan decided in 1975 to require auto emission standards comparable to the United States standards. Japan	Highly concentrated population and industry make Japan one of the most heavily polluted nations, giving rise in recent years to widespread public complaints and litigation. Widespread health damage is reported from air pollution.

Table I (*Continued*)

Nation	*Comprehensive pollution control ministry*	*Clean air law*	*Status of problems and programs*
Japan (*Continued*)		is now the first nation to set emission standards for some older vehicles, requiring retrofit controls.	
The Netherlands	Ministry of Public Health and Environmental Hygiene is responsible for most pollution control, but policy decisions come from the Physical Planning Council.	Air Pollution Act (1970) establishes regulatory system based on licenses for polluting facilities, administered by central, provincial, and municipal governments.	Considered by many to be the most pollution-conscious nation in Europe with strong antipollution commitment at all levels of government. Topography and climate help to reduce air pollution, except in several industrialized areas where it is severe.
Sweden	National Environment Protection Board.	Environmental Protection Act (1969) requires operating licenses for new or modified facilities in 38 industrial categories, with emission limits based on technical and economic feasibility. Royal ordinances restrict sulfur content in fuel oil and lead in gasoline. Emission standards for new motor vehicles from 1971 models. From 1976 models, Sweden has adopted the United States 1973 standards.	Despite a recently initiated national control program, air pollution remains significant in localized areas. About half of deposited sulfur drifts into Sweden from other nations. Vast pulp and paper industry accounts for most industrial problems.
United Kingdom	Department of Environment.	The Alkali and Works Regulation Acts (1863, 1906, 1966) and the Clean Air Acts (1956, 1968) require	Pollution control laws have a long history, but controls are still sought more by persuasion than

		annual registration of about 3000 designated industrial processes; firms must demonstrate that they are using the "best practicable means" determined on a firm-by-firm basis to control emissions. Controls also apply to motor vehicles.	enforcement and are generally implemented on a firm-by-firm technical–economic basis. Air pollution from domestic heating has decreased considerably but is still high.
Union of Soviet Socialist Republics	None. Hydrometeorological Service has overall monitoring and control function for air, soil, and water pollution. Various ministries (e.g., Agriculture, Health, Reclamation, and Water Management) have standard-setting roles for different aspects of the environment. State Committee for Science and Technology has coordinative role and with Academy of Sciences runs interagency council responsible for overview of environment. Individual republics in many cases have environmental policy bodies.	No comprehensive air law. There is a Law on Air Pollution in Moscow of January 8, 1963, and some broad legislation under the heading of the Protection of Health (1969). Part of the "control" strategy is the "sanitary clearance" or buffer zone established between industrial facilities and residential areas.	Environmental problems being given increasing attention. Special session of Supreme Soviet led to the Decree on Strengthening the Protection of Nature and Improving the Utilization of Natural Resources of December 28, 1972, raising the priority of environmental concerns within the Soviet system, calling for annual and long-range environmental plans as part of the planning system, and assigning more definite responsibilities for environmental protection among the various agencies of the Soviet government. There is also growing concern about air pollution.
United States	Environmental Protection Agency (EPA) (1970)	Clean Air Act (1963) with amendments (1965, 1970, 1977), National Environmental Policy Act (1970). Under the provisions of the Clean Air Act Amendments of 1977, EPA is authorized to pro-	Air quality emerged as a major national issue in the 1960s as the public became keenly aware of pollution problems from motor vehicles and industry. The first federal program on air pollution was

Table I (*Continued*)

Nation	*Comprehensive pollution control ministry*	*Clean air law*	*Status of problems and programs*
United States (*Continued*)		mulgate ambient air standards. EPA establishes and enforces performance standards for new or modified stationary sources. EPA establishes and enforces federal emission standards for hazardous pollutants. EPA establishes and enforces emission standards for new motor vehicles and may regulate the manufacture or sale of fuels or fuel additives.	developed in 1955 by the Public Health Service of the Department of Health, Education, and Welfare. Up to 1970, the efforts were piecemeal; various government units and state governments reacted in their own way. The state implementation plans under the provisions of the Clean Air Act Amendments of 1970 and 1977 are an all-important and essential element in the national strategy to achieve clean air in the 1970s and 1980s.

In the Federal Republic of Germany, responsibility for certain air pollution questions rests with the Ministry of the Interior; the eleven provincial governments are, however, largely responsible in this field.

In Poland, the Ministry of Health and the Committee for Protection of the Environment play decisive roles in air pollution control. Provincial administrations play an increasing role.

In Sweden, ministries are much smaller than in other countries, but are assisted by boards with broad functions; the Ministry of Agriculture, assisted by the National Environment Protection Board, has major responsibility for air pollution questions. The twenty-four provincial administrations have sections with environmental responsibilities. Local authorities were strengthened through the reduction of communes, from about 1000 some years ago to 274 in 1974.

In October 1970, the United Kingdom was the first country to set up a Department of Environment, comprising three previous ministries—Housing and Local Government, Transport, Public Buildings and Works. The United Kingdom has a two-level governmental structure, national and local; a growing influence on environmental matters is exercised by its Regional Economic Planning Councils.

In Yugoslavia, major roles are played by the Ministry of Health and by the Federal Commission for Town and Regional Planning, but many important details are left to the six provincial governments. In some Yugoslav provinces, responsibilities for environmental matters are centralized in the Secretariats for Town Planning. Local responsibilities are of recent origin but of growing importance.

In spite of great differences in the institutional arrangements, the countries had similar problems. All five countries recognized the importance of strong coordination and the need for links between controls over environmental disfunctions and various types of planning. In the federal countries, overall coordination was limited by the strength of provincial administrations. Separate parts of government machinery like to maintain their spheres of influence. It appears unlikely that all environmental questions can be brought under one umbrella, even where ministries of environment are set up. The five countries found the best results were achieved by working toward full cooperation between authorities, industry, agriculture, and the public, rather than strict reliance on legal enforcement. However, in reorganizing administrative machineries some over-all recommendations can be made:

a. Governmental arrangements at the central level should be headed by a cabinet member or high-ranking official of equivalent importance, in order to provide an effective voice and leadership in the governmental decisions that are being made daily, many or most of which affect the environment.

b. While the central body itself should preferably coordinate and promote decisions, it should not be confined to an advisory role; it should have operational control and planning functions as well.

c. Such a central body should (i) have a comprehensive and systematic approach and both a short- and a long-term out-look; and (ii) have wide contacts with universities, research institutes, etc. so as to ensure a continuing flow of the necessary information and research results required in decision-making and also provide a feedback mechanism.

d. The central body should maintain close lines of communication with local and intermediate authorities so that all efforts will be well coordinated. Furthermore, the body should maintain close contacts with and encourage participation by citizen groups and the population at large.

e. Wherever applicable, the governmental arrangements covering the intermediate (regional) and local levels should be strong enough to cope comprehensively and systematically with specific environmental problems arising in individual regions and urban or rural areas. This need will best be served by establishing a special post (or organ) at the appropriate intermediate or local level to which would be delegated the authority and resources needed for bearing extensive environmental responsibilities, undertaking multidisciplinary studies and organizing and co-ordinating comprehensive environmental action on those levels.

Countries with an energetic policy in controlling air pollution have no common denominator in specific organizational and institutional arrangements. The United States, Canada, Japan, Australia, New Zealand, the Scandinavian nations, the Netherlands, Switzerland, and the Federal Republic of Germany seem to be the countries with the most stringent national air pollution control programs. The governments in these countries are much helped by favorable public opinion. What they have in common is a high material standard of living. Both the need for air pollution control and the ability to afford it tend to correlate with a high standard of living.

V. Balance between Central Control and Decentralized Functions

The balance between central control and decentralized functions is particularly important in fields such as the management of the environment, where many problems are local but require support and integration through strong national policies. The growing complexity of government and the increasing pressure for public participation point toward decentralization, but the planning process and the need for national policies tend to concentrate power at the center.

The administrative levels in the management of the environment are central, provincial, state, and local. It is generally recognized that central government alone can determine national strategy and must retain a continuing oversight of surveys, of monitoring, pollution control, and planning. The responsibility for control of pollution by regulation should be at the lowest level of government capable of dealing effectively with the problem in its entirety. Locally controlled programs make possible a quicker response to problems, a closer control over operations, and a mobilization of local resources. However, there are problems associated with local control. Local authorities with small populations are not often in a financial position to run an adequate program. Industrial and commercial interests sometimes have a very strong influence on the actions of local governments. Local agencies are often guided by the fear of losing industry to other regions that have weaker regulations or by the desire to attract new industrial enterprises by offering lenient regulations. This situation is leading to consolidation of pollution control activities in agencies of the provincial, state, or central governments.

It is not possible to have effective pollution control if it is planned and imposed in the isolation of an academic or administrative center. There must be continuing consultation with the various levels of government, with the industries that are the actual or potential polluters and, increasingly, with ordinary people. To succeed, there must be administrative flexibility. It is essential to take into account the wide regional variations in social and economic circumstances and the way these change.

A. Control of Individual Sources

Air pollution may be regulated differently according to the nature of the source: (a) stationary sources that individually contribute significantly to air pollution; (b) stationary sources that individually contribute less significantly to air pollution, but are built in large numbers; and (c) motor vehicles, aircraft, and mass produced fuel-burning appliances.

The most effective way for the control of air pollution from stationary sources is a permit system under which prior approval of plans, specifications, and other data for new construction or alterations is required. Approval is given only if the controls to be provided are, in the judgment of the agency, adequate to meet the air quality and emission requirements. Through this procedure, installations are avoided that are poorly designed, that would cause air·pollution thus requiring costly modifications or reconstruction, and that would cause air quality standards to be exceeded; i.e., "clustering" is thus avoided. This does, however, impose on the administration a requirement to have the necessary skilled engineering staff available. It is desirable to publish the emission and process standards that will be used in deciding whether installations will be approved. This reduces variations in strictness among agency staff and increases the possibility of plans being approved on first submission.

If a permit is made compulsory for a large number of installations, there is a risk that the expert staff available will be overloaded with work in administering the permit system. Other important matters may be neglected. To avoid this, it is well to exempt from construction permit requirements all sources belonging to group (b) above. For such sources, an alternate method is to establish a registration system. In such a system, those planning to build a new facility are required to submit only information on the location of the unit, a general description of what it will do, and information on the nature and location of expected emissions of pollutants. Reliance is placed on the applicant for meeting emission regulations after the unit goes into operation. The registration procedure does, however, give the agency knowledge of the installation so it can be checked after construction and an opportunity to call attention of the applicant to emission requirements.

A considerable number of environmental disfunctions can be ascribed to unfortunate siting, to an excessive concentration of certain activities, to inappropriate buildings, or to negligent operation and exploitation of natural resources. This shows the necessity of controlling the construction of buildings, factories, the utilization of the land and the exploitation of natural resources. Some regulations of this kind are found in virtually all countries in the legislation concerning urbanization and building permits. However, these regulations are not always properly applied.

One of the most common deficiencies of regulatory systems for the control of individual sources is that decisions concerning the location of industrial activities that are considered to be important for purposes of national physical planning are not made at a sufficiently early stage and on the basis of an all-around assessment. Often this is due to industry having chosen its location and gone very far in its planning and other preparations before central authorities are given the opportunity to make their own assessment. Different assessments are undertaken on the basis of a variety of laws, and coherence is lacking.

To remedy these deficiencies, decisions affecting the location of industrial or similar activities that are of major importance to the management of the country's natural resources should be taken by the government at an early stage. Before decision, a comprehensive assessment must be made relating considerations of environment protection and planning to, e.g., considerations of labor market policy, regional policy, and industrial policy. In order for the government's decisions to be as broadly based as possible, opinions must be requested from authorities, union bodies, trade associations, etc. It is also important to hold a public hearing at the proposed site. After the location issue has been decided by the government, the industry will have to apply for a permit in the same way as for other sources belonging to group (a) above.

In a regulatory system for control of individual sources, emission and process standards are favored by control agencies and widely used all over the world. There is a trend toward the development of national standards from considerations of processes and equipment and, where necessary, more stringent regional standards from air quality standard considerations. Emission standards derived from considerations of processes and equipment normally require the use of the best practicable means to control emissions. The word "practicable" implies best available and economically feasible control technology. To adopt the best available control technology is not only to copy the best installation within an industry, but also to borrow the technology of one industry to improve that of another. It is inherent in the term "feasible" that the cost will be within the reach of industry. This does not mean that it has to be within the reach of each individual firm. No doubt, however, what will be considered a feasible cost will depend greatly on the environmental effects involved and the political climate or public opinion. Therefore, air quality *is* considered, indirectly, even if there is no formal reference to air quality guides or standards. As it is generally more expensive to reach a certain control level with existing installations than with new installations, the adoption of the best practicable means leads to different requirements for new and existing plant. The best practicable means for new installations are normally arrived at by considering experience in all the technically advanced nations, but for existing installations, more weight has to be given to national conditions.

Motor vehicles, aircraft and mass-produced fuel-burning appliances should be subject to general provisions. Since mobile sources circulate across frontiers, uniform standards for their construction and operation should be established at an international level. Sources belonging to category (c) above must be constructed in such a way that, if properly maintained and used, they will conform to the regulations throughout their useful life. The responsibility of the manufacturer must be made clear in the legislation.

The most important functions of a central body in the control of individual sources are (a) to administer the permit system for stationary sources that individually contribute significantly to air pollution and for which the siting is of national interest or for which expertise is not available at the regional or local level; (b) to review periodically the list of categories of stationary sources that should be covered by a permit or registration system, work out process and emission standards, and publish manuals and codes of practice for these sources as an assistance to regional and local agencies; and (c) to administer to certificate system for mobile sources.

B. Systems Planning

As a general rule, prohibitions and restrictive measures do not attack the roots of the problem. Environmental legislation ought to go deeper into the problem and abandon its passive attitude. It ought to direct the activities of society so as to prevent them from becoming a danger to the environment. To achieve this, the environment has to be subjected to a certain amount of planning. The construction of buildings and factories, the performance of certain activities, and everything that may affect the existing environment ought to be planned. Without planning, environmental care will remain just conservation, conceptually. It is, of course, an illusion to believe that everything can be planned, but nevertheless, in certain countries, the planning of land development and urbanism affords an appreciable protection. In a number of countries, and more particularly in those that have adopted a planned economy, the environment is protected and improved in large measure by national plans of economic development.

Systems planning in areas such as energy supply and transportation are of vital importance for air quality. As is shown in Figures 2 and 3, environmental protection is only one of several items to be considered in these systems. Therefore, regulations concerning the composition and use of fuels, as well as emission standards for stationary and mobile sources remain indispensible but should be considered in a broader context. Regulation of fuels without planning of the whole energy supply system can never give an optimal solution to the problems. An illustration is given in Figure 4, which shows how the sulfur dioxide concentration in Stockholm, Sweden, would vary with different types of heating systems, as calculated from a dispersion model for the area.

In planning for energy supply, air pollution control requirements have

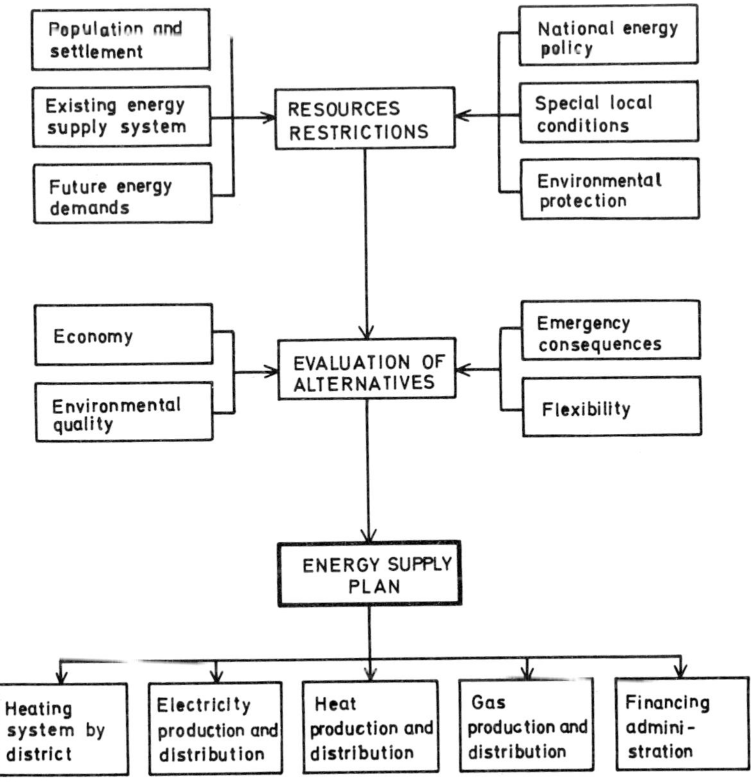

Figure 2. Development of an energy supply plan.

to be defined. In Stockholm, these were 60 $\mu g/m^3$ for sulfur dioxide as a 6-month average. As can be seen from Figure 4, district heating has a very marked effect on local air quality. The calculated reduction of the 6-month winter average is about 85% when individual heating is replaced by district heating from 50 MW stations. It is better than 95% for the most centralized case, although this means a 50% higher emission due to power production.

To implement a system planned with consideration for environmental quality, it will be necessary for the municipality to take responsibility for heat and electricity supply in much the same way it does for water supply, sewage treatment, and waste disposal. The free choice of a heating and cooling system by the individual probably can not be preserved in the future. New regulations will be required. The development of a planned energy supply system in an urban area is not a rapid process. This fact has to be reflected in the time schedule for air quality achievement. New areas can be included in a district heating or cooling scheme

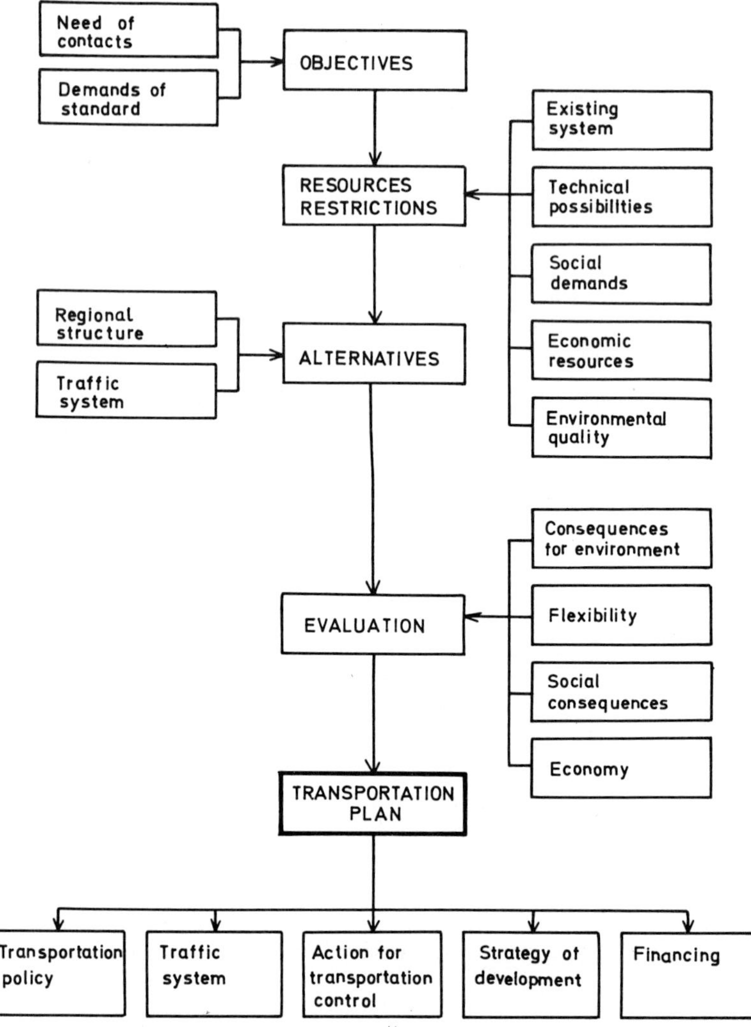

Figure 3. Development of a transportation plan.

without difficulty, but built-up areas will require a transition time of several years.

A strong argument for systematic transportation planning in urban areas is that air pollution, noise, and "space pollution" are equally important in most areas. A transportation plan must aim at solving all these problems.

Achieving a balance between central control and decentralized functions in systems planning of energy supply and transportation is not without problems. The development and establishment of such plans may be in conflict with the autonomy of local administrations. Legislation that is clear on the powers, duties, and responsibilities of the various authorities is required. The role of the central body is to develop the whole concept of systems planning in the fields where it is needed and to work out "model" plans to facilitate work at the local level. Assistance in financing is another important matter. The central body should have the power to require municipalities to develop such plans in a reasonable time period.

C. Enforcement and Supervision

The existence of laws does not alone provide an index of the effectiveness of control. Enforcement policies may be more important than the laws themselves. Laxity in enforcement has been common even in countries where legal machinery for control has been available. A prerequisite

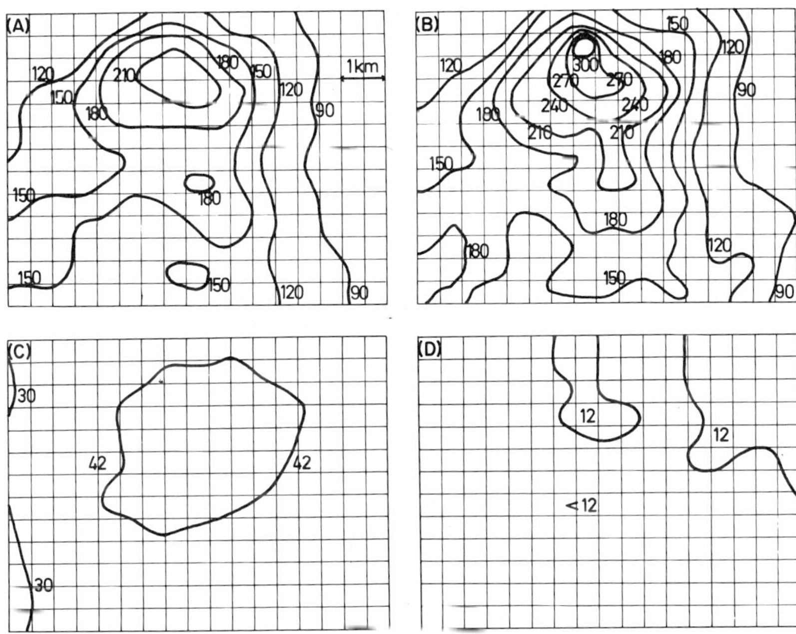

Figure 4. Calculated sulfur dioxide concentrations (6-month average in micrograms per cubic meter) in the city of Stockholm, Sweden, for different heating systems. The total emissions of sulfur dioxide are the same in alternatives (A), (B), and (C), but 50% higher in alternative (D) due to combined production of heat and power. (A) Actual heating system for the winter of 1969–1970; (B) no district heating; (C) district heating from a number of 50 MW (thermal) stations (stack height 60 m); (D) district heating from a number of 500 MW (thermal) stations for combined production of power and heat (stack height 140 m).

of enforcement is that regulations and air pollution control standards are clear and well defined.

In general, air quality standards are difficult to enforce. This is one reason that emission, process, and product standards are preferred by control agencies. But not even emission standards are always properly defined. Before numerical values are established, emission standards must be defined with regard to the conditions for which they are applicable. Figure 5, for example, shows the range of fluctuation of emissions from an industrial process, how the degree of control imposed by the standard will depend upon whether the numerical value has been related to normal conditions alone or to all operating conditions, and upon whether the degree of control is expressed as a short- or long-term average.

A system of supervision of stationary sources should include (a) initial inspection to ensure compliance with standards or specific requirements in the permit, (b) continuing check on the emissions by the plant personnel, (c) routine periodic inspections to insure continuous compliance with regulations, and (d) occasional inspections to investigate complaints or suspected violations. The initial inspection would normally be carried out within 6 months after a new or altered installation comes into operation. If a special permit to operate were required, such a permit would be issued if emissions were satisfactory. At complex in-

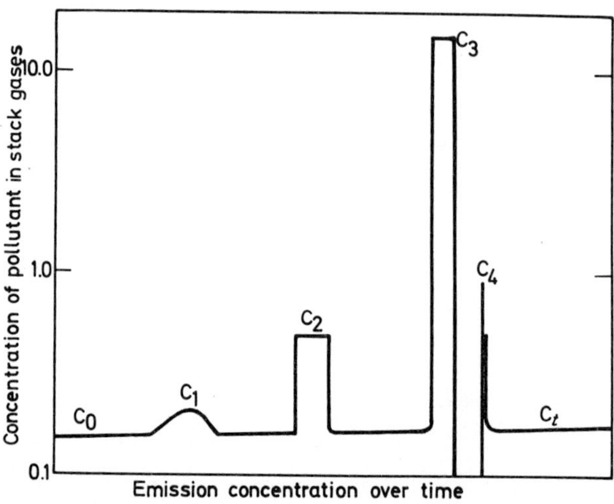

Figure 5. Variations of emission concentration with time (particulates controlled by electrical precipitator). Pollutant concentrations in stack gases are given during normal operation (C_0), during a process operation (C_1), with one section out of operation (C_2), with both sections out of operation (C_3), at start of operation (C_4), and during normal operation at time t (C_t).

dustrial sources, the initial inspection might require emission measurements during several days to cover different operational conditions. Continuous recording instruments should be calibrated and the operators instructed in their use and maintenance.

A very important part of the enforcement of emission standards is continuing check on emissions by the plant personnel. Not only emissions, but also parameters connected with the operation should be recorded. It is important that written instructions be available at each installation. These should describe the function of the various components and the reason for keeping these parameters within specified limits. The frequency of the routine periodic inspections may be the same for all kinds of plant, e.g., once per year; or more frequent inspections may be made of sources that have a tendency to cause excessive emissions. If a continuing check by plant personnel is made and reported in a satisfactory way, periodic inspections may be on a much smaller scale than initial inspections. Periodic inspections are also useful in maintaining interest of operating personnel in air pollution control matters, motivating them to maintain minimum pollutant emissions, and fostering exchange of information among those having similar source control problems.

The financing of the supervision system should be made clear in the legislation. The continuing check on emissions as well as the initial and periodic inspections can be considered as a part of the emission control program and the costs paid by the owner of the plant. The initial and periodic inspections can be made by the control agency or by authorized laboratories. In the latter case, every agency should at least be capable of doing occasional inspections or having them done by contract.

Supervision of stationary sources is normally left to local and regional authorities. This may cause problems with the more complicated process industries, because skilled personnel are often not available, at least at the local level. Supervision of scheduled premises through a central body like the British Alkali Inspectorate has definite advantages.

Duties of the central authority are to design the general structure of the supervision system, to work out model supervision programs for various categories of industries, to run a national center for testing analytical and sampling techniques and measurement equipment, and for the training of personnel in the use of the relevant techniques and equipment. If initial and periodic inspections are carried out by authorized laboratories, the central body has to keep good control of those laboratories through continuous intercalibration and exchange of information. Another important task is to summarize and evaluate the results from the supervision programs and make this information available to agencies administering permit and registration systems, to industries, and to other interested parties.

Monitoring new motor vehicles for compliance with current standards is a subject that requires a strong engagement of the central body. One problem immediately arises. Should it be the responsibility of the Ministry of Traffic or the Ministry of Environment? In the United States, it rests with the Environmental Protection Agency and in Canada, with the Department of the Environment, but in other countries with the Ministry of Traffic or central agencies within that ministry. Experience shows that an air pollution compliance monitoring program is likely to have a low priority if it is not a part of the environmental enforcement and supervision procedure.

A compliance monitoring program is normally based on all new motor vehicles offered for sale being certified by manufacturers that they conform to current emission standards. To make sure that production vehicles are in all respects of the same design as vehicles already tested and certified, production-line inspections are carried out. The third part of the program—and perhaps the most difficult one—is the testing and inspection of in-use vehicles, i.e., "used cars."

Durability tests are an important part of the certification procedure, but are very costly to perform in small countries where a comparatively small number of cars are sold of each model. It seems necessary, therefore, in these countries to be more lenient on the requirements for durability testing but instead monitor the emissions from a representative number of in-use vehicles for which emission standards are in force. As there is no representative short test available, time-consuming and expensive standardized test procedures have to be used. If a group of vehicles with excessive emissions is identified, the manufacturer may be ordered to recall the vehicles for modification. If new vehicles of the actual type are still being sold, the certificate may be withdrawn. In addition, more complete durability tests may be required for certification of new models. These actions are very expensive for the car manufacturer and would therefore deter him from putting defective exhaust control systems on the market. The difficulties involved in running different compliance monitoring programs in individual countries make internationally agreed upon procedures for testing and certification very desirable.

D. Financing

The "polluter pays principle" is recognized by all Organization for Economic Cooperation and Development (OECD) countries. This means that the accounts of private or public economic transactions must include the cost of environmental degradation resulting from production and consumption activities, so that the rational economic management of environmental resources that are now becoming scarce can be achieved. To put it in simple terms, anybody causing pollution must also bear the costs of reducing or eliminating it.

As the OECD Guiding Principles (*6*) recognize, trade distortions can result from differences in cost allocation. To the extent that a country provides government subsidies to firms, the prices of their goods and services will not fully reflect all related costs of protecting the environment. This will cause overconsumption of that country's products and lead to a missallocation of resources compared to that which would be attained if all domestic and international prices fully reflected the social costs of production and consumption.

A wide variety of deterrents or incentives are in effect or have been proposed (Table II) (*7*). The most common subsidy is a reduction in corporate profit taxes through liberalized depreciation allowances. Table III compares depreciation policy on pollution control equipment for seven nations that engage in substantial international trade (*8*). After the depreciation allowance, the next most popular method of subsidization is the low-interest or interest-free loan. Japan, for example, lends private industries up to 80% of the cost of equipment to control air, water, and noise pollution. France lends up to one-half the cost of such equipment. Canada also makes direct government loans for pollution abatement purposes. Borrowing from private sources is supported by federal guarantees in Germany and by interest rate subsidies in Belgium.

Direct subsidies are not generally accepted in the United States—either on policy grounds or (in some cases, at the state level) on constitutional grounds. Nevertheless there is use, in the United States, of tax-free industrial development bonds issued by local governments to make low-interest loans to companies to buy pollution control equipment.

In Sweden, the Government gives subsidies for the control of air pollution, water pollution, and noise from industry. Factories in existence before the Environment Protection Act came into force in 1969 can get a part of the investment costs paid by state money. The system applies only during a 6 year period. Perhaps the most interesting thing about the subsidy system is that it has been used to stimulate the labor market during periods with low economic activity. The total cost of the investments was estimated at approximately one billion Swedish krona. Of this sum, the government was expected to contribute 25%, or 250 million krona, allocated at the rate of 50 million krona per year.

A condition for allocation of a subsidy is that the purpose of the action

Table II Financial Deterrents or Incentives Proposed to Supplement or Replace Regulation *(7)*

Taxes	
Sales taxes	On fuel, fuel additives, ingredients in fuel, pollution producing equipment
Ultimate disposal taxes	On automobiles or other objects requiring ultimate disposal
Land-use taxes	For pollution producing activities
Tax remission	
Corporate income tax	For investment in or operation of pollution control equipment; accelerated write-off of pollution control equipment
Property taxes	For pollution control equipment
Fines, effluent charges, and fees	
Fines	For violation of regulations
Effluent charges	For permission to emit excessive quantities of pollution —paid after emission
Fees	For permission to emit excessive quantities of pollution —paid before emission
Subsidies	
Direct	Governmental production (e.g., nuclear fuel)
Grants-in-aid	For pollution control installations
Indirect (low-interest bonds and loans)	For pollution control process or equipment development
Import restraints	
Duties and quotas	On materials, fuels, pollution control, and pollution-producing apparatus
Domestic production restraints	
Quotas, land, and off-shore use restraints	On materials and fuels

must be important from the standpoint of the general public. Moreover, subsidies may be given only for industrial establishments expected to remain in existence for a sufficiently long time that the cost appears justifiable.

When the general level of economic activity began to show signs of stagnation in the autumn of 1971, it was decided by the Swedish Parliament in December of the same year that increased subsidies of up to 75% of the costs would be allocated for pollution-control measures in industry for work commenced between November 1, 1971, and June 30, 1972. A further stipulation was that the work should be at least 60% completed before June 30, 1972, and special time limits were set for the procurement of machinery and other equipment. To facilitate com-

Table III Depreciation Allowances for Pollution Control Equipment in Selected Countries *(8)*

Country	*Depreciation allowance (%)*	*Time (years)*	*Comments*
Canada	100	2	Equipment purchased 1965–1973
France	50	1	Facilities built 1968–1974, in addition to normal accelerated depreciation
Germany	50	5	In addition to straight-line depreciation
Japan	50	1	In addition to accelerated depreciation
Sweden	—		No accelerated depreciation
United Kingdom	—		No accelerated depreciation
United States	100	5	Equipment purchased after 1970 for plants built before 1969; 5-year period must be amortized on straight-line basis; no investment tax credit permitted

mencement of such work, the Environment Protection Board and the county administrations were authorized to allow work on a plant to be commenced on certain conditions despite the fact that environment control questions had not been finally settled. Otherwise, the same grounds were valid as for the 25% subsidy. For the increased subsidies during the winter of 1971–1972, the Parliament allocated a total of 400 million krona over and above the funds already reserved for the ordinary 25% subsidies. In 1972 and 1973, it was again decided that increased subsidies of up to 50% would be distributed to industry during the autumn and winter. For these increased subsidies, 100 and 70 million krona, respectively were allocated by Parliament. Forty percent of the total subsidies to industry have been for air pollution control.

A feature common to all socialist countries, and important from the point of view of financing measures for the control of environmental pollution, is the social ownership of the basic means of production, banks, building, and transport. Another matter of importance is the basing of their social and economic development on national economic plans.

In Bulgaria, administrative regulations under the 1963 law define the scope of the investments and technical equipment required in newly built industrial plants that might cause air pollution. It is assumed that any damage to the environment should be repaired at the expense of the economic sector that caused the damage. In accordance with these principles, much of the investment is financed from the industrial enterprises' own budgets.

In Czechoslovakia, the financial outlays for air pollution control are an integral part of the total cost of plant building. The costs of new installations for air protection are financed from an air protection fund made up of government subsidies and of fees paid by plants according to the quantity of pollutants emitted into the air over and above permissible norms. Higher penalties are imposed on plants located in tourist and health resort areas. Each industrial plant is assessed separately.

In the German Democratic Republic, industrial plants finance the building of installations for the protection of the air.

In Hungary, the financing of the program for improvement of the quality of the air in the region of the capital, Budapest, derives partly from government funds.

In Poland, the program for the building and development of installations related to the protection of air are carried out within the framework of the national annual and quinquennial plans and are financed mainly from sectoral investments and, to a smaller degree, from plants' investments. The implementation of sectoral investment is facilitated by bank credits, which must be repaid upon termination of the work and after settlement of accounts. The credit may be repaid partly from the enterprise investment fund and partly from the additional profit obtained during 5 years after the plant has begun work. The enterprise investments are financed from repair and investment funds. Sanitary control of air cleanliness and technical control of industrial plants, from the point of view of their pollution of the atmosphere, are financed in Poland from the state budget.

In Romania, funds for air protection installations are also an integral part of general investment funds for particular industrial projects. Financial investments for air protection, that is, for the building of installations for cleansing the emissions of gas and dust from industrial plants, come from the central budget.

In the Union of Soviet Socialist Republics, financial investments in this field are part of the cost of building new, or the development of existing, industrial plants.

In Yugoslavia, the industrial plants and communes finance investments in the area of air protection from their own resources or with the aid of bank credit.

One thing should be borne in mind when assessing, the effects of environmental measures and their financing on foreign trade and international competitiveness; the costs of environmental management is not particularly burdensome when compared, for instance, with increases in wages and cyclical fluctuations in prices of raw materials and energy.

E. Information, Education, and Training

For environmental protection efforts to be successful, the public must have both knowledge and involvement, which in turn, makes it essential that information about environmental matters be made generally available. Training for the administration of environmental protection programs or to enable people to take environmental aspects into consideration in the course of their work is also necessary. Information on environmental matters is for the most part disseminated through pamphlets, the press, radio, and television. It is an important duty of the central authorities to supply these sources with factual material. The authorities themselves at all levels must also participate in the public debate on environmental protection matters.

The objective of education in environmental protection is the integration of environmental protection subjects in general education. Elementary instruction in environmental protection matters should be provided as a part of education at all levels—preschool, primary school, secondary school—to create an enduring awareness of the limitations that the protection of the environment imposes on human activity. In view of the broad interdisciplinary character of environmental science, such educational programs should embrace not only simple biology and earth sciences, but also chemistry and physics, mathematics, history, the arts, and literature. One of the principal and necessary objectives of an integrated approach must be to overcome the existing dichotomy between the social sciences and technology.

The main purpose of introducing environmental education at university level is to equip young specialists with knowledge that will enable them, in their future work, whether in industry, agriculture, public health service, chemistry, civil engineering, or other land-linked professions, to solve the problems of conserving, restoring, and utilizing natural resources. At every university there should be a general course dealing with environmental protection as a complex problem. This course should also explain the ecological interrelations in nature, the consequences of disrupting biological balances, etc. Specialized university subjects should introduce a general survey of, and the correct approach to, the environment as a whole and show the importance of taking environmental problems and interactions into consideration. Environmental education should deal with the dynamics of natural complexes (ecosystems—equilibrium in nature) in relation to human activities. Natural processes have to be taken into consideration when dealing with the problems arising in the course of a large-scale economic development project. Ecological thinking can be introduced into university courses by lectures, seminars, and postgraduate study programs.

The objective of training in air pollution control is to develop and improve the knowledge and skills of staff employed in air pollution control activities. There is a need for basic courses and advanced courses ranging from one week to one or two years in duration. Several of the courses should also provide opportunities for extensive laboratory practice. Rotation of personnel between local, regional, and central authorities, as well as between industry and control agencies, is important. The production of basic material for information, education and training, as well as coordination, often rests with the central bodies for environmental protection, but active participation is needed at all levels.

F. Future Needs

The need to improve and strengthen the machinery for the management of the environment at all levels has already been acknowledged by governments. However, the patterns now in existence reflect their historical growth and are still mainly sectoral rather than comprehensive. The great variety of environmental processes and the fragmented character of institutional arrangements emphasize the importance of horizontal and vertical coordination in the machinery for management of the environment. Major coordinating links are needed between controls over environmental disfunctions and various types of planning. Environmental policies and measures also need to be coordinated with other, particularly economic, measures.

In the long term, sanctions, prohibitions and restrictions are not sufficient. However such measures are always necessary and may be adequate for solving some of the worst problems from industrial activities in the first years of a national air pollution control program. An adequate policy in fields like energy and transportation requires an active interference in all matters that are likely to affect the environment. On the other hand, neither declarations of intent nor sweeping legislation and planning in themselves can cure pollution or improve the environment. In the end, doing what is needed to protect the environment nearly always comes down to hard-slogging, unglamorous details and painstaking administration.

REFERENCES

1. "National Report on the Human Environment, United States of America" (prepared for the United Nations Conference on Human Environment in Stockholm, Sweden, 1972). Washington, D.C. 1972.

2. "Sweden's National Report to the United Nations on the Human Environment." Royal Ministry for Foreign Affairs, Royal Ministry of Agriculture, Stockholm, Sweden, 1971.
3. "Clean Air Act," 42 U.S.C. 1857 et seq. Public Law 91-604. United States Environmental Protection Agency, Washington, D.C., 1970 (U.S. Govt. Printing Office, Washington, D.C., 1971), and P. L. 95-95 (1977).
4. "Air Quality Criteria and Guides for Urban Air Pollution," W.H.O. Tech. Rep. Ser. No. 506. World Health Organization, Geneva, Switzerland, 1972.
5. "ECE Symposium on Problems Relating to the Environment" (Proceedings and Documentation of the Symposium organized by the United Nations Economic Commission for Europe and held in Prague from 2 to 15 May 1971). United Nations, New York, New York, 1971 (United Nations Publication. Sales No. E. 71. II.E.6).
6. "Guiding Principles Concerning the International Aspects of Environment Policies," Counc. Doc. C (72) 128. Organization for Economic Co-operation and Development. Paris, France, 1972.
7. A. C. Stern, M. S. Heath, and M. M. Hufschmidt, A Critical Review of the Role of Fiscal Policies and Taxation in Air Pollution Control, pp. 10–12. *In* "Proceedings of the Third International Clean Air Congress." VDI Verlag Gmbh, Dusseldorf, Federal Republic of Germany, 1973.
8. "The Effects of Pollution Abatement on International Trade," Public Law 92-500. First Report of Secretary of Commerce to the President and Congress in Compliance with Section 6 of Federal Water Pollution Control Act Amendments of 1972. U.S. Govt. Printing Office Washington, D.C. (GPO), 1972.

10

Manpower Development and Training

P. W. Purdom

I. Introduction

This chapter is concerned primarily with the manpower requirements in the technical and professional areas and will not consider administrative and clerical personnel, albeit they are essential to both public and private organizations. Attention will be focused on manpower requirements in public regulatory agencies and in private organizations that must comply with the government regulations. The data cited herein are only for the United States (*1*).

A wide range of technical and professional personnel are employed in air pollution control activities. These include professional personnel (engineers, meteorologists, chemists, and other scientists) and technical personnel, such as, operating "engineers," laboratory, electronic, mechanical, and other technicians, inspectors, etc. It is estimated that about 80,000 technical and professional personnel were needed in the public and private sectors to meet 1974 obligations, but the actual number employed may have been somewhat less than this because of inadequate state and local budgets. These numbers do not include some 400,000 operators of combustion equipment and an estimated 500,000 auto mechanics. This chapter examines these manpower requirements in detail, the education of such personnel, and manpower management and development.

II. Manpower Requirements

In viewing manpower requirements for air pollution control, there is a tendency to think primarily or exclusively of the governmental sector. The actual control of emissions is accomplished by the engineers in industry who design pollution-free manufacturing processes or, failing in this, design pollution control equipment; by the installers and operators of the pollution control processes and devices; and by the business officers and managers who commit funds for these purposes and set the policies for performance. If control technology is required, the inventors and manufacturers advance the state of the art of control. The foundation for air pollution control activities rests on the work of researchers who identify the effects of pollution on humans, sources of pollution, control alternatives, and the like. One may add to these personnel the individuals in education and voluntary organizations.

A. Regulatory Activities

The need for government personnel depends upon the activities to be conducted, not on the population or level of government. These activities will usually be mandated by a basic charter or constitution, by laws or ordinances, and by regulations. The financial ability of a unit of government to employ qualified personnel will depend in large part upon the population base of the political unit that will pay the taxes or fees to support the services.

A minimum population base to maintain a comprehensive program with reasonable taxes is about 100,000, although, nuisance-type inspections and some other limited activities may be conducted by smaller population units. The extent of activities seems to change markedly in jurisdictions at about 1,000,000 population and over.

Certain types of activities are more apt to be performed at different levels of government. For example, inspection, stack testing, and monitoring of individual sources of polluting emissions are apt to be local service, while research into the effects of pollution on populations will likely be performed or supported financially by the national government. A state agency may perform local-type activities where local services are lacking, and in the more populous states and provinces, research and other activities thought to be national may be conducted. The Los Angeles County, California, Air Pollution Control District is an example of a local agency that has conducted a full range of all types of activities, including research.

1. Federal Activities

In the United States, the Environmental Protection Agency (EPA) is charged with enforcing the provisions of the Clean Air Act. EPA is required to set national standards for air quality. State and local agencies establish programs to achieve these standards. EPA conducts a great deal of in-house research and maintains a staff of engineers, chemists, meteorologists, statisticians, physical and biological scientists, various technicians, and others for this purpose. In addition, field studies, technical and educational services to state and local agencies, and administrative and enforcement activities are performed. Manpower levels for EPA are given in Table I.

In addition to serving as a regulatory body through EPA, the United States government conducts many activities itself that produce air pollution. Executive Order 11507 requires federal departments and agencies to abate air pollution from their facilities. The Energy Research and Development Administration, the General Services Administration, the Defense Department, the National Oceanographic and Atmospheric Administration, and the Tennessee Valley Authority have air pollution personnel, but in smaller numbers than EPA.

2. State and Local Activities

Under the United States Clean Air Act, state and, in turn where approved, local agencies must formulate and implement programs to achieve air quality standards. In addition, state and local agencies may

Table I Air Pollution Manpower Needs of the United States Environmental Protection Agency (1)

Discipline	*1970*	*1974*
Engineers	208	752
Chemists	82	271
Meteorologists	24	102
Statisticians	36	102
Physical Scientists	17	59
Biological Scientists	70	170
Other Professional	46	138
Technicians	208	664
Administrative–clerical	309	642
Total	1000	2900

be independently charged by the laws of their own jurisdiction to perform certain activities. Local agencies, in this sense, include townships, cities, counties, and combinations of these in local districts, as well as districts or other regional subdivisions of states.

Routine activities may involve investigation of complaints, inspection and testing of stack emissions, review and approval of plans for new and modifications of old installations, sampling of community air, observation of atmospheric phenomena, analysis of fuels, and enforcement activities. State agencies and some local units may inspect mobile sources. In more populous areas, the air pollution control agency may be concerned with the impact on air quality of transportation systems, land-use patterns, energy sources and systems, and evaluation of alternative strategies in these areas insofar as they influence air quality. Epidemiological and demographic studies of populations may be conducted for long-range planning.

Actual employment in United States state and local agencies was estimated at 2837 in 1969, and 4662 in 1971. For 1974, estimates ranged from 7000 to 11,000. The higher figure is based on full budgeting. EPA estimates of state and local manpower needs by disciplines are shown in Table II. The educational background of professional personnel in state and local agencies, reported for 1971, are given in Table III, and the status position classification in Table IV. The data in Tables III and IV do not include operating personnel at facilities of state and local agencies who may be charged with reducing emissions from operations.

Table II Estimates of Manpower Needs in the United States for State and Local Governmental Air Pollution Control Agencies *(1)*

Discipline	*1969*[a]	*1971*[a]	*1974*[b]	*1974*[c]
Engineers[d]	692	1117	2156	2965
Chemists	232	335	539	741
Meteorologists	30	45	121	166
Sanitarians	187	353	459	631
Other Professionals	174	0	381	524
Inspectors	598	864	1571	2160
Technicians	371	581	964	1326
Administrative–clerical[e]	553	1367	1811	2490
Total	2837	4662	8002	11,003

[a] Historical data (from Report to Congress by the Secretary of Health, Education, and Welfare, 1970).
[b] EPA projection of 1974 personnel (made in 1969).
[c] EPA projection of 1974 personnel needs (revised 1973).
[d] Includes supervisors in agencies.
[e] Includes directors.

B. Private Sector

Air pollution personnel in the private sector are employed in industry to reduce emissions, in manufacturing to produce control systems and monitoring and testing equipment; and in service companies providing laboratory, research and development, engineering design, and maintenance services.

1. Industrial and Commercial Sector

Other than the air pollution from private autos and mobile sources, processes, control equipment, and operations at industries will have the

Table III Educational Background of Professional Personnel in United States State and Local Air Pollution Control Agencies (1971) *(1)*

Discipline	*Number*	*Percent*
Engineering	1204	39.66
Science and mathematics	952	31.36
Liberal arts, social sciences, etc.	182	5.00
Health professions	114	3.75
Other	584	19.24
Total	3036	100.00

Table IV Classification of Positions in United States State and Local Air Pollution Control Agencies (1971) *(1)*

Occupational title	*Number*	%
Director	232	5.0
Supervisor	313	6.7
Engineer I	346	7.4
Engineer II	458	9.8
Chemist I	135	2.9
Chemist II	200	4.3
Meteorologist I	14	0.3
Meteorologist II	31	0.7
Specialist I	171	3.7
Specialist II	182	3.9
Technician I (Trainee)	143	3.1
Technician II	245	5.3
Technician III	95	2.0
Inspector I (Trainee)	124	2.7
Inspectcr II	514	11.0
Inspector III	226	4.8
Aide I	41	0.9
Aide II	57	1.2
Administrative, clerical, all other	1135	24.3
Total	4462	100.0

greatest influence on air quality. It has been estimated that 3 to 5 engineering man-years are required for each million dollars of installed air pollution equipment *(2)*. Types of industrial (and commercial) operations will determine the kinds of problem encountered and, consequently, personnel needs. These operations may roughly be grouped as combustion, dusty, metallurgical, chemical, and nuclear processes.

Many industrial and commercial establishments will burn fossil fuels for heating and sometimes to produce their own electric energy or steam power. Incinerators may be used to reduce the volume of paper and other wastes. While some of these combustion operations may be individually a substantial source of pollution, such as smoke or fly ash and oxides of sulfur and nitrogen, by far the most significant potential source of such pollutants in a populous community is apt to be the fossil-fuel-fired electric generation station(s).* Of course, the fuel (gas, oil, or coal) and its quality (ash, sulfur) will determine the problems

* It is recognized that if private homes used unprocessed coal for heating this, too, could be significant, but gas, fuel oil, or electric energy is used in most homes in United States urban areas.

encountered. Small-capacity installations require trained technical personnel to operate and maintain the combustion processes and related particulate removal equipment. Larger installations may require mechanical engineers as well as technicians, not only for operation and maintenance, but also for design. Dusty processes, such as, grinding, crushing, and materials handling require similar personnel.

Chemical and metallurgical industries have the added complication that the control of the process itself may be the best method of preventing air pollution. A wider range of pollution control facilities may be encountered, including scrubbers, chemical exchange processes, high temperature and catalytic oxidation, and bag houses and other particulate removal equipment. A wider range of technical personnel, as well as chemists and chemical and other engineers, will be required.

Large operations with significant pollution potential may employ meteorologists to advise on adverse weather conditions that require unusual operating procedures or plant shut down.

Industries with extensive operations may employ staffs of personnel at the corporate level. These personnel may perform functions for the company similar to that of a regulatory governmental agency, like inspection, monitoring, sampling, and testing. In addition, the corporation may do much or all of its long-range planning, site selection, and engineering design of manufacturing process and pollution control. It is also necessary for corporations to employ or retain legal counsel familiar with environmental law.

Nuclear energy electric generating plants have unusual needs for personnel. In addition, to the anticipated needs for personnel to control atmospheric emissions, the company may be required to maintain an extensive environmental and biological surveillance activity.

Industrial personnel may not work exclusively on pollution control, let alone air pollution; hence, manpower estimates are in terms of man-years. Data for certain industries are shown in Table V.

2. Manufacturing of Air Pollution Control Equipment

It is estimated that there are less than a hundred major manufacturers of air pollution control equipment, including hardware like cyclone collectors, electrostatic precipitators, and scrubbers, with fifteen to twenty concerns supplying about 85% of the market. Many manufacturers in this field provide other products for industry, especially in the instrumentation field.

Professional and technical personnel constitute about one-third of the manpower of an equipment manufacturer. While Middlebrook's

Table V Estimated Manpower Needs for Air Pollution Control Personnel in United States Industry (in Man-Years) *(1)*

Industry	*Professional and executive*		*Technician*		*Operation and maintenance*		*Total*	
	1969	*1974*	*1969*	*1974*	*1969*	*1974*	*1969*	*1974*
Rock product processing								
Cement (portland)	75	135	[a]	175	300	450	375	760
Other (lime, glass and ceramic, gypsum)	71	91	7	7	75	110	153	208
Chemical[b]	950	1700	150	1200	1400	1750	2500	4650
Food and kindred products[c]	214[d]	306[d]	—	—	15	15	229	321
Steel and related manufacturing								
Integrated steel mills	400	790	175	200	1400	1800	1975	2790
Electric steel operations	20	35	—	—	75	150	95	185
Gray-iron foundry	40	40	—	—	50	1800	90	1840
Nonferrous metals								
Copper (primary smelting)	69	162	38	90	230	600	337	852
Lead and zinc (primary and secondary smelting)	121	152	28	38	750	1140	899	1330
Aluminum reduction	70	144	30	72	140	144	240	360
Nonferrous foundries	20	20	—	—	50	750	70	770
Petroleum								
Petroleum refining	650	740	200	260	2000	2600	2850	3600
Other (hot paving mix, asphalt-saturated felt)	[a]	[a]	—	—	—	140	[a]	140
Steam-electric power plants	200	2080	50	870	60	150	310	3100
Pulp and paper	240	550	150	220	210	330	600	1100
Rubber	20	60	[a]	40	—	—	20	100

[a] Nil.

[b] Includes alkalis and chlorine, intermediate coal-tar products, pharmaceutical preparations, paints and allied products, fertilizers, agricultural chemicals, soaps and detergents, synthetic rubber and fibers, plastics, other inorganic and organic chemicals.

[c] Includes meat, fish, canned and frozen foods, flours and cereals, other milling and allied operations, fats and oils.

[d] Includes 150 man-years for rendering operations.

committee *(1)* found no data on current employment in this field, it was estimated that about a thousand more engineers and technicians would be required in 1975 than in 1973. Over this same 2-year period, air pollution instrument manufacturers were estimated to require an additional three to four hundred, and auto manufacturers to require about sixteen hundred, with no estimate for expanding needs thereafter.

Services provided to industry and government agencies include research and testing laboratory services and consulting engineering, meteo-

Table VI Estimated Air Pollution Manpower Engaged in Services in United States (1969) *(1)*

Service	*Professional*	*Nonprofessional*
Contract research	125	125
Testing laboratories	30	30
Architectural and engineering	75	75
Consultants	200–500	—
Total	430–730	230

rological, biological, and chemical services. Consultants may provide designs, surveys, and operational consultation. Estimated manpower engaged in these services in 1969 are given in Table VI.

C. Institutional Sector

Air pollution control personnel are employed in colleges and universities in teaching and research associated with graduate and undergraduate degree programs. These personnel may be natural scientists, engineers, economists, social scientists, mathematicians, etc. Except for persons of exceptional ability and attainments, most university faculty must have a Ph.D. for entrance positions. However, other professional personnel may be employed as adjunct faculty and lecturers. In addition, universities employ technicians for laboratory and research work.

Two- and four-year colleges and technical institutes engaged in technician and technology training place more emphasis on the use of experienced technicians and professional personnel for practical training for operations and maintenance. In high schools and community colleges, personnel with knowledge of environmental issues, including air pollution, are required for informative programs.

Citizen activist groups and other voluntary organizations require persons with knowledge of air pollution issues for community educational and promotional activities

III. Educational Development

To produce the manpower described in the preceding paragraphs, a variety of educational programs is indicated. These, of course, depend on the job to be performed.

A. Graduate Education

At the first national conference dealing with the education of sanitary engineers, held at Harvard University, Cambridge, Massachusetts, in 1960, the first recognition of the desirability of specialized formal education for the field of air pollution control occurred. Prior to that time, the type of education of workers in the field depended on the stage of development of concepts of air pollution.

In the United States, the cities of Pittsburgh, Pennsylvania, and St. Louis, Missouri, mounted early successful programs to curb the nuisance created by the improper combustion of coal, especially soft coal, and also the fly ash produced with more complete combustion. Since these early efforts to combat air pollution were actually programs of smoke abatement, persons with mechanical engineering education tended to dominate the professional jobs, and the inspector jobs in government and the operating jobs in industry were filled with technicians, sometimes classified as "operating engineers."

While such jobs continue today, the air pollution episodes at Donora, Pennsylvania; London, England; and elsewhere, and the recognition of the Los Angeles, California, type photochemical smog, led to a broadened perspective of the nature of air pollution control activities. This led to chemical engineers entering the field to design processes that gave off less pollutants and to develop systems for controlling emissions. But the implementation of control programs to assure an adequate quality of clean air proved to be complex, and now meteorologists, epidemiologists, plant pathologists, economists, political and other social scientists, lawyers, and others are recognized as necessary.

The 1960 conference recognized that air pollution professionals might come from a variety of undergraduate disciplines. To be considered adequately trained for professional jobs, a graduate program of at least twelve months was recommended. Twenty-four months was considered to be more desirable.

The recommendation that a master's degree be considered the minimum for entry into a professional job was reiterated at a second national conference held at Northwestern University in 1967. True, both of these conferences dealt mainly with the education of engineers, but there have not been similar conferences dealing with other professions separately. In general, it appears that other professionals have also recognized the value of specialized training at the master's level for work in air pollution control. Sometimes persons who have obtained Ph.D.s in a discipline have sought and obtained special graduate education for air pollution work or other environmentally oriented programs.

The primary issue concerning the content of graduate programs at the master's level is the extent to which the material is specialized in technical subjects (e.g., meteorology, unit processes, toxicology), and the extent to which the curriculum covers related areas of environmental concern, like water pollution and solid wastes; areas that present problems for air resource management (e.g., energy conversion systems, transportation systems); and constraints (e.g., political, cultural, and economic factors). The tendency has been to emphasize the technical subjects, since there is a very limited amount of time in a twelve month program.

At the 1973 conference at Drexel University, the Task Force concerned with graduate education (*3*) reconsidered this subject and came to about the same conclusions. A typical program suggested for a master's degree program for engineers in air pollution was suggested (Table VII).

Doctoral programs tend to be rather specialized according to the interests of the individual professor guiding the student and the philosophy of the particular school. In general, doctoral programs are designed to prepare the student for a career in research rather than to perform a professional job with excellence. Even so, some schools will tend to favor applied research more than others. The present policies of the United States National Science Foundation in establishing RANN

Table VII Suggested Curriculum for Master's Degree for Air Pollution Engineers *(3)*

Area	*Emphasis* (%)	*Content*
Engineering	25	Air pollution dynamics, source factors, control systems, air resource management, water and waste management
Chemistry	10	Chemistry of pollutants, sampling and analysis, photochemistry
Biology	10	Cell and human biology; effects on vegetation, animals, and humans; air microbiology
Physics	10	Aerosol science and technology, optical properties of atmospheric aerosols
Meterology	10	Properties and dynamics of atmosphere, atmospheric transport and diffusion
Applied Mathematics	10	Statistics, computer programming
Environmental Health	10	Elements of physiology, toxicology, and epidemiology; history and organization of public health and sanitary engineering; current environmental health issues and problems
Social Sciences	15	Air resource management; electives in law, economics, business, urban planning

(Research Applied to National Needs) make applied research more respectable.

The task force on graduate education (*3*) surveyed the current status of graduate programs in air quality engineering in the United States. They obtained information from eight of fourteen institutions offering programs. Three schools offered only the master's degree, while five offered both the master's and a Ph.D.

Four programs were in a department of civil engineering, one in mechanical engineering, and another in a school of public health. Of the remainder, one was a cooperative program between two departments (civil and mechanical engineering), and two were within the chemical engineering department but included course offerings of other science and engineering departments.

These programs granted 160 degrees in 1970, including 140 master's degrees and 20 doctor's degrees. This output does not begin to meet the forecast needs, and the output may very well decline because the policy of the United States government enunciated in 1972 and 1973 to curtail much of the previously available direct financial support of graduate students.

The Appendix shows an extensive list of colleges and universities in the United States and Canada offering programs in some phase of environmental engineering and science developed from information prepared by the Air Pollution Control Association for the U.S. Environmental Protection Agency. Those schools that indicate that they offer general programs in the environment may well provide for elective specialization in air quality areas. There were forty-one schools that offer special programs related to air quality. In fiscal year 1973, the federal Environmental Protection Agency supported thirty-three programs, but this support is being phased out.

B. Undergraduate Education

The Graduate programs existing before 1970 took students from various disciplines, such as, civil or chemical engineering, chemistry, physics, or other science. The students were unequally prepared for graduate work in different aspects of basic education in science and engineering. The chemists may have had little concept of fluid mechanics; the civil engineers may not have had enough chemistry; both may have known little about human or plant physiology. The end result was that many graduate programs had one or more courses designed to make up some of the essential deficiencies. Consequently, not as much time was available to go into the depth desired in the specialized subjects. Also, knowledge of interacting environmental problem areas was omitted or neglected, and courses desirable for development of air resource management plans were left out or minimized.

The undergraduate education task force at the Third National Conference (*4*) suggested that a fundamental change in educational approach be taken. Looking at the definition of environmental engineering* adopted by the conference, this group proposed that a new identifiable undergraduate degree program be instituted. A few schools have already taken steps in this direction.

Under this plan, a broad education for environmental science and engineering would be provided in a 4-year undergraduate program, which would also provide the introductory courses in various areas of special concern such as air pollution. A person so prepared presumably would be better able to enter a field like air pollution as compared with a graduate of a single discipline program. Nevertheless, the person could still pursue a master's degree to secure more specialized knowledge in air pollution; but in this case the graduate work would not have to overcome undergraduate deficiencies and could concentrate on graduate-level work. To be well prepared for professional jobs, this combination of undergraduate and graduate work would be vastly preferable to the then existing system.

Even so, there would continue to be a need for the highly specialized person in science or engineering to work in concert with those persons who are trained in environmental science and engineering. The more broadly trained individuals would be the choice for any jobs involving planning and management.

The educational plan advanced as preparation for a career in environmental protection is possibly broader than that for any other engineering or science field. Consequently, the task group was reluctant to suggest a scientific curriculum, since they felt that flexibility was desirable and several approaches should be examined. Even so, a summary of the types of courses that should be included in a 4-year plan was advanced (Table VIII).

Undergraduate programs such as this are said to exist at Rensselaer Polytechnic Institute; Purdue, Cornell, and Vanderbilt Universities; and others, in conjunction with graduate programs which complement the undergraduate program (see Appendix). In these programs, material of

* Environmental engineering is that branch of engineering that is concerned with the application of scientific principles for (a) the protection of human populations from the effects of adverse environmental factors, (b) the protection of environments, both local and global, from the potentially deleterious effects of human activities, and (c) the improvement of environmental quality for man's health and well-being.

Table VIII Types of Courses That Should Be Included in a 4-Year Undergraduate Program in Environmental Protection *(4)*

Course	*Suggested credit hours*
Basic science	
Mathematics through differential equations	9
Physics, general (not more than 2 semesters)	6
Chemistry, inorganic and organic chemistry theory, physical chemistry including thermodynamics	9
Environmental biology (ecosystem studies including man) and/or microbiology	3
Computers and/or computer application	3
Total	30
Basic engineering	
Heat transfer	3
Fluid Mechanics (should include open and closed channel flow)	6
Engineering lab (may include project)	6
Mechanics (including strength of materials and structure analysis)	6
Total	21
Humanities	
Completely elective courses with no suggestions or recommendations that would intimidate an engineer to take a certain course because it is related to engineering. There may be certain categories such as literature, English, etc. It is considered desirable to have the humanities courses spread out fairly evenly over the 4-year program	24
Total	24
Environmental engineering	
Air resources	3
Water resources, including hydrology	3
Solid wastes	3
General environmental engineering (to include radiological health, noise, conservation, health, and safety)	3
Resources, energy and the environment	3
Monitoring and analysis	3
Unit operations in environmental engineering	3
Unit processes in environmental engineering	3
Environmental engineering laboratory and systems design	4
Total	28
Optional or elective	
Probability and statistics	
Environmental systems management	
Environmental design (prerequisite is monitoring and analysis)	
Impact and law	
Management	
Writing and public speaking	
Engineering economics	
Geology or groundwater hydrology	
Suggested Total Credits	17

low priority has been deleted in order to provide the desired broad undergraduate education in science and engineering pertinent to environmental quality programs.

One problem posed by such an approach is the university administrative structure to be used to implement the plan. It was felt that several models could be employed. These included administration as an independent department, as a division or program within an existing department, as a multidisciplinary program administered by a faculty who are members of many departments, and as a program under committee supervision. The choice will probably depend upon historical precedents and the politics of the individual university.

The major difference between the previously existing and the proposed combination of undergraduate and graduate programs is where specialization takes place and for what purpose. In the proposed undergraduate degree program for environmental careers, the broad courses, such as chemistry, biology, and unit processes, would take place at the undergraduate level, with more specialized courses (e.g., aerosol technology or photochemistry) at the graduate level. Under the previously existing system, the graduate program has been a mixture of graduate work with some undergraduate material necessary to broaden the student beyond the previous undergraduate specialty.

C. Technical Education

In recent years, formal educational programs have been established for technicians and technologists. Such programs have been developed for environmental personnel and a task force at the Third National Conference considered these programs (*5*).

The technician receives an associate degree. A technician is defined as a person who has fulfilled academic requirements for an associate degree in engineering technology and is able to assist professional engineers and engineering technologists in the conduct of engineering work under direct supervision and apply established methods and procedures. A technician should be able to perform specified observations, tests, measurements, field sampling, inspections, detections, laboratory analysis and instrumentation, etc. Also, such a person could, under supervision, design, operate, maintain, measure, and evaluate performance effectiveness of devices and processes employed in pollution prevention, as well as diagnose and correct equipment problems. The technician might also supervise other maintenance and operational personnel. The associate degree is usually a 2-year program and may be a terminal degree. Some schools, which have both an associate degree and a technology degree, may make provision for the associate to continue on to the 4-year technology degree.

The engineering technologists complete a 4-year program leading to a baccalaurate degree in engineering technology. The technologist is able to undertake responsibility for the conduct of engineering work under the direct supervision of a professional engineer or independently when applying established practices and procedures. A technologist might review plans, reports, and proposals for approval; process permits; act as an administrative assistant; and otherwise assist in enforcement activities. In industry, the technologist might do the above and also serve as a safety officer, surveillance supervisor, or supervisor of the operation of pollution control facilities.

The use of trained technicians will help insure that the most efficient performance of control facilities will be obtained. They can also relieve professional personnel of many chores that do not require a high degree of advanced education. For environmental engineering activities, the task force recommended the following 2-year technician program:

1. At least the equivalent of one-half academic year of basic sciences, about half of which is mathematics and of which the mathematics include carefully selected topics suited to each curriculum from appropriate areas of college algebra and trigonometry.
2. At least the equivalent of one academic year of technical courses.
3. At least the equivalent of one-fourth academic year of nontechnical subjects, such as oral and written communications and humanistic–social subjects, exclusive of courses in physical education or military reserve officer training.
4. The specifications listed in 1–3 total less than the minimum period of two academic years required to achieve an integrated and well-rounded engineering technician curriculum. This additional time is available for the implementation of the educational objectives of the individual and the institution.
5. A minimum of 60 semester hour (90 quarter hour) credits for a 2-year associate degree curriculum.

There is sharp division of opinion concerning the wisdom of advocating a 4-year technology program. The essence of the argument against such programs is that in four years a student could pursue a program leading to a B.S. degree in one of the disciplines, which would not have the deficiencies of the technology degree and would not be as likely to be limited in future opportunities for job advancement or further graduate work. However, advocates of such programs consider their practical orientation an asset to the employer. One major difference in technology and regular B.S. programs appears to be the amount of mathematics required. The task force recommended that a 4-year technology program for an environmental engineering technologist include the following as a minimum:

1. At least the equivalent of three-fourths academic year of basic sciences, with the proportion of mathematics to be dependent on the needs of the particular technology.

2. At least the equivalent of one and one-half academic years of technological courses, including technical science, technical specialty, and technical electives.

3. At least the equivalent of three-fourths academic year in communications, humanities, and social sciences (not to include physical education or military reserve officer training courses).

4. The specifications listed above total three years of the four years required for a baccalaureate degree in a well-rounded engineering technology curriculum. The additional time is available for the implementation of the educational objectives of the individual and the institution.

5. A minimum of 120 semester hour (180 quarter hour) credits for a baccalaureate degree curriculum.

6. The total program should reflect adequate work and student progress throughout the four years to warrant the conferring of a baccalaureate degree.

D. Continuing Education

The process of education is not completed simply by mastering a formal course of instruction, whatever the time period allocated to the program of study. Knowledge is expanded, new technology is developed, and the demands of society for environmental quality requirements change. Furthermore, no program of study should be expected to completely prepare an individual for all of the practices and procedures applicable to a particular job. Consequently, programs of continuing education are essential to maintain individual performance and to fulfill organizational objectives.

Continuing education activities have four principle purposes—upgrading, updating, diversifying, and broadening. One might add to this list the practice of providing orientation to the job for in-house in-service programs. Such programs of continuing education are sometimes conducted in-house by either industry or government for its own personnel or by any of several sponsors for the benefit of others. In the United States, federal agencies, such as EPA and its predecessors, have over a period of years conducted short courses at their centers for local and state government personnel and for industry. Many professional societies do likewise. Private interests and educational institutions also have a variety of programs. Some colleges and universities have not done as much in this field as they could because their faculty members may not recognize such activities as important in decisions regarding pay, promotion and tenure.

The form of continuing education varies from lectures to formal courses extending over a period of weeks and months. Some of the more formal and extended efforts may provide for academic credits toward graduate degrees and others may award a certificate of proficiency.

To be useful, the particular project must be tailored to fit the need of the worker. Some air pollution control agencies have found it desirable to conduct such programs to qualify inspectors to give judgmental testimony in court. For example, compliance with opacity standards for smoke plumes may depend upon a visual observation unassisted by instruments. Special training in making such observations may be accepted as evidence of proficiency by the court.

Care should be exercised that the faculty providing continuing education undertake activities only for those projects in which they have competency. There has been a suggestion that an accreditation system be established for assuring the quality of continuing education programs offered to the public.

The Air Pollution Training Institute of the United States Environmental Protection Agency conducts short courses of 2–5 days duration each at its headquarters located at Research Triangle Park, North Carolina, as well as selected locations throughout the United States. These courses have included such titles as are in Table IX.

Table IX Typical Course Titles—Air Pollution Training Institute, United States Environmental Protection Agency, Research Triangle Park, North Carolina

Air quality monitoring systems	Environmental training simulations
Air pollution administration	Gas chromatographic analysis of air pollutants
Air pollution control orientation	Measurement of atmospheric metals
Air pollution control technology	Meteorological instrumentation in air pollution
Air pollution field enforcement	Principles and practice of air pollution control
Air pollution meterology	Regional planning for air pollution control officers
Air pollution microscopy	Sampling and identification of pollen and fungus spore aeroallergens
Air pollution principles for planners	Source sampling for particulate pollution
Air pollution systems management	Special "reference methods" laboratory courses
Analytical methods for air quality standards	Statistical evaluation of air pollution data
Atmospheric sampling	Visible emissions evaluation
Combustion evaluation	Workshop on environmental quality planning
Control of gaseous emissions	
Control of particulate emissions	
Determination of polycyclic aromatic hydrocarbons	
Diffusion of air pollution	
Effects on vegetation	

IV. Manpower Development and Management

The management and development system for an organization's manpower will, to a large degree, determine its capacity to function, the quality of its work, the attainment of its objectives, and its response to changing conditions. While, to be effective, a personnel plan must be part of an overall program plan, the best formulated program plan can falter on failure of the personnel to function as an effective team.

A. Personnel Practices

The personnel plan begins with an allocation of work and a division of responsibilities. It is under such a system that the efficiency of specialization of tasks is realized. It is necessary to examine each job to be performed, determine the skills and knowledge necessary, and the performance units to be expected. One should bear in mind that a job a little beyond the capacity of an individual is totally beyond his capability. Placing an individual in such a situation can only lead to frustration for the worker and the supervisor.

Where formal training is a necessary prerequisite, this should be specified. There is a tendency to specify a higher degree of training than necessary in order to persuade the personnel director to allocate a higher pay range, especially where recruitment is difficult. Again, this practice leads to a frustrated employee, and the question of adequate pay should be met squarely. In cases of scarce personnel, however, one should carefully examine the possibility of separating out the tasks that can readily be performed by less highly skilled individuals.

To insure proper placement of personnel, most organizations will give some kind of examination. It does little good to give an examination that does not relate to the job. If the job requires manual skills or dexterity, then a practical demonstration of performance is called for. Where technical information must be applied, a written examination is appropriate. If public contacts and impressions created thereby are important, an oral examination is indicated.

In some circumstances, the person recruited is not expected to immediately perform the tasks for which he was hired. Instead the individual brings certain basic qualifications to the job, and a plan of in-service education is provided before the individual is expected to reach full performance level. In some cases, this in-service education even involves a one-year graduate program at some university.

It is generally conceded undesirable to establish dead-end jobs in an organization. In an organization that is only large enough to have one inspector, or even one that has one chemist and one engineer, this condition may not be avoidable. The question might be raised in such cases whether the basic geopolitical base for the program is adequate. Even in large organizations where it is possible to maintain a full range of services, there may be a need for some specialized skills. The possibility of using consultants should be considered, especially if the need is not for the full time of the required specialist. Where at all possible, a promotional ladder should be provided. Where both technical and professional personnel are employed, this may be considered difficult, as the technician may not possess the requisite education for the higher job. In-service training programs are frequently provided to not only help the employee to do a better job where he is, but also to help prepare the individual for further advancement. Many government and industrial organizations provide such programs to varying degrees.

Most frequent is the training program provided in-house for which the employee is paid for time devoted to the activity. Many organizations will pay expenses and tuition for seminars and short courses provided by external organizations. These are usually directly job related. It is difficult, however, to get away from the job more than one week, or two at the most. The more forward-looking organizations will also encourage formal academic pursuits, both undergraduate and graduate. In such cases, they may pay tuition and fees, usually requiring some level of academic performance. In other cases, they may provide paid time off from work to attend courses in a part-time program; but this is limited to those locations where opportunities exist within commuting distance. There are also organizations that will select employees for graduate work at full pay and tuition, usually for periods not to exceed one year. Of course, there are financial aid programs available to individuals who exercise the initiative to seek them and are willing to risk taking leave or resigning from their job. Organizations that allow paid or unpaid educational leave usually restrict this to one year.

B. Transfer of Training and Retraining

Because of changing technology or because of change in social priorities, it sometimes becomes necessary to redirect program efforts. This may have implications for manpower requirements or utilization. In such cases, it is desirable to provide for the retraining or reorientation of personnel. Organized in-service programs, as discussed above, are suitable for this, both in-house and at other locations.

In periods of scarcity in a particular field, such as might occur in a rapidly expanding air pollution control effort, it may be useful to recruit persons trained for other endeavors, particularly engineering and scientific personnel. For professional jobs, it is considered desirable and most efficient to provide at least a one year program of graduate study at the master's level. Such individuals have much to offer from prior experience after they have had the basic master's training in air pollution control.

C. Multidisciplinary Teamwork

The development of extensive areawide or industry-wide air pollution control efforts requires persons from many disciplines to develop workable solutions. Not only is it necessary to solve the particular technical problem, but the impact of the proposed solution on other aspects of the environment, other systems necessary to sustain the quality of life, and economic and cultural vitality must be considered. This means than an organizational setting must be provided to encourage interaction, and personnel must be conditioned to work with their allies from other fields. Such personnel may include meteorologists, chemists, engineers, lawyers, management and program analysts, life scientists, economists, behaviorists, and planners.

D. Leadership Development for Planning and Management

Most of the jobs associated with air pollution control require technical competence, and training efforts are aimed primarily at improving this competence. Nevertheless, there is a necessity for some one or a group to take an overall and comprehensive view to develop and implement a sound air quality management program. Few educational or training efforts are designed to meet such a need.

Planning is an operation that is part of management, not apart from it. To assure acceptance and implementation, the planning operation must include the persons who allocate the organization's resources and direct activities—the decision makers.

A prerequisite for good planning for air quality management is an understanding of the technical factors involved—the science and engineering considerations. The graduate programs now offered and the combination undergraduate and graduate programs proposed provide such a foundation. This is just a beginning, however. Grafted onto this basic preparation should be a program of study that includes decision-making sciences, behavioral sciences, and other courses to develop an understanding of economic, legal, political, and cultural factors and processes that limit the options available to the administrator or that must be solved as separate problems before plans can be implemented.

Air quality management programs also depend upon developments in related fields. Notable among these are energy systems, transportation systems, industrial location and regional economic development, and patterns of land use and population distribution. Thus, an effective manager must be familiar with land-use planning and the key issues in these other fields as they impact on air quality. The ability to appraise the environmental impact of alternative strategies in these essential fields is necessary.

Program planning of necessity must take place within the existing political system. The manager may find that existing political institutions cannot effectively control the air quality (possibly because they cannot command sufficient resources to execute the plan), they cannot legally control sources and operations that produce air pollutants that influence materially the quality of air, or their geographic area is too large to be manageable for controlling individual sources or is too small to contain enough sources to have any impact on air quality, if controlled. An understanding of the process of program planning is essential. Such a process includes problem definition, setting of priorities, establishment of long- and short-term objectives, organization of personnel and other requisites, budgeting, management of activities, communication, and evaluation of program achievements related to postulated objectives. In connection with these, an understanding of organizational controls is essential.

V. International Manpower Development and Training

Since its inception, the World Health Organization (WHO) has paid considerable attention to the problem of development of environmental health manpower (*6*). Over the years, a number of individual projects in member states have contributed to the education and training of engineers and technical personnel engaged in environmental health. In response to an increasing number of requests for assistance from member states, and taking also into account the ever-changing and broadening scope of environmental health, the organization has on many occasions restructured its program of assistance in this domain. In recent years, the WHO Regional Office for Africa has undertaken a survey of existing training institutions in the region. The Region for the Americas, with headquarters in Washington, D.C., possesses a wealth of information on this subject. Perhaps the thrust of the WHO effort is in the recent study initiated by the European Regional Office of WHO in Copenhagen on manpower requirements in environmental health. For this purpose, several questionnaires have been prepared and studies are being carried out in selected areas in five countries of the European region. This study is the first of its kind and will undoubtedly provide the type of information that will not only affect education and training programs for environmental health personnel, but also will contribute to ensuring the linkage between the basic education and training programs and the manpower concept. An example of WHO activity, is the assistance to the government of Morocco in training sanitary engineers and technical personnel to meet the needs of some thirty countries where French is the commonly spoken language.

VI. Summary

In this chapter, air quality management requirements have been considered from the standpoint of the kinds of personnel required, the education and training of these personnel, and manpower development and management. Manpower requirements in government depend upon mandated activities and responsibilities. In industry, the manpower required depends upon the problems to be solved and an effective response to government and social pressures. Others are needed to produce air pollution control equipment and testing devices, as well as staff service organizations. Still others are needed in educational activities.

Educational programs are necessary to produce the variety of manpower required. These needs vary from technician and technologist programs to undergraduate and graduate programs. To maintain competence continuing education is also required.

Effective use of manpower requires a personnel system that places the qualified person in an appropriate job and updates and upgrades the manpower through in-service training. A resilient organization will also assist personnel in adjusting to changes in technology and priorities and in advancing to higher level jobs. Above all, there is a need for developing a comprehensive and overall view of the planning and management responsibility.

REFERENCES

1. J. E. Middlebrooks, chairman, "Task Force on Manpower Needs," 3rd Natl. Conf. Environ. Eng. Educ. Drexel University, Philadelphia, Pennsylvania, 1973.
2. W. L. Faith, personal communication to Joe Middlebrooks.
3. G. Hanna, Jr., chairman, "Task Force on Graduate Education," 3rd Natl. Conf Environ. Eng. Educ. Drexel University, Philadelphia, Pennsylvania, 1973.

4. D. Aulenbach, chairman, "Task Force on Undergraduate Education," 3rd Natl. Conf. Environ. Eng. Educ. Drexel University, Philadelphia, Pennsylvania, 1973.
5. J. H. Austin, chairman, "Task Force on Technician and Technologist Education," 3rd Natl. Conf. Environ. Eng. Drexel University, Philadelphia, Pennsylvania, 1973.
6. G. Etienne, personal letter 6 March 1974.

GENERAL REFERENCES

American Society for Engineering Education, "Engineering Technology Education Study," Final Report. Washington, D.C., 1972.

Anonymous, Back to school for millions of adults. *U.S. News World Rep.* April 2 (1973).

Anonymous, "Conclusions and Recommendations," 1972 FEANI/UNESCO Seminar on Continuing Education for Engineers. Helsinki, Finland, 1972.

Anonymous, "Continuing Engineering Studies," A Report of the Joint Advisory Committee, American Society for Engineering Education, Engineers Council for Profession Development, Engineers Joint Council, National Society of Professional Engineers, Engineers Council for Professional Development, New York, New York, 1965.

Anonymous, Final report-goals of engineering education. *J. Eng. Educ.* (1968).

Anonymous, "Guidelines to Professional Employment for Engineers and Scientists." American Society of Civil Engineers, American Society of Mechanical Engineers, Engineers Council for Professional Development, Engineers Joint Council, etc., New York, New York, 1973.

Anonymous, "Newsletter of the State of Wisconsin." Examing Board of Architects, Professional Engineers, Designers and Land Surveyors, Madison, Wisconsin, 1973.

Anonymous, Recommendations of Commission on Non-Traditional Study. *Chron. Higher Educ.* (1973).

C. M. Berry, What is an industrial hygienist? *Natl. Saf. News* **108,** No. 2, 69 (1973).

W. R. Brode, Manpower in science and engineering, based on a saturation model. *Science* (1971).

Carnegie Commission on Higher Education, "The Fourth Revolution—Instructional Technology in Higher Education." McGraw-Hill, New York, New York, 1972

K. A. Doyle, Manpower for pollution control. *Environ. Rep.* **1,** 37 (1971).

Engineers' Council for Professional Development, "40th Annual Report." New York, New York, 1972.

Environmental Science and Technology, Manpower for environmental protection. *Environ. Sci. Technol.* 316–317 (1971).

W. L. Faith, "Air Pollution Control Manpower Needs in the Private Sector." National Air Pollution Control Administration, United States Department of Health, Education, and Welfare, Washington, D.C., 1969.

E. F. Gloyna and D. E. Griffith, "Continuing Engineering Education for Professional Practice—A University Viewpoint." Texas Section Meeting, American Society of Civil Engineers, New York, New York, 1973.

R. C. Graber, F. K. Erickson, and W. B. Parsons, Manpower for environmental protection. *Environ. Sci. Technol.* **5,** No. 4, 342 (1971).

N. Hackerman, "What Business and Community Leadership Should Know About Campus Unrest," Proc. 32nd Annu. Manage. Conf University of Texas, Austin, Texas, 1970.

W. J. Kaufman and E. J. Middlebrooks, "An Evaluation of Sanitary Engineering Education." American Association of Professors in Sanitary Engineering and Environmental Engineering Intersociety Board, 1970.

E. J. Middlebrooks, M. S. Ettelstein, R. G. Snider, and L. N. Spiller, Manpower needs of consulting engineering firms. *J. Water Pollut. Contr. Fed.* **44,** No. 10, 1865–1883 (1972).

National Sanitation Foundation, "Staffing and Budgetary Guidelines for State Solid Waste Control Agencies." Ann Arbor Sci. Publ., Ann Arbor, Michigan, 1973.

Olympus Research Corporation, "Evaluation of Manpower Needs in the Pollution Control Field." Salt Lake City, Utah, 1972.

"Projection of Education Statistics to 1979–80." Office of Education, United States Department Health, Education, and Welfare, Washington, D.C., 1970.

"Report on the Second National Conference on Environmental and Sanitary Engineering Graduate Education," sponsored by the Environmental Engineering Intersociety Board and the American Association of Professors in Sanitary Engineering, Northwestern University, Evanston, Illinois, 1967.

R. S. Roberts, "Pesticide Information: Federal Environmental Pesticide Control Act of 1972." Utah State University Extension Services, Logan, Utah, 1973.

Secretary of Health, Education, and Welfare, "Manpower and Training Need for Air Pollution Control" (Report to Congress). Washington, D.C., 1970.

Senate Resolution 218, California Legislative, October, 1071.

R. G. Snider, "Professional Manpower Needs." Office of Water Programs, U.S. Environmental Protection Agency, Washington, D.C., 1971.

"Summary of Discussions of Environmental Engineering Committee." American Association of Professors of Sanitary Engineering, Denver, Colorado, 1971.

United States Congress, "Public Law 92–516, Federal Environmental Pesticide Control Act of 1972." Washington, D.C., 1972.

United States Environmental Protection Agency, 1971. "Preliminary Report of a State and Local Air Pollution Control Agency Manpower and Training Survey." Office of Air Programs, U.S. Environmental Protection Agency, Research Triangle Park, North Carolina, 1971.

United States Environmental Protection Agency, "United States Environmental Protection Agency: A Progress Report." Washington, D.C., 1972.

United States Environmental Protection Agency, "A Report to Congress on Water Pollution Control, Manpower Development and Training Activities." Office of Water Programs, Washington, D.C., 1972.

United States Environmental Protection Agency, "Solid Waste Management Manpower Profile and Analysis." Washington, D.C., 1973.

Appendix: Directory of Colleges and Universities

The listing of colleges and universities offering curricula in air pollution is alphabetical according to state and then alphabetical by school with city and ZIP number. The line under the name and address shows first the type of degree or degrees offered in the environmental program by the school, then the department or de-

partments in which courses are available, within the parenthesis is the type of environmental program and following the parenthesis is the number of courses. A dash (—) indicates that data were unavailable.
The letters used are coded as follows:

Degree awarded		*Department*	
A	Associate Degree	CE	Civil Engineering
B	Bachelor Degree	Ch E	Chemical Engineering
M	Masters Degree	H	Public Health
D	Doctorate Degree	Env. Eng	Environmental Engineering
		Env. Sci	Environmental Science
		ME	Mechanical Engineering

Environmental program

A*	Air Quality
W	Water Quality
E	Environmental General Science
H	Health
S	Sanitation

Using the explanation above, the training offered at California Polytechnic State University, San Luis Obispo, California (M; Env. Eng; (A,W) 13) would read as follows: "Master of Science degree from the Department of Environmental Engineering, specializing in either air or water pollution control, with thirteen courses available on air quality.

Asterisk indicates United States Environmental Protection Agency support of air quality program in Fiscal Year 1973.

Note: Since programs vary from school to school, and from time to time, it is necessary to obtain a school catalogue to determine the type and number of environmental courses offered.

Data constructed from material prepared by the Air Pollution Control Association (4400 Fifth Avenue, Pittsburgh, Pennsylvania.

CALIFORNIA

California Polytechnic State University
San Luis Obispo; 93401
M; Env Eng; (A, W) 13

Calif. State University at Long Beach; 90801
B M; CE; (A, W) 5

University of California—Riverside Campus
Riverside; 92502
B; Env Sci; (A, E) 3

University of Southern California
Los Angeles; 90007
M; Eng; (A) 8

CONNECTICUT

Yale University
New Haven; 06520
M; H; (A, H, E) 8

DELAWARE

University of Delaware
Newark; 19711
B M D; CE, Ch E; (A, W) 8

* Only colleges and universities having this identification are listed in this directory. Other colleges and universities, not listed herein, may offer courses having air quality or air pollution control subject content under the E, H, or S categories.

DISTRICT OF COLUMBIA
Howard University
Washington, DC; 20001
M; —; (A, W) 5

FLORIDA
University of Florida
Gainesville; 32601
B M D; Env Eng; (A, H, S, W, E) 13

GEORGIA
Georgia Institute of Technology
Atlanta; 30314
M D; Ch E; (A, H, S, W, E) 7

INDIANA
Purdue University
West Lafayette; 47907
M D; CE, Env; (A, H, W, E) 13

Rose-Hulman Institute of Technology
Terre Haute; 47803
B; CE, Ch E, M E; (H, A) 1

KENTUCKY
University of Kentucky
Lexington; 40506
B M D; Ch E; (A, W) 10

LOUISIANA
Louisiana State University
Baton Rouge; 70803
—; Eng; (E, A, W) 2

Tulane University
New Orleans; 70118
M D; CE; (A, E) 2

MARYLAND
Johns Hopkins University
Baltimore; 21218
M D; Env Eng, H; (A, W, E, H) 5

University of Maryland
College Park; 20742
B M D; CE, Ch E; (A) 11

MASSACHUSETTS
Harvard University
Cambridge; 02138
M D; H; (A, H, W, E) 7

University of Massachusetts
Amherst; 01002
M; Ch E, CE; (A, W, E) 5

MICHIGAN
Genesee Community College
Flint; 48503
A; Sci; (A, E) 4

University of Michigan
Ann Arbor; 48104
M D; Env H; (A, H, W, S, E) 13

NEW JERSEY
Rutgers University—
New Brunswick; 08903
B M D; Env Sci; (A, W, E) 8

NEW YORK
City University of New York-City College
New York; 10031
M; CE; (A, E) 8

New York University
New York, 10003
M D; Ch E, CE, Meteorology; (A, H, W, E)

Sullivan County Community College
South Fallsburg; 12779
A; Nat Sci; (A, W) 4

NORTH CAROLINA
North Carolina State University
Raleigh; 27607
M D; Graduate; (A, W, E) 11

University of North Carolina at Chapel Hill
Chapel Hill; 27514
M D; Env Sci, H (Graduate); (A, H, W, E) 16

OHIO
University of Cincinnati
Cincinnati; 45221
B M D; CE, Env; (A, H, W, E) 21

OREGON
Oregon Technical Institute
Klamath Falls; 97601
A; Env H; (H, A) 7

PENNSYLVANIA
Drexel University
Philadelphia; 19104
M D; Env Sci; (A, S, W, E) 13

Pennsylvania State University-Berks Campus
Reading; 19610
A; Eng; (A) 4

Pennsylvania State University—University Park Campus
University Park; 16802
M; Graduate; (A, W, E) 14

Temple University
Philadelphia; 19122
—; —; (A, H, E) —

University of Pittsburgh
Pittsburgh; 15213
M D; H, Ch E; (A, H, W, E) 5

TENNESSEE
Vanderbilt University
Nashville; 37240
B M D; Env Eng; (A, W, E) 12

TEXAS
University of Texas at Austin
Austin; 78712
B M D; Env Eng; (S, E, A)—

UTAH
University of Utah
Salt Lake City; 84112
M D; Ch E; (E, A, W) 4

WASHINGTON
University of Washington
Seattle; 98195
M D; CE; (A, H, S, W, E) 20

Washington State University
Pullman; 99163
M D; CE; (E, A, W) 10

WEST VIRGINIA
West Virginia University
Morgantown; 26506
—; —; (A, S, W, E) —

WISCONSIN
University of Wisconsin at Madison Campus
Madison; 53706
—; —; (A, W)—

CANADA
ONTARIO
University of Toronto
Toronto; 181
B M D– Env Eng; (S, E, A)—

Part C

AIR POLLUTION STANDARDS

11

Air Quality Standards

Vaun A. Newill

I. Introduction

The preferred sequence of development of air quality standards is as follows:

a. Prepare air quality criteria, which are analyses of the relationship between pollutant concentrations in the air and the adverse effects associated therewith. The World Health Organization calls these "guides."

b. From the air quality criteria, develop air quality goals, which are the concentrations of pollutants with which we believe we can live without adverse effects on health and welfare.

c. From the air quality criteria, develop air quality standards, which are the concentrations of pollutants that we intend to achieve in the immediate future but that may fall short of our air quality goals because the standards must give consideration to feasibility of achievement within the immediately foreseeable future.

d. In order to develop the above standards, there must be standards for measurement and testing of the ambient air and air pollution effects.

II. Air Quality Criteria or Guides

A. *Health Effects*

In developing statements of the relationship between air pollution levels and the effects caused by exposure to these levels, many decisions must be made as to the existence or extent of cause-and-effect relationships. Some of the issues with which environmental health scientists must deal in the assessment of these relationships with regard to their usefulness as air quality criteria are:

1. Spectrum of Response (*1*)

Air pollutants can affect the health of individuals or communities over a broad range of biological responses. One can conveniently think of five biological response stages of increasing severity as illustrated in Figure 1: (a) a tissue pollutant burden not associated with other biological changes, (b) physiological or metabolic changes of uncertain significance, (c) physiological or metabolic changes that are clear-cut disease sentinels, (d) morbidity or disease, and (e) mortality or death. Boundaries between categories may occasionally overlap. Furthermore, each category shows a range of responses rather than a simple all-or-none phenomenon.

At any point in time, more severe effects, such as death or chronic disease, will be manifest in relatively small proportions of the population. In very few cases can death or disease be attributed directly and solely to pollutant exposure. Death and disease are end products of repeated cumulative insults (cumulative risks) from sources such as diet,

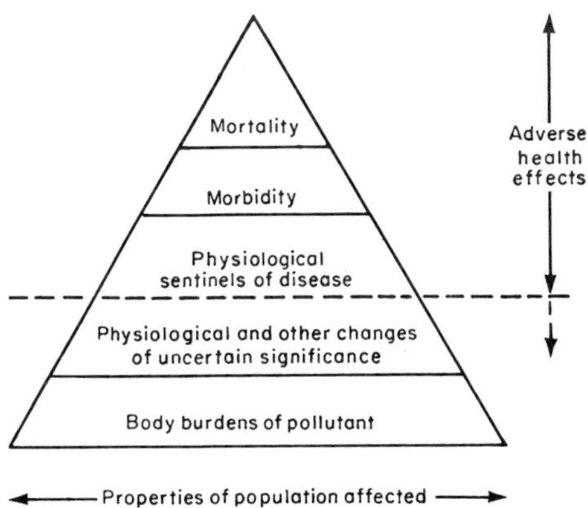

Figure 1. Schematic spectrum of biological response to pollutant exposure (1).

cigarette smoking, physical inactivity, infectious challenges, and accidental injury. In general, the role of environmental contaminants in the mortality or morbidity experience of a community is difficult to quantify, because so many other determinants of death and disease cannot be adequately measured.

The lower levels of the response spectrum shown in Figure 1 are subclinical manifestations of pollutant exposures. Larger portions of the population are effected at these levels. Pathophysiological responses, such as impaired mucociliary clearance and bronchoconstriction and physiological changes of uncertain significance, such as neurobehavioral responses, are more adaptable to experimental studies on animals or humans than is the case with acute or chronic disease, and they can be more readily associated with specific pollutant exposures. Pollutant burdens are tissue residues resulting from pollutant exposure. Pollutant burdens are highly specific effects of exposure, can be readily quantified in population studies, and may be used as indicators of environmental quality. If the bridge between the lower and higher levels of the response spectrum can be established, the disease risk associated with pollutant burdens or physiological changes can be shown, and ultimately, the role of pollutant exposure in the total community morbidity and mortality experience can be defined.

Some groups within the population may be especially susceptible to environmental factors. Notably, these include the very young, the very old, and those affected by a disease. Thus, susceptibility may be temporary or permanent. Inherited abnormalities, such as alpha antitrypsin deficiency and abnormal hemoglobins, are examples of permanently altered sensitivity. Temporarily increased sensitivity may be associated with periods of growth, with weight reduction, with pregnancy, and with reversible illnesses.

Diseases commonly result from complex causal webs rather than single factors (*2*). Environmental pollution may contribute a number of strands to such webs. Other strands may arise from such diverse origins as genetic heritage, nutritional status, and personal habits. Moreover, pollutant exposure may alter the severity of disease without altering its frequency.

To use such information in the regulatory decision-making process, one needs to distinguish those effects that should be labeled as adverse. Adverse effects include both the aggravation of pre-existing diseases and an increased frequency of a health disorder. In addition, good preventive medicine dictates that evidence for an increased risk of future disease is an adverse health effect.

2. Health Disciplines Utilized

The health effects information useful for regulatory decision making comes from the integrated output from several health disciplines. The major ones involved are epidemiology, clinical research, and the whole set of experimental studies that will be included under the term "toxicology" (*1*). Each of these discipline approaches has useful and, in some instances, unique information to provide.

a. Epidemiology. Epidemiology can provide studies of population exposure in real-life settings. The advantages of epidemiology are the natural exposure, no need for extrapolation of data to the human, the most vulnerable groups in the population can be studied, and both current and long-term low-level exposures can be evaluated. The major problems relate to quantifying the exposure, to dealing with the many covariates, to obtaining dose–response data, and to deciding association versus causation.

b. Clinical Research. Clinical research studies can be used to gather human data on either normal or diseased persons regarding absorption, metabolism, and excretion of pollutants and can be used for in-depth studies of humans accidentally exposed to high levels of pollutants where new study parameters and response indicators can be identified. The advantages of clinical studies are that the pollutant exposure is controlled so that improved dose measurements are obtained; since each person can usually be used as his own control, covariates are well controlled; vulnerable subjects can be included in the studies; cause–effect relationships are more easily ascertained; there is no need for extrapolation to humans; and, thus, the derived information gives maximum input into criteria. The problems are that the exposure is artificial; there can be no long-term exposures, thus, only acute effects are determined, and there can be real hazard to the exposed person.

c. Toxicological Studies. Toxicological studies can use many response systems, such as the whole animal, organs, cells, or biochemical systems. The advantages of toxicological studies are maximum dose–reponse data can be obtained, though this is incomplete at the low end of the curve; data acquisition is rapid; cause-effect relationships are more sure; and mechanism of response studies, such as kinetics of pollutant absorption, distribution, metabolism, and excretion, can be performed. While known quantitative exposure requirements are most easily satisfied under controlled or experimental exposure settings, appropriate human disease models in animals have not been available and thus have not been evaluated in this manner. Neither can such experimental studies readily identify delayed responses or chronic cumulative effects of exposure. Laboratory experimental studies cannot provide complete assurance of dose effects, because community exposures cannot be duplicated in the laboratory. Furthermore, the difficulties of extrapolating animal data to the human remains, particularly in estimating the threshold of human response.

3. Pollutants and Human Exposure Response

The effects of pollutants on human health depend on the physical and chemical properties of the pollutant; on the duration, concentration, and route of exposure; and on the human uptake and metabolism of the pollutant. Man's biological response is likewise a function of occupational, psychosocial, and climatological factors and is tempered by the phenomena of tolerance and adaptation. These exposure factors underlie attempts to understand the impact of pollutant exposure on human health (*1*).

The physical and chemical properties of pollutants determine their potential as a health hazard. These properties—including size, density, viscosity, shape, electrical charge, volatility, solubility, and chemical reactivity—all affect the absorption, retention, and toxicity of the pollutants. Many pollutants do not retain their exact identities after entering the environment. Thermal, chemical, and photochemical reactions occur when pollutants move through the environment from source to receptor. These factors affect their final physical and chemical state at the point of human exposure and help determine the toxic potential of the pollutants (*1*).

4. Exposure Response Matrix

The duration and concentration of pollutant exposure are measures of the total dose to the human. It must be remembered that the rate at which the total dose is received may influence response. Health effects of an environmental pollutant may be either short-lived (acute) or relatively permanent and irreversible (chronic). Acute or chronic effects may occur after a single exposure to a hazardous substance. This can be illustrated by acute radiation exposure, which can cause acute radiation gastroenteritis and chronic leukemia. An air pollution episode may have similar effects, though the permanent sequelae of acute episodes have never been adequately studied. Likewise, acute and chronic effects can result from long-term exposure. For example, excess acute respiratory illness and chronic respiratory disease have been repeatedly demonstrated in high-exposure situations. Acute effects and short-term exposures are less difficult to study than chronic effects or long-term exposure. Moreover, the effects of dose rate, i.e., large dose in a short interval versus repeated small doses over a long period, have seldom been investigated systematically. Little effort has been expended on the monitoring of long-term exposure and disentangling the causal webs underlying chronic effects of pollutant exposure. Such effort has been hampered by methodological difficulties.

5. Estimation of Exposure

Population exposure to individual pollutants or pollutant mixes changes rapidly with time; they vary with season, day, and hour. Also, within short-time frames, our very mobile population moves from indoor to outdoor environments and from one neighborhood to another. Approximately 20% of the United States population changes residence each year. Many cross-sectional studies of communities have shown that one-third of the population have resided at their current address for only 2–3 years.

Attempts to derive estimates of long-term integrated exposure, even of small populations in a single neighborhood, are fraught with difficulties in quantifying personal exposure. Quiet, tolerable, and small-scale instruments for personal monitoring are being developed. However, when developed they will not reduce the need for stationary monitors even for community research. Control programs will still need to be managed by stationary monitoring systems, and integrated exposures from personal monitors will need to be related to the measurements at stationary sites.

Stationary monitors have inherent drawbacks due to variation in performance over time and between instruments, interferences caused by variations in temperature or in concentration of pollutants, and simple instrument malfunction. Continuous monitoring, required to assess the effect of short-term exposures, is very costly and technologically complex. Furthermore, monitoring equipment is too often dissociated in time and space from measured health effects, especially where chronic effects are considered.

Studies of disease frequency in a large city or metropolitan area often rely on exposure estimates based on one or a few stationary monitoring units. These stations are usually not representative of community-wide exposure and often provide erroneous estimates of exposure in residential areas. Conclusions based on such results may imply higher exposures than actually occurred in residential areas having pollutant-associated disease excess. On the other hand, ascribing excess chronic disease to pollutant concentrations currently measured may imply lower than actual exposure since, for example, air pollutant levels in most of the world's larger cities tended to be considerably higher in the 1940s and 1950s than in the 1970s, by which time air pollution control measures had become more prevalent.

Information is seldom available to cope adequately with these methodological problems. In establishing exposure response relationships, the weakest links in the association are quantitative estimates of exposure. However, sound criteria require as precise data about exposures as about effects. Actions based on poorly defined criteria may be either overrestrictive or underprotective of health.

6. Evaluating Exposure–Response Relationships

Given adequate characterization of exposures and health effects, many additional considerations bear on the scientific validity of the exposure–response relationship. Hill (*3*) has given an exposition of criteria that can be used to judge whether an observed exposure–disease relationship is causal. Hill's criteria were developed as guides for occupational health studies. With minimal modification, however, they can be applied to general population studies. These criteria are consistency of observed associations, coherence of results, plausibility of the association, and strength of association. In addition, exposure–response gradients, intervention, and control of covariates need to be considered.

a. Consistency of Observed Association. Consistency of observed association is perhaps the most important criterion. Does the health effect occur in various age, sex, and race groups? Has the effect been repeatedly observed in different places, circumstances, and times? Even small differences that are not quite statistically significant bear great weight in standard-setting when the criterion of consistency is met. When the same effect is observed in a variety of population groups under varying conditions and at different times, the likelihood of a constant error or fallacy becomes progressively less.

b. Coherence of Results. When animal studies, experimental human exposures, and epidemiological data are coherent, i.e., they all demonstrate the same or similar health effects of exposure, the bits and pieces of evidence, when brought together, form a mosaic of health intelligence. One study supports another; experimental exposures identify pathophysiological pathways by which effects observed in epidemiological surveys become biologically explainable. The three disciplinary approaches discussed earlier are needed to form this interlocking mosaic—animal experimentation, controlled human exposures, and epidemiological studies. Crosswalks between each discipline can be readily identified. Through animal studies, metabolic pathways, organs of response, and effects identified in humans can be verified across a broad range of pollutant exposures. In well controlled human exposure studies (as employed extensively in tests of new drugs), results from animal studies can be made more relevant to human exposure at ambient levels. Also, effects observed in population studies can be studied under well controlled human exposures from near zero to ambient levels at varying exposure times. Such results have high immediate utility for short-term air quality standards. Finally, epidemiological surveys of communities before and after introduction of environmental quality control measures provide unique data for evaluating the adequacy of established standards and demonstrating the health benefits of control. Results from epidemiological studies provide the impetus to develop animal disease models from which more complete dose–response curves may be developed. By reinforcing results obtained in one approach with studies in another, a coherent health intelligence system will provide scientifically strong and readily acceptable guides for acceptable air quality.

c. Plausibility of the Association. Initial studies may uncover unexpected relationships between exposures and effects. Mere statistical associations must be rejected when no reasonable biological explanation, based on experimental evidence, can justify the association. On the other hand, when experimentation points to a disturbed physiological process that may lead to some clinical manifestation, subsequent epidemiological studies designed to test such hypotheses are well grounded in biological plausibility. New exposure–response associations require support from other disciplines before casual inferences can be readily defended.

d. Strength of the Association. When disease frequency is nine- to tenfold greater in exposed than in nonexposed populations, the exposure–effect relationship is extremely strong. Relationships of this magnitude have been found in studies of cigarette smoking and respiratory disease, including lung cancer and bronchitis. However, pollutant exposures are generally less intense and less reactive than cigarette smoke inhaled deeply into the lungs. Differences in exposure between high- and low-pollution neighborhoods seldom exceed two- or threefold concentration gradients. At present, ambient air concentrations, relative differences in effects between high- and low-exposure areas are unlikely to be as striking as ratios observed in smokers versus nonsmokers. However, for common and frequent disease events, such as acute respiratory disease, relatively small differences in disease experience can have a large and costly impact.

e. Exposure–Response Gradient. When a stepwise increase in exposure can be associated with a stepwise increase in the frequency of the adverse health effect, the evidence is strong for a cause–effect relationship. Linear relationships over an exposure gradient become increasingly difficult to explain by third intervening variables. Response gradients can be investigated in relation to exposure gradients across geographic areas, differences in length of residence in high exposure areas, and migration gradients constructed from various combinations of childhood and adult exposures of the same individuals. Human exposures occur naturally over an exposure gradient, especially over time and across areas. Exposure–response gradients obviously can be easily created in experimental settings.

f. Intervention. Protection of health requires society to intervene in public exposures to air pollution. This intervention is largely based on observed exposure–health effects associations. As desirable air quality is achieved, will the frequency of adverse health effects be affected? If so, the causal nature of the exposure–response association is strongly supported. The high cost of air pollution control warrants a national program of community health and environmental surveillance in those places where achievement of required air quality will require vigorous abatement efforts. Such a program has another ancillary benefit in that newly developing air pollution problems can be identified.

g. Control of Covariates. The causal nature of an exposure–response association is convincingly identified when, after the effects of known covariates are first displayed, increased disease risk within covariate classes can be clearly demonstrated in high exposure populations. For example, in any studies of chronic bronchitis prevalence, smokers and males show more disease than nonsmokers and females, respectively. An air pollution–chronic bronchitis study should reveal the above smoking–sex differences in prevalence rates, thereby giving assurance that the study has internal consistency. If smoking–sex-specific groups in high-exposure neighborhoods have excess chronic bronchitis, the hypotheses that air pollution exposure causes excess chronic bronchitis is considerably more convincing than if the smoking–sex covariates were not analyzed. Most epidemiological studies of air pollution require similar analysis of excess disease risk within covariate categories, with particular attention given to age, sex, smoking, socioeconomic level, and duration of residence at current location. When covariates are systematically analyzed for relationships to the health indicator under study, residual excesses in disease frequency can be attributed to area differences in pollution exposure with a reasonable degree of confidence.

B. Other Biological Effects

Deleterious effects of air pollutants in biological species other than man also must be considered as air quality criteria for regulatory purposes. In natural settings, air pollutant exposure can result in producing significant effects in plants and animals. When these effects occur at atmospheric concentrations of a pollutant that are less than the lowest level at which known human effects occur, then these levels become the basis for establishing a long-term goal or for determining the secondary ambient air quality standards as defined in the United States Clean Air Act [Section 109(b)(2)].

Such effects are somewhat easier to identify and to establish as a cause-and-effect relationship than is true for human effects. Ethics of experimentation are not nearly as much of a constraint, so that experimental studies can be performed by directly exposing the plants or animals to a pollutant or combination of pollutants in ranges to which they are likely to be exposed. Somewhat higher levels of exposure will also be used initially to identify the specific effects to be documented. These statements do not imply, however, that such studies are simple or inexpensive to perform. Actually, these studies must be meticuously carried out under a wide variety of conditions, and very subtle effects, such as the detection of a small decrement in yield of offspring or crop, must be searched for. Furthermore, the investigator must be able to assure that the effects found in the plant did not arise from inadequate light, insect infestation, or less than optimum temperature, moisture, nutrients, or soil conditions. (See also Chapter 4, Volume II).

C. Criteria for Physical Effects

Pollution also affects the physical properties of the atmosphere. The determination of what constitutes a physical aberration is as difficult as the similar decision making with respect to biological deviations. The physical properties of the atmosphere affected include its electrical properties and its ability to transmit radiant energy; to convert its water vapor to fog, cloud, rain, and snow; to deteriorate; and to soil surfaces. Our concern with atmospheric transmission of radiant energy encompasses the entire radiomagnetic spectrum, but of particular concern is the infrared region, as it affects the terrestrial heat balance; the ultraviolet region, as it affects both biological processes and photochemical reactions in the atmosphere; and the visible spectrum, as it affects both our ability to see things and our need for artificial illumination.

The physical property most apparent to people is diminution of visibility through the atmosphere. In assessing loss of visibility, we must bear in mind that under certain conditions there will be loss of visibility resulting from fog, wind-blown particles of natural origin, and hazes created by the photochemical reaction in the atmosphere between naturally occurring substances. We therefore cannot use unlimited visibility as the norm for unpolluted air. However, the fact that nature does not provide 100-mile visibility as the norm is no argument for our being willing to accept one-mile visibility in our urban communities.

Exposure of materials to air pollutants also can cause effects that will shorten their useful lives or alter their appearance. For example, ozone can cause fabrics to be weakened and discolored, works of art to lose their color, and rubber tires to crack, and other pollutant mixes can cause surface coatings of paint or metal to change appearance, iron to deteriorate more rapidly, electrical contacts to corrode, and other effects to occur (See also Chapter 2, Volume II). Knowledge of the dose–exposure relationships with these effects are useful data as criteria and can serve as the basis for air quality standards.

D. Conversion of Effects Data to Criteria

This is a three step process. The first step consists of identifying the relevant published literature containing appropriate effects information (described previously in this chapter and by Goldsmith and Friberg in Chapter 7, Volume II). As a second step, this literature should be systematically and critically reviewed and the pertinent information, adequately qualified as to its usefulness, summarized and assembled into a background document (*5, 6*). The third step is to further summarize into a coherent criteria document that can serve as the legal basis for the judgemental standards established and promulgated by a regulatory agency. These need not be separate steps and can be compressed into a single operation. Interposing the separate preparation of a background document is useful, however, because it gives an opportunity to have the job done by expert scientists most knowledgeable regarding a specific pollutant and its effects and still retains for the appropriate government regulatory agency the opportunity to prepare and issue the criteria document. Such criteria documents also can be adequately prepared by an international agency such as the World Health Organization (*7*). Preparation of a background document by a group of expert scientists has other benefits as well, e.g., it identifies gaps in knowledge and thus serves the additional useful purpose of identifying high-priority research needs.

E. Examples of Background and Criteria Documents

1. United States Environmental Protection Agency (USEPA)

Section 102(a)(2) of the United States Clean Air Act (*4*) calls upon the USEPA Administrator to issue air quality criteria that shall ". . . accurately reflect the latest scientific knowledge useful in indicating the kind and extent of all identifiable effects on public health or welfare which may be expected from the presence of such pollutants in the ambient air, in varying quantities." Six such criteria documents have been published, namely, for particulate matter, sulfur oxides, carbon monoxide, nitrogen oxides, hydrocarbons, and photochemical oxidants (*8*). Each of these documents are currently under review and should be reissued in the 1978–1982 time span.

2. North Atlantic Treaty Organization (NATO)

In 1969, the North Atlantic Treaty Organization established the Committee on the Challenges of Modern Society (CCMS), which sponsored four air pollution studies, for which the United States was designated as the pilot nation and the Federal Republic of Germany and the Republic of Turkey as copilot nations. One of these studies related to air pollution criteria documents. Using the initial United States air quality criteria documents noted above as input, additional material, particularly from the literature of countries other than United States, was assessed and included where pertinent. Preparation procedures were altered somewhat after the completion of the sulfur oxides and particulate matter criteria documents, so that those for carbon monoxide, nitrogen oxides, and photochemical oxidants and related hydrocarbons are based less on the United States air quality criteria documents, and more on new data assembled by the NATO member nations (*9*).

3. United States National Research Council/National Academy of Sciences

The National Research Council/National Academy of Sciences (NRC/NAS) has published a series of documents that review existing knowledge concerning specific substances, including at the time of writing this chapter, asbestos, chlorine and hydrogen chloride, chromium, copper, fluorides, lead, manganese, nickel, particulate polycyclic organic matter, ozone and other photochemical oxidants, vanadium, vapor-phase organic pollutants and selenium (*10*). Reports are planned for publication on aeroallergens; arsenic; carbon disulfide; infectious aerosols; the platinum group of heavy metals; sulfides, including hydrogen sulfide and sulfides of ammonia, barium, and iron; and zinc. The Advisory Center for Toxicology, NRC/NAS, has prepared a document outlining a philosophy for short-term air quality limits and followed it with short-term air quality limit reports for chlorine; hydrazine, monomethylhydrazine, and 1,1-dimethylhydrazine; hydrogen chloride; hydrogen fluoride; and oxides of nitrogen (*11*).

4. World Health Organization*

A resolution adopted by the 23rd World Health Assembly in 1970 expressed the wish that due consideration should be given to the effect of water, soil, food, air pollution, noise, and other environmental factors harmful to human health and to the need for the establishment of environmental health criteria, guidelines for preventive measures, and methods of determining priorities and allocating resources based on health problems and needs in both developing and developed countries. As a result, WHO convened an Expert Committee on Air Quality Criteria and Guides that met in April, 1972, and prepared the report entitled *Air Quality Criteria and Guides for Urban Pollutants* (*7*). This activity was

* Most of the methods and references in this chapter are from the Western part of the world. Those who would like to explore the methods used in the USSR should refer to the Report of a Working Group convened by the Regional Office for Europe of the World Health Organization entitled, "Methods for Studying Biological Effects of Pollutants (A Review of Methods Used in USSR)." This 1975 report is available only through the Regional Office for Europe in Copenhagen.

the forerunner of the present *WHO Environmental Health Criteria Programme* (*5*).

The United Nations Conferences on the Human Environment, Stockholm, Sweden (1972) specifically recommended that WHO establish primary standards for the protection* of the human organism, especially for pollutants that are common to air, water, and food, and stressed the need to develop agreed procedures for setting derived working limits† for common air and water contaminants.

Under the stimulus of the World Health Assembly resolutions and the United Nations Conference recommendations, the WHO program underway was expanded. Work is proceeding on the development of criteria documents for asbestos, cadmium, carbon disulfide, germanium, lead, manganese, mercury, molybdenum, mycotoxins, nitrates–nitrites–nitrosamines, and oxides of nitrogen, selenium, tellurium, tin, and titanium.

III. Air Quality Goals

The obvious goal for urban areas is air of the same quality as air from the hinterlands. This air can never have lower concentrations of any substance than the global background concentration of that substance. Depending upon the substance, the global background concentration may be either constant, decreasing, or increasing. A constant or decreasing concentration results when there are sinks to consume the substance at a rate equal to or greater than its rate of introduction to the atmosphere. Otherwise, there would be an increasing global background concentration. Some estimates of global background concentrations are given in Table I.

Air pollution goals will vary from country to country and within a country from one time period to another. To decide on the appropriate emphasis to be placed on air pollution control to improve health status requires consideration of the other competing health problems within the country. Certainly, where the major health problems are arising from the need for adequate and clean water supply or there are other major sani-

* A "primary protection standard" is an accepted maximum level of a pollutant for its indicator in the target or some part thereof, or an accepted maximum intake of a pollutant or nuisance into the target under specified circumstances.

† "Derived working levels" (or limits) are maximum acceptable levels of pollutants in specified media other than the target designed to insure that under specific circumstances a primary protection standard is not exceeded. Derived working levels are known by a variety of names, including environmental or ambient air quality standards, maximum permissible limits, and maximum allowable concentrations. When derived working levels apply to products such as food or detergents, they may be known as product standards.

Table I Natural Source and Background Concentrations for Pollutants for Which Ambient Air Quality Standards Have Been Established (7)

Pollutant	*Natural source*	*Background concentration ($\mu g/m^3$)*
SO_2	Volcanoes	1–4
H_2S	Volcanoes; biological decay	0.3
NO	Bacterial action in soil; photodissociation of N_2O and NO_2	0.3–2.5
NO_2	Bacterial action in soil; oxidation of NO	2–2.5
NH_3	Biological decay	4
CO	Oxidation of methane; photodissociation of CO_2; forest fires; oceans	100
O_3	Tropospheric reactions and transport from stratosphere	20–60
Hydrocarbons	Biological processes in swamps	CH_4 = 1000; non-CH_4 <1
Reactive hydrocarbons	Biological processes in forests	<1

tation needs, less emphasis should be placed on the air pollution problems than in a highly industrialized country where the air pollution exposure is contributing a larger proportion of the total health problem.

Table II Recommended Long-Term Goals[a]

Pollutant and measurement method	*Limiting level*	
Sulfur oxides[b]—British Standard Procedure[c]	Annual mean	60 $\mu g/m^3$
	98% of observations[d] below	200 $\mu g/m^3$
Suspended particulates[b]—British Standard Procedure[c]	Annual mean	40 $\mu g/m^3$
	98% of observations[d] below	120 $\mu g/m^3$
Carbon monoxide—nondispersive infrared[c]	8-Hour average	10 mg/m^3
	1-Hour maximum	40 mg/m^3
Photochemical—oxidant as measured by neutral buffered KI method expressed as ozone	8-Hour average	60 $\mu g/m^3$
	1-Hour maximum	120 $\mu g/m^3$

[a] The committee specifically urged that this table should not be considered independently of the accompanying text. (7)

[b] Values for sulfur oxides and suspended particulates apply only in conjunction with one another.

[c] Methods are not those necessarily recommended but indicate those on which these units have been based. Where other methods are used an appropriate adjustment may be necessary.

[d] The permissible 2% of observations over this limit may not fall on consecutive days.

The 1972 World Health Organization Expert Committee report on air quality guides and criteria (7) supplies guides useful in determining short-term goals, and after taking into consideration all the evidence available regarding sulfur oxides, suspended particulates, carbon monoxides, and photochemical oxidants, reached the consensus that, in the light of present knowledge, Table II could be offered as long-term goals intended to prevent undesirable effects from the air pollutants under discussion. It was emphasized, however, that these tentative recommendations were subjected to change as and when more data on dose–response relationships within different populations became available.

IV. Air Quality Standards

A. Definition

Air quality standards are legal limits placed on levels of air pollutants in the ambient (outdoor) air during a given period of time. As such, they characterize the allowable level of a pollutant or a class of pollutants in the atmosphere and thus define the amount of exposure permitted to the population and/or to ecological systems. Air quality standards are expressions of public policy and thereby requirements for action. Thus, they are not based solely on air quality criteria but are also based on a broad range of economic, social, technical, and political considerations. When considered in the framework of definitions proposed for the United Nations Conference on the Human Environment, air quality standards are derived working levels (See page 458). Air quality standards have evolved differently in different countries, depending on exposure conditions, the socioeconomic situation, and the importance of other health-related problems. Table III shows the great variability in these standards in different countries.

Figures 2 and 3 show air quality standards from Table III for selected countries plotted for comparison purposes. With the exceptions of the USSR and Czechoslovakia, the data bases for each of these standards are the same, yet the actual concentration chosen for a standard is very different. This again emphasizes that standards are political rather than scientific decisions.

B. Single or Multiple Standards

A major decision that must be taken in the adoption of air quality standards is whether there should be one standard for an entire jurisdic-

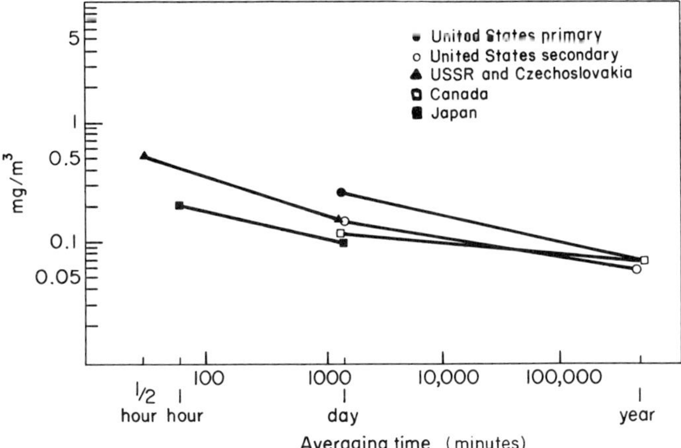

Figure 2. National ambient air quality standards (mg/m^3) for suspended particulates for selected countries for various averaging times (Table III).

tion or different standards for different areas within the jurisdiction. Our present concept for coping with this problem is the "air shed" or air quality region comprising that geographic area that encompasses both the area's air pollution sources and its receptors. All political jurisdictions within an air quality region would have the same air quality standards. In general, air quality regions would not be contiguous but would be separated by nonurban hinterland. If two separated regions were to ex-

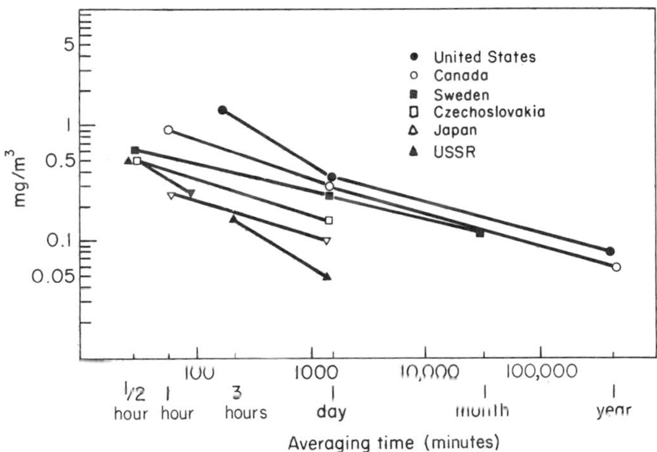

Figure 3. National ambient air quality standards (mg/m^3) for sulfur dioxide for selected countries for various averaging times (Table III).

Table III. National Ambient Air Quality Standards

Substance and country	*Long-term standard[a]*			*Short-term standard[a]*			*Notes[b]*
	mg/m³	*ppm*	*Averaging time (hours)*	*mg/m³*	*ppm*	*Averaging time (minutes)*	
Acetaldehyde							
Bulgaria, Yugoslavia	—	—	—	*0.01*	0.005	30	—
East Germany	*0.01*	0.005	24	*0.03*	0.015	30	
USSR	*0.01*	0.005	24	*0.01*	0.005	30	—
West Germany (VDI 2306)	*4.0*	*2.0*	½	*12.0*	*6.0*	30	3, 4
Acetic acid							
Bulgaria	—	—	—	*0.2*	0.08	30	—
East Germany, USSR	*0.06*	0.024	24	*0.2*	0.08	30	1, 2
West Germany (VDI 2306)	*5.0*	*2.0*	½	*15.0*	*6.0*	30	3, 4
Acetic anhydride							
Bulgaria	—	—	—	*0.1*	0.025	30	—
East Germany, USSR	*0.03*	0.0075	24	*0.1*	0.025	30	1, 2
Acetone							
Bulgaria, Hungary, USSR, Yugoslavia	*0.35*	0.15	24	*0.35*	0.15	30	2, 5
East Germany	*0.35*	0.15	24	*1.0*	0.42	30	1
Hungary	*12.0*	5.0	24	*180.0*	75.0	30	—
Israel	7.2	*3.0*	24	24.0	*10.0*	30	6
Romania	*2.0*	0.83	24	*5.0*	2.1	30	—
West Germany (VDI 2306)	*120.0*	*50.0*	½	*360.0*	*150.0*	30	3, 4
Acetophenone							
Bulgaria	*0.35*	0.07	24	*0.35*	0.07	30	—
East Germany	*0.003*	0.0006	24	*0.01*	0.002	30	1
USSR, Yugoslavia	*0.003*	0.0006	24	*0.003*	0.0006	30	2

Acrolein								
Bulgaria, Czechoslovakia, Hungary, Romania, Yugoslavia	*0.1*	0.04	24	*0.3*	0.12	30		—
East Germany	*0.01*	0.004	24	*0.02*	0.008	30	1	
Israel	0.1	*0.04*	24	0.25	*0.1*	30	6	
USSR	*0.03*	0.012	24	*0.03*	0.012	30		—
West Germany (VDI 2306)	*0.01*	*0.005*	½	*0.025*	*0.01*	30	3, 4	
Ammonia								
Bulgaria, Hungary, USSR, Yugoslavia	*0.2*	0.28	24	*0.2*	0.28	30	5	
Czechoslovakia, East Germany, Romania	*0.1*	0.14	24	*0.3*	0.43	30	1	
Hungary	*0.5*	0.71	24	*1.5*	2.14	30		—
Amyl acetate								
Bulgaria, Hungary, USSR, Yugoslavia	*0.1*	0.019	24	*0.1*	0.019	30	5	
East Germany	*0.1*	0.019	24	*0.3*	0.057	30	1	
Hungary	*30.0*	5.7	24	*90.0*	17.1	30		—
Israel	5.25	*1.0*	24	15.75	*3.0*	30	6	
West Germany (VDI 2306)	*30.0*	*5.0*	½	*90.0*	*15.0*	30	3, 4	
Amyl alcohol								
West Germany (VDI 2306)	*20.0*	*5.0*	½	*60.0*	*15.0*	30	3, 4	
Amylene								
Bulgaria, USSR, Yugoslavia	*1.5*	0.5	24	*1.5*	0.5	30	2	
East Germany	*1.0*	0.33	24	*1.5*	0.5	30	1	
Aniline								
Bulgaria, Czechoslovakia, East Germany, USSR, Yugoslavia	*0.03*	0.008	24	*0.05*	0.013	30	1	
Romania	*0.02*	0.005	24	*0.05*	0.013	30		—
West Germany (VDI 2306)	*0.8*	*0.2*	½	*2.4*	*0.6*	30	3, 4	
Arsenic								
Bulgaria, Czechoslovakia, USSR	*0.003*	—	24	—	—	—	7	
East Germany	*0.003*	—	24	—	—	—		—
Israel	*0.006*	—	24	—	—	—	6, 7	

Table III (Continued)

Substance and country	Long-term standard[a] mg/m³	ppm	Averaging time (hours)	Short-term standard[a] mg/m³	ppm	Averaging time (minutes)	Notes[b]
Poland	*0.003*	—	24	*0.01*	—	20	9
	0.002	—	24	*0.005*	—	20	10
Romania	*0.01*	—	24	*0.03*	—	30	—
Yugoslavia	*0.003*	—	24	—	—	—	8
Benzene							
Czechoslovakia, Romania	*0.8*	0.25	24	*2.4*	0.75	30	—
East Germany, Hungary, Yugoslavia	*0.8*	0.25	24	*1.5*	0.46	30	1, 5
Hungary, West Germany (VDI 2306)	*3.0*	0.94	24	*10.0*	3.12	30	3, 4
Israel	1.6	*0.5*	24	4.8	*1.5*	30	6
Poland	*0.3*	0.09	24	*1.0*	0.31	20	9
	0.1	0.03	24	*0.2*	0.06	20	10
Benzene (high alkyl)							
West Germany (VDI 2306)	*5.0*	—	½	*15.0*	—	30	3, 4
Benzine							
East Germany	*0.03*	0.007	24	*0.05*	0.012	30	1, 13
Hungary	*80.0*	20.0	24	*240.0*	60.0	30	—
Hungary, USSR	*1.5*	0.38	24	*5.0*	1.25	30	5, 12
Israel	3.3	*0.8*	24	*10.0*	*2.4*	30	6, 12
Poland	*0.75*	0.19	24	*2.5*	0.63	20	10, 12
Romania	*2.0*	0.48	24	*6.0*	1.45	30	—
West Germany (VDI 2306)	*80.0*	*20.0*	½	*240.0*	*60.0*	30	3, 4, 12
Yugoslavia	*1.5*	0.38	24	*5.0*	1.25	30	11
Benzine (from shale)							
Bulgaria, USSR	*0.05*	0.012	24	*0.05*	0.012	20	11

Benzine (low sulfur)								
Bulgaria	*1.5*	0.38	24	*5.0*	1.25	30	11	
East Germany, Yugoslavia	*1.5*	0.38	24	*5.0*	1.25	30	1	
Beryllium								
Israel, Yugoslavia	*0.00001*	—	24	—	—	—	6	
Butane								
Bulgaria, USSR, Yugoslavia	—	—	—	*200.0*	85.0	30		—
East Germany	*50.0*	21.0	24	*200.0*	85.0	30	1	
Butanol								
Bulgaria, Yugoslavia	—	—	—	*0.3*	0.1	30		—
East Germany	*0.1*	0.03	24	*0.3*	0.1	30	1	
USSR	—	—	—	*0.1*	0.03	30		—
West Germany (VDI 2306)	*15.0*	5.0	$\frac{1}{2}$	*45.0*	15.0	30	3, 4	
n-Butyl acetate								
Bulgaria, USSR, Yugoslavia	*0.1*	0.021	24	*0.1*	0.021	30		—
East Germany	*0.1*	0.021	24	*0.3*	0.063	30	1	
Israel	4.7	*1.0*	24	14.0	*3.0*	30	6	
West Germany (VDI 2306)	*25.0*	*5.0*	$\frac{1}{2}$	*75.0*	*15.0*	30	3, 4	
Butylene								
Bulgaria, USSR, Yugoslavia	*3.0*	1.2	24	*3.0*	1.2	30	2	
East Germany	*2.0*	0.8	24	*3.0*	1.2	30	1	
Butyric acid								
Bulgaria, USSR, Yugoslavia	*0.01*	0.003	24	*0.015*	0.004	30		—
Cadmium								
Yugoslavia	*0.003*	—	24	*0.01*	—	30		—
Caproic acid								
USSR, Yugoslavia	*0.005*	0.001	24	*0.01*	0.002	30		—
Caprolactam								
Bulgaria, USSR, Yugoslavia	*0.06*	0.013	24	*0.06*	0.013	30	14	
East Germany	*0.06*	0.013	24	*0.1*	0.022	30	1	
Caprylic acid								
Bulgaria, East Germany	*0.005*	0.001	24	*0.01*	0.002	30	1	

Table III (*Continued*)

Substance and country	*Long-term standard*[a]			*Short-term standard*[a]			*Notes*[b]
	mg/m^3	*ppm*	*Averaging time (hours)*	*mg/m*^{3}	*ppm*	*Averaging time (minutes)*	
Carbon disulfide							
Bulgaria, Czechoslovakia, Romania, Yugoslavia	*0.01*	0.0033	24	*0.03*	0.01	30	—
East Germany	*0.003*	0.001	24	*0.03*	0.01	30	1
Israel	0.15	*0.05*	24	0.45	*0.15*	30	6
Poland	*0.015*	0.005	24	*0.045*	0.015	20	9
USSR	*0.005*	0.0016	24	*0.03*	0.01	30	2
Carbon monoxide							
Argentina	11.5	*10.0*	8	57.7	*50.0*	60	—
Bulgaria, East Germany, Hungary, USSR, Yugoslavia	*1.0*	0.9	24	*3.0*	2.7	30	1, 2, 5
Canada—Desirable level	*6.0*	*5.0*	8	*15.0*	*13.0*	60	15, 16
—Acceptable level	*15.0*	*13.0*	8	*35.0*	*30.0*	60	15, 17
Czechoslovakia	*1.0*	0.9	24	*6.0*	5.4	30	—
Finland	*10.0*	*9.0*	8	*40.0*	*35.0*	60	71
Hungary, Romania	*2.0*	1.8	24	*6.0*	5.4	30	—
Israel	11.5	*10.0*	8	35.0	*30.0*	30	—
Italy	23.0	*20.0*	8	57.7	*50.0*	30	18
Japan	11.5	*10.0*	24	—	—	—	19
	23.0	*20.0*	8	—	—	—	19
Poland	*0.5*	0.45	24	*3.0*	2.7	20	9
Spain	*15.0*	13.0	8	*45.0*	39.0	30	20
USA, West Germany	*10.0*	8.6	8	*40.0*	35.0	60	21
Carbon tetrachloride							
East Germany, USSR	*2.0*	0.33	24	*4.0*	0.66	30	1, 2
Romania	*1.0*	0.16	24	*3.0*	0.5	30	—
West Germany (VDI 2306)	*3.0*	*0.5*	½	*10.0*	*1.5*	30	3, 4

Chlorine								
Bulgaria, Czechoslovakia, East Germany, Hungary, USSR, Yugoslavia	*0.03*	0.01	24	*0.1*	0.03	30	1, 5	
Hungary	*0.3*	0.1	24	*0.6*	0.2	30		—
Israel	0.1	*0.03*	24	0.3	*0.1*	30	6	
Italy	—	—	—	*0.58*	*0.2*	30	18	
Poland	*0.03*	0.01	24	*0.1*	0.03	20	9	
	0.01	0.003	24	*0.03*	0.01	20	10	
Romania	*0.1*	0.033	24	*0.3*	0.1	30		—
Spain	*0.05*	0.016	24	*0.3*	0.1	30	20	
West Germany	*0.3*	*0.1*	½	*0.6*	*0.2*	30	22	
m-Chloroaniline								
East Germany	*0.01*	0.003	24	*0.03*	0.01	30	1	
USSR	*0.01*	0.003	24	—	—	—		—
Yugoslavia	—	—	—	*0.04*	0.013	30		—
p-Chloroaniline								
Bulgaria	—	—	—	*0.04*	0.008	30		—
East Germany, USSR	*0.01*	0.002	24	*0.04*	0.008	30	1	
Chlorobenzene								
Bulgaria, USSR, Yugoslavia	*0.1*	0.02	24	*0.1*	0.02	30		—
East Germany	*0.1*	0.02	24	*0.3*	0.06	30	1	
West Germany (VDI 2306)	*5.0*	*1.0*	½	*15.0*	*3.0*	30	3, 9	
Chloroform								
West Germany (VDI 2306)	*10.0*	*2.0*	½	*30.0*	*6.0*	30	3, 4, 23	
m-Chlorophenyl isocyanate								
Bulgaria, USSR, Yugoslavia	*0.005*	—	24	*0.005*	—	30	2	
East Germany	*0.003*	—	24	*0.005*	—	30	1	
p-Chlorophenyl isocyanate								
Bulgaria, East Germany, USSR, Yugoslavia	*0.0015*	0.0002	24	*0.0015*	0.0002	30	1, 2	
Chloroprene								
Bulgaria, USSR, Yugoslavia	*0.1*	0.028	24	*0.1*	0.028	30		—
East Germany	*0.05*	0.014	24	*0.1*	0.028	30	1	
Israel	0.14	*0.04*	24	0.5	*0.14*	30	6	

Table III (*Continued*)

Substance and country	*Long-term standard*[a]			*Short-term standard*[a]			*Notes*[b]
	mg/m^3	*ppm*	*Averaging time (hours)*	*mg/m*^{3}	*ppm*	*Averaging time (minutes)*	
Chlorotetracyclin							
East Germany	*0.03*	—	24	*0.05*	—	30	1, 24
USSR	*0.05*	—	24	*0.05*	—	30	25
Chromium							
Romania	*0.0015*	—	24	*0.0015*	—	30	26
Chromium (hexavalent)							
East Germany	*0.001*	—	24	*0.0015*	—	30	1, 27
Israel	*0.0015*	—	24	—	—	—	6, 27
USSR	*0.0015*	—	24	*0.0015*	—	20	27
Yugoslavia	*0.0015*	—	24	*0.0015*	—	30	27
Cresol (all isomers)							
West Germany (VDI 2306)	*0.2*	*0.05*	½	*0.6*	*0.15*	30	3, 4
Cyclohexane							
East Germany	*1.0*	0.3	24	*1.4*	0.4	30	1
USSR	*1.4*	0.4	24	*1.4*	0.4	30	—
Cyclohexanol							
Bulgaria, USSR, Yugoslavia	*0.06*	0.015	24	*0.06*	0.015	30	—
East Germany	*0.06*	0.015	24	*0.15*	0.037	30	1
Cyclohexanone							
Bulgaria, Hungary, Yugoslavia	*0.04*	0.01	24	*0.04*	0.01	30	5
East Germany	*0.04*	0.01	24	*0.1*	0.02	30	1
Hungary	*10.0*	2.5	24	*30.0*	7.5	30	—
USSR	—	—	—	*0.04*	0.01	30	—
West Germany (VDI 2306)	*10.0*	*2.0*	½	*30.0*	*6.0*	30	3, 4

Cyclohexanon oxime								
East Germany	*0.04*	0.01	24	*0.1*	0.025	30	1	
USSR	—	—	—	*0.1*	0.025	30		—
Dichloroethane								
Bulgaria, East Germany, Romania, USSR, Yugoslavia	*1.0*	0.25	24	*3.0*	0.75	30	1	
Israel	2.0	*0.5*	24	6.0	*1.5*	30	6	
West Germany (VDI 2306)	*8.0*	*2.0*	½	*25.0*	*6.0*	30	3, 4	
2-3-Dichloro-1-4-naphthaquinone								
Bulgaria, East Germany	*0.02*	—	24	*0.05*	—	30	1	
USSR, Yugoslavia	*0.05*	—	24	*0.05*	—	30		—
Diethylamine								
Bulgaria, Romania, USSR, Yugoslavia	*0.05*	0.016	24	*0.05*	0.016	30		—
East Germany	*0.02*	0.008	24	*0.05*	0.016	30	1	
West Germany (VDI 2306)	*0.03*	*0.01*	½	*0.05*	*0.02*	30	3, 4	
Diethyl ether								
West Germany (VDI 2306)	*65.0*	*20.0*	½	*155.0*	*60.0*	30	3, 4	
Diketene								
Bulgaria, USSR, Yugoslavia	—	—	—	*0.007*	0.002	30		—
East Germany	*0.002*	0.001	24	*0.007*	0.002	30	1	
Dimethylamine								
East Germany	*0.005*	0.003	24	*0.015*	0.0075	30	1	
USSR	*0.005*	0.003	24	*0.005*	0.003	30		—
West Germany (VDI 2306)	*0.02*	*0.01*	½	*0.06*	*0.03*	30	3, 4	
Dimethylaniline								
Bulgaria, Yugoslavia	—	—	—	*0.0055*	0.001	30		—
East Germany	*0.005*	0.001	24	*0.015*	0.003	30	1	
USSR	*0.0055*	0.001	24	*0.0055*	0.001	30		—
Dimethyl Disulfide								
Bulgaria, USSR	—	—	—	*0.7*	0.18	30		—
East Germany	*0.2*	0.05	24	*0.7*	0.18	30	1	
Yugoslavia	—	—	—	*0.07*	0.018	30		—

Table III (*Continued*)

Substance and country	Long-term standard[a] mg/m³	Long-term ppm	Long-term Averaging time (hours)	Short-term standard[a] mg/m³	Short-term ppm	Short-term Averaging time (minutes)	Notes[b]
Dimethylformamide							
Bulgaria, USSR, Yugoslavia	*0.03*	0.01	24	*0.03*	0.01	30	—
East Germany	*0.01*	0.003	24	*0.03*	0.01	30	1
Israel	0.018	*0.006*	24	0.06	*0.02*	30	6
Dimethyl sulfide							
Bulgaria, USSR, Yugoslavia	—	—	—	*0.08*	0.03	30	—
East Germany	*0.03*	0.01	24	*0.08*	0.03	30	1
Dinitrobenzene							
West Germany (VDI 2306)	*0.035*	*0.005*	½	*0.1*	*0.015*	30	3, 4
Dinyl							
Bulgaria, Romania, USSR, Yugoslavia	*0.01*	0.0015	24	*0.01*	0.0015	30	2, 28
East Germany	*0.003*	0.0045	24	*0.01*	0.0015	30	1, 28
Dioxane							
West Germany	*20.0*	*5.0*	½	*60.0*	*15.0*	30	3, 4, 48
Divinyl							
Bulgaria, East Germany, USSR, Yugoslavia	*1.0*	0.4	24	*3.0*	1.2	30	1
Epichlorohydrin							
Bulgaria, USSR, Yugoslavia	*0.2*	0.05	24	*0.2*	0.05	30	—
East Germany	*0.06*	0.016	24	*0.2*	0.05	30	1
Ethanol							
Bulgaria, USSR, Yugoslavia	*5.0*	2.5	24	*5.0*	2.5	30	2
East Germany	*5.0*	2.5	24	*15.0*	7.5	30	1
West Germany (VDI 2306)	*100.0*	*50.0*	½	*300.0*	*150.0*	30	3, 4

Ethyl acetate							
Bulgaria, USSR, Yugoslavia	*0.1*	0.029	24	*0.1*	0.029	30	—
East Germany	*0.1*	0.029	24	*0.3*	0.085	30	1
Israel	14.0	*4.0*	24	42.0	*12.0*	30	6
West Germany (VDI 2306)	*75.0*	*20.0*	½	*225.0*	*60.0*	30	3, 4
Ethylbenzene							
East Germany	*0.02*	0.005	24	*0.06*	0.014	30	1
USSR	*0.02*	0.005	24	*0.02*	0.005	30	—
Ethylene							
Bulgaria, USSR, Yugoslavia	*3.0*	2.3	24	*3.0*	2.3	30	2
East Germany	*2.0*	1.53	24	*3.0*	2.3	30	1
Israel	0.26	*0.2*	24	0.65	*0.5*	30	6
Ethylene oxide							
Bulgaria, East Germany, USSR, Yugoslavia	*0.03*	0.015	24	*0.3*	0.15	30	1
West Germany (VDI 2306)	*4.0*	*2.0*	½	*12.0*	*6.0*	30	3, 4
Ethylenimine							
East Germany	*0.001*	0.0005	24	*0.003*	0.0015	30	1
USSR	*0.001*	0.0005	24	*0.001*	0.0005	30	—
Flourides (as F)							
Bulgaria, East Germany, Romania	*0.005*	0.002	24	*0.02*	0.01	30	1, 29
Czechoslovakia, Hungary, Israel	*0.01*	0.005	24	*0.03*	0.015	30	5, 6, 29
Hungary	*0.03*	0.015	24	*0.1*	0.05	30	—
Italy, Spain	*0.02*	0.01	24	*0.06*	0.03	30	18, 20
Fluorides							
Bulgaria, Poland	*0.01*	—	24	*0.03*	—	20	9, 31, 33, 35
East Germany, Yugoslavia	*0.01*	—	24	*0.03*	—	30	1, 31, 33, 35
Hungary	*0.02*	0.015	24	*0.02*	0.015	30	32
	0.0013	0.001	24	*0.005*	0.004	30	5, 32
Netherlands	*0.01*	0.008	24	—	—	—	—
Poland	*0.003*	—	24	*0.01*	—	20	10, 35
Spain, USSR	*0.01*	0.008	24	*0.03*	0.022	30	2, 20, 32, 33

Table III (*Continued*)

Substance and country	*Long-term standard*[a] *mg/m*^{3}	*ppm*	*Averaging time (hours)*	*Short-term standard*[a] *mg/m*^{3}	*ppm*	*Averaging time (minutes)*	*Notes*[b]
USSR	*0.005*	0.002	24	*0.02*	0.01	20	2, 29, 30
West Germany	*0.002*	0.001	$\frac{1}{2}$	*0.005*	0.004	30	32
Yugoslavia	*0.005*	0.004	24	*0.02*	0.015	30	32
Fluorides (insoluble)							
Yugoslavia	*0.03*	—	24	*0.2*	—	30	—
Fluorides (sparingly soluble)							
East Germany, USSR	*0.03*	—	24	*0.2*	—	30	1, 34
Formaldehyde							
Bulgaria, East Germany, Hungary, USSR, Yugoslavia	*0.012*	0.01	24	*0.035*	0.025	30	1, 5
Czechoslovakia	*0.015*	0.01	24	*0.05*	0.033	30	—
Hungary	*0.03*	0.02	24	*0.07*	0.05	30	—
Israel, West Germany (VDI 2306)	*0.03*	*0.02*	24	*0.07*	*0.06*	30	3, 4, 6
Poland	*0.02*	0.014	24	*0.05*	0.033	20	9
	0.01	0.007	24	*0.02*	0.014	20	10
Romania	*0.01*	0.007	24	*0.03*	0.02	30	—
Furfural							
Bulgaria, USSR, Yugoslavia	*0.05*	0.013	24	*0.05*	0.013	30	2
East Germany, Romania	*0.05*	0.013	24	*0.15*	0.04	30	1
Israel	0.08	*0.02*	24	0.25	*0.06*	30	6
West Germany (VDI 2306)	*0.08*	*0.02*	$\frac{1}{2}$	*0.25*	0.06	30	3, 4
Hexachlorocyclohexane							
East Germany	*0.01*	—	24	*0.03*	—	30	1
USSR	*0.03*	—	24	*0.03*	—	30	—

Hexamethylenediamine							
Bulgaria, USSR	*0.001*	—	24	*0.001*	—	30	—
East Germany	*0.001*	—	24	*0.003*	—	30	1
Yugoslavia	*0.01*	—	24	*0.01*	—	30	—
Hydrocarbons (total)							
Israel	2.0	*3.0*	24	5.0	*7.5*	30	6
Italy	26.6	*40.0*	24	53.3	*80.0*	30	18, 36
United States	*0.16*	*0.24*	3	—	—	—	21, 37
Hydrochloric acid							
Bulgaria	*0.2*	0.14	24	—	—	—	39
Bulgaria, USSR, Yugoslavia	*0.006*	—	24	*0.006*	—	30	38
Czechoslovakia	—	—	—	*0.01*	0.007	30	39
	—	—	—	*0.01*	—	30	38
East Germany	*0.015*	0.01	24	*0.05*	0.035	30	1, 39
Hungary	*0.7*	0.5	24	*1.4*	1.0	30	2, 39
Hungary, USSR	*0.2*	0.14	24	*0.2*	0.14	30	2, 5, 39
Israel	0.4	*0.3*	24	1.4	*1.0*	30	6, 39
Italy	0.04	*0.03*	24	0.28	*0.2*	30	18, 39
Poland	*0.1*	0.07	24	*0.2*	0.14	20	9, 39
	0.02	0.014	24	*0.05*	0.035	20	10, 39
Romania	*0.1*	0.07	24	*0.3*	0.21	30	39
West Germany	*0.05*	0.035	$\frac{1}{2}$	*0.15*	0.1	30	39
Yugoslavia	—	—	—	*0.2*	0.14	30	39
Hydrogen cyanide							
East Germany	*0.005*	0.004	24	*0.015*	0.014	30	1
USSR	*0.01*	0.009	24	—	—	—	—
Hydrogen sulfide							
Bulgaria, Czechoslovakia, Hungary, USSR, Yugoslavia	*0.008*	0.005	24	*0.008*	0.005	30	2, 5
East Germany	*0.008*	0.005	24	*0.015*	0.01	30	1
Finland	*0.05*	0.03	24	*0.15*	0.1	30	71
Hungary	*0.15*	0.1	24	*0.3*	0.2	30	—
Israel	0.045	*0.03*	24	0.15	*0.1*	30	—
Italy	0.04	*0.03*	24	0.1	*0.07*	30	18

Table III (*Continued*)

Substance and country	*Long-term standard*[a] *mg/m³*	*ppm*	*Averaging time (hours)*	*Short-term standard*[a] *mg/m³*	*ppm*	*Averaging time (minutes)*	*Notes*[b]
Poland	*0.02*	0.013	24	*0.06*	0.04	20	9
	0.008	0.005	24	*0.008*	0.005	20	10
Romania	*0.01*	0.006	24	*0.03*	0.02	30	—
Spain	*0.004*	0.0025	24	*0.01*	0.006	30	20
West Germany	*0.02*	0.013	$\frac{1}{2}$	*0.05*	0.03	30	—
Intrathion (M-81)							
USSR	*0.001*	—	24	*0.001*	—	30	—
Isooctanol							
East Germany	*0.05*	—	24	*0.15*	—	30	1
USSR	—	—	—	*0.15*	—	30	—
Isopropanol							
East Germany	*0.6*	0.24	24	*2.0*	0.82	30	1
Isopropyl benzene							
Bulgaria, USSR	*0.014*	—	24	*0.014*	—	30	2
East Germany	*0.014*	—	24	*0.05*	—	30	1
Isopropyl benzene (hydroperoxide)							
Bulgaria, USSR	*0.007*	—	24	*0.007*	—	30	2
East Germany	*0.007*	—	24	*0.02*	—	30	1
Lead							
Bulgaria, Czechoslovakia, East Germany, USSR, Yugoslavia	*0.0007*	—	24	—	—	—	2, 42
Hungary	*0.001*	—	24	*0.002*	—	30	—
	0.0007	—	24	*0.0007*	—	30	5
Israel	*0.005*	—	24	—	—	—	—
Italy	*0.01*	—	8	*0.05*	—	30	18

Poland	*0.001*	—	24	—	—	—	9
	0.0005	—	24	—	—	—	10
Romania	*0.001*	—	24	—	—	—	—
Lead sulfide (as Pb)							
Bulgaria	*0.0007*	—	24	—	—	—	—
East Germany, USSR, Yugoslavia	*0.0017*	—	24	—	—	—	—
Israel	*0.0035*	—	24	—	—	—	6
Malathion							
Bulgaria, USSR, Yugoslavia	—	—	—	*0.015*	—	30	45
Maleic Anhydride							
Bulgaria, East Germany, USSR, Yugoslavia	*0.05*	0.012	24	*0.2*	0.05	30	1, 2
Manganese							
Bulgaria, Czechoslovakia, East Germany, Yugoslavia	*0.01*	—	24	—	—	—	43
Israel, Romania	*0.01*	—	24	*0.03*	—	30	6
USSR	*0.01*	—	24	—	—	—	—
Mercury							
Bulgaria, East Germany, Hungary, USSR, Yugoslavia	*0.0003*	—	24	—	—	—	—
Israel, Romania	*0.001*	—	24	—	—	—	6
Mesidine							
Bulgaria, Yugoslav	—	—	—	*0.003*	—	30	44
USSR	*0.003*	—	24	*0.003*	—	30	44
Methanol							
Bulgaria, Czechoslovakia, East Germany, Hungary, USSR, Yugoslavia	*0.5*	0.38	24	*1.0*	0.77	30	1, 2, 5
Hungary	*15.0*	10.0	24	*40.0*	27.0	30	—
Israel	1.5	*1.0*	24	4.5	*3.0*	30	6
Romania	*1.0*	0.77	24	*3.0*	2.3	30	—
West Germany (VDI 2306)	*15.0*	*10.0*	$\frac{1}{2}$	*40.0*	*30.0*	30	3, 4
Methyl acetate							
Bulgaria, USSR, Yugoslavia	*0.07*	0.023	24	*0.07*	0.023	30	—
East Germany	*0.07*	0.023	24	*0.2*	0.066	30	1
Israel	3.0	*1.0*	24	9.0	*3.0*	30	6
West Germany (VDI 2306)	*15.0*	*5.0*	$\frac{1}{2}$	*45.0*	*15.0*	30	3, 4

Table III (*Continued*)

Substance and country	*Long-term standard*[a]			*Short-term standard*[a]			*Notes*[b]
	mg/m^3	*ppm*	*Averaging time (hours)*	*mg/m*^{3}	*ppm*	*Averaging time (minutes)*	
Methyl acrylate							
Bulgaria, Yugoslavia	—	—	—	*0.01*	0.003	30	—
East Germany	*0.01*	0.003	24	*0.03*	0.009	30	1
USSR	*0.01*	0.003	24	*0.01*	0.003	30	—
Methylaniline							
USSR	*0.04*	0.01	24	*0.04*	0.01	30	—
Yugoslavia	—	—	—	*0.04*	0.01	30	—
Methyl ethyl ketone							
West Germany (VDI 2306)	*30.0*	*10.0*	½	*90.0*	*30.0*	30	3, 4
Methyl isobutyl ketone							
West Germany (VDI 2306)	*20.0*	*5.0*	½	*65.0*	*15.0*	30	3, 4
Methyl mercaptan							
Bulgaria, USSR, Yugoslavia	—	—	—	9×10^{-6}	—	30	—
East Germany	—	—	—	10^{-5}	—	30	1
Methyl methacrylate							
Bulgaria, USSR Yugoslavia	*0.1*	0.025	24	*0.1*	0.025	30	—
East Germany	*0.1*	0.025	24	*0.3*	0.075	30	1
Israel	*0.2*	0.05	24	*0.6*	0.15	30	6
Methylparathion							
Bulgaria, USSR, Yugoslavia	—	—	—	*0.008*	—	30	46
Methylene chloride							
West Germany (VDI 2306)	*20.0*	*5.0*	½	*55.0*	*15.0*	30	3, 4
α-Methylstyrene							
Bulgaria, USSR, Yugoslavia	*0.04*	0.01	24	*0.04*	0.01	30	—
East Germany	*0.03*	0.0075	24	*0.05*	0.0125	30	1

Monoethylamine							
East Germany	*0.01*	0.005	24	*0.03*	0.015	30	1
West Germany (VDI 2306)	*0.02*	0.01	½	*0.06*	0.03	30	3, 4
USSR	*0.01*	0.005	24	*0.01*	0.005	30	—
Monomethylaniline							
Bulgaria	—	—	—	*0.04*	0.009	30	—
East Germany	*0.03*	0.007	24	*0.05*	0.01	30	1
Naphthalene							
East Germany	*0.001*	0.0002	24	*0.003*	0.0006	30	1
USSR	*0.003*	0.0006	24	*0.003*	0.0006	30	—
West Germany (VDI 2306)	*2.5*	*0.5*	½	*7.5*	*1.5*	30	3, 4
α-Naphthaquinone							
Bulgaria, USSR, Yugoslavia	*0.005*	0.001	24	*0.005*	0.001	30	2
East Germany	*0.002*	0.0004	24	*0.005*	0.001	30	1
Nitric acid							
Bulgaria, USSR, Yugoslavia	*0.006*	0.0024	24	*0.006*	0.0024	30	2, 3
Bulgaria, Yugoslavia	—	—	—	*0.4*	0.16	30	47
Czechoslovakia	—	—	—	*0.01*	0.004	30	38
East Germany	*0.06*	0.024	24	*0.14*	0.056	30	1
Hungary	*1.3*	0.5	24	*2.6*	1.0	30	—
	0.4	0.16	24	*0.4*	0.16	30	5
Israel	0.42	*0.17*	24	1.3	*0.5*	30	6
USSR	*0.4*	0.16	24	*0.4*	0.16	30	47
West Germany (VDI 2106)	*1.3*	*0.5*	½	*2.6*	*1.0*	30	3, 40
Nitrobenzene							
Bulgaria	—	—	—	*0.04*	0.008	30	—
East Germany	*0.005*	0.001	24	*0.01*	0.002	30	1
Hungary	*0.3*	0.06	24	*0.85*	0.17	30	—
	0.008	0.0016	24	*0.08*	0.016	30	5
USSR, Yugoslavia	*0.008*	0.0016	24	*0.008*	0.0016	30	—
West Germany (VDI 2306)	*0.3*	*0.005*	½	*0.85*	*0.15*	30	3, 4

Table III (*Continued*)

Substance and country	*Long-term standard*[a]			*Short-term standard*[a]			*Notes*[b]
	mg/m^3	*ppm*	*Averaging time (hours)*	*mg/m*^{3}	*ppm*	*Averaging time (minutes)*	
o-Nitrochlorobenzene							
East Germany	*0.004*	—	24	*0.008*	—	30	1
p-Nitrochlorobenzene							
East Germany	*0.004*	—	24	*0.008*	—	30	1
o- and *p*-Nitrochlorobenzene							
USSR	*0.004*	—	24	—	—	—	—
Nitrogen dioxide							
Argentina	—	—	—	0.85	*0.45*	60	—
Bulgaria, Hungary, USSR, Yugoslavia	*0.085*	0.045	24	*0.085*	0.045	30	2, 5
Canada—Desirable level	0.06	0.03	1 yr	—	—	—	15.16
Acceptable level	0.1	0.05	1 yr	0.4	0.21	60	15.17
Acceptable level	0.2	0.11	24	—	—	—	15.17
Czechoslovakia, Romania, West Germany	*0.1*	0.05	24	*0.3*	0.16	30	—
Finland	*0.2*	0.1	24	*0.56*	0.3	30	71
Hungary	*0.15*	0.08	24	*0.5*	0.27	30	—
Japan	0.04	*0.02*	24	—	—	—	19
Nitrogen monoxide							
West Germany	*0.4*	—	½	*0.8*	—	30	—
Nitrogen oxides							
Argentina	0.9	*0.45*	1	—	—	—	49
East Germany	*0.004*	0.002	24	*0.1*	0.06	30	1, 49
Hungary	*0.15*	0.075	24	*0.5*	0.25	30	49
	0.05	0.025	24	*0.15*	0.075	30	5, 49
Israel	0.6	*0.3*	24	1.0	*0.5*	30	49

Italy	*0.2*	*0.1*	24	0.6	*0.3*	30	18, 49
Poland	*0.2*	0.1	24	*0.6*	0.3	20	49
	0.05	0.025	24	*0.15*	0.075	20	10, 49
Spain	*0.2*	0.1	24	*0.4*	0.2	30	20, 49
United States	*0.1*	0.05	1 yr	—	—	—	21, 37, 49
West Germany (VDI 2105)	*1.0*	0.5	½	*2.0*	1.0	30	3, 41, 49
Nitrogen pentoxide							
Yugoslavia	*0.1*	—	24	*0.3*	—	30	—
Oxidants							
Argentina	—	—	—	0.2	*0.1*	60	51
Canada—Acceptable level	*0.05*	*0.025*	24	*0.16*	*0.08*	60	15, 17, 51
Acceptable level	*0.03*	*0.015*	1 yr	—	—	—	15, 17, 51
Desirable level	*0.03*	*0.015*	24	*0.1*	*0.05*	60	15, 16, 51
Israel	0.2	*0.1*	8	0.4	*0.2*	30	51
Japan	—	—	—	0.12	*0.06*	60	50
Romania	*0.05*	0.015	24	*0.1*	0.05	30	51
United States	—	—	—	*0.16*	*0.08*	60	21, 37, 51
Ozone							
Israel	0.1	*0.05*	24	0.2	0.1	30	6
Pentane							
Bulgaria, East Germany, USSR, Yugoslavia	*25.0*	8.5	24	*100.0*	33.9	30	1
Perchlorethylene							
West Germany (VDI 2306)	*35.0*	*5.0*	½	*110.0*	*15.0*	30	3, 4
Phenol							
Bulgaria, Hungary, Yugoslavia	*0.01*	0.0026	24	*0.01*	0.0026	30	5
Czechoslovakia	*0.1*	0.026	24	*0.3*	0.079	30	—
East Germany	*0.01*	0.0026	24	*0.03*	0.0079	30	1
Hungary	*0.2*	0.052	24	*0.6*	0.16	30	—
Israel	0.1	*0.025*	24	0.3	*0.075*	30	6
Poland	*0.01*	0.0026	24	*0.02*	0.0052	20	9
	0.003	0.0008	24	*0.01*	0.0026	20	10

Table III (*Continued*)

Substance and country	*Long-term standard*[a]			*Short-term standard*[a]			*Notes*[b]
	mg/m^3	*ppm*	*Averaging time (hours)*	*mg/m*^{3}	*ppm*	*Averaging time (minutes)*	
Romania	*0.03*	0.0079	24	*0.1*	0.026	30	—
USSR	*0.01*	0.0026	24	*0.01*	0.0026	20	2
West Germany (VDI 2306)	*0.2*	*0.05*	½	*0.6*	*0.15*	30	3, 4
Phosphoric acid							
Romania	*0.1*	—	24	*0.3*	—	30	—
Phosphoric anhydride							
East Germany	*0.05*	0.0085	24	*0.15*	0.026	30	1
Israel	0.1	*0.017*	24	0.05	*0.0085*	30	6
Phosphorus pentoxide							
USSR, Yugoslavia	*0.05*	0.0085	24	*0.15*	0.026	30	—
Phthalic anhydride							
Bulgaria	*0.1*	0.015	24	*0.2*	0.03	30	—
East Germany	*0.03*	0.005	24	*0.1*	0.015	30	1
USSR	*0.1*	0.015	24	*0.1*	0.015	30	2, 14
Yugoslavia	*0.2*	0.03	24	*0.4*	0.06	30	—
Propane-2-ol							
USSR	*0.6*	—	24	*0.6*	—	30	—
Propanol							
Bulgaria	—	—	—	*0.3*	0.12	30	—
East Germany	*0.3*	0.12	24	*1.0*	0.36	30	1
USSR, Yugoslavia	*0.3*	0.12	24	*0.3*	0.12	30	—
West Germany (VDI 2306)	*50.0*	*20.0*	½	*150.0*	*60.0*	30	3, 4
Propyl–isobenzene hydroxide							
Yugoslavia	*0.007*	—	24	*0.007*	—	30	—

Propylene							
Bulgaria, USSR	*3.0*	1.5	24	*3.0*	1.5	30	2
East Germany	*2.0*	1.0	24	*3.0*	1.5	30	1
Pyridine							
Bulgaria, USSR, Yugoslavia	*0.08*	0.023	24	*0.08*	0.023	30	—
East Germany	*0.03*	0.009	24	*0.08*	0.023	30	1
Romania	*0.05*	0.014	24	*0.15*	0.04	30	—
West Germany (VDI 2306)	*0.7*	*0.2*	½	*2.1*	*0.6*	30	3, 4
Silica							
Italy	*0.02*	—	24	*0.1*	—	120	18
Soot							
Bulgaria, Czechoslovakia, East Germany, Romania, USSR	*0.05*	—	24	*0.15*	—	30	1
Hungary	*0.1*	—	24	—	—	—	—
	0.05	—	24	—	—	—	5
Israel	*0.1*	—	24	*0.3*	—	30	6
Styrene							
Bulgaria, Hungary, USSR, Yugoslavia	*0.003*	0.0007	24	*0.003*	0.0007	30	5
East Germany	*0.003*	0.0007	24	*0.01*	0.0023	30	1
Hungary	*20.0*	4.6	24	*50.0*	11.7	30	—
West Germany (VDI 2306)	*20.0*	*4.6*	½	*65.0*	*15.16*	30	3, 4
Sulfur dioxide							
Argentina	*0.67*	*0.03*	30 days	—	—	—	—
Belgium, Spain	*0.15*	*0.06*	1 yr	—	—	—	20, 76
Bulgaria, USSR	*0.05*	0.02	24	*0.5*	0.2	30	2
Canada—Acceptable level	*0.06*	*0.02*	1 yr	—	—	—	15, 17
Acceptable level	*0.3*	*0.11*	24	*0.9*	*0.34*	60	15, 17
Desirable level	*0.03*	*0.01*	1 yr	—	—	—	15, 16
Desirable level	*0.15*	*0.06*	24	*0.45*	*0.17*	60	15, 16
Columbia	*0.07*	0.03	1 yr	—	—	—	70
Czechoslovakia, East Germany, Hungary, West Germany, Yugoslavia	*0.15*	0.06	24	*0.5*	0.2	30	1, 5

Table III (*Continued*)

Substance and country	*Long-term standard*[a]			*Short-term standard*[a]			*Notes*[b]
	mg/m^3	*ppm*	*Averaging time (hours)*	*mg/m*^{3}	*ppm*	*Averaging time (minutes)*	
Finland	*0.25*	0.1	24	*0.72*	0.28	30	71
	0.18	0.07	1 yr	—	—	—	71
France	*1.0*	0.38	24	—	—	—	—
Hungary	*0.5*	0.2	24	*1.0*	0.38	30	—
Israel	0.26	*0.1*	24	0.75	*0.3*	30	—
Italy	0.38	*0.15*	24	0.75	*0.3*	30	18
Japan	0.1	*0.04*	24	0.26	*0.1*	60	19
Netherlands	*0.075*	0.03	24	—	—	—	52, 53
	0.25	0.1	24	—	—	—	52, 54
	0.35	0.13	24	—	—	—	56, 57
	0.125	0.05	24	—	—	—	56, 58
	0.275	0.1	24	—	—	—	56, 59
Netherland, Turkey	*0.15*	0.06	24	—	—	—	55, 56, 63, 64
Poland	*0.35*	0.13	24	*0.9*	0.35	20	9
	0.075	0.03	24	*0.25*	0.1	20	10
Romania	*0.25*	0.1	24	*0.75*	0.3	20	—
Spain	*0.4*	0.15	24	*0.8*	0.3	30	20
	0.256	0.1	30 days	—	—	—	20
Sweden	*0.25*	*0.1*	24	*0.625*	*0.25*	30	60
	0.125	*0.05*	30 days	—	—	—	60
Switzerland	*0.75*	0.3	24	*1.25*	0.5	30	62
Switzerland, West Germany (VDI 2108)	*0.5*	0.2	24	*0.75*	0.3	30	3, 40, 61
Turkey	*0.30*	0.12	24	—	—	—	63, 65
United States	*0.08*	0.03	1 yr	—	—	—	66

	0.365	0.14	24	—	—	—	37, 66
	1.3	0.5	3	—	—	—	37, 67
West Germany	*0.4*	0.15	½	*0.75*	0.3	30	—
Sulfuric acid							
Bulgaria, Romania, USSR, Yugoslavia	*0.1*	—	24	*0.3*	—	30	2, 68
Bulgaria, Yugoslavia	*0.006*	—	24	*0.006*	—	30	38
Czechoslovakia	—	—	—	*0.01*	—	30	38
East Germany	*0.02*	—	24	*0.05*	—	30	1
Hungary, Israel	*0.1*	—	24	*0.3*	—	30	6
Poland	*0.1*	—	24	*0.3*	—	20	9
	0.05	—	24	*0.15*	—	20	10
USSR	*0.002*	—	24	*0.006*	—	30	2, 38
Suspended particulates							
Argentina	*0.15*	—	30 days	—	—	—	—
Bulgaria, Czechoslovakia, East Germany, Finland, Romania, USSR	*0.15*	—	24	*0.5*	—	30	1, 71, 73
Canada—Acceptable level	*0.07*	—	1 yr	—	—	—	15, 17, 69
Acceptable level	*0.12*	—	24	—	—	—	15, 17
Canada (Desirable level), United States,	*0.06*	—	1 yr	—	—	—	15, 16, 67, 69
Colombia	*0.1*	—	24	—	—	—	70
Hungary	*0.2*	—	24	—	—	—	15, 17
Hungary, Turkey, United States	*0.15*	—	24	—	—	—	5, 63, 64, 67
Israel	*0.2*	—	24	—	—	—	—
Israel, United States	*0.075*	—	1 yr	—	—	—	66
Italy	*0.3*	—	24	*0.75*	—	120	18
Japan	*0.1*	—	24	*0.2*	—	60	19
Poland	*0.2*	—	24	*0.6*	—	20	9, 72
	0.075	—	24	*0.2*	—	20	10, 72
Spain	*0.13*	—	1 yr	—	—	—	20
	0.202	—	30 days	—	—	—	20
	0.3	—	24	*0.6*	—	30	20

Table III (*Continued*)

Substance and country	Long-term standard[a]			Short-term standard[a]				Notes[b]
	mg/m³	ppm	Averaging time (hours)	mg/m³	ppm	Averaging time (minutes)		
Sweden	—	—	—	0.1	—	60	73	
United States	0.26	—	24	—	—	—	66	
West Germany	—	—	—	0.48	—	30	73	
	0.1	—	½	0.3	—	30		—
Tar								
Israel	1.0	—	24	3.0	—	30	6	
Tetrachloromethane								
Bulgaria	—	—	—	4.0	—	30		—
Tetrahydrofuran								
East Germany	0.2	0.07	24	0.6	0.21	30	1	
USSR	0.2	0.07	24	0.2	0.07	30		—
West Germany (VDI 2306)	30.0	10.0	½	90.0	30.0	30	3, 4	
Thiophene								
Bulgaria, USSR, Yugoslavia	—	—	–	0.6	0.17	30		—
East Germany	0.2	0.06	24	0.6	0.17	30	1	
Toluene								
Bulgaria, East Germany, Hungary, USSR, Yugoslavia	0.6	0.16	24	0.6	0.16	30	1, 5	
Hungary	20.0	5.3	24	50.0	13.3	30		—
West Germany (VDI 2306)	20.0	5.0	½	60.0	15.0	30	3, 4	
Toluene diisocyanate								
Bulgaria, East Germany, Romania, USSR, Yugoslavia	0.02	0.0029	24	0.05	0.0071	30	1	
West Germany (VDI 2306)	0.009	0.001	½	0.021	0.003	30	3, 4	
Tributyl phosphate								
Bulgaria	—	—	—	0.01	—	30	1	—
USSR	0.01	—	24	0.01	—	30		—

Trichlorfon							
USSR	0.02	—	24	0.04	—	30	74
Trichloroethane							
West Germany (VDI 2306)	20.0	5.0	½	90.0	15.0	30	3, 4
Trichloroethylene							
Bulgaria, East Germany, USSR, Yugoslavia	1.0	0.18	24	4.0	0.74	30	1
Hungary	30.0	5.6	24	50.0	9.3	30	—
	0.2	0.04	24	0.2	0.04	30	5
West Germany (VDI 2306)	30.0	5.0	½	90.0	15.0	30	3, 4
Triethylamine							
East Germany	0.05	0.012	24	0.14	0.035	30	1
USSR	0.14	0.035	24	0.14	0.035	30	—
West Germany (VDI 2306)	0.04	0.01	½	0.12	0.03	30	3, 4
2,4,6-Trimethylaniline							
East Germany	0.003	—	24	0.01	—	30	1, 75
Turpentine							
West Germany (VDI 2306)	25.0	5.0	½	75.0	15.0	30	3, 4, 41
n-Valeric acid							
Bulgaria, East Germany, USSR, Yugoslavia	0.01	0.003	24	0.03	0.008	30	1
Vanadium pentoxide							
Bulgaria, East Germany, USSR	0.002	—	24	—	—	—	—
Czechoslovakia, Yugoslavia	0.003	—	24	—	—	—	—
Vinyl acetate							
Bulgaria, Czechoslovakia, Yugoslavia	0.2	0.006	24	0.2	0.006	30	—
East Germany	0.15	0.0045	24	0.4	0.012	30	1
Israel	4.0	1.0	24	12.0	3.0	30	6
USSR	0.15	0.0045	24	0.15	0.0045	30	—
West Germany (VDI 2306)	20.0	5.0	½	60.0	15.0	30	3, 4
Xylene							
Bulgaria, Hungary, USSR, Yugoslavia	0.2	0.05	24	0.2	0.05	30	5
East Germany	0.2	0.05	24	0.6	0.14	30	1
Hungary	20.0	4.6	24	50.0	11.5	30	—
West Germany (VDI 2306)	20.0	5.0	½	60.0	15.0	30	3, 4

[a] Italicized concentrations represent the standards listed in promulgated regulations; those not italicized are approximate conversions.

[b] NOTES:

1. Short-term averaging time 10–30 minutes for East Germany.
2. In USSR:

   A. If several substances with synergistic toxic properties are present in the air, the maximum permissible concentration (MPC) of the mixture is calculated from the formula $X = (A/M_1) + (B/M_2) + (C/M_3)$, where X is the (relative) MPC; A, B, C are the concentrations of the substances in the mixture, and M_1, M_2, M_3, are their respective maximum permissible concentrations.

   B. If this formula is applied to the following two, three, or four component systems, the value X should not exceed 1.0 for (a) acetone and phenol; (b) sulfur dioxide and phenol; (c) sulfur dioxide and nitrogen dioxide; (d) sulfur dioxide and hydrogen fluoride; (e) sulfur dioxide and sulfuric acid aerosol; (f) hydrogen sulfide and "dinyl"; (g) isopropyl benzene and isopropyl benzene hydroperoxide; (h) furfural, methanol, and ethanol; (i) strong mineral acids (sulfuric, hydrochloric and nitric—concentrations expressed as H^+); and (j) ethylene, propylene, butylene and amylene. The value X should not exceed 1.3 for acetic acid and acetic anhydride, and should not exceed 1.5 for (a) acetone and acetophenone, (b) benzene and acetophenene, (c) phenol, and acetophenone.

   C. If (a) hydrogen sulfide and carbon disulfide; (b) carbon monoxide and sulfur dioxide; (c) phthalic anhydride, maleic anhydride and α-naphthoquinone are present in the mixture, the MPC values of individual substances should not be exceeded.

   D. If *p*-chlorophenyl isocyanate is present together with *m*-chlorophenyl isocyanate, the MPC is determined by the presence of the more toxic substance, i.e., of *p*-chlorophenyl isocyanate.
3. VDI = Verein Deutscher Ingenieure—Kommission Reinhaltung der Luft, VDI-Verlag GmbH, Duesseldorf, Federal Republic of Germany.
4. Short term = Short-term exposure limit, not to be exceeded more than once in any 4 hours in West Germany.
5. Highly protected and protected areas in Hungary.
6. Tentative standards in Israel.
7. Also the inorganic compounds, except arsine, AsH_3.
8. As AsH_3.
9. Protected areas in Poland.
10. Specially protected areas in Poland.
11. As C.
12. Listed in Regulations as <10% aromatics.
13. Also benzine from oil shale.
14. Fumes and aerosols.
15. National Air Quality Objectives in Canada.
16. Desirable level in Canada.
17. Maximum acceptable level in Canada.
18. Once in 8 hours in Italy.
19. Average of hourly means for 24 hour value in Japan.
20. Proposed standard in Spain.
21. Primary and secondary ambient air quality standards in the United States.

22. 0.6 mg/m^3 once as a 30 minutes average in a time period of 8 hours in West Germany.
23. Also called trichloromethane.
24. Also called aureomycin.
25. For mixing with animal feed.
26. As Cr_6.
27. As CrO_3.
28. Diphenyl plus its oxide.
29. HF, SiF_4.
30. As F, gaseous compounds.
31. Gaseous plus salt combined.
32. As HF.
33. NaF, Na_2SiF_6.
34. AlF_3, Na_3AlF_6, CaF_2.
35. Readily soluble inorganic fluoride.
36. As hexane, for hydrocarbons emitted by oil refineries.
37. Not to be exceeded more than once a year.
38. As H^+.
39. As HCl.
40. Short-term standard not to be exceeded more than once in 2 hours in West Germany.
41. Short-term standard = short-term exposure limit, not to be exceeded more than once in any 8 hours.
42. Lead and its compounds, except tetraethyllead.
43. As MnO_2.
44. 2-Amino-1,3,5-trimethylbenzene.
45. Also called Carbophos.
46. Also called Metaphos.
47. As HNO_3.
48. Diethylene dioxide.
49. As NO_2.
50. By KI.
51. As O_3.
52. For areas with low smoke level.
53. Percentile of the cumulative frequency distribution of consecutive 24 hour samples: 50%.
54. Percentile of the cumulative frequent distribution of consecutive 24 hour samples: 38%.
55. Soot level <0.03 mg/m^3, frequency 50% in Netherlands.
56. Interim limit value in Netherlands for areas designated by commission.
57. Soot level <0.09 mg/m^3, frequency 98%.
58. Soot level <0.04 mg/m^3, frequency 50%.
59. Soot level <0.125 mg/m^3, frequency 98%.
60. Guideline.
61. Summer—March 1–October 31 guideline in Switzerland.
62. Winter—November 1–February 28/29 guideline.
63. Recommended standard in Turkey.
64. Residential areas in Turkey.
65. Industrial areas.
66. Primary standard in United States.
67. Secondary standard in United States.
68. As H_2SO_4.
69. Annual geometric mean.
70. Reference level.
71. Not national legal norms, communal health councils can enforce them.
72. Particle size <20 μm.
73. Basis for stack height calculation.
74. Also called Chlorophos.
75. Also called Mesidine.
76. Protected area in Belgium.

pand in area in the future until they became contiguous, they would presumably then be merged into one larger air quality region. Since each air quality region would adopt its own air quality standards, they could differ among the several regions. However, there should be a floor below which no region's air quality standards should drop.

Because the residential, commercial, and industrial zones of an urban area are all surmounted by the same air mass, diverse air quality standards within the same jurisdiction or air quality region would seem to be very difficult to administer. One of the banes of air pollution control is the juxtaposition of a community with little or no control to one with stringent control standards. When this happens there is objection on the grounds that air pollution does not recognize jurisdictional boundaries. Yet the use within one jurisdiction of different standards for contiguous land-use zones creates the very thing found objectionable when it involves more than one jurisdiction.

The foregoing argument is most valid for relatively small homogeneous jurisdictions such as the city, the county, and the air quality region, but it is less valid when applied to a state or the nation. As has been done in New York, a state may wish to preserve the superior air quality presently existing in large areas of the state, e.g., the Adirondack and Catskill Mountain State Park areas, by setting for these regions more restrictive standards of air quality than elsewhere in the state. It may also choose, as New York State has, to have less restrictive standards for its major metropolitan areas, on the basis of allowing, as an interim measure, that which is feasible to be done now. Where this is done, there should be a procedure to escalate these standards after a predetermined number of years until the standards for such regions approach more closely the state's basic air quality standard.

In American zoning practice, it is relatively easy to keep a lower-class use out of a higher-class zone. However, it is much more difficult to enforce regulations that keep residences out of lower-class zones. Courts are more prone to uphold restraints upon industrial and commercial location than upon the right of an individual to live where he chooses.

C. Economic, Social, and Political Considerations in Setting Air Quality Standards

Setting air quality standards discommodes no one. It is only their implementation, by enforcing emission, fuel, or equipment restrictions to achieve the air quality standard, that does so. Where such restrictions cause economic, social, or political problems, the alternatives include: (a) immediate adoption of air quality goals as standards, but the time phasing of adoption of the restrictions to achieve them; (b) immediate adoption and implementation of restrictions to achieve air quality standards that fall short of air quality goals, but time phasing for later adoption and implementation of air quality standards more closely approaching air quality goals; and (c) immediate adoption of both air quality standards meeting air quality goals and restrictive limitations sufficient to achieve them, but time phasing of the application of these limitations. As a further alternative to time phasing with respect to pollutants, one can time phase with respect to classes of sources, fuels, equipment, etc. The technological feasibility of meeting such standards is a prime consideration in selection of the appropriate alternative.

Economic considerations are sometimes cited as the argument for allowing the air quality standard to be relaxed in cities or industrial areas. The economic consideration of primary concern is the overall economy of the community, region, or nation. However, many things that benefit our overall economy cause hardship to certain individuals, industries, or regions. Thus, lowering tariffs on certain imports may benefit our general economy while placing domestic producers of these products at a competitive disadvantage. The same is true of the application of air pollution limitations. Those who have to spend money to achieve a common standard are disadvantaged with respect to those who do not. However, such economic differences may be remedied by means other than the adoption of air pollution standards less stringent than required for the general welfare. These other means include subsidies and tax relief to ease the financial impact on a minority, as well as other means for spreading the cost to all the segments of the economy that are the ultimate beneficiaries. When the British create smokeless zones, they simultaneously provide a subsidy to cover part of the cost of the equipment changeover needed to achieve smokeless operation. The tax laws of several nations and states provide tax relief on expenditures for air pollution control.

There is a widespread delusion that stringent air pollution standards repel new industry, whereas relaxed standards attract new industry. This is a fallacy. Los Angeles County, California, the county with the most stringent air pollution regulations in the world, has experienced an unprecedented growth. If anything, the promise of a future of clean air attracts rather than repels both employers and employees.

D. General Approaches Used for Setting Air Quality Standards

Several quite different approaches have been considered for setting air quality standards (*17*). One such approach was to use another community's air as the standard. In this approach, community A would indicate, "We will be satisfied if our air quality is as good as that which now exists in community B." Knowledge of air quality in community B thus provides a standard for community A. A second approach was to use as the standard for air quality that air that existed at an earlier time and for which it was believed that adverse effects were either nonexistent or tolerable by the community. In this approach, the community says, "We will be satisfied if our air quality is as good as it was here in 1940." If there were objective data on air quality for the earlier data—which is usually not the case—this would provide a standard. A variant of this approach would be to compute from present-day data what the air quality most likely was at the earlier date. A third approach would be to use as the standard for air quality that air that exists in the community on certain days of good ventilation. In this approach, the community says, "We will be satisfied if our air quality every day is as good as it now is on certain days." Knowledge of air quality on these certain days thus provides the standard.

A fourth approach would be to consider health protection–control cost relationships. Pollution abatement to reduce or negate health risks requires financial expenditures and possibly even lifestyle changes. Figure 4 is a simplified schematic diagram that presents several of the principal factors in the decision-making process concerning air pollution standard setting. It indicates that the degree of health protection attained is a function of costs of pollution control mandated by standards. The minimum acceptable level of health protection is, at the very least, the level necessary to protect from death, and the immediate regulatory action program adopted for air pollution control should certainly protect from illness as well (*7*).

The degree of health protection to be selected above the minimum acceptable level from Figure 4 is a matter of political decision. The appropriate authorities must decide on the level of health protection desirable for their society. Increments of health protection above the minimum acceptable level are generally purchased at ever increasing increments in control costs. The zone in which increased health protection (benefit) is obtained at increasing control costs (the cross hatched area in Figure 4) can also be thought of as the region of social decision making. In deciding where to utilize limited resources, the total set of risks that man must bear needs to be considered and the greatest efforts should be expended to reduce the largest risks, be they environmental or other societal problems.

Each regulatory action is taken for society's benefit. However, each action also has a societal disbenefit because it will consume a certain quantity of the society's resources for its accomplishment. Obviously,

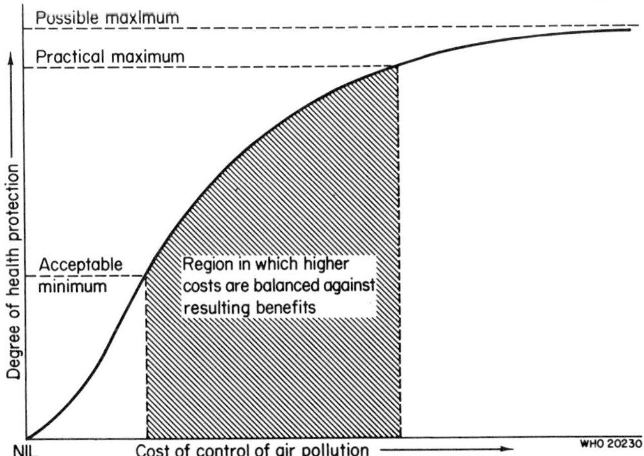

Figure 4. Relationship between degree of health protection and cost of control of air pollution (7).

there must be some give and take between the societal benefits and disbenefits. Decisions involving this process are difficult, because the groups to benefit usually are different from the groups that will receive the disbenefits, and thus, it is difficult to make this an equitable process.

Another important concept follows from further consideration of Figure 4. Infinite health protection is unattainable; thus, either society must decide what level of health risk is acceptable—a decision that will be reflected in the societal investment in environmental pollution control—or the decision will be made indirectly, since the societal investment in environmental pollution control will dictate the level of health protection to be achieved.

The process of standard setting requires policy decisions concerning the protection of public health at least cost to society. Sound quantitative exposure response data allows the policy maker to make clear-cut decisions that can be defended. When range estimates are broad due to inadequate health intelligence, one of two possible kinds of error may be made. The resulting standards could be too lenient and not protective enough or they could be too strict and excessively costly for the real benefit derived. Thus, society must pay a price for not generating adequate environmental health information. It must also be remembered that generation of health information itself can be costly and that in some instances it may be less costly to over-control than to develop the specific information required. Margins of safety built into standards are judgements required to bridge the gap between the inadequacy of health intelligence and the need to stop continued exposure to hazardous pollutant levels. Greater degrees of uncertainty generate larger safety margins. Unfortunately, control costs increase exponentially as exposure is limited more severely. The health benefit from the additional degree of control may be insignificant or actually negative if overly stringent controls cause health disbenefits, as could occur if scarce resources are diverted from health care or protection to unnecessary pollution abatement.

The function of a health intelligence system in the standard setting process can be graphically displayed in the form of marginal cost curves (Fig. 5). The true social cost of pollution is equal to the sum of the marginal cost of control and the health and welfare cost of exposure. Health costs of disease, and more especially of physiological dysfunction, cannot be readily estimated with current scientific knowledge.

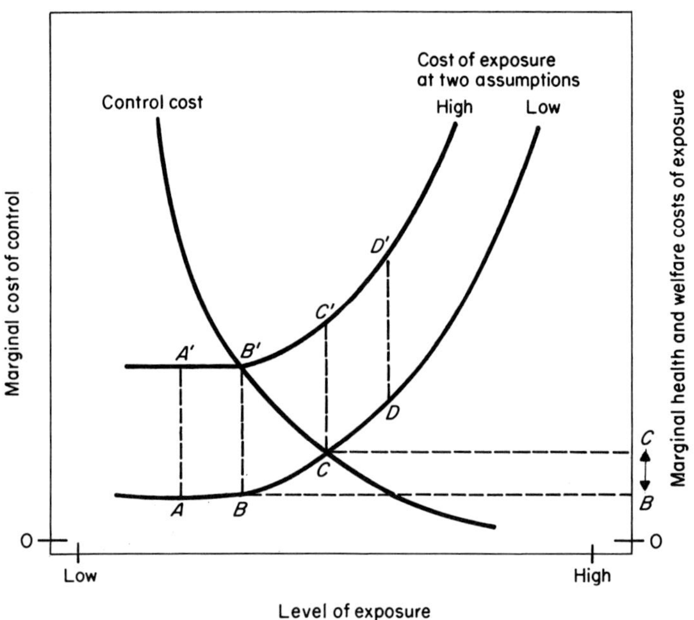

Figure 5. True social costs of pollutant exposure.

Possible exposure limit		
At low assumption	*At high assumption*	*Resulting standard*
A, *B*	*A′*	Excessively stringent
C	*B′*	Least cost
D	*C′*, *D′*	Too lenient

Health and welfare costs increase with higher exposures, while control costs increase as exposures are lowered. True social costs of pollution are minimal where the two marginal cost curves intersect. If society will tolerate some pollutant–associated excess health costs, curve AD may represent this social preference. On the other hand, if society places a very high premium on preventable disease or reducing risk of disease, curve $A'D'$ may well represent this attitude. On both curves, the points B or B' represent the lowest pollutant exposure associated with adverse health (or) welfare effects. Under the low health cost assumption (curve AD), society will accept the health costs (and associated adverse health effects) represented by the vertical distance between B and C. Under the high cost assumption, society will not allow any adverse health effect caused by exposure and will require control to point B', where high marginal cost of control is equated with derived health benefit.

Excessively stringent standards are shown at points A and B under the low cost assumption and at point A' under the high assumption, while too lenient standards are reflected in point D under the low assumption and C' and D' under the high assumption. The least cost standard, in terms of true social costs, is at point C under the low health cost assumption and at point B' under the high assumption. Least cost standards protective of health require adequate quantitative exposure–response information, knowledge of control costs, and range estimates of health cost of pollution effects.

Even range estimates of pollutant-related health costs are difficult to derive. Many direct and indirect health costs must be considered, including immediate and delayed health effects of short-term and long-term exposure; the contribution of pollutant exposure to the occurrence and severity of major public health problems, such as acute and chronic respiratory diseases, heart disease, cancer, and congenital defects; and adverse health effects of control strategies. While these estimates are formidable tasks, failure to make the estimates will prevent even the possibility of achieving least social cost standards, except entirely fortuitously. Annual reports of air pollution control costs as required by legislation in the United States clearly warrant better and more quantitative health effects data and concerted efforts to convert health effects to health costs.

E. Kinds of Air Quality Standards

1. Ambient Air Quality Standards

These are the legal limits placed on the concentration of air pollutants in a community where people and things are exposed. Enforcement of ambient air quality standards is difficult. To know if the actual concentration are in compliance requires an air monitoring program. However, if the monitoring program reveals that the ambient air quality standard is being exceeded, there is no direct action that can be taken to reduce the concentrations. The enforcement action to bring an area into compliance requires a reduction of the air pollution emissions.

Ambient air quality standards need to be distinguished from threshold limit values (TLV) for workroom atmospheres, e.g., those published by the American Conference of Governmental Industrial Hygienists (*12*). These latter values are the permissible exposure levels for healthy adult workers for 8 hours per day, 5 days per week. In contrast, air quality standards are permissible exposures of all living and nonliving things for 24-hours per day, 7 days per week. The two may not be used interchangeably and will not show a constant mathematical relationship to each other.

In the United States, the ambient air quality standard based on air quality criteria has been the legislatively mandated approach. Because of all the uncertainties in the effects research results and because of a desire to see the air pollution control program stimulated, the Clean Air Act restricts the USEPA Administrator to a precise set of steps to be taken to arrive at an air quality standard. The Act states that national primary ambient air quality standards are those which ". . . in the judgment of the Administrator, based on such criteria and allowing for an adequate margin of safety, are requisite to protect public health" [Section 109(b)]. Thus, air quality criteria reflect scientific knowledge, while primary ambient air quality standards involve a judgment as to how that knowledge must be used in regulatory action to protect public health. In addition, the legal definition of the primary ambient air quality standard does not permit consideration of the availability of control technology to achieve the standard or of risk benefit analysis.

The above definition of the national primary ambient air quality standard implies that the lowest level at which an effect occurs should be identified and, after allowing an adequate margin of safety, the standard established. The only flexibility allowed in this standard is the determination of what an effect is and what magnitude of safety factor is adequate.

The United States Environmental Protection Agency (USEPA) has compiled all ambient air quality standards from various governmental jurisdictions within the United States (*13*). Table IV lists specific substances for which there are state air quality standards.

In 1951, the Union of Soviet Socialist Republics (USSR) established maximal permissible concentrations for ten pollutants and thus launched

Table IV Specific Substances for Which There Are State Air Quality Standards in the United States

Substance	*States having air quality standards*
Asbestos	New Mexico
Beryllium	Idaho, Montana, New Mexico, New York, Pennsylvania
Calcium oxide	Oregon
Carbon monoxide	All
Fluorides	Kentucky, New Hampshire, Pennsylvania, South Carolina, Tennessee, Washington
Heavy metals	New Mexico
Hydrocarbons	All
Hydrogen sulfide	California, Kentucky, Minnesota, Montana, New Mexico, New York, North Dakota, Pennsylvania
Lead	California, Montana, Pennsylvania
Nitrogen dioxide	All
Oxidants	All
Sulfur dioxide	All
Sulfuric acid mist	Idaho, Louisiana, North Dakota
Suspended particulate matter	All
Sulfate	Montana, North Dakota, Pennsylvania

their air pollution program on an ambient air quality approach. Since then, the number of such standards established has exceeded a hundred (*14*). These standards are included in Table III. According to the approach used in the Soviet Union, the air quality standards are based on hygienic, sanitary, and technological considerations, which are defined as follows.

> Hygienic standards are requirements that meet the interests of man as a biological species. They are aimed at providing environmental conditions completely favorable for man from the physiological point of view. Not only do they serve to protect him, they are to ensure his biological comfort or the hygienic optimum.
>
> Sanitary standards are hygienic standards corrected for factors in the current situation: technical and economic feasibility. These standards represent a concession to practical possibilities and are therefore of a temporary nature. They may differ in different countries and at different times in the development of an economy.
>
> Technological standards, the third group, are aimed at preventing, insofar as this is justified by economic considerations, the discharge of harmful substances into the atmosphere (*14*).

The hygienic standards are based on tests that measure human odor perception, human reflex response, and effects of animal chronic exposure. More detail regarding the Soviet approach to establishing maximal permissible concentrations is presented by Izmerov (*14*, *15*).

2. Other Air Quality Standards

There are other types of air quality standards:

a. Quasi-emissions standards, called "point of impingement standards," which are the limit on specific pollutants in the ambient air at ground level required by national, state, or local regulations to be used in diffusion computations to determine limits of emissions from specific sources.

b. Standards for fluoride are based upon the fluoride content of vegetation, especially forage.

c. Standards for particulate matter deposited by sedimentation, washout, dust fall, or soot fall onto or into exposed receptacles.

d. Soiling index, which is the measurement of transmitted or reflected light through or from a spot of particulate matter collected on a filter for a prescribed period of time.

e. Sulfation rate is a measure of the conversion of lead oxide to lead sulfate by exposure of candles or plates covered with a paste of this material.

f. Odor standards.

g. Visibility standards.

h. Episode action standards.

Table V shows that national point of impingement standards have been set in only three countries, namely, France, Italy, and the Phillipines (*13*). Specific substances for which there are state point of impingement standards in the United States are given in Table VI. National deposited particulate standards have been established in seven countries and are presented in Table VII (*13*). The only national standards set on the basis of a soiling index are in Israel. The soiling index is expressed as the number of coefficients of haze (COH) per "30 m" of air and is measured by the transmittance of light through the deposit collected on a filter. The 24 hour standard is an average value of 1.0 COH. The 2 hour standard is 2.0 COH.

Since weather conditions are natural occurrences and are not under man's control, effects resulting from episode conditions can be tolerated that would not be tolerated normally. Significant endangerment to health is the upper bound of the pollutant concentration that can be tolerated. This is the concentration where a significant health effect can be identified clearly among an exposed population. The strategy for dealing with such episodes is to designate several levels of pollutants, such as alert, warning, emergency and significant harm, to trigger (initiate) various degrees of regulatory action designed to keep the actual air quality con-

Table V National Point of Impingement at Ground Level Standards

Substance and country	Standard: Original units	Standard: Averaging time (hours)	Notes[a]
Ammonia			
Philippines	20.0 ppm	1	—
	10.0 ppm	24	—
Carbon Monoxide			
France	50.0 ppm	8	1
	100.0 ppm	Peak	1
Italy	57.24 mg/m^3	½	2
	22.89 mg/m^3	8	—
Philippines	100.0 ppm	1	—
	30.0 ppm	24	—
Chlorine			
Italy	0.58 mg/m^3	½	2
Philippines	1.0 ppm	1	—
	0.2 ppm	24	—
Ethylene			
Philippines	5.0 ppm	½	—
	0.2 ppm	24	—
Fluorides			
Italy	4.5 μg/m^3	½	2
	0.02 mg/m^3	24	—
Hydrogen chloride			
Italy	0.30 mg/m^3	½	2
	0.05 mg/m^3	24	—
Philippines	1.0 ppm	1	—
	0.5 ppm	24	—
Hydrogen sulfide			
Italy	0.10 mg/m^3	½	2
	0.04 mg/m^3	24	—
Philippines	0.2 ppm	1	—
	0.1 ppm	24	—
Lead			
Italy	0.05 mg/m^3	½	2
	0.01 mg/m^3	8	—
Nitrogen dioxide			
Italy	0.56 mg/m^3	½	2
	0.19 mg/m^3	24	—
Nitrogen Oxides			
France	200.0 mg/m^3	24	3
Philippines	2.0 ppm	1	—
	0.3 ppm	24	—
Organic substances			
Italy	80.0 ppm	½	4
	40.0 ppm	24	—

Table V (*Continued*)

Substance and country	*Standard* Original units	*Averaging time (hours)*	*Notes*[a]
Ozone			
Philippines	0.3 ppm	1	—
	0.1 ppm	24	—
Silica (free)			
Italy	0.10 mg/m^3	2	2, 5
	0.02 mg/m^3	24	5
Sulfur dioxide			
France	250.0 μg/m^3	24	3
Italy	0.79 mg/m^3	$\frac{1}{2}$	2
	0.39 mg/m^3	24	—
Philippines	2.0 ppm	1	—
	0.3 ppm	24	—
Suspended particulate			
France	150.0 μg/m^3	24	3
Italy	0.75 mg/m^3	2	2
	0.30 mg/m^3	24	—
Philippines	900.0 μg/m^3	1	6
	300.0 μg/m^3	24	6
	600.0 μg/m^3	1	7
	200.0 μg/m^3	24	7

[a] NOTES: (1) Underground parking lots. (2) Peak concentration allowable once in 8 hours. (3) Basis for stack height calculations. (4) As hexane—derived from refineries. (5) As SiO_2. (6) Industrial area. (7) Residential area.

Table VI Specific Substances for Which There Are State Point of Impingement at Ground Level Standards in the United States

Substance	*States*
Beryllium	Texas
Coefficient of haze	Missouri
Dustfall	Arkansas, Mississippi
Fluorides	Montana, Texas, Washington
Hydrogen sulfide	Missouri, Montana
Sulfur dioxide	Colorado, Missouri, Oregon, Oklahoma, Texas, Washington
Sulfuric acid mist	Arkansas, Missouri, Texas
Suspended particulate matter	Arkansas, Indiana, Kansas, Missouri, Pennsylvania, Texas

Table VII National Deposited Particulate Matter Standards

Countries	Original units
Argentina	1.0 mg/cm²/month
Colombia[a]	0.5 mg/cm²/month
Finland	
Lead	10.0 mg/m²/month
Chromium	10.0 mg/m²/month
Vanadium	10.0 mg/m²/month
Total	10.0 g/m²/month
Hungary, Romania	200.0 tons/km²/year
Hungary[b]	150.0 tons/km²/year
Poland[b]	250.0 tons/km²/year
Poland[c]	40.0 tons/km²/year
Poland[c]	6.5 tons/km²/month
Spain[d]	200.0 mg/m²/day
West Germany[e]	0.35 gm/m²/day
West Germany[f]	0.65 gm/m²/day

[a] Set by PAHO and adopted by Colombia.
[b] Protected Areas.
[c] Specially protected areas.
[d] Proposed standard.
[e] Yearly average of the 12 monthly averages.
[f] Monthly average.

centration from reaching the significant harm level. Both the actual concentration established for each level of episode action and the actual pollution emissions restrictions imposed are dependent on the set of pollution sources, the topographic contour of the land and the local meterological conditions. In the United States such episode action standards are required as part of the implementation plan to achieve air pollution control. Table VIII lists the National Emergency Procedure Concentrations that have been established in three countries, namely, Argentina, Israel, and Japan (*13*) (See also Chapter 3, Vol. V, pp. 72–74.)

F. Expression of Air Quality Standards

Air quality standards perforce must be expressed in the same terms as are used to categorize air quality data. Air quality data, its measurement, and the manner in its presentation are discussed in detail in other sections of this treatise. At the risk of some repetition of this material, Figure 6 is included here to show the relationship between atmospheric concentration of a pollutant, the averaging time, and the frequency of

Table VIII National Emergency Procedure Concentration Levels[b]

Country	*Sulfur dioxide (ppm)*	*Suspended particulate (coefficient of haze)*	*Carbon monoxide (ppm)*	*Nitrogen dioxide (ppm)*	*Oxidants (ppm)*
		Alert Levels			
Argentina	1.0 (1 hr) 0.3 (8 hr)	—	15.0 (8 hr) 100.0 (1 hr)	0.6 (1 hr) 0.15 (24 hr)	0.15 (1 hr)
Israel[a]	3.5 (24 hr) 1.5 (6 hr) 2.0 (6 hr)	10.0 (24 hr) 2.5 (6 hr)	—	—	—
Japan	0.2 (3 hr) 0.3 (2 hr) 0.5 (1 hr) 0.15 (48 hr)	2.0 mg/m^3 (2 hr)	30.0 (1 hr)	0.5 (1 hr)	0.14 (1 hr)
		Alarm Levels			
Argentina	5.0 (1 hr)	—	30.0 (8 hr) 120.0 (1 hr)	1.2 (1 hr) 0.3 (24 hr)	0.25 (1 hr)
Israel[a]	Operational Status A				
	5.0 (24 hr) 2.0 (6 hr) 3.0 (6 hr)	10.0 (24 hr) 2.5 (6 hr)	—	—	—
	Operational Status B				
	7.5 (24 hr) 3.0 (6 hr) 4.5 (6 hr)	10.0 (24 hr) 2.5 (6 hr)			
		Emergency Levels			
Argentina	10.0 (1 hr)	—	50.0 (8 hr) 150.0 (1 hr)	0.4 (24 hr)	0.4 (1 hr)
Israel[a]	12.5 (24 hr) 7.5 (6 hr)	20.0 (24 hr)	—	—	—
Japan	0.5 (3 hr) 0.7 (2 hr)	3.0 mg/m^3 (3 hr)	50.0 (1 hr)	1.0 (1 hr)	0.5 (1 hr)

[a] If stagnation period is forecast for an additional 12 hours.
[b] For United States levels see Tables I and II, Chapter 3, Vol. V.

occurrence of each concentration-averaging time pair for a specific pollutant at a specific measuring site (*16*).

If there is only one source of the pollutant being measured, its plume will intercept the measuring site for only one wind direction; whereas if there are many sources of the pollutant uniformly distributed around the site, the measuring site will be subjected to the pollutant for all wind directions. Similarly, for two sites with the same source–site relationship,

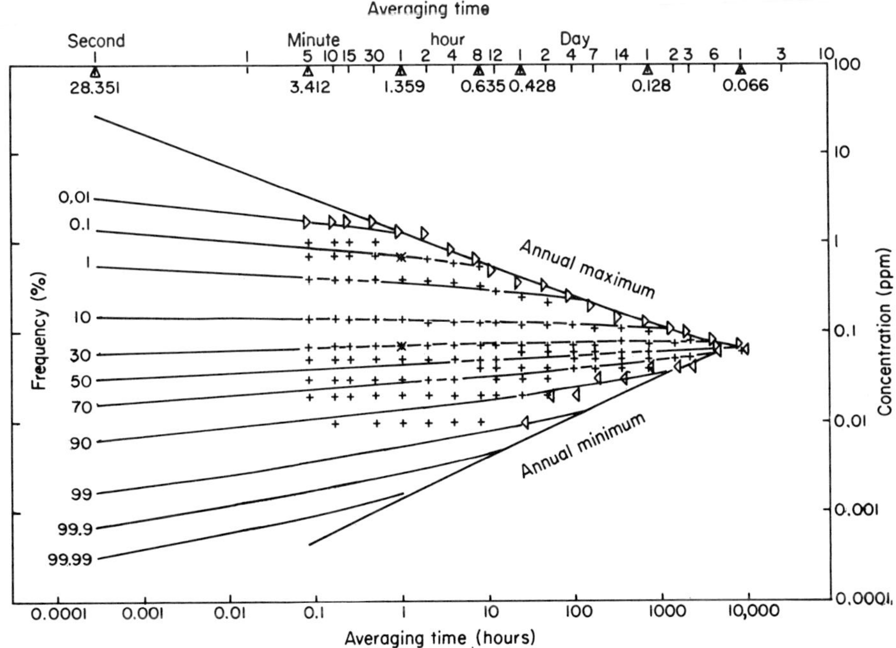

Figure 6. Arrowhead chart for concentration versus averaging time and frequency for nitrogen oxides in Washington, D.C.—December 1, 1961–December 1, 1964. Expected annual maximum concentration in ppm. The geometric mean for 1 hour (average) is 0.044 ppm. The standard geometric deviation is 2.46. Seventy percent of the hours have data available.

the one with high winds and much vertical mixing will be subject to a different pollutant concentration distribution than the one with low wind velocity and little vertical mixing. These factors, plus variations in source strength, result in differences in slope and ordinate of the lines on charts of the Figure 6 type for different pollutants and different sites.

If a community wished its present air quality to be its air quality standard, a chart such as Figure 6 for each pollutant could be its standard for the pollutant, since it describes that air quality. Alternatively, it could use one line from the figure, such as the maximum line or the 1% line, as its standard. If the community wished to set its standard at concentrations lower than its present air quality, its new air quality standard could be represented by shifting the ordinate scale of a Figure 6 type representation of its present air quality. This is not identical to, but has the same effect as, the drawing of a line parallel to and below the ambient air concentration of a pollutant frequency of occurrence plot on log probability coordinates of data representing only one averaging time, which has been used in Nashville, Tennessee, St. Louis, Missouri, and other communities as the basis for proposed air quality standards.

Any point on a figure similar to Figure 6, defined in terms of all three parameters—concentration, averaging time, and frequency of occurrence—may be used to describe the atmosphere and hence may be used as an air quality standard. An air quality standard that does not include frequency of occurrence—in other words, that states a concentration-averaging time pair that shall not be exceeded—defines the point on the "maximum" line for the specified averaging time. An air quality standard that includes neither averaging time nor frequency of occurrence—in other words, that states a concentration that shall not be exceeded—defines the point on the "maximum" line for an averaging corresponding to that of the device or method specified to measure the concentration. If no device or method is specified, it must be assumed to refer to the averaging time of the measuring device or method most likely intended to be employed.

It should be recognized that when a standard states "not to be exceeded more than A times in any period of B hours" and when the averaging time C in minutes is either stated or assumed, it is the same as stating not more than CA minutes in $60B$ minutes, or less than $[100(60B - CA)]/60B$ percent. For example, a standard of not more than once in 8 hours (30-minute averaging time) computes as 94%. It should be noted that a concentration that is not allowed 94% of the time is allowed 6% of the time and would therefore be the point on the 6 percentile line at 30-minute averaging time.

It should be noted that the arithmetic mean concentration of all the data is the same for all averaging times less than the span of the data record. The geometric mean (50% value on Figure 6) of all the data is essentially the same for all averaging times. Therefore, standards stated in terms of annual mean—either arithmetic or geometric—require no further definition and yet specify an atmosphere with just as much specificity as other forms of definition.

If for the same pollutant and jurisdiction, the several air quality standard values in a sequence such as

0.02 ppm, annual average

0.10 ppm, 24-hour average, not to be exceeded over 1% of the days in any 3-month period

0.20 ppm, not to be exceeded more than 1 hour in any 4 consecutive days

0.50 ppm, not to be exceeded for more than 5 minutes in any 8-hour period

are derived from different concentration–effect considerations, it may turn out that they are mutually inconsistent, i.e., that the values will not all fall on a figure of the type of Figure 6 for any one monitoring site in the jurisdiction. However, this does not invalidate the use of such a sequence to set limits on all monitoring sites in the jurisdiction. Consideration should be given to having the values in the sequence as consistent with the concentration–averaging-time–frequency-of-occurrence relationships in the real atmosphere of the jurisdiction as is possible without violating the air quality criteria on which the standards were based.

In the foregoing, 50% occurrence is an annual average, 2% is essentially any 1 week per year, and the 0.3% occurrence any 1 day per year. Intermediate values are determined by two straight lines connecting these three points plotted on log probability coordinates.

REFERENCES

1. Message from the President of the United States transmitting the Report of the Department of Health, Education, and Welfare and the Environmental Protection Agency on the Health Effects of Environmental Pollution, pursuant to Air Quality Act of 1967 P.L. 90-148 (1967), the Clean Air Amendments of 1970—P.L. 91-604 (1970) and the Clean Air Act of 1970.
2. B. MacMahon, T. F. Pugh, and J. Ipsen, "Epidemiologic Methods," p. 18. Little, Brown, Boston, Massachusetts, 1960.
3. A. B. Hill, *Proc. Roy. Soc. Med.* **58,** 295 (1965).
4. Clean Air Act (42 U.S.C. 1857 et seq.) includes the Clean Air Act of 1963 (P.L. 88-206), and amendments made by the Motor Vehicle Air Pollution Act—P.L. 89-272 (1965), the Clean Air Act Amendments of 1966—P.L. 89-675 (1966), the Air Quality Act of 1967—P.L. 90-148 (1967), the Clean Air Amendments of 1970—P.L. 91-604 (1970), the Clean Air Act of 1977, and Clean Air Amendments of 1977 (P.L. 95-95, 1977).
5. World Health Organization Environmental Health Criteria Programme, "Report of a Meeting on Environmental Health Criteria and Standards, Geneva, Switzerland, 20–24 November, 1972," EP 73/1," World Health Organization, Geneva, Switzerland, 1973.
6. A. C. Stern, and Y. Horie, "Incorporation of Population Exposure Concepts in Air Quality Criteria Documents (Proceedings of the International Symposium of Pollution). Published by the Commission on the European Communities, Vol. IV, pp. 2143–2151. General Directorate Scientific and Technical Information Management, Luxembourg, 1975.
7. World Health Organization, "Air Quality Criteria and Guide for Urban Air Pollutants," Tech. Rep. Ser. No. 506, World Health Organization, Geneva, Switzerland, 1972.
8. U.S. Department of Health, Education, and Welfare, Air Quality Criteria. Particulate Matter (AP-49), Sulfur Oxide (AP-50) (1969); Carbon Monoxide (AP-62), Photochemical Oxidant (AP 63), Hydrocarbons (AP 64) (1970). Public

Health Service, Washington, D.C. U.S. Environmental Protection Agency, Nitrogen Oxides (AP-84) Research Triangle Park, North Carolina, (1971).

9. North Atlantic Treaty Organization/Committee on Challenges to Modern Society, Brussels, Belgium. Air Quality Criteria, Sulfur Oxide (N-7), Particulate Matter (N-8), Carbon Monoxide (N-10), Nitrogen. Oxides (N-15), Photochemical Oxidants and Related Hydrocarbons (N-29). U.S. Environmental Protection Agency, Research Triangle Park, North Carolina.

10. Division of Medical Sciences, "Biological Effects of Atmospheric Pollutants," Fluorides ISBN-0-309-01922-2 (1971); Asbestos ISBN-0-309-01927-3, Lead ISBN-0-309-01941-9, and Particulate Organic Matter ISBN-0-309-02027-1 (1972); "Medical and Biological Effects of Environmental Pollutants," Manganese ISBN-0-309-02143-X (1973), Chromium ISBN-0-309-02217-7, Vanadium ISBN-0-309-02218-5, and Nickel ISBN-0-309-02314-9 (1974). National Research Council—National Academy of Sciences, Washington, D.C., Vapor phase Organic Pollutants—Volatile Hydrocarbons and Oxidation Products (NTIS-PB 249 357) (1975, Arsenic (NTIS—PB 262 167) 1976, Chlorine and Hydrogen (NTIS—PB 253 96) Ozone and Other Photochemical Oxidants (PB 260 570–571), Selenium (NTIS—PB 251 318 (1976).

11. Advisory Center on Toxicology, "Short-term Limit Reports," Oxides of Nitrogen PB-199 903, Basis for Guides PB-199 904, Hydrogen Chloride PB-203 464, and Hydrogen Fluoride PB-203 465 (1971); Ammonia PB-244 336 (1972); Carbon Monoxide PB-244 338 and Chlorine PB-244 339 (1973); and Hydrazine, Monomethyl Hydrazine, and 1,1-dimethylhydrazine PB-244 337 (1974). National Research Council—National Academy of Sciences, Washington, D.C.

11a. K. H. Rasmussen, M. Taheri, and R. L. Kabel, *Water Air Soil Pollut.* **4**, 33–64 (1975).

12. American Conference of Governmental Industrial Hygienists, "Threshold Limit Values for Chemical Substances and Physical Agents in the Workroom Environment with Intended Changes for 1976," p. 94. ACGIH, Cincinnati, Ohio, 1976.

13. W. Martin and A. C. Stern, The World's Air Quality Management Standards. 2 Vols. Vol. 1: "The World including U.S. Federal Standards." Misc. Series—EPA—650/9-75-001a, 383 pp. Vol. 2: U.S. Misc. Series—EPA-650/9-75-001b, 373 pp. U.S. Environmental Proetction Agency, Office of Research and Development, Washington, D.C. (1974).

14. N. F. Izmerov, *World Health Organ., Chron.* **28**, 255 (1974).

15. N. F. Izmerov, Control of Air Pollution in the U.S.S.R., Public Health Paper #54, World Health Organization, Geneva, Switzerland, 157 pp. (1973).

16. Continuous Air Monitoring Program in Philadelphia, 1962–63–64–65. Public Health Service, U.S. Department of Health, Education and Welfare, Cincinnati, Ohio, 1968.

17. A. C. Stern, *J. Occup. Med.* **18**(5), 297–303 (1976).

12

Emission Standards for Mobile Sources

John A. Maga

I. Introduction

Generally, mobile sources include motor vehicles, aircraft, ships, and railroad locomotives. These categories are subdivided for control purposes; e.g., motor vehicles are subdivided into automobiles, trucks, buses, and motorcycles, and these, in turn, are subdivided into those powered by gasoline engines and those powered by diesel engines.

Railroad locomotives and ships were among the very first targets of smoke abatement regulatory action in the latter half of the nineteenth century and have remained an element of air pollution control programs of nations the world over to the extent that their steam locomotives and ships have not been displaced by diesel- or electric-powered equipment. In those nations where the conversion to diesel- or electric-powered locomotives has been essentially complete, the contribution of railroads to total air pollution has been considered so low that no special regulatory standards have been developed for diesel locomotives. Usually, they still have only to meet the emission requirements of steam locomotives, which, in general, they do quite well. Similarly, no new coal-fired seagoing ships have been built in years, and, as a disappearing breed, they have ceased to be the smoke and cinder problem they once were at seaports. Bridges, tunnels, and oil-fired steam or diesel equipment have displaced coal-fired ferries, tugs, and freshwater ships. In general, no new emission standards have arisen for ships to replace those originally developed to control smoke from coal-fired vessels.

Emission standards for motor vehicles have been one of the most significant developments in the field of air pollution control. For many years prior to the development of emission standards for motor vehicles, little attention was given to pollutants emitted from this source. If there was a concern about such emissions, it would have been about excessive smoke, which was usually evaluated using standards or regulations developed for evaluating smoke emissions from railroad locomotives or stationary sources. The lack of early attention to emission standards for motor vehicles was due to several factors. Smoke from gasoline-powered automobiles was not considered to be a serious problem; hydrocarbons, carbon monoxide, and oxides of nitrogen were not early identified as important atmospheric pollutants (photochemical air pollution had not yet been described), and the legal and administrative authority to deal with emissions from motor vehicles did not exist. This situation has changed markedly since 1960 as more was learned about motor vehicle created air pollution and as concern grew over air pollution from all sources. Control of emissions from mobile sources is now recognized as one of the most important elements in air pollution control.

II. Need for Emission Standards for Mobile Sources

A. Railroad Locomotives and Ships

Railroads generally go through the heart of the community. Freight yards are frequently near the center of the city. In fact, in many communities, the railroad was there first and the community grew up around the railroad station. Because of this, if the locomotives used for through service or local switching and freight classification service produce smoke and cinder emission, they need to be regulated by emission standards designed for the purpose.

Most of the major cities of the world are on navigable waterways, and their most desirable industrial, commercial, and residential sites tend to be on or near the water. Because of this, ships that emit smoke and cinders need to be regulated by emission standards.

B. Motor Vehicles

In the early 1950s, Dr. A. J. Haagen-Smit reported on his now famous studies that showed that organic compounds and oxides of nitrogen reacted in sunlight to produce the air pollution symptoms that were closely associated with those experienced in Los Angeles, California (*1*) As emissions of both organic compounds and oxides of nitrogen are attributable to automobiles, a large number of which were in operation in Los Angeles, his findings provided the first evidence that pollutants from automobiles were causes of community-wide pollution and needed to be controlled.

Subsequent to Dr. Haagen-Smit's report, the contribution of automobiles to communitywide air pollution was further defined, as other researchers confirmed his findings and broadened our knowledge of the nature and quantity of emissions from motor vehicles. These findings led to development and later refinements of emission standards for carbon monoxide, hydrocarbons, and oxides of nitrogen. Persistent public complaint against smoke from diesel engines during this period led to development of emission standards for these engines.

Hydrocarbons at concentrations that can be experienced in the atmosphere have few direct effects; they are not known to affect human health. Ethylene at concentrations occurring in the atmosphere can, however, damage some ornamental plants. Control of hydrocarbons is based on the need for preventing photochemical reactions that can produce pollutants that do have health effects and cause widespread damage to vegetation. The measures to prevent photochemical air pollution can be expected to reduce ethylene and direct effects on ornamental plants. Nitrogen dioxide, one of the oxides of nitrogen in the atmosphere, is a participant and a product of the photochemical reactions and may cause health effects at concentrations that can occur in the atmosphere.

Figure 1 shows a typical concentration pattern of the main chemical species involved in the photochemical reaction. The primary pollutants, hydrocarbons and nitric oxide (NO), which are discharged from motor vehicles and other sources, undergo complex photochemical reaction to produce oxidant (mainly ozone), nitrogen dioxide (NO_2), aerosols, and many other compounds. These products, in turn, can cause health effects, eye irritation, vegetation damage, and reduced visibility. Motor vehicle

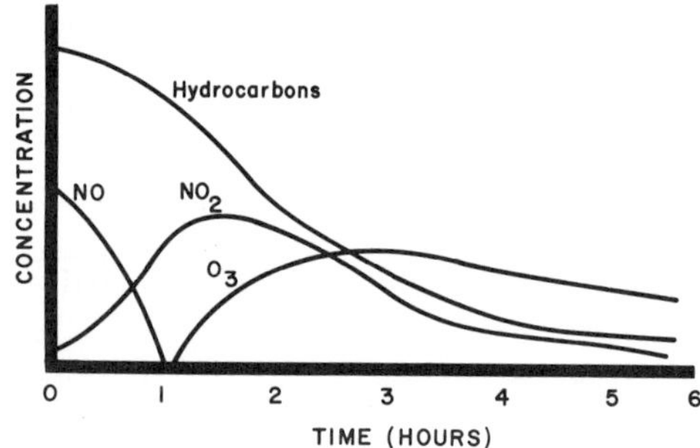

Figure 1. Concentration patterns of main participating chemical species in typical photochemical reactions (not drawn to scale). (NO, nitric oxide; NO_2, nitrogen dioxide; O_3, ozone.)

Table I Relative Contribution of Carbon Monoxide, Hydrocarbons, and Oxides of Nitrogen by Motor Vehicles and Stationary Sources

		Contribution (%)					
		Carbon monoxide		*Hydrocarbons*		*Oxides of nitrogen*	
City	*Year*	*Vehicular*	*Stationary*	*Vehicular*	*Stationary*	*Vehicular*	*Stationary*
Chicago, Illinois	1967	94	6	81	19	35	65
Los Angeles, California	1966	95	5	72	28	73	27
New York, New York	1965	96	4	84	16	38	62
Philadelphia, Pennsylvania	1967	70	30	47	53	27	73
Houston, Texas	1967	75	25	58	42	43	57
Pittsburgh, Pennsylvania	1967	80	20	70	30	29	71
St. Louis, Missouri	1967	77	23	80	20	48	52

emission standards represent the limits for emissions of primary pollutants intended to reduce the photochemical reaction products to levels that no longer produce undesirable effects.

Carbon monoxide has not been shown to be an important component in photochemical air pollution. It does, however, combine with the hemoglobin in the blood to form carboxyhemoglobin, thereby reducing the amount of hemoglobin available to carry oxygen. Emission standards for carbon monoxide are designed to prevent atmospheric concentrations that can result in undesirable amounts of carboxyhemoglobin in the blood.

Motor vehicles as a group produce large amounts of hydrocarbons, carbon monoxide, and oxides of nitrogen, both because of the nature of combustion in engines and because of the numerous vehicles in use. Figure 2 shows motor vehicle registration in the United States from 1910 to 1975. It can be seen that the number of cars has increased from a few million in 1910 to more than 106 million in 1975. During the same period, trucks increased from less than one million in 1910 to almost 26 million (*2*). These vehicles are concentrated in the metropolitan areas, where most people, industry, and commercial activities are also located. This situation leads to the development of severe air pollution in areas where a large number of people can be exposed.

The relative contribution of motor vehicles to the air pollution problem of a community depends, of course, on how much of these same contaminants are emitted from other sources. It can be seen from Table I that in many large cities motor vehicles contribute over 90% of the carbon

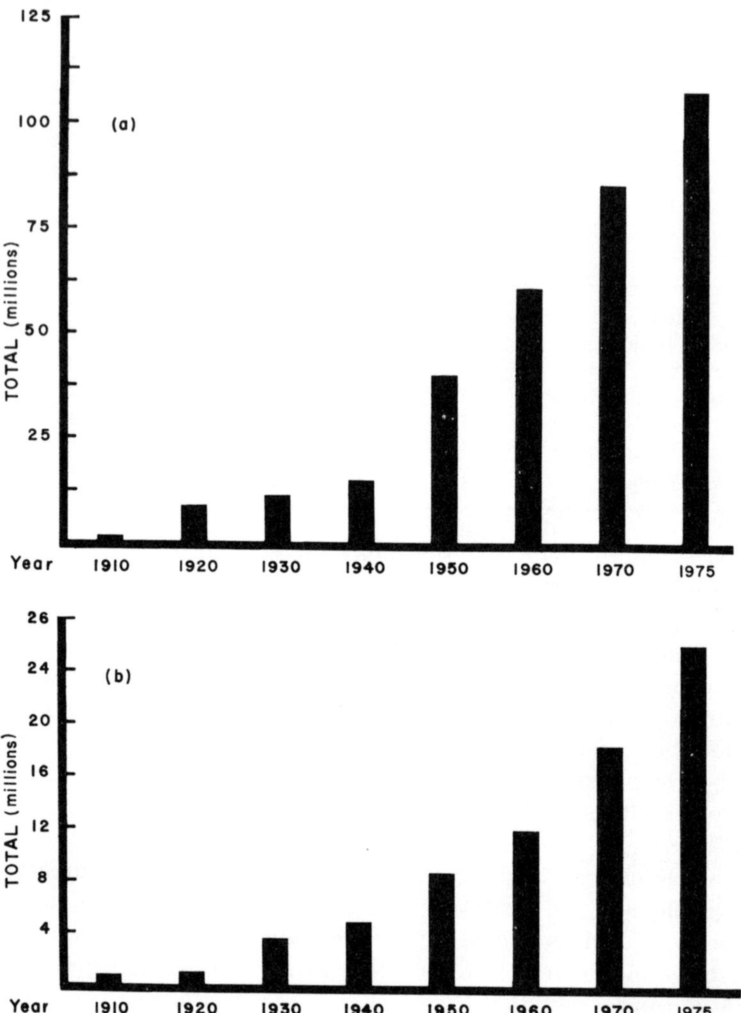

Figure 2. Motor vehicle registration in United States; (a) cars, (b) trucks.

monoxide, and seldom in any city is the contribution less than 75% (*3*). The relative contribution of hydrocarbons from automobiles may constitute 75% or more of the total emission from all sources. Motor vehicles do not contribute as high a percentage of oxides of nitrogen as they do hydrocarbons and carbon monoxide. In some communities, less than 50% of the oxides of nitrogen are attributable to motor vehicles. In such communities, it is the fossil fuel power-generating plants that account for the largest quantity of oxides of nitrogen emitted.

C. Aircraft

The introduction of jets in the late 1950s marked the beginning of widespread public complaint that aircraft were a cause of air pollution. These complaints were directed at the dark smoke that was so clearly visible in the skies near all major airports. Public awareness of the dark plumes and the odor associated with the burning of jet fuel increased with the extraordinary growth in commercial air service, conversion to jet aircraft, and expansion of major airports. Aircraft emissions are of particular concern in areas adjacent to large airports, where emissions from jets concentrate.

The United States Clean Air Act, as amended in 1977, further focused attention on aircraft emissions by requiring the United States Environmental Protection Agency (USEPA) to set standards for these emissions and requiring the states to adopt implementation plans to achieve stringent national air quality standards. In some metropolitan regions, the degree of emission control required in their state's implementation plan was so high that it was necessary to include control of pollutants from aircraft. Although the contribution of aircraft, in terms of relative quantity, is low (a few percent or less of the area's total emissions), its significance will increase as the other sources are more strictly controlled and aircraft use expands.

III. Regulatory Considerations

A. Railroad Locomotives and Ships

The usual regulatory statement applying emission opacity standards prohibits an "emission of a density of *A* or greater for more than *B* minutes in any period of *C* minutes." The principal consideration for establishing the time period *C* is the mobility of the source. A moving locomotive or ship can be observed from a stationary observation point on the ground for a short time before it travels out of the observer's view. The concepts of regulations based upon time *B* within a period *C* and of allowances for smoke released when starting fires had their historical basis in the nature of smoke emission from hand-fired steam locomotives and ship's furnaces and represent a compromise to meet the objections of operators of that class of equipment to more stringent regulations. Even when the reasons for these compromises have disappeared along with the class of equipment they were intended to regulate, their legacy remains embedded in regulatory legislation. The hand-fired coal-burning steam locomotive and ship's furnace can be, and in some jurisdictions has been, regulated out of existence by setting emission standards too stringent for this class of equipment to meet but that can be met by oil-fired, diesel, or electric equipment.

B. Motor Vehicles

The severe photochemical smog problem in Los Angeles, California, together with the recognition that motor vehicles were a major cause of this type of pollution led the California Legislature in 1959, to require that the Department of Public Health establish air quality and motor vehicle emission standards (*4*). The standards established by that department were the first such standards and marked the beginning of a control effort that would extend far beyond California and eventually threaten the very existence of the internal combustion engine as now used in automobiles.

The emission standards set by the Department of Public Health were followed in 1960 by legislation that mandated the control of motor vehicle emissions in California. During the decade that followed, California adopted emission standards for crankcase vapors, fuel tank and carburetor evaporative losses, and exhaust carbon monoxide, hydrocarbons, and oxides of nitrogen. Standards were also extended to include heavy-duty gasoline- and diesel-powered vehicles.

By mid-1960, motor vehicle-related air pollution became a matter of national concern in the United States, and the Congress enacted laws establishing a federal program to control vehicular emissions. The period from 1968 through 1970 saw a new direction in the type of legislation to solve vehicular air pollution. Both the United States Congress and the California Legislature placed motor vehicle emission standards into the law, rather than directing administrative agencies to set the standards, as had been the case theretofore.

This new legislative policy brought about a significant change in the manner by which emission standards are determined and enforced. By placing the standard into the law, the legislative bodies clearly indicate their intent and limit the authority of the administrative agencies to modify the standards. The use of the legislative rather than administrative process for establishing standards has had the effect of applying much more pressure on the automobile manufacturers, but less emphasis on technical considerations in setting standards. It had been recognized that the technical problems of emission standards and control systems were complex and controversial. The setting of standards directly through legislation is even more controversial, because technical attainability appears to have been assumed when many problems remain unsolved.

An example of legislated standards are those specified in the United States Clean Air Act as amended in 1970 and as it has been further amended in 1977. In the 1970 act, the Congress required that the emissions of hydrocarbons and carbon monoxide in 1975 automobiles be reduced to one-tenth of those emitted by 1970 vehicles, and that oxides of nitrogen in 1976 automotive emissions be reduced to one-tenth of that emitted by 1971 vehicles. More than any other step, this action by the Congress stimulated a large research and development effort by the automobile industry to produce the needed control systems. At the same time, however, it produced much controversy over the basis, feasibility, costs, need, and benefits of the legislated standards.

The United States federal law pre-empts the field of motor vehicle emission control. States that had adopted emission standards prior to 1967 are, however, permitted to enact their own standards. California is the only state to which this provision of the act applies. The law allows California to act only if it obtains a waiver from USEPA. To obtain the necessary waiver, California must convince USEPA that it has an extraordinary and compelling problem, that its standards will be more stringent than the federal ones, and that its standards will be technologically feasible. These requirements have resulted in a situation where some United States emission standards apply in all fifty states and some standards in California are different from those in the remaining forty-nine states. As the federal standards become more stringent, there will be less reason for separate ones that are applicable only in California.

C. Aircraft

The public complaints about smoke from jet aircraft led to action being taken in the late 1960s by several state and local agencies in the United States. This action included lawsuits against commercial airlines and laws establishing opacity standards for aircraft emissions. These actions were aimed at forcing the airlines to use new jet engines that produce little smoke and to replace the combustors in existing engines with types that do not smoke.

In the 1970 amendments to the United States Clean Air Act and the amendments to the Act adopted in mid-1977 just before this chapter went to press, the Congress requires the USEPA to establish aircraft emission standards. In contrast to motor vehicle emission control, the Congress does not permit any state or local government to adopt and enforce aircraft emission standards different from those set by USEPA. Congress concludes that only one standard should be permitted, because aircraft generally operate between states, and it would not be practical to have state and local agencies in the United States enforce different standards.

IV. Basis for Emission Standards

A. Railroad Locomotives and Ships

Emission standards for smoke from railroad locomotives and ships rely upon the Ringelmann chart. The use of this chart for stationary sources is discussed in Chapter 13 of this volume. The standards for locomotives and ships were developed in parallel with the development of smoke emission limits for stationary sources and have represented what could be accomplished by careful firing of properly designed and maintained equipment with fuel of appropriate quality. With hand-fired coal-burning equipment, it has been customary to allow a specified relaxation of emission standards for specific time periods for cleaning of fires, start-up, and soot blowing. Some ordinances have allowed relaxation for the latter two causes in oil-fired steam boiler furnaces in locomotives and ships. Since fossil-fuel-fired furnace emissions from locomotives and ships are highly dependent upon fuel, operation, and maintenance, there has never been prototype certification as in the case of motor vehicles and aircraft. Although it has not yet been done, there is no reason why prototype emission certification could not be used for diesel locomotives.

Even though cinder emission was a recognized problem from coal-fired steam locomotives and ship's furnaces, the only effective emission standard that evolved to cope with the problem was to treat a ship tied up to a pier as a stationary source and apply stationary source emission limits to it. No cinder emission standards have been devised for moving locomotives or ships.

B. Motor Vehicles

Emission standards prescribe limits of contaminants discharged into the atmosphere, so that when the standards are met, adverse effects from air pollution will be minimized or eliminated. One way of defining adverse effects to be expected is through establishment of air quality standards that relate the undesirable effects to the concentration of pollutants and the duration of exposure. Once air quality standards are established, such standards can be used as the basis for formulating emission standards. The emission standards, therefore, represent emission levels not to be exceeded if air quality goals are to be achieved.

The air quality standards for hydrocarbons, oxides of nitrogen, carbon monoxide, and oxidant (ozone) are particularly relevant in the control of vehicular air pollution. The first three of these pollutants are discharged directly from motor vehicles; oxidant is produced in photochemical reactions and is the most frequently used index for this type of air pollution.

Although air quality standards have been established by a number of governmental agencies in the United States, the ones that have been used to determine motor vehicle emission standards are those adopted by California and USEPA (*5, 6*). Table II shows the USEPA air quality standards for hydrocarbons, carbon monoxide, nitrogen dioxide, and oxidant and atmospheric concentrations measured during severe episodes of vehicular air pollution.

In principle, the calculation of motor vehicle emission standards is straightforward (*7*). The steps are as follows:

a. Description of the quantitative relationships between the undesirable effects of pollution and atmospheric concentration of the pollutant (air quality criteria), and establishment of air quality standards.

b. Determination of current concentration of contaminants on days of meteorological conditions most unfavorable to the dispersal of pollutants.

Table II United States Air Quality Standards for Hydrocarbons, Carbon Monoxide, Nitrogen Dioxide, and Oxidant, and Typical Concentrations of These Pollutants during Severe Air Pollution Episodes

	Air quality standards		*Typical high atmospheric concentration for duration specified*
Pollutant	*Concentration*	*Duration*	*(ppm)*
Hydrocarbons[a]	160 μg/m³ (0.24 ppm)	3 hours (6–9 A.M.)	6
Carbon monoxide	10 mg/m³ (9 ppm)	8 hours	40
	40 mg/m³ (35 ppm)	1 hour	50
Nitrogen dioxide	100 μg/m³ (0.05 ppm)	1 year	0.1
Oxidant	160 μg/m³ (0.08 ppm)	1 hour	0.6

[a] Excluding methane.

c. Determination of the total emissions from motor vehicles and nonvehicular sources and the average quantity of emissions from vehicles.

d. Projection of the emission data to a selected future date, allowing for increases due to growth and decreases due to controls.

e. Calculation of the control of emissions (reduction factors) required to achieve the air quality standards on days of most unfavorable dispersal conditions.

f. Application of the reduction factors to the average exhaust contaminant concentrations, with appropriate regard to the reductions also to be achieved by nonvehicular sources.

In practice, the calculation of emission standards is very complicated and the results are somewhat unreliable because of the many variables and unknowns involved. These include growth and changing conditions in a community, uncertainties in the quantitative relations between the concentration and effects of pollutants, lack of sufficient data on atmospheric concentrations and source emissions, and uncertainties about the relationship between amount of emissions and atmospheric concentrations. The most complicated situation is with photochemical air polution and use of the oxidant air quality standard as the index for this type of pollution. This is because oxidant is a product of a reaction that is not adequately understood and because of the complex atmospheric conditions that exist in a large metropolitan area.

The United States air quality standard for hydrocarbons was established for the purpose of controlling atmospheric oxidant, because hydrocarbons are precursors to oxidant formation in photochemical reactions. The hydrocarbon standard is applied to the concentration of hydrocarbon between 6–9 A.M., the period of high concentrations in the central city. This period coincides with early morning commuter traffic and is just before bright sunlight becomes available, which is important, because sunlight is one of the essential ingredients in photochemical reactions. Time is required for photochemical reactions to complete. The maximum oxidant, therefore, is found usually from midmorning to late afternoon at an area downwind from the area of emission of the pollutants. The time when maximum concentrations occur depends upon the period required for the reaction, movement of the air mass, and injection of additional pollutants into the air mass. This dynamic situation introduces uncertainties as to the period and location of hydrocarbon emissions that produce peak oxidant values. Midmorning oxidant concentrations are expected to be largely the result of the 6–9 A.M. emissions. However, it is not clear that the afternoon oxidant concentrations are due mainly to 6–9 A.M. emissions. Although oxides of nitrogen are also involved in the photochemical reactions, the federal standard does not specify the 6–9 A.M. oxides of nitrogen concentration. Instead, the United States ambient air quality standard for nitrogen dioxide (NO_2), which is stated as an annual average, is based upon health effects.

Most of the information used to describe atmospheric photochemical reactions comes from laboratory studies, where reactions are produced in experimental chambers under artificial sunlight. These studies can produce typical photochemical reactions, but the product yields from given concentrations of hydrocarbons and oxides of nitrogen frequently differ. To further complicate the situation, data obtained in the atmosphere are much more variable than those obtained from laboratory experiments. As a result, laboratory data alone cannot reliably describe the reactions occurring in the atmosphere.

Diffusion models would be useful for calculating emission standards. Such models, however, have not been developed to a point where they can be used for this purpose. The use of such models requires the availability of definitive data on inversion heights, wind flow, topographic features, traffic, emissions from all sources, and photochemical reactions.

A simple model, the proportional or "rollback" model, was followed in developing the first motor vehicle emission standards in 1959 (*7*). This method has since been widely used in many state implementation plans for calculating emission standards needed for all sources. This model assumes that the emissions and atmospheric concentrations of contaminants are linearly related, i.e., that a given percent reduction in emission will result in a similar percent reduction in atmospheric concentrations. The maximum concentrations that have been measured in the atmosphere are assumed to represent the limiting dispersal conditions.

The assumption that a given reduction in the emission of a contaminant will produce a proportional reduction in atmospheric concentrations is expected to be more valid for carbon monoxide than for other gaseous contaminants, because carbon monoxide is more stable than hydrocarbons and oxides of nitrogen and is not consumed appreciably in photochemical reactions.

The simple model when applied to mobile sources assumes that emissions of hydrocarbons, carbon monoxide, and oxides of nitrogen from stationary sources will also be reduced. If the emissions from one class of sources are not reduced by the required percentage, then those from other sources must be reduced correspondingly more. In the case of carbon monoxide, motor vehicles are the only important source in most communities, so the model should be fairly reliable for determining the degree of control required for motor vehicle emissions. In some places, motor vehicles are also the only important source of hydrocarbons and oxides of nitrogen, but such places are exceptions rather than the rule. One proportional model is that described by Larsen (*8*), in which

$$R = \frac{g(P) - D}{g(P) - B} \times 100 \tag{1}$$

where R = required reduction in percent emissions, g = growth factor in the emissions, P = present air quality, D = air quality standard, and B = background concentration. To determine g, it is necessary to select a year in the future when it is expected that the emission standards will be applied to all vehicles. Virtually all vehicles are no longer operating after they are 15 years old. Fifteen years would then be a suitable projection period. In selecting a growth factor for motor vehicle emissions, it could be assumed that the emissions will increase in proportion to expected increase in vehicle registration. If it is assumed that the number of vehicles will grow at an average of 3% per year (noncompounded), the growth factor is 1.45; should it be assumed that the average growth is 5%, the growth factor is 1.75. The selection of a growth factor is, however, much more complicated.

Crowded streets in large cities already carry about as much traffic as these streets can carry. Few additional vehicles will travel in these areas in the future. It could be argued, therefore, that vehicle growth is near unity in the built-up older parts of large cities, where the greatest amount of emissions and highest atmospheric concentrations now occur. In rapidly growing areas, the growth factor will exceed the average value based on increased vehicle registration, but the atmospheric concentrations will be lower in these areas while they are growing. When these areas are fully developed, their concentrations will approach those now found in the crowded mature areas. Wind transport of pollutants from mature areas can add pollutants to receptor sites over and above that contributed by the traffic in the receptor sites. Therefore, concentration growth factor may be greater than expected from the traffic growth factor in the receptor site.

C. Aircraft

The emission standards for aircraft have largely been established on the basis of known control technology. This is in contrast to the motor vehicle emission standards, which often have mandated very stringent controls in future years, even though the control systems needed to meet the standards were not known at the time. As a result, the degree of control required of aircraft is much lower than that for motor vehicles. Aircraft emission standards also do not consider quantitatively the degree of control of these emissions to meet air quality standards. The use of standards largely based on known technology recognizes that aircraft generally are not major sources of pollutants in metropolitan areas and that aircraft safety might have to be compromised if engines were to be designed primarily to meet stringent emission limits.

V. Sources of Pollutants Covered by Emission Standards

A. Railroad Locomotives and Ships

All the pollutants resulting from combustion of coal or oil in stationary furnaces, gas turbines, and internal combustion engines can be found in the emissions of such equipment when it is used in mobile sources. However, the space constraints and the accelerations associated with a locomotive or ship make it difficult to apply as sophisticated combustion controls as can be used in a stationary plant. The combustion products and the equipment operating conditions associated with them are discussed in Chapters 10 and 11 of Volume IV for other than the internal combustion engine, and in Chapter 14 of Volume IV for internal combustion engines, including the diesel.

B. Motor Vehicles

The sources of pollutants from motor vehicles include the exhaust, crankcase, fuel tank, and carburetor. Table III shows the relative emissions of hydrocarbons, carbon monoxide, and oxides of nitrogen from these sources in a typical gasoline-powered passenger vehicle without emission control systems. It can be seen that significant amounts of hydrocarbons are emitted from the exhaust, crankcase, fuel tank, and carburetor. Exhaust, however, is the only significant source of carbon monoxide and oxides of nitrogen.

Table III Relative Contribution of Pollutants from the Crankcase, Fuel System, and Exhaust of an Uncontrolled Vehicle[a]

Emission source	*Emission (percent of total emission of the contaminants)*		
	Carbon monoxide	*Oxides of nitrogen*	*Hydrocarbons*
Crankcase	1–2	1–2	25
Fuel system	0	0	10
Exhaust	98–99	98–99	65

[a] Source: California Air Resources Board.

In the United States, emission standards have been mainly concerned with vehicles whose engines were over 50 inch^3 displacement (CID) or 820 cm^3 displacement. It was not until 1972 that USEPA included engines of less displacement in the emission standards for automobiles (*9*). The emissions from smaller engines are similar in nature to those from the larger ones. One difference, however, is that some of the smaller cars may have two-cycle engines in contrast to the four-cycle engines in the larger cars. In general, two-cycle engines have higher hydrocarbon and carbon monoxide emissions.

Lack of attention to the engines with less than 50 CID (820 cm^3) was due to the very small number of these vehicles in the United States. Their emissions, therefore, have not been of much significance in the motor-created air pollution problems of the United States, although this situation is likely to change in the future as engines in United States vehicles get smaller. In other countries, automobiles with smaller engines are much more prevalent, and emissions from such automobiles comprise a higher percentage of total vehicular pollutants. In these countries and in the United States in the future, it will be important to have motor vehicle emissions standards for the small automobile.

Motor vehicle emission standards for motorcycles were the last to be considered. Again, this was due in large part to the relatively small number of such vehicles as compared to cars and trucks. Unlike passenger cars, two-cycle engines are widely used in motorcycles.

1. Gasoline-Powered Vehicles

a. Crankcase Emissions. These emissions occur when some of the air–fuel mixture within the cylinder is forced past the piston rings; this is often called "blowby." As can be seen from Table III, crankcase vapors are about 25% of the total hydrocarbon emissions in an uncontrolled vehicle. The hydrocarbon concentrations do not vary widely among vehicles and among operating modes, because blowby is mainly carbureted air–fuel mixture plus a small quantity of exhaust products. The volume of blowby varies over a wide range due to the change in pressure in the cylinder according to operating mode. Blowby rates are highest during the compression and power strokes, low rates are experienced at deceleration and idle. The average volume of blowby under typical city driving conditions was found to be from 0.5–5 feet^3/minute (0.3 to 3 liters/second) with the composite average of about 1 cf^3/minute (0.6 liters/second) (*10*).

b. Evaporative Losses. In uncontrolled vehicles, gasoline vapors escape from the fuel tank and carburetor. How much of these vapors do escape depends upon the fuel composition, engine operating temperature, and ambient temperature. Evaporative losses from the fuel tank are strongly influenced by the ambient temperature and exposure of the tank to the sun, atmosphere, and pavements. These losses can be very high when the ambient air temperature approaches the boiling point of the gasoline.

The evaporative losses from the carburetor take place mainly during the "hot soak" period when the carburetor is heated by the hot engine. Ambient temperature has less effect on evaporative losses from the carburetor than on those from the fuel tank.

Emission standards for evaporative loss must reflect the temperature experience of the fuel tank and the number of "hot soaks" during the period under consideration, normally one day.

c. EXHAUST EMISSIONS. As can be seen from Table III, the exhaust emissions are the source of most of the hydrocarbons and practically all of the oxides of nitrogen and carbon monoxide. The exhaust products are the result of combustion conditions in the engine under high temperature and pressure and at an air–fuel ratio that is often less than stoichiometric. Sulfur dioxide, lead, lead scavenger compounds, oxygenated compounds, particulate matter, and other compounds are also present in the exhaust. However, emission standards have yet to be adopted for any of these, although concern has been expressed over lead, particulate matter, and oxygenated hydrocarbons. Fuel composition standards have, however, been set for lead and sulfur in a number of countries. USEPA has adopted limits on the amount of lead that can be added to gasoline in the United States. For vehicles equipped with catalytic devices, the USEPA regulations prescribe a "lead-free" gasoline (maximum of 0.01 gm of lead per liter). The average lead content of all gasoline, including the unleaded grade, was limited by the USEPA (*11*) to the following:

USEPA Limitations

Average lead content		
(grams/gallons)	*(grams/liter)*	*Date*
1.7	0.45[a]	After January 1, 1975
1.4	0.37[a]	After January 1, 1976
1.0	0.26[a]	After January 1, 1977
0.8	0.21[b]	After January 1, 1978
0.5	0.13[c]	After January 1, 1979

[a] Never enforced because of deferral pending outcome of litigation.

[b] Required only if a refiner does not show good faith effort to meet 1979 standard (41FR42675).

[c] Compliance date changed to Oct. 1, 1979 (41FR42675).

A law suit challenging the above standards was filed by some of the affected industries but was finally settled in USEPA's favor in 1976. This delayed the application of the standards. USEPA found that complience with the above schedule could lead to a gasoline shortage. As a result, it amended the schedule to leave in effect the 0.8 gm/gallon (0.21 gm/liter) to begin on January 1, 1978 and 0.5 gm/gallons (0.13 gm/liter) beginning October 1, 1979 (*12*). Because of the delays that occurred in the U.S. program, California adopted standards on the average lead content in gasoline (*12a*) prior to the time the suit against USEPA was settled. California's requirements in grams of lead per liter of gasoline were as follows:

California's Limitations

Average lead content		
(grams/gallon)	*(grams/liter)*	*Date*
1.4	0.37	After January 1, 1977
1.0	0.26	After January 1, 1978
0.7	0.18	After January 1, 1979
0.4	0.11	After January 1, 1980

Both the USEPA and California provided small oil refineries more time to comply.

The type and quantity of pollutants emitted through the exhaust is highly dependent on the engine operating modes. The wide variation in the concentration of carbon monoxide, hydrocarbons, and oxides of nitrogen that occur under selected operating modes is shown in Table IV (*13*). The volume of the exhaust is also highly dependent on the operating modes. In small passenger cars, it will be less than 6 feet^3/minute (3 liters/second) at idle. It can be over 200 feet^3/minute (110 liters/second) for large trucks operating under full load. Motor vehicles standards for exhaust emissions must, therefore, specify the operating modes and the time the engine is operated in these modes if meaningful measurements are to be made. The significance of driving modes will be discussed further in the section dealing with operating cycles.

2. Diesel Engines

The type of pollutants from diesel engines are similar to those from gasoline-powered vehicles. There are several important differences, however, in the operating characteristics of the two types of engines and in

Table IV Concentration of Exhaust Emission by Selected Mode of Operation of Uncontrolled Light-Duty Vehicles

Operating mode (*miles/hour*)	(*km/hour*)	*Carbon monoxide* (*volume %*)	*Hydrocarbons*[a] (*ppm by volume*)	*Nitrogen oxides* (*ppm by volume, as NO_2*)
Cruise				
(50)	(80)	2.8	445	890
Deceleration				
(30–15)	(48–24)	5.0	1650	—[b]
Cruise				
(15)	(24)	4.1	560	205
Acceleration				
(15–30)	(24–28)	2.8	500	1700
Cruise				
(50)	(80)	2.4	320	1580
Deceleration				
(50–20)	(80–32)	4.2	3495	—[b]
Idle				
		5.8	895	—[b]

[a] Measured by hexane-sensitized nondispersive infrared (NDIR) analyzer.
[b] Very low, not measured.

the fuels they use. Therefore, emission standards for diesels differ from those established for gasoline engines.

There are no carburetor evaporative losses from diesel vehicles, because the fuel in diesel engines is ignited by injection into the cylinders, which contain compressed air. Diesel fuel is much less volatile than gasoline; tank evaporative losses are, therefore, much lower than those from gasoline tanks. The diesel engines compress air before the fuel is injected. Blowby from these engines, therefore, is mainly air and contains only small amounts of hydrocarbons.

Exhaust emissions are more like those from gasoline-powered vehicles. The high temperature and compression ratios in diesel engines cause large amounts of oxides of nitrogen to be formed. Because diesels usually operate above the stoichiometric ratio, carbon monoxide and hydrocarbon emissions are lower than those from gasoline engines of comparable size. Excess fuel, however, is often used at full load, resulting in the black smoke so characteristic of diesel engines.

As a result of the above conditions, fuel evaporative and crankcase emission standards are not needed for diesel engines. However, diesel exhaust emission standards have been adopted for smoke, hydrocarbons, carbon monoxide, and oxides of nitrogen.

The number of diesel-powered vehicles is less than 1% of all vehicles in the United States. A diesel vehicle, however, use much more fuel than does a passenger car. It has been estimated that, in the United States, diesel fuel consumption is about 5% of the fuel consumed by all motor vehicles. Because of the difference in fuel and blowby, lower exhaust hydrocarbons and carbon monoxide, and fewer diesels the promulgation of emission standards for these vehicles did not receive as high a priority as those for gasoline-powered engines. Standards for diesels have received much more attention since 1970, reflecting the general concern over air pollution and the more stringent emission standards that were being established for gasoline engines.

The relatively small percentage of diesels in the United States is in large part due to the large number of gasoline-powered passenger cars and the greater reliance on automobiles for individual transportation. In other countries, diesels are generally a much higher percentage of all motor vehicles. Where there is less reliance on passenger cars as a means of transportation, diesels may be the predominant type of motor vehicle. In these countries the main concern with motor vehicle-created air pollution usually is over the smoke from the diesel engines, rather than carbon monoxide and photochemical air pollution.

3. Motorcycles

Information on emissions from motorcycles is much more limited than for automobiles. Also, the number of motorcycles in the United States is not accurately known, because many are recreational vehicles that need not be registered. Since 1960, motorcycle sales in the United States have increased rapidly. It has been estimated that in 1972, there were approximately 5,400,000 motorcycles in the United States and that 30% of these were unregistered recreational cycles (*14*). This would represent slightly less than 5% of all motor vehicles (passenger cars and trucks).

The type of emissions from motorcycles are similar to those from automobiles and include those from the crankcase, fuel, and exhaust systems. However, the emissions have been measured from only a few motorcycles. Crankcase emissions from one four-cycle engine were found to be largely hydrocarbons, with smaller quantities of carbon monoxide and oxides of nitrogen. The amount of blowby hydrocarbons was about 20% of those from the exhaust (*14*). This is very much like the situation for automobiles. Evaporative emissions from motorcycles are largely from the diurnal losses of the fuel tank. Carburetor hot-soak losses are much less than in automobiles, because the motorcycle engines are smaller and the carburetors are not mounted on the engines.

Table V Representative Emissions from Motorcycles

Engine type and displacement (cm^3)	*Emissions*					
	Hydrocarbons[a]		*Carbon monoxide*		*Oxides of nitrogen*	
	(gm/mile)	*(gm/km)*	*(gm/mile)*	*(gm/km)*	*(gm/mile)*	*(gm/km)*
Four-stroke						
0–50	2.2	1.4	19	12	0.53	0.33
51–90	2.3	1.4	21	13	0.35	0.22
91–190	2.6	1.6	24	15	0.32	0.20
191–290	3.4	2.1	32	20	0.17	0.11
over 290	4.8	3.0	46	29	0.11	0.07
Two-stroke						
0–50	5	3	5	3	0.24	0.15
51–90	7	4	7	4	0.26	0.12
91–190	10	6	12	7	0.16	0.10
191–290	18	11	30	19	0.04	0.02
over 290	25	16	50	31	0.04	0.02

[a] Includes 20% for crankcase losses in four-stroke engines.

In contrast to passenger cars, many motorcycles have two-stroke engines. The quantity of hydrocarbons emitted from four- and two-stroke engines appear to be very different. The emissions from seven motorcycles (three 2-stroke and four 4-stroke) were measured under the seven-mode and CVS-1 procedures*; the results of these tests were plotted as functions of engine displacements to indicate the range of values to be expected *(14)*. Emission points were then assumed to represent selected categories and the mass emissions determined as a function of engine size and type are shown in Table V. It is cautioned that the data in this table should not be used as actual emissions from motorcycles, since a much larger sample would have to be used for a good statistical analysis.

Table V shows that (a) motorcycles can have high emissions of hydrocarbons and relatively high emissions of carbon monoxide; (b) some four-stroke engines comply with the United States automobile emission standards for 1973–1974, but not for 1975, and (c) two-stroke engines appear to have hydrocarbons emissions considerably above United States passenger car standards for 1973 models.

* See Chapter 16, Vol. III.

C. Aircraft

Engines of piston aircraft use similar fuels and have the same general characteristics as gasoline-powered engines for automobiles. The sources and pollutants are therefore similar to those from motor vehicles. Emissions standards, however, have been adopted by USEPA only for the exhaust.

Gas turbine engine (jet) emissions and fuel are more like those of diesel engines, where smoke is probably the major concern and where exhaust hydrocarbons, carbon monoxide, and oxides of nitrogen must also be considered. An emission from aircraft gas turbine engines that differs from that in motor vehicles is the "fuel-venting emission." Each time a turbine engine is started up or shut down, some fuel already in the system is drained to a dump. It had been the practice to dump this fuel after take-off.

Aircraft emissions occur on the ground in idle, taxi, start-up and shut-down operations as well as in flight. Also, there are emissions from engine overhaul test cells, which are regulated by standards applicable to stationary sources.

VI. Emission Standards for Railroad Locomotives and Ships

When air pollution control programs were first directed at the emissions from locomotives and ships, the main objective was to prevent the discharge of smoke. The source of these pollutants was the combustion of fuel, and the fuel often was coal. It was natural, therefore, for the emission standards to take forms that were similar to those being developed for stationary combustion sources, but to provide observation periods that reflected the fact the locomotives and ships were moving sources.

With the change from coal to oil, the concern over emissions from locomotives and ships has decreased. On the other hand, the increased attention being devoted to the control of air pollution has led to more stringent limitation of visible emissions from all combustion sources. Regulation of emissions from locomotives and ships has therefore continued, although at a much reduced level from when coal was burned.

The visible emission standards for locomotives and ships have been based on the Ringelmann chart. Table VI shows the standards for some jurisdictions in the United States and other nations (*15*). As can be seen; the standards have a similar form in that they do not permit smoke above a specified Ringelmann number during a given period of time. Many of the standards are set at Ringelmann number 2 and a few are

Table VI Visible Emission Standards for Smoke from Marine Vessels and Locomotives (*15*) (United States and Other Nations)

Place	*Category*	*Standard*: *Ringelmann number*	*Standard*: *Not to exceed (time)*	*Remarks*
Australia	—	1	—	National guidelines for new sources
Queensland	Vessels	2	5 min/hr	Continuous emissions
		4	6 min/hr	Aggregate emissions
South Australia	Vessels	2	6 min/hr	Main fuel burning equipment, at no time more than Ringelmann 3
		2	3 min/hr	Auxilary fuel-burning equipment, at no time more than Ringelmann 3
Western Australia	Vessels	2	10 min/hr	Vessels not under way
	Vessels	2	20 min/hr	Vessels under way
Mexico	Locomotives	3	30 sec/30 min	—
United States				
Colorado, Boulder City	Locomotives	2	10 sec	—
Illinois, Chicago	Vessels and locomotives—yard-duty	2	—	Special allowances for start-up
		2	—	—
		3	1 min/15 min	Switching and transfer
Iowa				
Cedar Rapids, Linn County	Locomotives	2	40 sec	Special Allowance for start-up
Maryland, Baltimore	Vessels and locomotives	2	1 min/hr	—
Massachusetts	Vessels	1	6 min/hr	Ringelmann 2 maximum
New Jersey	Locomotives and vessels	2	10 sec	—
				—
Ohio				
Cincinnatti	Vessels and locomotives	1	5 sec	Special allowance for start-up
Toledo	Locomotives	2	30 sec	Special allowance for start-up
Pennsylvania Philadelphia	Locomotives	2	3 min/hr	Ringelmann 3 maximum
Yugoslavia	Trains and ships	3	3 min/hr	—

at Ringelmann number 1, reflecting the change from coal as a fuel. Special allowances are also often made for start-up and temporary operations by permitting the emission of more smoke during such events.

VII. Motor Vehicle Operating Cycles—United States

Motor vehicles are usually separated into light-duty and heavy-duty for the purpose of emission standard and test procedures. Light-duty vehicles are those having a gross vehicle weight (GVW) of 6000 pounds (2722 kg) and under. In 1979 California will include a separate class of medium-duty vehicles which will cover all heavy-duty vehicles having a manufacturers' gross vehicle-weight rating of 8500 pounds (3855 kg) or less. Starting in 1979 USEPA will define all trucks under 8500 pounds (3855 kg) as light-duty trucks.

A. Light-Duty Vehicles

1. Exhaust Emissions

As has been mentioned, motor vehicles emit through their exhaust pollutants that vary with driving conditions. Motor vehicle exhaust emission standards must therefore be related to operating modes specified in test procedures. A change in the test procedures will result in a change in emissions so the standards and the test procedures specified in them are inseparable.

Motor vehicles are driven in many ways. Some trips are short, others are very long; some are a combination of these. Trips are also made for many purposes, e.g., commuting to work, shopping, business, and recreation. Drivers will often operate vehicles differently when driving under similar traffic conditions. One test procedure cannot represent all of the possible operating modes in each trip or the driving habits of different drivers. It is necessary, therefore, to select a test cycle that is believed to represent the traffic that produces the bulk of the emissions in areas where the most severe motor vehicle air pollution is experienced. In Los Angeles, California, where the photochemical air pollution problem is severe, heavy traffic occurs in the downtown area during morning commuting hours. Studies of automobile travel in this area, as well as in other cities in the United States, show that most trips are short ones and are made at relatively low average speeds. The driving cycle used for evaluating exhaust emissions, therefore, emphasizes urban-type driving (short trips at relatively low speeds). It must also provide for start-up of cars to reflect emissions when the engine is cold and the choke is operating.

The earliest comprehensive study of automobile driving modes and emissions was made in Los Angeles in 1956 by the Traffic Survey Panel of the Automobile Manufacturers Association participating as a member of the Coordinating Research Council (*16*). The findings of this survey were used by California to formulate the first motor vehicle emission standards, which were adopted in 1959. These standards were expressed in terms of eleven separate vehicle operating modes, including various vehicle speeds.

The measurement of emissions to determine compliance with exhaust standards needs to be done on a dynamometer, because a large number of tests are required to be made under identical conditions. The vehicle operating modes determined by the Coordinating Research Council in 1956 were not suitable for testing on a dynamometer, because they could not be combined into a closed cycle. It was necessary to modify these modes to obtain a cycle that could be used conveniently to operate a vehicle on a chassis dynamometer. The selected test cycle included seven of the eleven modes used in the standards and approximated the time distribution of driving modes in the standards (*17*). The test cycle also has three extra events not found in the eleven modes. These extra events were needed to connect the seven modes into a continuous cycle. The seven modes and three extra events are shown in Table VII. This cycle was utilized for certifying exhaust control systems. The emissions

Table VII Seven-Mode Dynamometer Test Cycle

Mode				
(miles/hour)	*(km/hour)*	*Time in mode (seconds)*	*Cumulative time (seconds)*	*Weighting factor*
1. Idle	Idle	20	20	0.042
2. 0–25	0–40	11.5	31.5	0.244
25–30[a]	40–48	2.5	34	Not used
3. Cruise–30	Cruise–48	15	49	0.118
4. 30–15	48–24	11	60	0.062
5. Cruise–15	Cruise–24	15	75	0.050
6. 15–30	24–48	12.5	87.5	0.455
30–50[a]	48–80	16.5	104	Not used
7. 50–20	80–32	25	129	0.029
20–0[a]	32–0	8	137	Not used

[a] Extra events used to connect the seven modes into a continuous cycle.

measured during seven modes (the interconnecting three were omitted) were weighted in arriving at the overall emission. This method of testing has been referred to as the "seven-mode method." This same procedure was incorporated in 1968 in the United States vehicle control regulations and used as the basic test procedure until 1972.

Because the cycle was based on operating modes determined in a study made in 1956, questions were soon raised about the adequacy of the test cycle for changed traffic conditions. As a result, a study was made of the driving modes associated with early morning traffic in the downtown Los Angeles area in 1966 (*18*). This type of traffic was selected because it was believed to be the major contributor to the peak pollution concentrations that later led to the photochemical reactions. The downtown Los Angeles driving route thus developed (often referred to as street route LA-4) was used by USEPA in the federal test procedure for 1972 model vehicles (*19*). It simulates a trip of 7.5 miles (12.1 km) over a period of 23 minutes, quite similar to the operating conditions prescribed in the seven-mode procedure. The revised federal procedure contains, however, some features that are not in the seven-mode procedure. The chassis dynamometer in the new procedure simulates an urban trip instead of a limited number of operating modes. The mass emissions during a test are determined by a constant-volume sampler. This procedure is often referred to as CVS-1. (See also Chapter 16, Vol. III.)

The CVS test collects all of the exhaust gases and uses a variable volume of dilution air to obtain a volumetric sampling rate that remains constant throughout the test (dilution air + exhaust gases). The contaminant concentrations in the diluted mixture is proportional to the mass of the contaminant (mass = concentration × volume × density). The air and exhaust mixture is passed through a heat exchanger to maintain uniform gas temperature. A constant volume of the exhaust–dilution air mixture is pumped by means of a constant-volume pump into a sample bag. Contents of the bag are analyzed for the pollutants of interest.

A new federal test procedure (CVS-2) was adopted for 1975 models (*20*). The main differences between CVS-1 and CVS-2 is that CVS-2 proportionately weights the hot and cold engine operation portions of the test according to the average number (4.8) of trips per day. The CVS-2 procedure (Fig. 3) is similar to the CVS-1 (1972 procedure), except that the first 8 minutes in the CVS-2 test are repeated after a 10-minute soak. The CVS-2 procedure consists of the 23-minute cold start CVS-1 test, 10-minute soak, and 8-minute hot portion (total 41 minutes). Three bags are required in the CVS-2 procedure to collect the gases under the separate portions of the test.

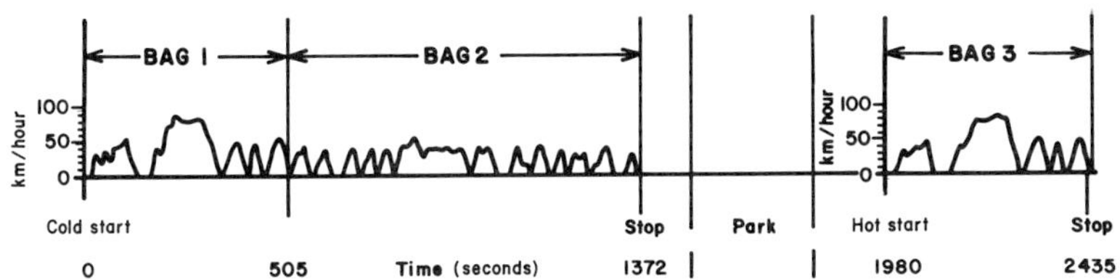

Figure 3. Driving cycle under CVS–2 test procedure. (1 mile = 1.609 km.)

The test procedure is a complicated one, involving a constant-volume pump, heat exchanger, multiple sample collection bags, and an assortment of instruments. The test is also lengthy (12-hour pretest soak and 41-minute test duration) and costly. The complexity of the procedure, nature of exhaust emissions, and other factors give rise to difficulties in reproducing the emissions from individual vehicles. A number of tests are therefore required to accurately establish a specific mass emission value from one vehicle.

2. Evaporative Losses

The USEPA procedure to determine compliance with the evaporative emission standards is designed to measure the vapors lost from the fuel tank, from the fuel system, and from the carburetor during operation of the vehicle in an urban trip. The procedure also must simulate the heating of the fuel in the fuel tank as the ambient temperatures increases during the day, and the hot soaking of the carburetor by the hot engine after the trip ends (*20*).

During the test, the evaporative losses are collected in activated carbon traps. The fuel in the tank is first heated to 84°F (51°C) over a period of 1 hour (diurnal breathing losses); the vehicle is then operated on a chassis dynamometer to simulate an urban trip (running losses); and finally, the hot engine is allowed to soak for 1 hour (hot soak losses). Upon completion of the test, the activated carbon traps are weighed to determine the quantity of vapors lost.

Studies by California in 1975 showed that the carbon trap test procedure did not accurately measure the evaporative emissions. As a result, the Air Resources Board adopted a Sealed Housing for Evaporative Determinations (SHED) procedure to become effective in 1978. In this method the vehicle is placed in a sealed enclosure and the evaporative emissions are determined by measuring the hydrocarbons in the enclosure. The USEPA has also adopted the SHED procedure (*21*).

3. Crankcase Emissions

No test procedure is specified because the USEPA emission standard requires a closed system for returning the crankcase vapors to the intake system of the vehicle.

B. Heavy-Duty Vehicles

1. Exhaust

Operating modes for heavy-duty vehicles have not been studied as extensively as those for automobiles. The test cycles in the test procedure for such vehicles are therefore based on much less data. Operating modes that have been used were mainly developed from a study in 1963 by the Ethyl Corporation under contract with the United States Public Health Service of truck operating patterns in a metropolitan area (*22*). The dynamometer cycle for gasoline-powered heavy-duty vehicles is shown in Table VIII and that for a diesel vehicle is shown in Table IX.

2. Smoke

The test procedure for evaluating smoke emissions from diesel vehicles is designed to produce dynamometer cycles representative of operating modes that causes the most smoke. A less complicated cycle can therefore be used than the one used to measure carbon monoxide, hydrocarbons, and oxides of nitrogen. The cycle includes only acceleration and lugging modes, and smoke is measured by means of the USEPA smoke meter (*20*).

Table VIII United States Dynamometer Test Procedure for Heavy-Duty Gasoline-Powered Vehicles[a]

Mode	*Manifold vacuum (mm of Hg)*	*Time (seconds)*	*Cumulative time (seconds)*	*Weighting factor*
Idle	—	70	70	0.232
Cruise	406	23	93	0.077
Part-throttle acceleration	254	44	137	0.147
Cruise	406	23	160	0.077
Part-throttle acceleration	483	17	177	0.057
Cruise	406	23	200	0.077
Full load	76	34	234	0.113
Cruise	406	23	257	0.077
Closed throttle	—	43	300	0.143

[a] A similar procedure is used by California.

Table IX United States Dynamometer Test Procedure for Heavy-Duty Diesel-Powered Vehicles[a]

Mode number	*Engine speed*	*Percent load*
1	Low idle	0
2	Intermediate	2
3	Intermediate	25
4	Intermediate	50
5	Intermediate	75
6	Intermediate	100
7	Low idle	0
8	Rated[b]	100
9	Rated	75
10	Rated	50
11	Rated	25
12	Rated	2
13	Low idle	0

[a] A similar procedure is used by California.
[b] Rated speed by the manufacturer for that engine.

C. Aircraft

Although the sources and type of pollutants from aircraft are quite similar to those from motor vehicles, the modes of operation are very different. Aircraft operations can be divided into those on the ground at the airport and those in flight after take-off and in landing. The ground operations include start-up and idle, taxi to the runway, idle before take-off, and the reverse processes—landing, taxi, and idle. In-flight operations cover climb-out to about 3000 feet (914 m) above the ground and descend from 3000 feet (914 m) to the ground. The 3000 feet (914 m) elevation is assumed to be the top of the temperature inversion below which atmospheric dispersion would be poor. Pollutants below this level would most likely affect the community near the airport, whereas above 3000 feet (914 m) the pollutants are dispersed more thoroughly.

Emission tests to determine compliance with the standards are made on an engine dynamometer or thrust-measuring stand as appropriate for the engine under test. Gaseous pollutants (hydrocarbons, carbon monoxide, and oxides of nitrogen) are measured under operating cycles which simulate engine operations at major airports (*23*). Smoke measurements for gas turbine engines are under operating modes very much the same as the gaseous pollutants. The density of smoke emitted from the engine is determined by conducting a representative portion of it through a filter. The smoke particles deposited on the filter are measured by means of a reflectometer. From the reflectance reading, the smoke number is calculated (*23*).

VIII. Motor Vehicle Emission Standards—United States

A. Light-Duty Vehicles

1. Crankcase Emissions

The United States federal standard for crankcase emissions, which applies in all fifty states, specifies that no pollutants from this source may be discharged into the atmosphere (*20*). Test procedures are not needed for determining compliance with the federal standard, because completely closed control systems are needed to comply with it.

2. Evaporative Emissions

The first standard for evaporative losses was established in California for the 1970 model light-duty vehicles. This standard limited evaporative emissions to 6 gm per test under the carbon trap method. A United States standard was established for 1972 models and limits evaporative emissions to 2 gm of hydrocarbons per test under the carbon trap procedure (*20*).

The California evaporative emission standard under the Sealed Housing for Evaporative Determinations (SHED) procedure is 6.0 gm per test. Although this appears to be a more lenient standard, it is not the case because the SHED method gives considerably higher values than does the carbon trap procedure. The USEPA also adopted an evaporative emission standard of 6 gm per test for light-duty vehicles and light-duty trucks in 1978 (*21*). California and USEPA use similar test procedures.

3. Exhaust Emissions

a. LIGHT-DUTY VEHICLES. The exhaust emission standards for light-duty vehicles have been the subject of the greatest attention in standard-setting efforts. Attention was centered on these vehicles because they constitute the largest percentage of all vehicles.

Exhaust emission standards have undergone much evolution since they were first promulgated in 1959. There has been a continuing trend to make the standards more stringent; the method of their expression has changed from a "concentration" to a "mass" basis, and test procedures and instrumentation have been modified. The first exhaust emission standards specified emissions on the basis of concentration (hydrocarbons 275 ppm by volume and carbon monoxide 1.5%). Hydrocarbons were required to be measured by nondispersive infrared analyzers (NDIR) (*7*). The NDIR is not a completely satisfactory method, because only a portion of the hydrocarbons is detected. However, no other method available at the time was suitable.

The baseline emission values (i.e., emissions from the population of precontrolled cars) were obtained from a field study by the Coordinating Research Council (CRC) in Los Angeles, California (*24*). There was considerable uncertainty, however, about the exact emissions values obtained; future studies were to show that the CRC investigation reported hydrocarbon concentrations that were too high.

It was recognized that emission standards based on exhaust pollutant concentrations are not equitable. Such standards do not reflect the lower mass emissions from small cars with low exhaust volumes and the higher mass emissions from large cars with high exhaust volumes. The lower volumes from small vehicles was partially taken into account in exhaust emissions standards for 1968 model light-duty vehicles. These standards, as shown in Table X, prescribed higher concentrations for the smaller cars. Although the lower exhaust volumes from small cars was recognized in these standards, the fact that exhaust flow rates vary within the displacement class and that very large vehicles have very high exhaust flow rates was not.

The belief by many individuals that the emission values determined by the CRC in 1956 were not representative of the true values led to a new study by the automobile industry, the United States Government, the State of California, Los Angeles County Air Pollution Control District, and the Automobile Club of Southern California (*13*). The study, called the Los Angeles Test Station Project, showed that the concentrations of hydrocarbons as determined in the earlier CRC project were too high. These findings were used by California to establish new exhaust emission standards for 1970 models. These new standards were 180 ppm hydrocarbons and 1.0% carbon monoxide.

Table X United States and California Motor Vehicle Emission Standards for 1968 and 1969, Reflecting Lower Exhaust Volumes for Smaller Vehicles

Engine displacement		*Exhaust emission standards*	
(in.^{3})	*(cm^3)*	*Hydrocarbons (ppm)*	*Carbon monoxide (%)*
50–100	820–1639	410	2.3
101–140	1640–2290	350	2.0
140 and above	2290 and above	275	1.5

California also adopted an emission standard for oxides of nitrogen at 350 ppm. This was the first time oxides of nitrogen had been included in emission standards. These standards for 1970 models were soon modified by action of the state legislature and by United States government test procedures.

A significant change in the manner in which the standards were expressed was made by the United States government when it specified the emission in terms of mass of pollutants (grams per miles) for 1970 models. These standards were developed to give the same degree of control as the more stringent ones that had been adopted for the 1970 models in California, but were specified in terms of mass emission rate: 2.2 gm/mile (1.4 gm/km) hydrocarbons and 23 gm/mile (14 gm/km) carbon monoxide.

These mass emission standards were still based on the seven-mode test procedure. However, the emissions were calculated by using an empirical equation that prescribed exhaust volumes according to the weight of the vehicle (*25*). The calculation included the use of a multiplying factor of 1.8 to convert the hydrocarbons measured by NDIR to total hydrocarbons measured by the flame ionization detector. This factor was an average value determined by comparing hydrocarbons as measured using NDIR, which measures hydrocarbons selectively, and by the flame ionization detector, which measures nearly all hydrocarbons. These new federal emission standards more nearly reflected mass emissions but did not do this very well, because the flow rates were assumed to be constant for vehicles of the same weight class but were not measured for each vehicle tested.

In 1968, the California Legislature enacted a law that placed exhaust emission standards into state statutes. As mentioned previously, this was a change from the previous practice by which administrative agencies adopted the standards. Hydrocarbon and carbon monoxide values for 1970 were similar to those adopted by the Department of Public Health, and the standards in the law became more stringent with time for the 1970–1974 models (see Table XIII). Another basic change was that the law required that the standards had to be met as a condition of registration of new cars in the state. Prior to this time California emission standards became effective when suitable control devices were found to be available and were given state approval.

The USEPA constant volume sampler test procedures for 1972 models was designed to correct the inaccuracies of the empirical formula, which related flow rate to vehicle weight. The procedures also included a new driving cycle and the use of the flame ionization detector instead of NDIR for measuring hydrocarbons. The effect of these new procedures was to change the numerical values of previously adopted exhaust emission standards even though the new procedure was not intended to increase the degree of control. The change in values was due to the new procedures and the use of the LA-4 driving route, which from the standpoint of emissions is more severe than the original CRC modes upon which the seven-mode test was based. The flame ionization detector measured all hydrocarbons, so there no longer was a need to apply the factor of 1.8 to NDIR readings.

1970 was the year when low emission standards were imposed on the, then forthcoming, 1975–1976 models. It was in that year that California adopted its low emission standards and United States Congress required that hydrocarbons and carbon monoxide in 1975 models be one-tenth the emissions of 1970 cars and oxides of nitrogen in 1976 models be one-tenth that of 1971 cars (Table XI).

At the time of adopting these 90% reduction requirements, the Congress recognized that there might be a problem in developing the necessary emission control systems within the time required. As a result, it permitted the administrator of USEPA to grant the automobile manufacturers a delay of one year in achieving the standards if he found that the manufacturers could not comply, in which case the administrator could establish interim standards for that one year.

Because of the questions raised by the feasibility of meeting the emission standards, Congress asked USEPA to contract with the National Academy of Sciences for a comprehensive study of this problem. The Academy established a Committee on Motor Vehicle Emissions to undertake this study. In turn, the committee formed eight panels of consultants on the more important subjects to be investigated. The Academy submitted its report to Congress and USEPA in February 1973 (*26*). Separate reports on the subject studies by the panels of consultants were also published by the Academy in 1973. The Academy was not able to

Table XI Comparison of United States and California Standards Adopted in 1970 for 1975 Model Light-duty Vehicles

	United States (1975)[a]		*California (1975)*[b]	
Pollutant	*(gm/mile)*	*(gm/km)*	*(gm/mile)*	*(gm/km)*
Hydrocarbons	0.41	0.25	0.9	0.6
Carbon monoxide	3.4	2.1	17	11
Oxides of nitrogen	3.1[c]	1.0	1.5	0.9

[a] Required by Congress, but later waived by USEPA (Table XII).
[b] Adopted by California Air Resources Board, later modified by USEPA waiver to California (Table XII).
[c] Was to become 0.4 gm/mile (0.25 gm/km) in 1976.

give an unqualified answer of "yes" or "no" to the question "Are the emission standards required by Congress technically feasible?" It stated that "achievement of the 1975 standards may be technologically feasible and achievement of the 1976 standards is likely but may not be attainable on the established schedule." Thus, the study by the Academy was unable to resolve the issue of the feasibility of the standards. It did, however, provide much valuable information on the many uncertainties and complex problems that are involved in the control of motor vehicle emissions and in achieving the standards.

After lengthly hearings in 1973, USEPA found that the 1975 standards legislated by the United States Congress could not be met, delayed the standards for one year, and set interim standards for 1975 and 1976 models as shown in Table XII.

In permitting delay for the oxides of nitrogen standard until 1977, USEPA stated it had re-evaluated the analytical procedure for measuring ambient nitrogen dioxide concentrations and found that the procedure had given values that were too high. As a result, the reported atmospheric concentrations were higher than they should have been. USEPA concluded that the emission standard for oxides of nitrogen was too severe and recommended that the 1970 Clean Air Act be amended to permit USEPA to establish emission standards for this pollutant.

Most of the automobile manufacturers announced that they would have to use catalytic control systems on all California cars and most of the cars in the other states if the 1975 interim emission standards were to be met. A new controversy developed over the selection of this control system because of its cost, need for unleaded gasoline, durability of the catalyst, increased emissions of sulfates, and other factors. Further con-

Table XII United States Motor Vehicle Exhaust Emission Standards for 1975 and Later Passenger Vehicles

	1975–1976				*1977 and later years*[a]			
	California		*Other states*		*California*		*Other states*	
Pollutant	*(gm/mile)*	*(gm/km)*	*(gm/mile)*	*(gm/km)*	*(gm/mile)*	*(gm/km)*	*(gm/mile)*	*(gm/km)*
Hydrocarbons	0.9	0.6	1.5	0.9	0.41	0.25	1.5	0.9
Carbon monoxide	9.0	5.6	15	9.3	9.0	5.6	15	9.3
Oxides of nitrogen	2.0	1.2	3.1	1.9	1.5	0.9	2.0	1.2

[a] See Table XIII for standards after 1977.

troversy arose because, in light of a nationwide shortage of gasoline, the fuel penalties that would result if the legislated exhaust emissions requirement for 1976–1977 were imposed became more significant.

Because of the continuing controveries, the United States Congress requested the National Academy of Sciences to undertake additional studies on both the feasibility of these emission standards and the reasonableness of the air quality standards adopted by USEPA in 1970 for the pollutants emitted by motor vehicles (Table II).

In a report to the United States Senate Committee on Public Works, a coordinating committee of the National Academy of Sciences and the National Academy of Engineering stated that the committee found no substantial basis for changing the USEPA air quality standards. The coordinating committee also reported that the United States air quality standard for carbon monoxide could probably be met even with some relaxation in the statutory emission standard and that the statutory emission standards for oxides of nitrogen may be more stringent than needed to achieve the air quality standard for nitrogen dioxide (*27*). The committee, however, was careful to point out the lack of adequate data on which to determine the validity of some of the air quality and vehicular emission standards.

In 1974, the Congress amended the United States Clean Air Act to continue the USEPA 1975 interim hydrocarbon and carbon monoxide standards for the 1976 models and required an interim standard of 2.0 gm/mile (1.2 gm/km) of oxides of nitrogen for the 1977 models. The subjects of what the exhaust emission standards for passenger cars should be and when such standards should go into effect remained controversial issues within the United States government, California, the automobile industry, and technical and environmental groups. Because of the delays in the application of the standards in the 1970 Clean Air Act, California adopted more stringent standards for the 1977 and later model passenger cars as shown in Tables XII and XIII (*28*).

In 1977, the United States Clean Air Act was again amended (*28a*) with respect to emissions from vehicles for model 1977 and beyond (Table XIII). The carbon monoxide standard for model years 1981–1982 may be waived up to 7.0 grams per mile by the United States Environmental Protection Agency for any model line if the Administrator finds for such vehicles that (1) public health does not require attainment of the statutory standard, (2) the waiver is in the public interest, (3) good-faith efforts have been made to meet the statutory standard, (4) control technology to meet the statutory standard is not available, and (5) a National Academy of Sciences study concurs with this assessment of availability of technology. The oxides of nitrogen standard was waived for specified small manufacturers for model years 1981–1982 and for all manufacturers for not more than 5% of a manufacturer's production or 50,000 vehicles, whichever is greater, to 1.5 grams per mile oxides of nitrogen to allow use of innovative technology. A similar waiver was made available for model year 1981–1984 light-duty vehicles using diesel engines, if necessary to permit the use of diesels in such vehicles.

A summary of the United States and California exhaust emission

Table XIII Summary of United States and California Motor Vehicle Emission Standards for Light-duty Vehicles, 1966–1981 Models[a]

			Hydrocarbons			*Carbon monoxide*			*Oxides of nitrogen*	
Year	*Standard*	*Test*	*(ppm)*	*(gm/mile)*	*(gm/km)*	*(%)*	*(gm/mile)*	*(gm/km)*	*(gm/mile)*	*(gm/km)*
1966–67	California	7-mode	275	—	—	1.5	—	—	—[b]	—[b]
1968–69	California	7-mode	—	—	—	—	—	—	—	—
	and United	50–100CID	410	—	—	2.3	—	—	—[b]	—[b]
	States	101–140CID	350	—	—	2.0	—	—	—[b]	—[b]
		Over 140CID	275	—	—	1.5	—	—	—[b]	—[b]
1970	California and United States	7-mode	—	2.2	1.4	—	23	14	—[b]	—[b]
1971	California	7-Mode	—	2.2	1.4	—	23	14	4	2.5
	United States	7-mode	—	2.2	1.4	—	23	14	—[b]	—[b]
1972	California	CVS-1	—	3.2	2.0	—	39	24	3.2[c]	2.0
	United States	CVS-1	—	3.4	2.1	—	39	24	—[b]	—[b]
1973	California	CVS-1	—	3.2	2.0	—	39	24	3	1.9
	United States	CVS-1	—	3.4	2.1	—	39	24	3	1.9
1974	California	CVS-1	—	3.2	2.0	—	39	24	2	1.2
	United States	CVS-1	—	3.4	2.1	—	39	24	3	1.9
1975–76	California	CVS-2	—	0.9	0.6	—	9.0	5.6	2.0	1.2
	United States	CVS-2	—	1.5	0.9	—	15	9.3	3.1	1.9
1977–79	California	CVS-2	—	0.41	0.25	—	9.0	5.6	1.5	0.9
	United States	CVS-2	—	1.5	0.9	—	15	9.3	2.0	1.2
1980–81	California	CVS-2	—	0.41	0.25	—	9.0	5.6	1.0	0.6
	United States	CVS-2	—	0.41	0.25	—	7.0	4.4	2.0	1.2
1981[d] and later	United States	CVS-2	—	0.41	0.25	—	3.4	2.1	1.0	0.6
1982 and later	California	CVS-2	—	0.41	0.25	—	9.0	5.6	0.4	0.25

[a] Same standards applies to passenger cars and light-duty trucks through 1974. After 1975, standards apply only to passenger cars.

[b] No standard.

[c] Hot CVS test.

[d] See text regarding waiver.

standards are shown in Table XIII for light-duty vehicles in the period 1966 through 1982. Also shown are the test procedures used to determine compliance with the standards. From 1966 through 1974, the standards apply to both passenger cars and light-duty trucks. After 1975, they apply only to passenger cars.

b. Light-Duty and Medium-Duty Trucks. Exhaust emissions standards for light-duty vehicles had always included passenger cars and light-duty trucks. This was changed in 1973 by a lawsuit brought by one of the truck manufacturers. As a result of this suit, the court ordered USEPA to provide separate standards for light-duty trucks. The argument for separate standards is that many of the trucks are designed to operate differently from passenger cars; the need for heavier duty operation makes it more difficult for these trucks to meet emission standards than for passenger cars to meet them. This action set a new policy in which passenger cars and small trucks are covered by different standards.

Questions were, however, raised as to which standards were appropriate for the trucks. Many light-duty trucks (pickups) operate very much like passenger cars and are used as a substitute for such cars. On the other hand, larger trucks in this class may be operated differently. Also, trucks between 6000 and 8500 pounds (2722 and 3856 kg), which had been defined as heavy-duty vehicles, are operated more like light-duty vehicles.

As a result of these considerations, California defined trucks from 6000 to 8500 pounds (2722 to 3856 kg) as medium-duty trucks and established separate exhaust emission standards for this new class but retained the basic light-duty vehicle test procedures (*20*). The USEPA, on the other hand, has proposed that trucks up to 3867 kg be classified as light-duty trucks. The applicable California exhaust emission standards for light-duty trucks (under 6000 pounds) (2722 kg) are shown in Table XIV. California's standards for medium-duty vehicles are given in Table XV. USEPA standards for light-duty trucks are listed in Table XIVA.

B. Heavy-Duty Vehicles

1. Crankcase Emissions

As in the case of light-duty vehicles, the United States crankcase emission standards for heavy-duty vehicles apply in all fifty states. These standards cover gasoline-powered vehicles and permit no discharge of crankcase vapors into the atmosphere (*20*). Because of the design of diesels, no crankcase emission standards are needed (see Section V,B,2).

Table XIV California Exhaust Emission Standards for Light-Duty Trucks, 1975 and Later Model Years[a]

	Equivalent inertia weight		*Hydrocarbons*		*Carbon monoxide*		*Oxides of nitrogen*	
Year	*(pounds)*	*(kg)*	*(gm/mile)*	*(gm/km)*	*(gm/mile)*	*(gm/km)*	*(gm/mile)*	*(gm/km)*
1975	All	All	2.0	1.2	20	12	2.0	1.2
1976–78	All	All	0.9	0.6	17	11	2.0	1.2
1979–80	0–3999	0–1813	0.41	0.25	9.0	5.6	1.5	0.9
	4000–6000	1814–2722	0.50	0.31	9.0	5.6	2.0	1.2
1981 and later	0–3999	0–1813	0.41	0.25	9.0	5.6	1.0	0.6
	4000–6000	1814–2722	0.50	0.31	9.0	5.6	1.5	0.9

[a] 6000 pounds (2722 kg) GVW rating or less.

Table XIVA United States Exhaust Emission Standards for Light-Duty Trucks, 1975 and Later Model Years[a]

	Hydrocarbons		*Carbon monoxide*		*Oxides of nitrogen*	
Year	*(gm/mile)*	*(gm/km)*	*(gm/mile)*	*(gm/km)*	*(gm/mile)*	*(gm/km)*
1975–78	2.0	1.2	20	12	3.1	1.9
1979 and later	1.7	1.1	18	11	2.3	1.4

[a] 1975–1978 light-duty trucks are 6000 pounds (2722 kg) GVW rating or less. Starting in 1979 these trucks are 8500 pounds (3856 kg) GVW rating or less.

Table XV California Exhaust Emission Standards for Medium-Duty Vehicles, 1978 and Later Years[a]

	Hydrocarbons		*Carbon monoxide*		*Oxides of nitrogen*	
Year	*(gm/mile)*	*(gm/km)*	*(gm/mile)*	*(gm/km)*	*(gm/mile)*	*(gm/km)*
1978–80	0.9	0.6	17	11	2.3	1.4
1981 and later	0.60	0.4	9	5.6	2.0	1.2

[a] For years prior to 1978, medium-duty truck standards were the same as those for heavy-duty vehicles (Table XVII).

2. Evaporative Emissions

At the time of writing this chapter, standards for evaporative losses from heavy-duty vehicles had been adopted only by Cailfornia. No standards were set for pre-1973 model heavy-duty vehicles. From 1973 to 1977 models, the standard limits these emissions to 2 gm per test based on the carbon trap test procedures. In 1978, the standard becomes 6.0 gm per test based on the Sealed Housing Evaporative Determination (SHED) method (*29, 21*). The test procedures permit an engineering evaluation to determine compliance of control systems with the standards if the systems are an extension of those approved for light-duty vehicles. Evaporative standards are not needed for diesel engines because the fuel is less volatile than gasoline and fuel injectors are used instead of carburetors (see Section V,B,2).

3. Exhaust Smoke Emissions

Most of the concern over smoke from vehicles is with diesel engines. These engines are the only ones covered by standards for smoke emissions. The United States smoke emission standards apply in all fifty states and are shown in Table XVI. As can be seen from this table, the standard is based only on acceleration and lugging modes, since these are the two which produce the most smoke.

4. Exhaust Gaseous Emissions

The original California emission standards made no distinction between light- and heavy-duty vehicles (*4*); however, a different schedule was developed for implementing the exhaust emission standards. The exhaust control systems were required on light-duty vehicles in 1966 models and on heavy-duty gasoline-powered vehicles in 1969 models. The lag in implementation of the exhaust standards for the heavy-duty heavy vehicles occurred because emphasis was placed first on controlling emissions from passenger cars. Automobiles are present in much greater

Table XVI United States Emission Standards for Smoke from Diesel Powered Heavy-Duty Vehicles, 1970 and Later Years

Model year	*Opacity of smoke emission (%)*		
	Acceleration mode	*Lugging mode*	*Peak values*
1970–1973	40	20	—
1974 and later	20	15	50% during peaks in either modes

numbers, and control systems for these vehicles can usually be extended to the larger vehicles. Also, very little was known in the early 1960s about the emissions and operating modes of large vehicles.

As was the case for light-duty vehicles, the first emission standards for heavy vehicles were expressed as the concentration of the pollutant (parts per million by volume). The change to mass emission standards on a grams per kilometer basis for the light vehicle has not been followed for the heavy vehicles. Because of their size and method of operation, large heavy-duty vehicles have much higher exhaust volumes and would not be able to reduce the concentration of pollutants to levels needed to comply with the recent stringent mass standards for automobiles. Instead, emission standards for large vehicles have been expressed on the basis of grams per horsepower-hour. This standard reflects the mass of emissions while producing a given amount of work. It is calculated by dividing the mass rate (grams per hour) by the weighted brake horsepower values for the cycle (*20*). The emission levels specified in the standards for heavy vehicles were generally calculated to represent the same degree of emission control as required for light-duty vehicles.

Another departure from light-duty vehicle standards was the expression in the standards for 1974 of emissions in terms of hydrocarbons plus oxides of nitrogen. This was done to permit greater flexibility in the heavy vehicles meeting the standards under full-load operations. The California exhaust emission standards for hydrocarbon, carbon monoxide, and oxides of nitrogen are shown in Table XVII, United States exhaust emission standards are given in Table XVIII. The USEPA has proposed more stringent emission standards for 1979.

C. Motorcycles

As was discussed earlier in this chapter (Section V,B,3), motorcycles have been one of the last class of vehicles to be considered for emissions standards and controls. With the finding that motorcycles may emit large amounts of hydrocarbons and carbon monoxide, interest developed in the United States for also applying emission standards to these vehicles. The USEPA felt that motorcycles should comply with the exhaust emission standards for automobiles that were included in the 1970 Clean Air Act (*30*). However, more time would be allowed for compliance since adequate lead time would need to be provided for developing the control systems for motorcycles. Because USEPA delayed the adoption of its proposed standards, the California Air Resources Board established it's own exhaust emission standards (*31*). The California emission standards for motorcycles were limited to hydrocarbons and are shown in Table

Table XVII California Exhaust Emission Standards for Heavy-Duty Vehicles[a]

Year[b]	*Carbon monoxide* (%)	*Carbon monoxide* (gm/BHP-hour)[c]	*Hydrocarbons* (ppm)	*Hydrocarbons* (gm/BHP-hour)[c]	*Oxides of nitrogen* (gm/BPH-hour)[c]	*Hydrocarbons plus oxides of nitrogen* (gm/BPH-hour)[c]
1969–71	1.5	—	275	—	—[d]	—[d]
1972	1.0	—	180	—	—[d]	—[d]
1973–74	—	40	—	—	—	16
1975–76	—	30	—	—	—	10
1977–78[e]	—	25	—	—	—	5
	—	25	—	1.0	7.5	—
1979[e]	—	25	—	—	—	5
	—	25	—	1.5[f]	7.5	—
1980–82[e]	—	25	—	—	—	5
	—	25	—	1.0[f]	—	6.0
1983 and later[e]	—	25	—	0.5[f]	—	4.5

[a] From 1968 to 1972 standards apply to gasoline-powered vehicles only. After 1973 Standards apply to both gasoline- and diesel-powered vehicles.
[b] Effective in 1978; standards do not apply to medium-duty vehicles.
[c] Grams per brake horsepower-hour.
[d] No standards.
[e] Manufacturers may use either set of standards.
[f] Flame ionization detector required in 1979 and later years. This results in higher readings for hydrocarbons.

Table XVIII United States Exhaust Emission Standards for Heavy-Duty Vehicles over 6000 Pounds (2722 kg)

Model year	*Type of vehicle*	*Carbon monoxide*	*Hydrocarbons*	*Oxides of nitrogen*
Through 1973	Gasoline	1.5%	275 ppm	None
1974 and later	Gasoline and diesel	40 gm/BHP-hour[a]	16 gm/BHP-hour[a,b]	

[a] Grams per brake horsepower-hour.
[b] Hydrocarbons plus oxides of nitrogen.

XIX. In 1977, USEPA prescribed the emission standards for motorcycles, which are shown in Table XIXA. It also required that there be no crankcase emissions from these vehicles (*31a*).

D. Control Systems—Used Vehicles

California has established emission standards for the purpose of requiring exhaust emission control systems on used vehicles. These stan-

Table XIX California Exhaust Emission Standards for Hydrocarbons from Motorcycles

Date of manufacture	*Emission standard for hydrocarbons (gm/km)*
After January 1, 1978	10.0
After January 1, 1980	5.0
After January 1, 1982	1.0

Table XIXA United States Exhaust Emission Standards for Motorcycles

Year	*Engine displacement (cm^3)*	*Carbon monoxide (gm/km)*	*Hydrocarbons (gm/km)*
1978–79	50–170 (3.1–10.4 CID)[a]	17	5.0
	170–750 (10.4–45.8 CID)[a]	17	$5.0 + 0.0155 \times (D - 170)$[b]
	over 750 (over 45.8 CID)[a]	17	14
1980	All	12	5.0

[a] CID = cubic inch displacement.
[b] D = engine displacement in cubic centimeters (cm^3).

dards cover two groups of light-duty vehicles; the 1955–1965 models (*32*) and the 1966–1970 models (*33*). The 1955–1956 vehicles were manufactured prior to the installation of any exhaust emission control systems and are considered uncontrolled vehicles. The 1966–1970 models when new were equipped with exhaust control systems for hydrocarbons and carbon monoxide, but not for oxides of nitrogen (since this latter pollutant was first controlled in California with the 1971 models).

The exhaust emission standards for approval of control devices on used vehicles in California are shown in Table XX.

Table XX California Exhaust Emission Standards for Approval of Control Devices on Used Vehicles

Model Year	*Test*	*Hydrocarbons (ppm)*	*Carbon monoxide (%)*	*Oxides of nitrogen*
1955–1965	Seven mode	350	2.0	800 ppm
1966–1970	Modified CVS	—	—	42% reduction

The standards for 1955–1965 vehicles were used to accredit emission control systems that are required to be installed on vehicles of these model years upon change of ownership or first registration from out-of-state in the larger metropolitan areas of California. The standards for 1966–1970 models led to the accreditation of control systems that are required to be installed on 1966–1970 models upon change of ownership or first registration from out-of-state throughout California.

California, since 1965, has required crankcase control devices to be installed on vehicles that did not have such systems when they were first registered in the state. The crankcase control devices must be installed upon sale of the vehicle or first registration of a car brought into the state. This program covers the larger metropolitan regions. No emission standard is required for these devices, since they are required to collect all of the crankcase vapors.

E. Inspection Standards—Used Vehicles

1. Visible Emissions

The first emission standards applied to motor vehicles operating on streets and highways were for smoke. These standards were based on Ringelmann numbers or Ringelmann equivalent opacity. Smoke from automobiles has been decreasing because of improvements to engines and because of compliance with emission standards for hydrocarbons and carbon monoxide. As a result, there is not as great a need for enforcement programs directed at smoke from automobiles. However, excessive smoke from poorly maintained older cars can still be a problem that requires enforcement action.

Smoke from diesel-powered trucks and buses remains a much more serious problem than smoke from cars. As a result, smoke enforcement programs for motor vehicles are mainly directed at diesel vehicles. Since 1970, new diesel engines in the United States have had to comply with smoke emission standards (Table XVI). This will in time decrease the problem of smoke from diesels. However, smoke enforcement programs for these vehicles will still be needed; particularly where there are large numbers of diesels, the engines are old, and proper maintenance and operating procedures are not followed.

The standards that have been adopted follow a somewhat similar pattern by limiting smoke to a prescribed Ringelmann equivalent opacity over a given time or distance traveled. The main differences are in the amount of smoke and period time prescribed. In a few instances (Alaska, Illinois, and District of Columbia) the standards designate that there shall be no visible smoke (0% opacity) from automobiles; this requirement, however, is not extended to diesels. Most jurisdictions prohibit emissions in excess of 20–40% opacity for a period of 5–15 seconds. Instead of time, a few places use distance traveled by the vehicle.

The same standard (California, for example) is sometimes applied to both automobiles and diesels; in other instances, different standards are used for the two types of vehicles. When this is done, the standard for diesels is not as stringent. Several jurisdictions permit more smoke at higher elevations. This is more common for diesels and recognizes that the lower air–fuel ratio at higher elevation causes the vehicle to smoke more. Also, some of the regulations specify more stringent smoke limits for diesel engines manufactured after 1970–1973 to reflect the smoke standards applied to new engines. In some cases, such as Los Angeles, California, and Wayne County, Michigan, one standard is used for all stationary and mobile sources. Table XXI shows typical visible emission standards in the United States (*15*).

Table XXI Typical Visible Emission Standards for Operating Vehicles—United States (*15*)

Jurisdiction	*Vehicle*	*Ringelmann equivalent opacity (%)*	*Time or distance*	*Remarks*[a]
California	Autos and diesels	20	10 sec	Under 3000 feet (914 meters) and new vehicles after January 1, 1971
		40	10 sec	Under 3000 feet (914 meters) and new vehicles prior to January 1, 1974
Colorado	Autos	20	10 sec	
Colorado	Diesels	30	10 sec	Under 8000 feet (2438 meters)
Colorado	Diesels	40	10 sec	Over 8000 feet (2438 meters)
District of	Autos	0		
Columbia	Diesels	20	5 sec	
Oregon	Autos	10	7 sec	40% maximum
Oregon	Diesels	10	7 sec	Under 3000 feet (914 meters) (40% maximum)
Oregon	Diesels	20	7 sec	Over 3000 feet (914 meters) (60% maximum)
Utah	Autos	20	100 yards (91 meters)	
Utah	Diesels	20	100 yards (91 meters)	After January 1, 1973
Utah	Diesels	30	100 yards (91 meters)	Prior to January 1, 1973

[a] Elevation, effective date, and maximum opacity.

2. Gaseous Emissions

A number of components can affect the gaseous emissions from motor vehicles, and these components must be kept in proper adjustment and maintenance if the original emission standards are to be met by vehicles in private use. The emission control components, spark setting, carbureted air–fuel mixture, air cleaner, and spark plugs are some of the more important items that require continued adjustments and maintenance.

The need to comply with the transportation control requirement of the United States Clean Air Act and the dependence of low emissions upon maintenance and adjustment has led to vehicle emission programs by some state and local governmental agencies in the United States. These programs are designed to measure the emissions from vehicles in private use and to require appropriate repairs and adjustments if designated inspection standards are exceeded.

Inspection standards must be based upon short and relatively simple test procedures so that the costs and inspection times will be minimized. The procedures used to determine compliance with the emission standards for new vehicles are far too detailed and costly for inspection purposes (see Section VII,A). In addition, they require a cold start of the vehicle; this would be impractical for used vehicle emission inspection programs.

Because the inspection test must be very different, it is not possible to apply the same standards used to approve new vehicles. Separate inspection standards must therefore be developed to reflect the new test conditions. Inspection procedures under which emissions are to be measured generally fall into two types: (a) idle condition and (b) a few driving modes with vehicles on a dynamometer.

Inspection standards for gaseous pollutants have been adopted by several states and local agencies in the United States; among these are Arizona, California, New Jersey, New York, and Chicago, Illinois. The standards for New Jersey are shown in Table XXII (*34, 35*). These apply to light-duty gasoline-powered vehicles when the engine is running at idle. Vehicles are to be tested once a year.

The inspection, which was voluntary in New Jersey from July 1972 through January 1974, became mandatory in February 1974. As can be seen from Table XXII, the standards become more stringent with time to reflect the development of the program and the more limiting approval emission standards in late model new vehicles.

California has adopted inspections standards that are used in that state's random highway roadside inspection program. As in New Jersey, an idle test is used. The California idle emission inspection standards

Table XXII State of New Jersey Emission Inspection Standards for Light-Duty Vehicles[a]

	Emission standard[b]					
	Effective date Feb. 1, 1974		Effective date Aug. 1, 1975		Effective date Feb. 1, 1976	
Model year	*Carbon monoxide (%)*	*Hydrocarbons (ppm)*	*Carbon monoxide (%)*	*Hydrocarbons (ppm)*	*Carbon monoxide (%)*	*Hydrocarbons (ppm)*
Pre-1968	10.0	1600	8.5	1400	7.5	1200
1968–1969	8.0	800	7.0	700	5.0	600
1970–1974	6.0	600	5.0	500	4.0	400
1975 and later	—	—	3.0	300	2.0	200

[a] Voluntary compliance prior to February 1974, mandatory after that date.
[b] Carbon monoxide and hydrocarbons measured by nondispersive infrared instrument.

are given in Table XXIII (*36*). As can be seen, there are different standards depending upon the number of cylinders, vehicle model, and type of emission control system. In 1973, the California Legislature established an annual periodic inspection program for all vehicles in much of Southern California, where the state's most severe photochemical air pollution is experienced. One of the most significant requirements of the law is that a dynamometer cycle must be used as the inspection test.

Table XXIII California Highway Exhaust Emission Inspection Standards for Light-Duty Vehicles (Idle Condition)

Model year	*Number of cylinders*	*Hydrocarbons*[a] *(ppm)*		*Carbon monoxide (%)*	
1955–1965	4 or less	1900		8.0	
	5 or more	1200		8.0	
1966–1967	4 or less	1900		8.0	
		AI[b]	*Others*[c]	*AI*[b]	*Others*[c]
1966–1967	5 or more	400	500	5.5	7.0
1968–1970	4 or less	500	650	5.5	7.0
	5 or more	400	500	5.5	7.0
1971 and later	4 or less	450	600	3.5	5.0
	5 or more	250	350	3.0	4.0

[a] As hexane, measured with nondispersive infrared analyzer.
[b] Air injection emission control system.
[c] Emission control system other than air injection.

Table XXIV California Inspection Standards for Light-Duty Vehicles (Dynamometer Test Procedures)

Model year	Number of cylinders	Idle: Hydrocarbons[a] (ppm)		Idle: Carbon monoxide (%)		Low cruise: Hydrocarbons (ppm)	Low cruise: Carbon monoxide (%)	Low cruise: Oxides of nitrogen (ppm)	High cruise: Hydrocarbons (ppm)	High cruise: Carbon monoxide (%)
1955–1965	4 or less	1900		8.0		1200	7.0	2500	1200	6.5
	5 or more	1200		8.0		1000	6.0	2500	1000	5.5
1966–1967	4 or less	1900		8.0		1200	7.0	2500	1200	6.5
		AI[b]	Other[c]	AI[b]	Other[c]					
1966–1967	5 or more	400	500	5.5	7.0	500	4.5	2500	500	4.0
1968–1970	4 or less	500	650	5.5	7.0	600	5.0	2500	600	4.5
	5 or more	400	500	5.5	7.0	500	4.5	2500	500	4.0
1971 and later	4 or less	450	600	3.5	5.0	500	4.0	2500	500	3.5
	5 or more	250	350	3.0	4.0	400	3.0	2500	400	2.5

[a] Hydrocarbons measured as hexane with NDIR analyzer.
[b] Air injection control systems.
[c] Control systems other than air injection.

This was done because of the concern that an idle test samples only one mode and does not produce meaningful data on oxides of nitrogen.

The inspection modes prescribed by the State Air Resources Board in the dynamometer test are: (a) idle, (b) low cruise load, and (c) high cruise load. Inspection standards for the three conditions are given in Table XXIV (*36*). The dynamometer inspection program was in the pilot stage. Experience with the program will be used to decide if it should be extended to other parts of California. It will be interesting to learn if a loaded dynamometer procedure will be superior to an idle test in reducing emissions from vehicles in private use.

IX. Aircraft Emission Standards—United States

In 1973, USEPA promulgated emission standards for aircraft (*23*). These standards include smoke for new and in-use gas turbine engines and hydrocarbons, carbon monoxide, and oxides of nitrogen from new gas turbine and piston-propulsion engines. The standards for smoke are on the basis of smoke numbers; those for the gaseous contaminants are on the basis of kilograms of emissions per 454 kg-thrust hour per test cycle for gas turbines and kilograms per rated power per cycle for piston engines. The aircraft emission standards are in Table XXV; the definition and examples of aircraft engine classes covered by the standards are shown in Table XXVI. The smoke standards, expressed in terms of the dimensionless smoke numbers, for aircraft engines according to thrust and to horsepower are shown in Figures 4 and 5, respectively. Smoke densities are measured by means of a reflectometer (*23*).

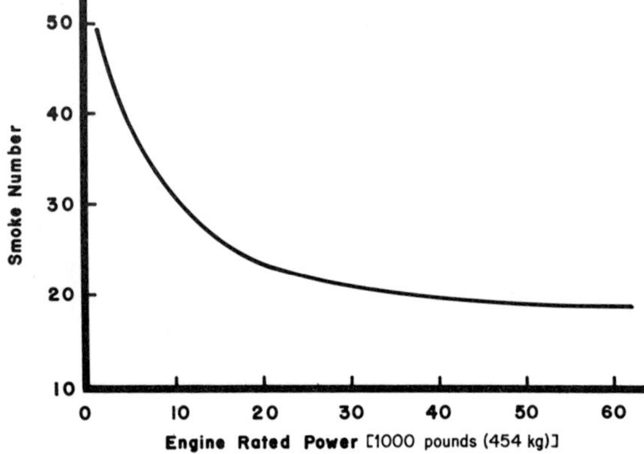

Figure 4. United States smoke standards for aircraft (engine power rated in terms of thrust).

Table XXV United States Smoke, Hydrocarbons, Carbon Monoxide, and Oxides of Nitrogen Emission Standards for Aircraft

Engine type[a]	*Hydrocarbons*	*Carbon monoxide*	*Nitrogen oxides*	*Smoke*[b]
New (in use) Class T4 gas turbines manufactured (in use) on or after January 1, 1974	None	None	None	Shall not exceed smoke number 30
New (in use) Class T2 gas turbines with greater than 13,106 kg thrust manufactured (in use) on or after January 1, 1976	None	None	None	Shall not exceed applicable smoke number from Figure 4
New (in use) Class T3 gas turbines manufactured (in use) on or after January 1, 1978	None	None	None	Shall not exceed smoke number 25
New Class T1 gas turbines manufactured on or after January 1, 1979	Not more than 1.6 lbs/1000 lbs (0.7 kg/454 kg) thrust hours/cycle	Not more than 9.4 lbs/1000 lbs (4.2 kg/454 kg) thrust hours/cycle	Not more than 3.7 lbs/1000 lbs (1.7 kg/454 kg) thrust hours/cycle	Shall not exceed applicable smoke number from Figure 4
New Class T2, T3, and T4 gas turbines manufactured on or after January 1, 1979	Not more than 0.8 lbs/1000 lbs (0.4 kg/454 kg) thrust hours/cycle	Not more than 4.3 lbs/1000 lbs (1.9 kg/454 kg) thrust hours/cycle	Not more than 3 lbs/1000 lbs (1.4 kg/454 kg) thrust hours/cycle	Shall not exceed applicable smoke number from Figure 4
New Class P2 gas turbines manufactured on or after January 1, 1979	Not more than 4.9 lbs/1000 lbs (2.2 kg/454 kg) thrust hours/cycle	Not more than 26.8 lbs/1000 lbs (12.1 kg/454 kg) thrust hours/cycle	Not more than 12.9 lbs/1000 lbs (5.8 kg/454 thrust hours cycle	Shall not exceed applicable smoke number from Figure 5
New Class T2, T3, and T4 gas turbines manufactured on or after January 1, 1981	Not more than 0.4 lbs/1000 lbs (0.2 kg/454 kg) thrust hours/cycle	Not more than 3 lbs/1000 lbs (1.4 kg/454 kg) thrust hours/cycle	Not more than 2.9 lbs/1000 lbs (1.4 kg/454 kg) thrust hours/cycle	Shall not exceed applicable smoke number from Figure 5
New piston propulsion engines manufactured on or after December 31, 1979	Not more than 0.00190 lbs (0.0009 kg)/rated power-cycle	Not more than 0.42 lbs (0.019 kg)/rated power/cycle	Not more than 0.0015 lbs (0.0007 kg)/rated power/cycle	None
New onboard power generating units manufactured on or after January 1, 1979	Not more than 0.4 lbs (0.2 kg)/1000 hp-hour of power output	Not more than 5 lbs (2.3 kg)/1000 hp-hour of power output	Not more than 3 lbs (1.4 kg)/1000 hp-hour of power output	None

[a] For definitions of engine classes, see Table XXVI.
[b] Smoke number is a dimensionless term that quantifies smoke emissions as determined by a reflectometer.

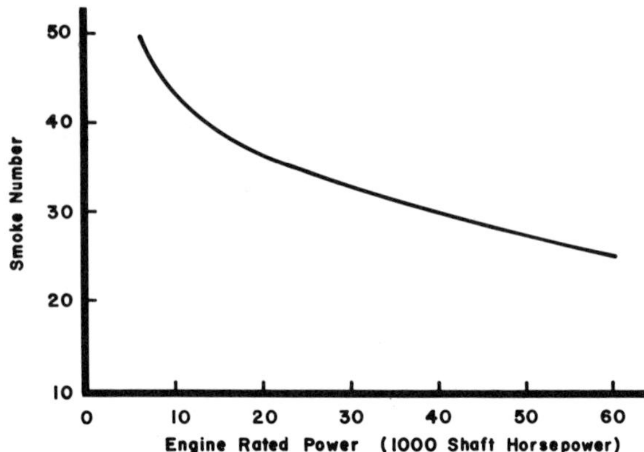

Figure 5. United States smoke standards for aircraft (engine power rated in terms of shaft horsepower).

Table XXVI Definitions and Examples of Aircraft Engine Classes Covered by Standards in Table XXV

CLASS T1—All aircraft turbofan or turbojet engines except engines of class T5 of rated power less than 8000 pounds (3628 kg) thrust. *Example aircraft;* Lear Jet, Grumman Gulstream, Lockheed Jetstar and Electra, Fairchild F27, and Convair 540.

CLASS T2—All aircraft turbofan or turbojet engines except engines of class T3, T4, and T5 of rated power of 8000 pounds (3628 kg) thrust or greater. *Example aircraft;* Boeing 707, 727, 737, and 747; McDonnell Douglas DC-8, DC-9, and DC-10; and Lockheed L1011.

CLASS T3—All aircraft gas turbine engines of the JT3D model family.

CLASS T4—All aircraft gas turbine engines of the JT8D model family.

CLASS T5—All aircraft gas turbine engines for propulsion of supersonic aircraft.

CLASS P1—All aircraft piston engines, except radial engines. *Example aircraft:* Cessna 150, Piper Cherokee 140, Beechcraft Queen Air, etc.

CLASS P2—All aircraft turboprop engines.

X. Motor Vehicle Operating Cycles and Emission Standards—Other than the United States

Interest in the adoption and enforcement of motor vehicle emission standards for gaseous components from new vehicles quickly extended to other countries after standards were first developed in the United States. Many of the factors that stimulated concern in the United States, such as the increasing number of vehicles and elevated ambient concentrations of vehicular pollutants in large cities, were present in these countries also. An added factor was that emission control systems had become available because of the regulatory programs in the United States.

As in the United States, it has been the practice in the other countries to require a higher degree of emissions control with time. In general, the standards are not as stringent as those in the United States for the same year of application, nor has as much attention been given to exhaust emissions from heavy-duty vehicles and evaporative losses. Japan has been particularly active in setting motor vehicle emission standards, and its requirements for 1975 and later model passenger cars are more like those of the United States in the degree of control required.

A. Economic Commission for Europe (ECE)

Western European countries working jointly through the Economic Commission for Europe (ECE) have carried out studies of driving conditions in selected European cities, established testing procedures, and have recommended uniform emission standards for adoption by the countries involved. These standards and test procedures are used, or were adopted with some modifications, by a number of countries.

There are three ECE test procedures—type I is for exhaust emissions from vehicles of 882–7716 pounds (400–3500 kg), type II is for carbon monoxide at idle, and type III is for crankcase vapors (*37*). The type I test includes a 6-hour soak and a fifteen-mode chassis dynamometer driving cycle, which is repeated four times without interruption. Total exhaust volume is sampled in a plastic bag, and the concentration of pollutants and accumulated volume of exhaust are measured. The hydrocarbons and carbon monoxide are determined by nondispersive infrared (NDIR) analyzers. The type I test cycle represents a shorter distance and lower average speed than the seven-mode or CVS cycles used in the United States. The equivalent distance traveled in the type I test is 2.5 miles (4.05 km), and the average speed is 11.8 miles/hour (18.7 km/hour), while the CVS cycle represents an equivalent distance of 7.5 miles (12.1 km) and an average speed of 19.7 miles/hour (31.6 km/hour). The lower average speed results in higher emissions of hydrocarbons and carbon monoxide. The ECE cycle is graphically represented in Figure 6, with the seven-mode cycle shown as a dashed line.

The type II test prescribes a warmed-up carbon monoxide measurement that is made by inserting a sampling probe in the tailpipe immediately after the fourth cycle in the type I test. Type III test is for crankcase emissions and includes idle and 31 miles/hour (50 km/hour) constant speed modes on a chassis dynamometer.

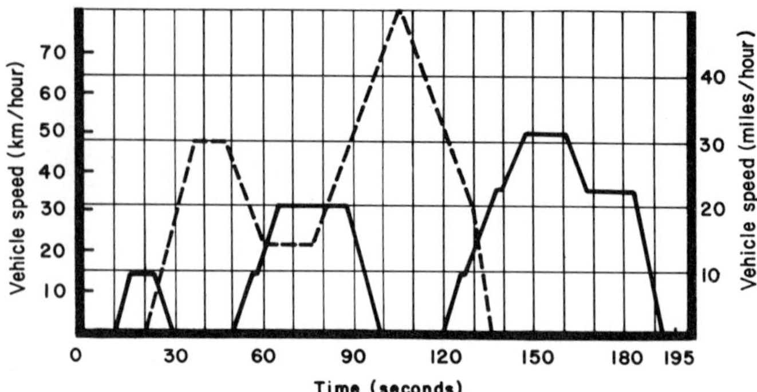

Figure 6. Comparison of Economic Commission for Europe type I (km/hour) (——) and seven-mode (miles/hour) (- - -) cycles.

Emission standards for new motor vehicles recommended by the Economic Commission for Europe have been adopted in a number of countries, and the standards, with some modification, have been used or proposed in additional countries. Standards for crankcase emission (ECE test III) and carbon monoxide at idle (test II) are shown in Table XXVII; the standards for exhaust hydrocarbons and carbon monoxide (ECE type I test) are shown in Table XXVIII. The countries that use the standards are listed in these tables.

As can be seen, the exhaust emission standards in Table XXVIII covers a range of values for vehicles from 882 to 7718 pounds (400 to 3500 kg) in weight. The specific value applied depends upon the size of vehicle as identified in the ECE type I test procedure. One set of standards is used for approval of prototype control systems; a more lenient second set is applied to production vehicles. The production standards for hydrocarbons permit 30% higher emissions than the comparable prototype approval

Table XXVII Economic Commission for Europe Emission Standards for Idle Carbon Monoxide (Type II Test) and Crankcase Vapor (Type III Test) for New Motor Vehicles over 882 pounds (400 kg) in Weight—Austria, Belgium, Federal Republic of Germany, France, Hong Kong, Italy, Luxembourgh, Netherlands, Norway, Spain, Switzerland, and United Kingdom

	Emission standard	
Test	*Carbon monoxide*	*Hydrocarbons*
Type II	4.5% by volume	
Type III		0.15% of supplied fuel

Table XXVIII Economic Commission for Europe Type I Test Exhaust Emission Standards for New Motor Vehicles 882–7718 Pounds (400–3500 kg) in Weight—Austria, Belgium, Federal Republic of Germany, France, Italy, Luxembourgh, Netherlands, Norway, Spain, Switzerland, and United Kingdom

		Prototype approval standards (gm/test)		*Production vehicle standards (gm/test)*	
Date	*Vehicle weight (kg)*	*Hydrocarbons*	*Carbon monoxide*	*Hydrocarbons*	*Carbon monoxide*
Prior to October 1, 1975	400–750	8.0	100	10.4	120
	750–850	8.4	109	10.9	131
	850–1020	8.7	117	11.3	140
	1020–1250	9.4	134	12.2	161
	1250–1470	10.1	152	13.1	182
	1470–1700	10.8	169	14.0	203
	1700–1930	11.4	186	14.8	223
	1930–2150	12.1	203	15.7	244
	2150–3500	12.8	220	16.6	264
After October 1, 1975	400–750	6.8	80	8.8	96
	750–850	7.1	87	9.3	105
	850–1020	7.4	94	9.6	112
	1020 1250	8.0	107	10.4	129
	1250–1470	8.6	122	11.1	146
	1470–1700	9.2	135	11.9	162
	1700–1930	9.7	149	12.6	178
	1930–2150	10.3	162	13.3	195
	2150–3500	10.9	176	14.1	211

standards; production standards for carbon monoxide allow 20% more. Different standards for production vehicles recognize that prototype engines and control systems can be made under more carefully controlled conditions than can those mass produced on assembly lines. The more stringent standards, which apply to vehicles after October 1, 1975 (Table XXVIII) restrict carbon monoxide emissions to 80% of levels allowed prior to that date and hydrocarbon emissions to 85% of previous standards (*38*).

In 1974, the ECE proposed oxides of nitrogen exhaust emission standards for light-duty vehicles as a part of the ECE type I test. These standards, which were proposed for 1977 application range from 10 to 16 gm of nitrogen dioxide per test, depending upon the vehicle inertia class. The measurement of oxides of nitrogen is by means of a chemiluminescent analyzer. Consideration is also being given to a possible reduction in 1980 of 30–50% in hydrocarbon and carbon monoxide exhaust emission standards (ECE type I test procedure). Vehicular smoke emission standards for France and Switzerland (countries otherwise conforming to ECE standards) are shown in Table XXIX.

B. Japan

Japan uses both a hot start ten-mode cycle and a cold start eleven-mode cycle for measuring the emissions from passenger cars with a capacity of ten or fewer persons and for other vehicles that have a gross weight of 5512 pounds (2500 kg) or less (*39, 40*). The exhaust under both cycles is collected by a constant volume sampling technique. Analytical instruments include the hydrogen flame ionization detector (FID) for hydrocarbons, the nondispersive infrared (NDIR) analyzer for carbon monoxide, and the chemiluminescence (CL) analyzer for oxides of nitrogen.

As in the case with the ECE type I test, the Japanese ten-mode cycle represents at a speed considerably less than the United States seven-mode or CVS cycles. The ten-mode cycle test, which is illustrated in Figure 7, has an average speed of 11 miles/hour (18 km/hour). The eleven-mode cycle is shown in Figure 8. It has an average speed of 19

Table XXIX Motor Vehicle Smoke Emission Standards for Australia, Brazil, Finland, France, Mexico, Sweden, and Switzerland (*15*)

		Ringel-mann no.	*Time*	*Bosch no.*	*Hartridge no.*	*Bacharach no.*
Australia	All cars	1	—	—	—	—
Brazil	All cars	2	—	—	—	—
Finland	Inspection	—	—	3.5	—	—
Finland	Roadway	—	—	4.5	—	—
France	Passenger	—	—	—	40	—
France	<6 tons	—	—	—	45	—
France	6–19 tons	—	—	—	50	—
France	>19 tons	—	—	—	60	—
Mexico	Gasoline	—	10 sec	—	—	—
Mexico	Diesels	2	15 sec	—	—	—
Mexico	Diesel locomotives	3	30 sec/30 min	—	—	—
Sweden	Diesel buses	—	—	2.5	30	—
Sweden	Diesel–other	—	—	3.5	45	—
Switzerland	<3 liters	—	—	6	—	6.5
Switzerland	3–5 liters	—	—	5.5	—	6.0
Switzerland	5–8 liters	—	—	5.0	—	5.5
Switzerland	>8 liters	—	—	4.5	—	5.0

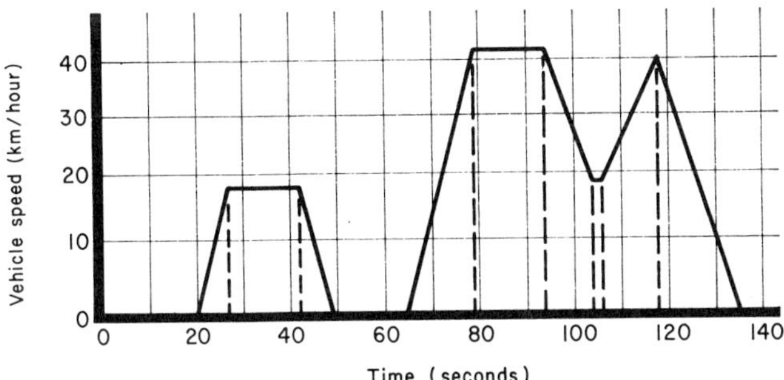

Figure 7. Ten-mode hot-start test procedure for exhaust emissions from light-duty vehicles in Japan. Incremental 10 km (6 mile) vehicle preconditioning followed by one warm-up test cycle; 135 second 10-mode test cycle repeated 5 times; constant volume sampling analysis. Total test cycle time, 675 seconds; average speed during test cycle, 18 km/hour (11 miles/hour); equivalent distance traveled—3.3 km (2.1 miles).

miles/hour (30.6 km/hour) which is almost as high as that of the United States CVS test (19.7 miles/hour). The ten-mode cycle was used for vehicles prior to 1975; after 1975, new light-duty vehicles in Japan were required to comply with the emission standards prescribed for both the ten- and eleven mode cycles. Exhaust emission for heavy-duty vehicles in Japan are measured under a chassis or engine dynamometer cycle which specifies engine speed in revolutions per minute, manifold vacuum, and driving time in six driving modes as shown in Table XXX.

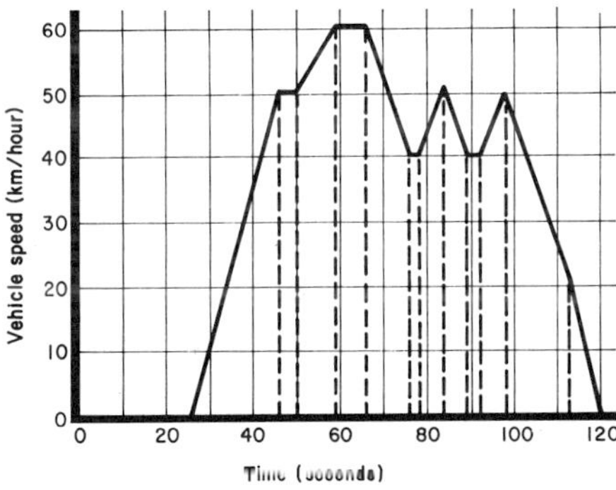

Figure 8. Eleven-mode cold-start test procedure for exhaust emissions from light-duty vehicles in Japan. No vehicle precondition requirements specified; 6 hour soak; cold start plus 25 second idle; 120 second 11-mode test cycle repeated 4 times; constant volume sampling analysis. Total test cycle time, 480 seconds; average speed during test cycle, 30.6 km/hour (19 miles/hour); equivalent distance traveled, 4.1 km (2.5 miles).

Table XXX Six-Mode Test Procedures for Exhaust Emissions from Heavy-Duty Vehicles in Japan

Driving mode	*Engine rpm*	*Intake manifold vacuum (mmHg)*	*Driving time*
1	Idling		3 minutes
2	2000 ± 100	125 ± 5	3 minutes
3	3000 ± 100	125 ± 5	3 minutes
4	3000 ± 100	200 ± 5	3 minutes
5	2000 ± 100	420 ± 5	3 minutes
6	2000 ± 100–1000 ± 100	Fully closed throttle	10 seconds

Japan's emission standards for 1975 and prior-year passenger cars and light-duty trucks are shown in Table XXXI and those for heavy-duty vehicles in Table XXXII. One feature of the Japanese standards that differs from the ECE or United States standards is the use of average and maximum limits. All certification test vehicles must meet the average specified in Japanese standards. No production vehicle can exceed the maximum standards and the average emission of the production vehicles must be within the average standard. It can be seen that there are separate standards for two-stroke engines. Japan also has adopted evaporative emission standards which limit emissions to 2.0 gm per test.

C. Australia

Australia, prior to July 1, 1976, used the ECE type I exhaust and ECE type II idle tests and emission standards (See Tables XXVII and XXVIII) and an evaporative emission standard of 2.0 gm of hydrocarbon per test (United States carbon trap test procedure) for all petrol passenger cars. In New South Wales the evaporative emission standard was also applied to trucks to 4.5 tons (4536 kg Gross Vehicle Weight).

After July 1976, the United States CVS-1 exhaust test procedure has been used. The associated exhaust emission standards with this procedure are tabulated below.

	Exhaust emission standards for light-duty vehicles	
	(gm/mile)	*(gm/km)*
Hydrocarbons	3.4	2.1
Carbon monoxide	39	24.2
Oxides of nitrogen	3.1	1.9

Table XXXI Exhaust Emission Standards for Light-Duty Motor Vehicles in Japan

		Emission standards					
		Ten-mode cycle (gm/km)[a]				Eleven-mode cycle (gm/cycle test)	
		1975		1973		1975	
Vehicle class	Pollutant	Maximum	Average	Maximum	Average	Maximum	Average
Passenger cars							
Gasoline and LPG[b]	Carbon monoxide	2.70	2.10	26.0	18.4	85.0	60.0
	Hydrocarbons	0.39	0.25	3.80	2.94	9.5	7.0
	Nitrogen oxides	1.60	1.20	3.00	2.18	11.0	9.0
Two-cycle gasoline	Carbon monoxide	2.70	2.10	26.0	18.4	85.0	60.0
	Hydrocarbons	0.39	0.25	22.5	16.6	9.5	7.0
	Nitrogen oxides	0.50	0.30	0.50	0.30	4.0	2.5
Light duty trucks							
Gasoline and LPG[b]	Carbon monoxide	17.0	13.0	26.0	18.4	130.0	100.0
	Hydrocarbons	2.70	2.10	3.80	2.94	17.0	13.0
	Nitrogen oxides	2.30	1.80	3.00	2.18	20.0	15.0
Two-cycle gasoline	Carbon monoxide	17.0	13.0	26.0	18.4	130.0	100.0
	Hydrocarbons	2.70	2.10	22.5	16.6	70.0	50.0
	Nitrogen oxides	0.50	0.30	0.50	0.30	4.0	2.5

[a] To calculate gm/mile, multiply gm/km by 1.609

[b] LPG stands for Liquid Petroleum Gas.

Table XXXII Exhaust Emission Standards for Heavy-Duty Gasoline-Powered Vehicles in Japan

	Emission standard	
Pollutant	Maximum	Average
Carbon monoxide (% by volume)	1.6	1.2
Hydrocarbons (ppm)	520	410
Oxides of nitrogen (ppm)	2200	1830

The 2.0 gm per test evaporative emission standard has been extended to vehicles over 4.5 tonnes (4535 kg Gross Vehicle Weight) after January 1, 1977 in New South Wales. After January 1, 1978, heavy-duty vehicle emissions standards will be applied in New South Wales only. The United States Heavy-Duty Engine Dynomometer Test will be used; the corresponding emission standards for vehicles over 4.5 tonnes (4535 kg Gross Vehicle Weight) are tabulated below.

Exhaust emission standards for heavy-duty vehicles	
Hydrocarbons	180 parts per million (ppm)
Carbon monoxide	1.0%

Australian vehicular smoke emission standards are shown in Table XXIX

D. Canada

Canada has exhaust emission standards for both light-duty and heavy-duty vehicles as well as evaporative emission standards for light-duty vehicles. Its test procedures have followed those of the United States, but its emission standards, in some cases, differ (*40*). The United States CVS-1, carbon trap, evaporative, and heavy-duty test procedures were used through the 1974 calendar year. The corresponding Canadian emission standards were:

Light-duty vehicles		
	Exhaust emissions	
	(gm/mile)	*(gm/km)*
Hydrocarbons	3.4	2.1
Carbon monoxide	39	24.2
Oxides of nitrogen	3.1	1.9
Evaporative emissions (gm/test)		
Hydrocarbons	2.0	
Heavy-duty vehicles		
Hydrocarbons	275 ppm	
Carbon monoxide	1.5% by volume	

After January 1, 1975, Canada used the United States CVS-2, carbon trap, evaporative, and heavy-duty vehicle tests; and the corresponding emission standards tabulated below:

Light-duty vehicles		
	Exhaust emissions	
	(gm/mile)	*(gm/km)*
Hydrocarbons	2.0	1.2
Carbon monoxide	25	16
Oxides of nitrogen	3.1	1.9
Evaporative emissions (gm/test)		
Hydrocarbons	2.0	
Heavy-duty vehicles		
Exhaust emissions (gm/brake horsepower-hour)		
Hydrocarbons plus oxides of nitrogen	16	
Carbon monoxide	40	

E. Mexico

Mexico adopted exhaust emission standards based on the United States CVS-1 test procedure for the 1975 model year and has proposed adding standards for oxides of nitrogen in the 1977 and 1978 model years passenger cars (*40*). The adopted and proposed standards are tabulated below:

	Emission standard					
	1975		*1977*		*1978*	
Component	*(gm/mile)*	*(gm/km)*	*(gm/mile)*	*(gm/km)*	*(gm/mile)*	*(gm/km)*
Hydrocarbons	3.4	2.1	—	—	—	—
Carbon monoxide	39	24.2	—	—	—	—
Oxides of nitrogen	—	—	3.6	2.2	3.0	1.9

Mexican vehicular smoke emission standards are shown in Table XXIX

F. New Zealand

New Zealand, after January 1, 1975, required compliance with the ECE type II idle test and emission standard. After January 1, 1977, it added the ECE type I exhaust test and the vehicles must meet the prototype standards as well as an evaporative emission standard of 2.0 gm of hydrocarbon per test, measured by the United States test procedure. All petrol fueled vehicles 882–7718 pounds (400–3500 kg) in weight must meet these standards, except that those manufactures in Australia, Japan, and the United States must meet the 1973 and 1974 United States federal Standards.

G. Sweden

Sweden, prior to October 1, 1975, used the ECE test I exhaust and type II idle tests. Its standards specified an idle carbon monoxide emission of 4.5% by volume and exhaust emission standards in grams per kilometer of hydrocarbons (2.2) and carbon monoxide (45). The test after October 1, 1975, has been the United States CVS-1 procedure and the corresponding exhaust emission standards for vehicles with gasoline engines from 0.8 liter or minimum horsepower of 30 horsepower through a maximum 2500 kg (GVW) are shown in the following tabulation (*40*).

	Standard	
Component	*(gm/mile)*	*(gm/km)*
Hydrocarbons	3.4	2.1
Carbon monoxide	39	24.2
Oxides of nitrogen	3.1	1.9

Swedish vehicular smoke emission standards are shown in Table XXIX.

H. Other Countries

Other countries have adopted or proposed motor vehicle emission standards for gaseous pollutants that use the ECE or United States test procedures but may specify different emission levels (*40*). In several cases the standard is limited to carbon monoxide at idle and in a few instances both the ECE and United States CVS cycles have been used or proposed.

Denmark and Israel limit their emission standards for gaseous pollutants to carbon monoxide at idle. Both countries follow the ECE type II test, vehicle designation, and emission limits (Table XXVII).

Motor Vehicle smoke emission standards for Brazil and Finland are shown in Table XXIX.

XI. Conclusion

Much progress has been made since 1959 in establishing emission standards for motor vehicles and in developing control systems to meet the standards. During this period, standards were adopted for crankcase, evaporative, and exhaust emissions; the standards were made increasingly stringent, test procedures and measurement methods were improved, and the method of expression was made more equitable (grams per mile). Aircraft emission standards were established for the first time in 1973. Little change has been made in the manner by which emissions from railroad locomotives and/or ships are regulated since pollutants from these sources continued to be controlled by standards developed long before motor vehicle and aircraft emission regulations came into being.

If history is an indication of the future, it can be expected that the major emphasis on emission standards for mobile sources will continue to be placed on automobiles. The reason for the emphasis in the past was the large quantity of hydrocarbons, carbon monoxide, and oxides of nitrogen emitted by motor vehicles, and this situation should remain for some years to come.

The present controversy in the United States over just how stringent the standards should be will probably continue and decisions on standards for motor vehicle emissions will continue to be made partly on the need to meet air quality goals, partly on the technology available to reduce emissions, and partly on political considerations. As a result, the subject of motor vehicle emission standards will involve complex issues and differing opinions on the best course of action to follow. In any event, it can be concluded that the future motor vehicle emission standards will impose a high degree of control.

Most mobile source emission standards are applied only to new motor vehicles. The full effects of a control program which is based on these standards will not be achieved until all motor vehicles are equipped with control systems which meet the standards. At the historical motor vehicle replacement rate, 12–15 years after standards first go into effect will have passed before the desired level of control is realized. Figure 9 shows how carbon monoxide emissions will be reduced under such a program. As can be seen, the total emissions of carbon monoxide increased until 1966 when control systems were first placed on new vehicles, and that under

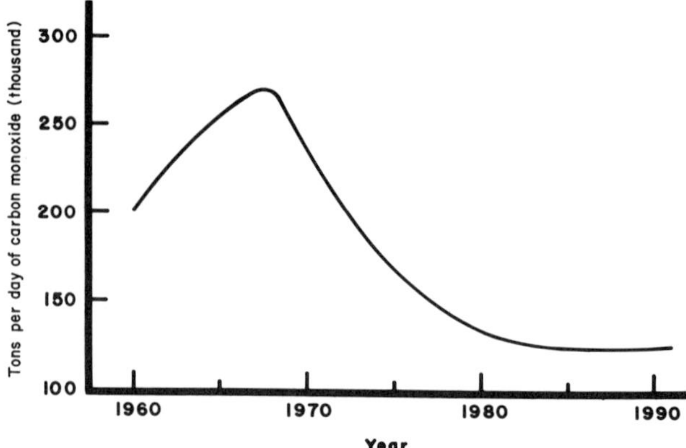

Figure 9. Estimates of carbon monoxide emissions from motor vehicles in United States (1960–1990).

current standards and a control program for new vehicles only, the minimum emission level for carbon monoxide from vehicles will not occur until after 1985. Similar curves have been developed for hydrocarbons and oxides of nitrogen.

It is clear that air quality goals will not be attained within a period of a few years if achieving the goals depends upon all or most of the vehicles in use complying with 1975 or later model emission standards. The air quality goals can be reached earlier only if the emissions from used vehicles are also controlled or if there is a reduction in vehicle usage. It is possible that air quality goals may not be reached in some areas with severe motor-vehicle created air pollution because of technological limitation on developing highly efficient emission control systems, regardless of the time permitted. In this case, the air quality goals can be met only by reducing vehicle usage.

The need to consider motor vehicle emission control measures other than standards for new vehicles was first dictated in the United States by the 1970 Clean Air Act Amendments. These amendments required the federal air quality standards to be met no later than 1977, although the minimum emissions from the standards included in the same act will not occur until after 1985. As a result, a number of measures have been proposed to reduce the emissions from the existing vehicle population. These include control systems for used cars, vehicle emission inspection and maintenance, incentives for using mass transit for transportation, disincentives for using private automobiles, and gasoline rationing.

Incentives for mass transportation include exclusive highway lanes for buses and car pools, and free fare for persons using public transit. Disincentives include measure to restrict vehicle use. It is yet too early to determine just how effective the disincentive/incentive measures are, or how acceptable they will be to the public.

Some of the controversy on the degree of control required is due to the uncertainties about the levels of oxidant, carbon monoxide, and nitrogen dioxide which should be specified as air quality standards. It is unlikely that new information will become available in the near future to establish the air quality standards much more firmly. As a result, this issue will continue to be a cause for differing opinions on how much reduction in automotive emissions is needed.

REFERENCES

1. A. J. Haagen-Smit, *Ind. Eng. Chem.* **44,** 1342 (1952).
2. United States Department of Transportation, Bureau of Public Roads, "Highway tistics." Washington, D.C., 1976.
3. B. J. Steigerwald, "Presentation to RECAT on August 30, 1970," Chart No. 3, p. 519–523. United States Environmental Protection Agency, Durham, North Carolina, 1970.
4. California Department of Public Health, "Technical Report on California Standards for Ambient Air Quality and Motor Vehicle Emissions." Berkeley, California, 1960.
5. California Air Resources Board, "Air Basins and Air Quality Standards," California Administrative Code, Title 17, Part III, Chapter 1, Subchapter 1. Sacramento, California, 1973.
6. United States, *Fed. Regist.* **36,** No. 84 Part II (1971).
7. J. A. Maga and G. C. Hass, *J. Air Pollut. Cont. Ass.* **10,** No. 11 (1960).
8. R. I. Larsen, "Future Air Quality Standards and Industrial Control Requirements," U.S. Pub. Health Publ. No. 1669. United States Government Printing Office, Washington, D.C., 1967.
9. United States, *Fed. Regist.* **37,** No. 221, Part II (1972).
10. H. Wong-Woo, "An Evaluation of Crankcase Emissions," California Department of Public Health, Berkeley, California, 1963.
11. United States, *Fed. Regist.* **38,** No. 6, (1973).
12. United States, *Fed. Regist.* **41,** No. 189, Part II (1976).

12a. California Air Resources Board, Resolution 76–3. Sacramento, California, 1976.

13. Los Angeles County (California) Air Pollution Control District, "Los Angeles Auto Exhaust Test Station Project," Joint Agency Report. Los Angeles, California, 1963.
14. United States Environmental Protection Agency, "Exhaust Emissions From Uncontrolled Vehicles and Related Equipment Using Internal Combustion Engines, Part 3, Motorcycles," APTD 1492. Washington, D.C., 1973.
15. W. Martin and A. C. Stern, "The World's Air Quality Management Standards," 2 Vols., Vol. I, The World, Including U.S. Federal Standards, Miscellaneous

Series—EPA-650/9-75-001a, 383 pp. Vol. 2. U.S.A. Miscellaneous Series EPA-65019-75-001-b, 373 pp. U.S. Environmental Protection Agency, Office of Research and Development, Washington, D.C. (1974).
16. D. M. Teague, "Los Angeles Traffic Pattern Survey," Vehicle Emissions, Tech. Progr. Ser., Vol 6 (PT–6). Society of Automotive Engineers, Detroit, Michigan, 1964.
17. G. C. Hass and M. C. Brubacker, *J. Air Pollut. Contr. Ass.* **12,** No. 11 (1962).
18. G. C. Hass, M. O. Sweeney, and J. N. Pattison, *SAE Tec. Pap.* **12,** Society of Engineers, Detroit, Michigan (1966).
19. United States, *Fed. Regist.* **36,** No. 128, Part II (1971).
20. United States, *Fed. Regist.* **37,** No. 221, Part II (1972).
21. United States, *Fed. Regist.* **41,** No. 164 Part II (1976).
22. Ethyl Corporation, "Survey of Truck and Bus Operating Modes in Several Cities," Pub Health Serv. Contract No. PH 86-62-12. U.S. Dept. of Health, Education, and Welfare, Washington, D.C.; 1963.
23. United States, *Fed. Regist.* **38,** No. 136, Part II (1973).
24. W. S. Fagley and G. Way, "Field Survey of Exhaust Gas Composition," Vehicle Emissions, *Tech. Progr. Ser.,* Vol. 6 (PT-6), Society of Automotive Engineers, Detroit, Michigan, 1964.
25. United States, *Fed. Regist.* **33,** No. 108, Part II (1968).
26. National Academy of Sciences, "Report by the Committee on Motor Vehicle Emissions." Washington, D.C., 1973.
27. United States Senate Committee on Public Works, "Air Quality and Automobile Emission Control, A Report by the Coordinating Committee on Air Quality Studies by the National Academy of Sciences and National Academy of Engineering." United States Government Printing Office, Washington, D.C., 1974.
28. Exhaust Emission Standards and Test Procedures–1975 through 1978 Model-Year Passenger Cars, California Administrative Code, Title 13, Section 1955.1, Register 76, No. 19-5-8-76, Sacramento, California, 1976.
28a. United States, "Amendments to the Clean Air Act," 95th Congress, Washington, D.C., 1977.
29. California Air Resources Board, "California Fuel Evaporative Emission Standards and Approval Procedures for 1973 and Subsequent Year Gasoline-Powered Vehicles over 6000 Pounds, Gross Vehicle Weight." Sacramento, California, 1976.
30. United States, *Fed. Regist.* **39,** No. 12 (1974).
31. Exhaust Emission Standards and Test Procedures-Motorcycles Manufactured Subsequent to January 1, 1968, California Administrative Code, Title 13, Section 1958, Register 76, No. 16-4-17-76, Sacramento, California, 1976.
31a. U.S., *Fed. Regist.* **42,** No. 3, Part II (1977).
32. California Air Resources Board, "California Exhaust and Fuel Evaporative Emission Standards and Test Procedures for Used Motor Vehicles Under 6001 Pounds, Gross Vehicle Weight." Sacramento, California, 1970.
33. California Air Resources Board, "California Oxides of Nitrogen Control Device Test Procedures for Used 1966 through 1970 Model Year Vehicles Under 6001 Pounds, Gross Vehicle Weight." Sacramento, California 1971.
34. New Jersey State Department of Environmental Protection and Pollution Control, "Control and Prohibitions of Air Pollution From Gasoline-Fueled Motor Vehicles," Chapter 15. Trenton, New Jersey, 1972.
35. A. J. Andreatch, J. C. Elston, and R. W. Lakey, *J. Air Pollut. Contr. Asso.* **21,** 757–763 (1971).

36. Highway and Mandatory Inspection Emission Standards, Air Resources Board, California Administrative Code, Title 13, Sacramento, California, 1977.
37. Economic Commission for Europe, "Uniform Provisions Concerning the Approval of Vehicles Equipped with a Positive-Ignition Engine with Regards to the Emission of Gaseous Pollutants by the Engine," Regulation 15. Geneva, Switzerland, 1970.
38. Economic Commission for Europe, "01 Series Amendments to Regulations 15." ECE, 1974.
39. A. Inakawa, Toyota Motor Sales, February 5, 1974 and February 11, 1974 (private communications).
40. L. V. Farago, Ford Motor Company, private communications, November 25 1974, June 8, 1976, and November, 18, 1976.

13

Emission Standards for Stationary Sources

Arthur C. Stern

I. Introduction

Stationary source emission standards are standards relating to a stationary site, facility, process, chimney, stack, or vent intended to help achieve desired air quality. They include standards for buffer zones, stack height, equipment design, and fuel composition, and those that directly limit the amount or concentration of a pollutant emitted from a source. Standards may be stated, subjectively, in terms of the appearance of an emission to the eye or its odor as determined by the nose; or, objectively, in terms of measures of weight or volume. Emission standards may be derived from process and equipment considerations, air quality considerations, or both. Emission standards sometimes reflect economic, sociological, and political considerations, in addition to those that are technological. In some cases, we have the technological ability to control certain pollutants but are not doing so for economic, sociological, or political reasons. However, despite considerable economic, sociological, and political pressure to more stringently regulate certain emissions, we are not doing so because we lack adequate technological ability.

A. Promulgated, Presumptive, and Informal Standards

Emission standards may be legally prescribed, officially promulgated, presumptive, or informal, i.e., "desk drawer" standards. Legally prescribed standards are rare.

Officially promulgated standards are those that have been published in the official bulletin, gazette, journal, or register prescribed for such publication by law or decree. The law or decree may also require that prior to official promulgation there be formal publication of intent to promulgate the standard, opportunity for public comment in writing, and, in some cases, formal public hearing. It may also require that official standards must be amended by the same process by which they are promulgated. The emission standards of the United States are mainly of this type.

Presumptive standards are those that may be obtained from the air pollution control agency in lists or publications freely available to the public and that the agency uses as its standards in reviewing plans and specifications for new installations and testing existing ones, but that have not gone through a formal promulgation procedure. The emission standards of a number of countries that do not promulgate air quality standards are of this type.

Informal or "desk drawer" standards are found mostly in countries whose laws or regulations permit them to promulgate air quality standards but do not permit them to promulgate emission standards. Such standards are not freely available to the public in lists or publications but represent the design criteria used by the air pollution control agency for reviewing plans and specifications of new installations and testing existing ones.

B. Uniform versus Regionally Variable Emission Standards (1, 2)

Uniform stationary source emission standards apply alike to all regions within a jurisdiction, be it a nation, a state or province, a county, or a city, regardless of their meteorology, topography, present air quality, population density, state of industrialization, or recreational values. Regionally variable standards apply one or more of the above list of factors as regional discriminators. It is preferable that a formal procedure be promulgated designating which of these factors are to be used as regional discriminators, defining the factors and specifying the manner of their use. Failing this, a jurisdiction can publish lists of its regions to which different emission standards apply, with or without supplying the public with details on the regional discriminators employed by the jurisdiction in making such determination.

II. Derivation of Emission Standards from Process and Equipment Considerations

A. Emission Standards

1. Direct Approach

The direct approach in setting an emission standard is to ask the question, what are the emissions from the best constructed and best operated plants (or, alternatively, well constructed and operated plants) in this category? and then rationalize that this performance should set the standard for all such plants. In an analogous approach, the first step is to determine for a range of plants from best to worst the pollutant loading of gases leaving the process at a point prior to their passing through any emission control equipment, such as a dust separator or gas scrubber. The next step is to compute what the emission from each would be if treated by an appropriate unit of emission control equipment. The result will depend upon the collection efficiency assumed for the control equipment. The old adage that "one can have as much control as he is willing to pay for" becomes the test in setting the assumed efficiency high or low. A good combination for such computation is to assume control equipment of better than average efficiency applied to a plant of average control equipment inlet loading.

2. Best Practicable Means Approach

One commonly used criterion for emission standards is to require the use of the best practicable means to control emissions. The word "practicable" implies not only technological practicability but also economic, sociological, and political practicability. Best practicable means derive not only from the technology within an industry but also from the borrowing of the technology of one industry to improve that of another. Thus, best practicable means utilize either the control actually achieved in the best plants in the industry or that which could be achieved by borrowing the best technology of other industries. Under this concept the emission standard may be made more stringent as the state of the art of control of the process improves over the years. The question as to the extent these more stringent standards should be made retroactive to installations that were in compliance prior to changing the standard, but not afterward, is provocative. It is more difficult to argue that emission standards be made retroactive when they are based upon process and equipment considerations than when they are based upon air quality considerations. The installation of improved controls in either an existing or a new plant may, by showing an improvement in best practicable means, serve to escalate the emission standard. Should it be shown that an emission standard based upon best practicable means does not result in air that meets an air quality standard, there is no force inherent in the best practicable means approach that is likely to bridge the gap.

The opposite situation also is possible, i.e., that application of best practicable means of control may result in community air quality considerably better than the air quality standard. In this case, best practicable means results in an emission standard more stringent than one derived from air quality considerations. There are those who call this "control for control's sake" or "overkill." This situation becomes particularly sticky when the existing air quality of an area is considerably better than the air quality standard and the application of best practicable means to a new source will allow this to continue but application of less than best practicable means will still not cause the air quality standard to be exceeded. This is the classic "nondegradation" situation where the introduction of any pollution to an area's atmosphere will cause it to degrade, even if initially it was cleaner than the air quality standard.

A problem with the best practicable means approach to emission standard setting is that an industry can avoid a more stringent standard by collectively refraining from installing or developing an improved process or control technology that would provide the example of better technology that could be used as proof of practicability. Such collective action need not be by collusion but could result from the unwillingness of each individual segment of the industry to be the one that would "rock the boat." In this situation, the government, as the representative of the people, must force the issue by demonstrating the better technology in its own or a contractor's pilot plant or must make a finding that an equivalent better technology exists in an analogous process or industry in which the advance of technology was not collectively impeded.

Where the government does force the issue by the pilot plant or equivalent process approach, the issue becomes one of full-scale practicability. If the industry is nationalized, the government can make the decision to make the necessary investment to prove or disprove full-scale practicability. If the industry is privately owned, it would seem that the government must be prepared to underwrite the cost of failure of the allegedly better technology in one or more demonstrations in industry-owned plants.

3. Sliding Scales

An alternate to requiring all plants to meet a best practicable means standard is to require only the largest plants to meet such a requirement while smaller plants meet less stringent standards. In comparing production installations within an industry, a large volume of stack discharge per unit time can usually be associated with a large production capacity and also with a high value for the installation. However, the large capacity implies a large number of production units to bear the cost of air pollution control equipment or processes. In general, the larger the production plant, the smaller the percentage of total cost associated with control equipment or processes. Thus, for the same percentage expenditure, a large plant could install relatively more expensive and presumably more efficient air pollution control equipment. If control practices within the industry vary from good to best, the tendency is to expect the large installations to perform in the higher end of the range and the smaller ones to conform only to good practice. Further, a large installation produces a greater absolute quantity of pollution than does a smaller one and thus imposes a greater burden upon the atmosphere when its control efficiency is no better than that of a small installation. This rationalization leads to the sliding scale emission standard that limits the emission per unit volume of stack gas to a greater extent on larger than on smaller installations.

The upper end of the range of a sliding scale of emission limits is usually the best feasible practice limit. The lower limit may be either the minimum or the good practice standard (Table I). An example of a standard, using a sliding scale and constrained by these upper and lower limits, would be as shown in Table II.

The scale in Table II shows an arbitrary five-unit increment for each exponential increase in productive capacity. In seeking a rational basis for establishing a scale, some form of emission allocation must be used. Emission allocations can be estimated on the basis of the potential emissions for all sources in each source size category. Allocations of the percentage of total emissions to be allowed for each source size category can be determined on the basis of technological, economic, social, and political considerations. With these percentages decided upon, standards intermediate between the minimum and best practice can be established. The minimum standard never should be relaxed, and control beyond the best practice is not feasible.

4. Options Available

In applying this procedure to the development of standards, several options are available for determining the area on which to base the

Table I Uniform National Standards *(1)*

Designation of standard	*Source characteristics*	*Remarks*
Minimum	Average installations in the United States	Readily achievable with little cost burden but of little real value
Good practice	Better installations in the world	Achieved in better installations
Best feasible practice	Best installations in the world	Highest achievable by current technology

Table II Application of a Sliding Scale Emission Standard *(1)*

Productive capacity of source[a] *(arbitrary units)*	*Emission limit*[a,b]	*Remarks*
1–99	25	Minimum standard equals 25
100–499	20	Good practice equals 18
500–999	15	—
1000–4999	10	Best feasible practice equals 10
5000 and larger	10	—[c]

[a] Continuous sliding scale between stated values.
[b] Per unit of productive capacity.
[c] The alternative approach would be to prohibit units larger than 5000.

estimated number of sources in each size category. One would be the nation as a whole. This could lead to anomalies in application if there were strong regional differences in the mix among size categories. The other options would follow a test of several state, provincial, or regional areas, incorporating the extremes in the mix variation, as well as the average of such variation. A size category allocation scheme that minimizes regional anomalies then could be selected or an alternative means for computing the emission standard between contrasting regions could be provided.

In countries that have not promulgated air quality standards but that have promulgated emission standards, the emission standards are usually based solely on best practicable means, i.e., upon process and equipment considerations. However, in countries that have promulgated both air quality and emission standards, the only guarantee that the latter are based upon the former is a statutory requirement that this be so, which is rarely the case. In the United States, federal emission standards take the form of new-source performance standards (*3*), and National Emission standards for Hazardous Air Pollutants (*3a*) which are based upon best practicable means. Publications giving the bases for these standards are listed in Tables IIA and IIB. However, state emission standards for existing sources of the same processes must at least in part be based on air quality considerations if the pollutant regulated thereby is one for which a national air quality standard has been promulgated.

In Canada, the national government has issued National Emission

Table IIA Publications Listing Background Information for United States Standards of Performance, Proposed New Source Performance Standards, and Establishment of National Standards of Performance for New Sources, U.S. Environmental Protection Agency, Research Triangle Park, North Carolina *(3)*

Steam Generation, Incineration, Portland Cement Plants, Nitric Acid Plants, Sulfuric Acid Plants (August 1971)—APTD 0711
Castor Bean Processing (July 1972)—APTD 1361
Asphalt Concrete Plants, Petroleum Refineries, Storage Vessels, Secondary Lead Smelters and Refineries, Brass and Bronze Ingot Production Plants, Iron and Steel Plants, and Sewage Treatment Plants
  Vol. 1. Main Text (June 1973)—APTD 1352A
  Vol. 2. Appendix: Summaries of Test Data (June 1973)—APTD 1352B
  Vol. 3. Promulgated Standards (February 1974)—EPA 450/2-74-003
 Primary Copper, Zinc, and Lead Smelters
  Vol. 1. Proposed Standards (October 1974)—EPA 450/2-74-002a
Electric Arc Furnaces in the Steel Industry
  Vol. 1. Proposed Standards (October 1974)—EPA 450/2-74-017a
  Vol. 2. Test Data Summary (October 1974)—EPA 450/2-74-017b
Electric Submerged Arc Furnaces for Production of Ferroalloys
  Vol. 1. Proposed Standards (October 1974)—EPA 450/2-74-018a
  Vol. 2. Test Data Summary (October 1974)—EPA 450/2-74-018b
Phosphate Fertilizer Industry
  Vol. 1. Proposed Standards (October 1974)—EPA 450/2-74-019a
  Vol. 2. Test Data Summary (October 1974)—EPA 450/2-74-019b
Primary Aluminum Industry
  Vol. 1. Proposed Standards (October 1974)—EPA 450/2-74-020a
  Vol. 2. Test Data Summary (October 1974)—EPA 450/2-74-020b
Coal Preparation Plants
  Vol. 1. Proposed Standards (October 1974)—EPA 450/2-74-021a
  Vol. 2. Test Data Summary (October 1974)—EPA 450/2-74-021b

Guidelines (*3b*) for the cement, asphalt paving, metallurgical, coke manufacturing, and secondary lead smelting industries that the Minister of the Environment states that the "provincial air pollution control agencies may wish to adopt . . . as minimum standards . . . within their juris-

Table IIB Publications Listing Background Information, Proposals for and Development of United States National Emission Standards for Hazardous Air Pollutants, U.S. Environmental Protection Agency, Research Triangle Park, North Carolina *(3a)*

Asbestos, Beryllium, and Mercury (December 1971)—APTD 0753
Asbestos, Beryllium, and Mercury (March 1973)—APTD 1503
Proposed Amendments to Standards for Asbestos and Mercury (October 1974)—EPA-450/2-74-009a
Standard Support and Environmental Impact Statement: Promulgated Emission Standard for Vinyl Chloride, Vol. 1 (October 1975)—EPA-450/2-75-009; Vol. 2 (September 1976)—EPA-450/2-75-009b

dictions." Publications giving the bases for these guidelines are listed in Table IIC.

B. Equipment Design Standards

The first standards to be used extensively in air pollution ordinances and regulations were design standards. The incorporation of such standards led to the requirement of the submission of plans and specifications for approval, i.e., for checking for compliance with design standards. Proponents of this approach have argued that it has been the only practicable preventive measure, in view of the difficulty in enforcing emission standards. Opponents raise the hypothetical issue of the plant erected in conformity with design standards but failing to comply with emission or air quality standards. Opponents also have held that design standards stifle engineering ingenuity by preventing designs that contravene the standards but that achieve the desired result in terms of emission or air quality. To avoid these pitfalls, design standards should never be used as a straitjacket on design, but rather as a guide to good design, and appeals procedures should be provided to allow fair consideration for nonstandard designs.

Design standards for stationary fuel-burning equipment relate to such factors as heat release per unit furnace volume, furnace refractory placement and arch design, chimney cross-sectional area and height, stoker setting, fuel oil viscosity and preheat temperature, etc. Being specific for the characteristics of the fuel used, which differ from region to region, such standards are perforce local and not universal. Design standards tend to become obsolete as technology changes and therefore should not be incorporated into legislation or regulation that is difficult to amend.

The testing and certification of prototypes has become accepted practice for control of mobile source emissions but has not been formally used for stationary sources. An equivalent practice has been informally used for years in refusal of staff members of municipal, county, state,

Table IIC Economic and Technical Review Reports Giving Background Information on Canadian National Emission Guidelines, Air Pollution Control Directorate, Environmental Protection Service, Department of the Environment, Ottawa, Ontario *(3b)*

Cement Industry (EPS 3 AP-74-3), vi + 50 pp. (1974)
Metallurgical Coke Manufacturing Industry (EPS 3 AP-74-6), viii + 74 pp. (1974)
Asphalt Paving Industry (EPS 3-AP-74-2), vi + 38 pp. (1975)
Secondary Lead Smelter and Allied Industries (EPS 3-AP-75-3), v + 30 pp. (1975)

or provincial air pollution control agencies to give construction or operating permits to applications or plans showing equipment they know from field experience will not work satisfactorily. It would be desirable to formalize this practice by national prototype testing and certification of classes of stationary air pollution producing equipment installed in large numbers nationally.

III. Derivation of Emission Standards from Air Quality Considerations

A. Single Sources

1. With No Buffer Zone

If one assumes all possible combinations of wind direction, wind velocity, atmospheric stability, and distance from a single elevated stack or a group of adjacent elevated stacks that may be considered to act as a single stack, there will be one combination for which (for a specific effluent emission rate and effective stack height) ground level concentration will be greatest. By setting this maximum ground level concentration equal to the air quality standard, the emission limit for the stack or group of stacks may be computed. Although meteorologists are working to improve the accuracy of this type of computation, the accuracy presently obtainable is good enough to allow use of this methodology for major single sources.

This procedure preempts all the allowable emission for this one stack. If another stack emitting the same pollutant has the proper combination of stack height and distance from the point of maximum ground level concentration of the one stack so that its point of maximum ground level concentration is the same as that for the one stack, these concentrations may be additive. If then the emissions for each stack were computed on the basis of maximum ground level concentration equaling the air quality standard, the resulting ground level concentration could be twice the air quality standard.

When a single source is built, there is no way of knowing in advance the number of additional sources that will ultimately share with it the capacity of the atmosphere to dilute pollutants to the air quality standard. There are several possible resolutions to this dilemma. One is the "flexible" emission standard that in the foregoing case allows the single plant an emission limit computed on the basis of its being the only source until the second plant is built, after which both plants must meet a revised emission limit half as large as the original limit, and so forth, as a third and fourth, etc. source is built. There is, in this approach, no "grandfather" clause protecting the right of the earlier-built plant or plants to maintain their earlier emissions, but rather they would, as each new source is built, have to decrease their emission. This is an unwieldy approach and difficult to administer but fits certain real situations that need resolution.

Another approach is to make the assumption that there will eventually be X sources; therefore the computation of emission standard should be based initially on the ground level concentration from any source not exceeding one-Xth of the air quality standard. This penalizes the first few sources but is easier to administer.

A third approach pre-empts a certain portion (e.g., 90%) of the air quality standard for multiple small sources, either existing or future, and allows the use of the remainder (e.g., 10%) as the basis for computations by the foregoing procedures for the few large sources.

All of these procedures result in a table or chart to describe the emission standard, rather than a single number. The reason for this is that the computations involved have as variables emission rate, effective stack height, distance, and meteorological values.

2. With Buffer Zone

These same considerations apply to setting emission limits on a large source surrounded by a buffer zone, where the computations are based on the air quality immediately beyond "the fence" surrounding the property, or the equivalent thereof where there is no actual fence. It is not sufficient that the land "within the fence" be owned by the owner of the source at the time of setting the emission limits. There must also be assurance that this buffer zone land cannot later be converted to other than buffer zone use without a concomitant change in applicable emission limits. In the case of emission limits based on ground level air quality in which the land where maximum ground level concentration occurs is in the same ownership as the source, it may be argued that the computation of emission limits should be based upon application of the air quality standard only to the land beyond such common ownership. However, it is use, not ownership, that is the important thing. If the use of the land contemplates the presence of people, the air quality standard should be applied to it for the computation of the emission limit. An exception would be reasonable only if by convenant, deed, zoning restrictions, or stipulation the land in question were converted to a true buffer zone. Since it is unreasonable to expect a control agency to recompute emission limits each time the use of a parcel of property is changed, the practicable procedure for incorporation in regulation is to assume the presence of people and buildings in all land surrounding a source unless such use is legally restricted. Essentially all of the stack height standards of Section VII are derived in the manner discussed in this section for single sources.

3. Point-of-Impingement-at-Ground-Level Standards

As an alternate to basing these computations upon some percentage of the air quality standard, a government can promulgate a set of ground level air pollutant concentrations specifically for use in such computations. These are listed in the tables of Chapter 11 of this volume as "point-of-impingement-at-ground-level standards." They are in the same form as air quality standards but can differ in numerical value from them by incorporation of the desired pre-emption percentage directly into them. Where there is no preemption for smaller sources or multiple large sources, the point-of-impingement-at-ground-level standard is numerically the same as the air quality standard. Whereas an air quality standard is constrained by air quality criteria considerations, a point-of-impingement standard can be administratively changed from time to time to reflect changes in control strategy, most particularly as reflected in changes in the pre-emption percentage.

4. Nondegradation

The resolution to the nondegradation situation in the United States has been to require all new plants in categories having federal new-source performance standards to meet such standards, regardless of the pollution of the area in which they are to be located, but to limit the approval of such new plants depending upon the nondegradation classification of the area (*4a*). These area classifications, set by the states with federal concurrence, are (I) areas described as "pristine", where practically any air quality deterioration would be considered significant; (II) described as allowing "moderate economic growth", where deterioration of air quality that would accompany moderate growth would not be considered significant; and (III) described as allowing "intensive economic growth," where intensive and concentrated major industrial growth is desired, permitting levels of air quality to rise to some designated ceiling.

5. Supplementary and Intermittent Control Systems (SCS/ICS)

Where a governmental agency allows a single source to be the sole polluter in an area, it can in return for that permission hold the source to be responsible for the pollution levels that occur in the area and for the damage caused thereby. This can be done either by law or by having the source stipulate to its assumption of such responsibility However, this is a hypothetical situation, since it is difficult to envision a situation where the large source would either locate where there were no people to draw upon as a labor force or, if there were initially no people, would not attract some to the area to provide it and them with labor and services. Thus, the real situation is where the large source is the predominant source in the area. Predominance can be computed as the percent contribution of the source to the maximum ground level pollutant concentration under specified meteorological conditions as compared with that from all other known or predicted sources in the area. Where predominance computed in this manner is 90% or greater, there is little disagreement but that the source may be treated as a sole source in the context discussed above if both the source and the government agrees. Likewise, where predominance is well below 50%, few would argue that the large source should be treated as a sole source. However, in the range from 50–90% predominance, one can argue that although ground-level pollution concentrations are closely tied to contributions from the single large source, there is serious question as to whether that source can be legally held liable for ground-level concentrations and damages or should be required to make such a stipulation as a condition to its operation.

This situation arises in decision-making as to whether or not to allow a large single source to use a "supplementary" or "intermittent" control system (SCS/ICS) rather than a "continuous control" system. SCS/ICS are systems of "meterological" control, which adjust emissions to atmospheric diffusive capacity to maintain acceptable ground level air quality by computational procedures similar to those discussed in this section of this chapter. As long as compliance is based solely on the ground-level concentrations measured, there is an unwillingness of large-single sources to accept or stipulate to sole responsibility in the 50–90% predominance range and an unwillingness of governmental agencies to allow them to use SCS/ICS without such acceptance or stipulation.

A compromise position would be to require large single sources permitted to use SCS/ICS in lieu of continuous control to show proof that their emissions were at all times within the limits computed by their approved SCS/ICS. This would require prior approval of the SCS/ICS hardware and computer software package (*5*) and of the instrumentation on emissions and plant operation to establish compliance required by the SCS/ICS computation. Repetitive failure of the area to meet air quality standards would be cause for review of, and possibly required upgrading of, the SCS/ICS hardware and software, but if the large source could show compliance with its approved SCS/ICS, it would not be held legally responsible for the air quality standard excess or for the damages resulting therefrom.

B. Multiple Sources

Emission standards can be derived from air quality standards by the use of multiple source urban diffusion models. The structure of these models is discussed in Chapter 10 of Volume I. The meteorological input to these models is discussed in Chapter 12 of Volume I. The emission inventory input to these models is discussed in Chapter 17 of Volume III. The use of these models in air pollution control strategies is discussed in Chapter 1 of this volume. The procedure of deriving emission standards in this manner has come to be known as the air quality management approach, as contrasted to the best practicable means approach discussed in Section II. The air quality management approach has not been much used outside the United States. It has been extensively used in the United States, because its use was required by the federal Clean Air Acts of 1970 and 1977. Its greatest use in the United States has been in the setting of the regional boundaries of interstate air quality control regions and the development of state implementation plans. A number of state and local emission and fuel standards were derived in this effort and by this procedure.

One problem with the state implementation plan effort was that the short time frame during which the necessary computations had to be made and the limited meteorological and emission inventory data base available restricted the number of areas in each state that could be modeled. The result was that the worst situations were chosen for modeling. The emission and fuel standards thus derived were then, in many states, applied state-wide, resulting in "overkill" in less impacted areas of these states. In the United States, the result of this overkill was to create a demand for clean fuels that proved to be in excess of immediately available supply, particularly after the advent of the energy "crisis" of the mid-1970s. This led to a federal move to get the states to relax their emission standards (specifically for sulfur dioxide) in those areas where such overkill had occurred.

IV. Derivation of Emission Standards from Rollback or Distance Proportional Considerations

A. Rollback Considerations

Rollback procedures to establish emission standards assume three points in time, an earlier time of lower (and presumably acceptable) pollution, the present time, and a future time. A computation estimates the number of source units at the three time periods. Assuming uniform emission per unit source for the several time periods, an estimate based on the present pollutant level is made of the past and future levels. Finally, the reduction in emission per unit of source necessary to roll back the future pollutant level to that of the past is computed. This has been the procedure used in setting motor vehicle emission standards in the United States. It is discussed in detail in Chapter 12 of this volume.

B. Distance Proportional Considerations

The distance proportional approach is used where there is a single dominant source and the fall-off of either pollutant concentration or its effects with distance from the source is known. First, it is determined that the concentration or effects at distance D_2 from the source are tolerable, but those at D_1, which are closer to the source, are not. On this basis, the required decrease in emission and the new emission standard are computed by the known mathematical relationship between pollution concentration and distance from the source.

V. Subjective Standards

A. Smoke and Soot Emission Standards

1. Methodology

Historically, the first emission standards were subjective. They prohibited the emission of "black," "dense," or "dark" smoke or offensive odors. A means to give definition to "black," "dense," or "dark" smoke was devised by Maximillian Ringelmann in 1898 (Fig. 1) and has been the accepted standard ever since. The original Ringelmann chart was a reflectance chart that the observer viewed at a considerable distance so that the black lines and white squares optically merged to shades of gray. Modern versions of the reflectance chart designed to be held at arms length use either a photoreduction of the original chart or halftone shades of gray. Modern transmittance charts, held close to the eye, use photographic film with gradations of optical density that correlate with the Ringelmann chart scale. In the United States, the courts have accepted the concept that emissions of colors other than black or gray, e.g., white, may be graded by Ringelmann chart numbers as being of "equivalent opacity." It is now common practice in the United States to

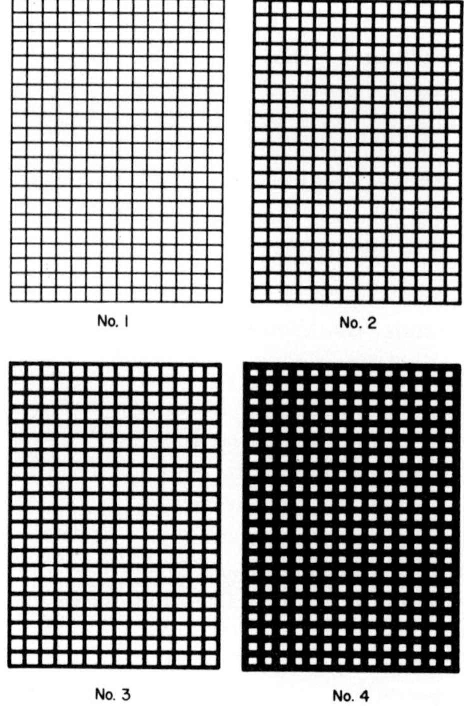

Figure 1. Ringlemann smoke chart.

SPACING OF LINES ON RINGELMANN CHART

Ringelmann chart no.	*Width of black lines (mm)*	*Width of white spaces (mm)*	*Percent black*
0	All white		0
1	1	9	20
2	2.3	7.7	40
3	3.7	6.3	60
4	5.5	4.5	80
5	All black		100

send air pollution inspectors to a "smoke school," where they are trained and certified as being able to read the Ringelmann density of black and white plumes to an accuracy that is acceptable for court testimony (Figs. 2 and 3).

In addition to subjective observation of smoke density, systems have been developed to objectively measure by means of a photocell the decrease in intensity of a beam of light projected through a plume prior

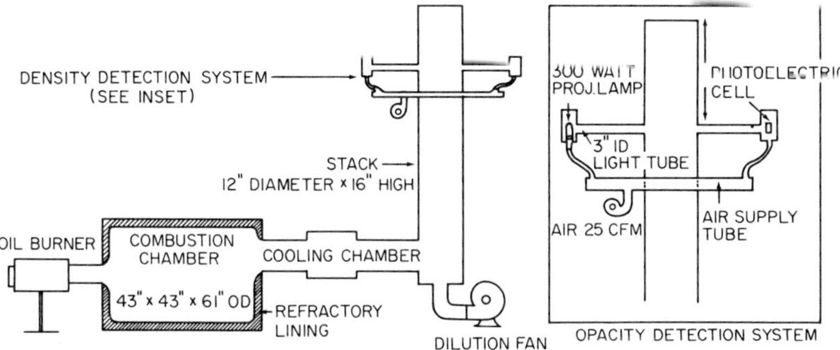

Figure 2. Black smoke generator for training people in reading Ringelmann smoke density. OD, outside diameter; ID, inside diameter; Proj., projection; CFM, cubic feet per minute.

to its emission, and to measure by photometer the blackness of a white paper filter through which smoke-containing gases have been passed. The blackness scales for the filtration instruments are supplied by the instrument makers and identified by their names, e.g., Bacharach, Bosch. (See also Chapter 15 of Volume III and Chapter 12 of this volume.)

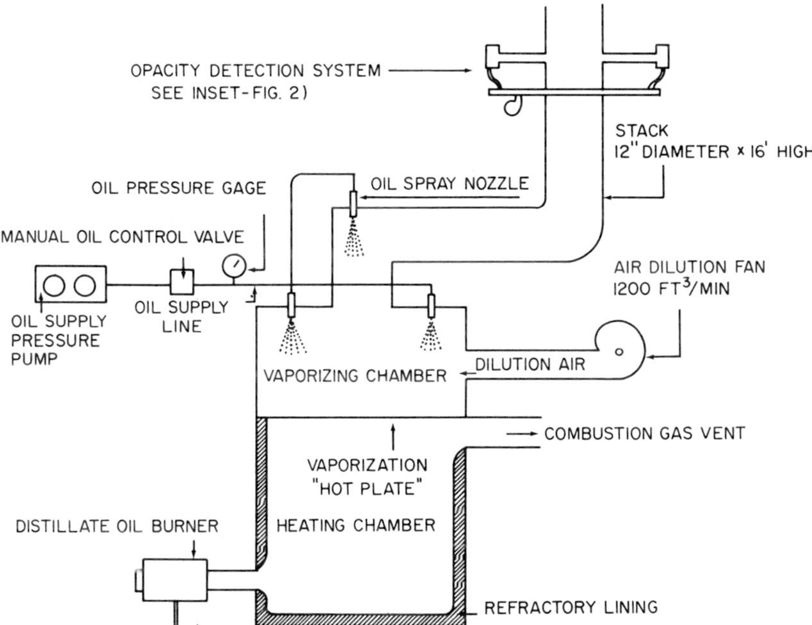

Figure 3. White smoke generator for training people in reading equivalent opacity in terms of Ringelmann smoke density.

The usual regulatory statement applying emission opacity standards prohibits an emission of a density of *A* or greater for more than *B* minutes in any period of *C* minutes. The principal consideration for establishing period *C* is the mobility of the source. A stack of a building or factory is stationary and can be observed over a long time, whereas a moving locomotive, marine vessel, or automotive vehicle can be observed from a stationary observation point on the ground for only a short time before it travels out of the observer's view. For stationary sources, the longer the time interval used for *C*, the stricter the regulation, particularly if the emission time *B* is cumulative within time period *C*. The concepts of regulations based upon time *B* within time *C* and of time allowances made for smoke released when starting fires had their historical basis in the nature of smoke emission from steam locomotives and hand-fired soft coal industrial boiler furnaces, and represented a compromise to meet the objections of operators of that class of equipment to more stringent regulations. Although the reasons for these compromises have largely disappeared along with the class of equipment they were intended to regulate, their legacy remains embedded in regulatory legislation. When time interval *C* is long, as it should be for present-day conditions of continuous fuel feed to fuel-burning devices, allowed emission time *B* becomes essentially an allowance for operating emergencies. A typical combination of *A*, *B*, and *C* as found in the Cleveland, Ohio, ordinance is shown in the following tabulation.

Source	*A* Emission prohibited equal to or darker than Ringelmann no.	*B* Time permitted (minutes)	*C* Observation time
New process or fuel burning source	2	3	1 hour
	2	9	8 hours
Existing process or fuel burning source	3	5	1 hour
	3	15	8 hours
Diesel locomotive or steamship	3	$\frac{1}{2}$	3 minutes
	3	4	15 minutes
Refuse incinerators	1	—	—

2. Smoke Density Regulations

In the early history of smoke regulation, only industrial and locomotive stacks were regulated. Later, when all stacks were regulated, the householder was accorded less stringent limits because it was felt that, with hand-fired coal-burning equipment, he would have more difficulty meeting strict limits than would the industrial plant. The modern approach to varying limits for different categories is to recognize the present availability of smokeless methods for heating homes and hot water and to make the smoke emission limits for domestic and commercial installations extremely restrictive (Table III). National stationary source smoke emission limits, expressed as Ringelmann number or percent opacity, have been adopted by a number of countries (Tables IV and IVA) (*6*). They also are in use in all states of the United States, in the provinces of five countries (Table V) (*6*) and in cities around world too numerous to mention. Stationary source emission standards based upon the Bacharach shade of samples of flue gas taken from the stack or breeching are used nationally in nine countries (Table VI) (*6*), and in addition, there are provincial regulations of this type in New South Wales, Australia, and Northrhine-Westphalia, Federal Republic of Germany, and municipal regulations in Paris, France; Madrid, Spain; Zurich, Switzerland; and Sarajevo and Zagreb, Yugoslavia.

Several United States communities have adopted "equivalent opacity" regulations, i.e., regulations that apply opacity measurements based on the Ringelmann scale to emissions other than smoke (as customarily defined) and of colors other than black (or gray). While the opacity of a smoke emission is a poor measure of its gravimetric particle loading, the opacity of a smoke-free fly ash emission correlates quite well with its gravimetric loading (Table VII). A special Bay Area (California) provision allows power plant stacks (oil-fired) that exceed the opacity standard of not more than 3 minutes per hour of no. 1 Ringelmann to alternatively meet a gravimetric emission standard of $[0.06 \times (\text{stack area in square feet})^{1/2}]$ grains/ft^3 at 60°F.

Table III Percent of United States Local Regulations Prohibiting Visible Emissions of Various Opacities from Stationary Sources

Emission greater than percent opacity prohibited	*Percent of regulations prohibiting visible emissions*				
	1940	*1950*	*1960*	*1965*	*1975*
60	81	69	41	32	3
40	19	31	59	66	33
20	0	0	0	2	56
0	0	0	0	0	8

Table IV National Emission Standards for Effluent Opacity of Gas from Stationary Sources[a] *(6)*

Country	*Category*	*Ringelmann number*	*Not to be exceeded more than (minutes/hour)*	*Notes*[b]
Australia	All stationary fuel-burning sources	1	—	1, 2
Belgium	Incinerators	2	—	—
Canada	Asphalt paving plants	1	—	8
	Arctic mining plants	1	—	—
Columbia	—	3	—	3
Great Britain	—	2	—	—
Guam	—	4	3	—
Hong Kong	All furnaces or ovens	2	—	4
Ireland	Chimney with 2 furnaces	2	16 min/8 hr	—
	Chimney with 3 furnaces	2	22 min/8 hr	—
	Chimney with 4 furnaces	2	27	—
	Private dwelling	4	4 min	—
	Private dwelling	2	8 min/8 hr	—
	Private dwelling	4	2 min/30 min	—
Italy	Smoke stack height up to 50 m	3	5	—
	Smoke stack height over 50 m	4	—	—
	—	2	5	—
	—	3	—	—
Malta	—	4	4	—
Mexico	Incinerators	2	3	—
	Other existing	2	5	—
	Other new	2	3	5
New Zealand	Clean air zones	1	—	—
	Other zones	2	4	—
Philippines	—	3	2	—
Singapore	—	2	5	6
West Germany	Solid fuel	1	—	7
	Graphite electrode manufacture	1	—	—
	Nonferrous alloys (other than aluminum)	1	—	—
	Aluminum reduction	1	—	—
	Refuse disposal	1	—	—
	Charging coke ovens	3	—	—

Country	*Category*	*% opacity*	*Not to be exceeded more than (minutes/hour)*	*Notes*[b]
United States	Asphalt concrete plants	20	—	—
	Coal preparation plants			
	Thermal dryers, processing, conveying, storing, transferring and loading	20	—	—

Table IV ***(Continued)***

Country	*Category*	*% opacity*	*Not to be exceeded more than (minutes/hour)*	*Notes*[b]
United States	Pneumatic cleaning equipment	10	—	—
	Ferroalloy production	15	—	—
	Electric arc furnaces—			
	From control devices	3	—	—
	From shop roofs	0	—	—
	Except while charging	20	—	—
	Except while tapping	40	—	—
	Fossil fuel fired boiler furnaces	20	—	—
	Iron and steel mills dust-handling equipment	10	—	—
	Nitric acid plants	10	—	—
	Oil fired boiler furnaces	40	2	—
	Petroleum refineries	30	3	—
	Portland cement kilns	20	—	—
	Primary aluminum reduction plants			
	Anode baking	20	—	—
	Potrooms	10	—	—
	Primary copper smelters			
	Drying, roasting, smelting, converting	20%	—	—
	Primary lead smelters			
	Blast, reverboratory or electric furnaces, sintering machines and converters	20	—	—
	Primary zinc smelters			
	Roasters and sintering machines	20	—	—
	Secondary brass and bronze smelters			
	Blast and electric furnaces	10	—	—
	Secondary lead smelters	20	—	—
	Sewage treatment plants	20	—	—
	Sulfuric acid plants	10	—	—

[a] See also Table IVA for standards for Spain.

[b] Notes: (1) No. 3 Ringelmann acceptable for lighting-up or soot-blowing; (2) national guideline for new plants; (3) except for 15 minutes in 24 hours; (4) not to exceed 6 minutes/4 hours or 3 minutes continuously at any one time; (5) start-up; (6) not more than three times per day; (7) also during soot-blowing, but not hand-firing coal; (8) national guideline.

Table IVA Emission Opacity Standards—Spain *(6a)*

Process	*Ringlemann number*	*Percent opacity*	*Remarks*
Asphalt plants	1	20	—
Cement kilns	0.5	10	—
Organic fertilizer	1	20	Except #2 for 3 minutes/hour
Petroleum refining	1	20	Except 3 minutes/hour for 2% of time/year
Reheating and heat treatment furnaces	1.5	30	—
Sludge drying	1	20	—
Power plants			
Coal	1	20	Except #2 <2 minutes/hour
Domestic oil	1	20	Except 3 times/day <10 minutes
Heavy #1 oil	2	40	Except 3 times/day <10 minutes
Heavy #2 oil	2.5	50	Except 3 times/day <10 minutes

Table V Provinces for Which There Are Provincial Standards for Smoke in Effluent Air or Gas from Stationary Sources *(6)*

Country	*Province*
Australia	New South Wales,[a] Queensland, South Australia, Victoria, Western Australia
Brazil	Guanabara, São Paulo
Canada	Alberta, British Columbia, Manitoba, Ontario, New Brunswick, Newfoundland
West Germany	North Rhine-Westphalia
Yugoslavia	Serbia

[a] Also provincial standard for soot in effluent air and gas from stationary sources.

B. Odor Emission Standards

The odor control regulations of the states of the United States have been reviewed by Leonardos (*7*) (Table VIII) (*7–9*). Subjective evaluation of odor emission is made difficult by the phenomenon of odor fatigue, which means that after a person has been initially subjected to an odor, he loses the ability to perceive the continued presence of low concentrations of that odor. Therefore, all systems of subjective odor evaluation rely upon preventing olfactory fatigue of the observer by letting him breath odor-free air for a sufficient time prior to his breathing the odorous air and evaluating its odor content. Usually, an activated

Table VI National Emission Standards for Soot in Effluent Air or Gas from Stationary Sources *(6)*

Country	*Source*	*Bacharach shade* (≤)	*Other units*	*Notes*[a]
Australia	Any boiler or furnace burning oil or gas	3	—	1, 2
Belgium	Domestic heating			
	≤ 300 therms/hr fuel oil	3	—	—
	> 300 < 2000	4	—	—
	>2000 therms/hr fuel oil	5	—	—
	Central stations—fuel oil	4	—	—
France	Space heating smokeless zones 1 and 2	6	—	—
Italy	Thermal installations	8	—	—
Philippines	—	3	0.92 gm/m^3	—
Spain	Power plants			
	Domestic oil	2	—	—
	Heavy #1 oil	4	—	—
	Heavy #2 oil	5	—	—
Sweden	New gas turbines operating more than 500 hours/year	3	—	3
	New gas turbines operating less than 500 hr/year	5	—	3
	Oil burners greater than 50 MW	3	1.0 kg/ton oil	4
	New oil burners less than 50 MW	3	1.0 kg/ton oil	5
	Existing oil burners less than 50 MW	3	1.5 kg/ton oil	5
Switzerland	Space heating <200 kg/hour	2	—	6
	New Heating <200 kg/hour	1	—	6
West Germany	Oil heating using extra light and light oil	2	—	—
	Oil heating using medium and heavy oil	3	—	—

[a] NOTES: (1) other than for lighting-up or soot-blowing; (2) national guidelines for new plants; (3) maximum values during steady-state operation—operating time is median for 5-year period; (4) flue gas velocity greater than 8 m/second at minimum load; (5) same as (4) if residual oil is burned with burner capacity at least 0.5 MW each; (6) other space heating standards in Table XVI in mg/m^3.

carbon bed is used to clean up the air to provide the odor-free air required by the observer. The odor observer generally uses a five-step scale of odor intensity, quite similar in concept to the Ringelmann scale for smoke density. (See also Chapter 8, Volume III.)

Table VII Relationship between Fly Ash Loading and Emission Opacity

Fly ash loading			*Emission opacity*	
Stack gases at 60°F (16°C)		*Pounds/1000 lb of stack gases at 50% excess air*		
grains/ft³	*mg/m³*		*Obscuration*	*Ringelmann number*
0.22	500	0.41	Visible plume	2
0.11	250	0.21	Light haze	1
0.03	68	0.06	Invisible effluent	0

Table VIII Odor Emission Regulations of the States of the United States *(7)*

Odor control regulations	*States*
No specific regulations	Arizona, Georgia, Hawaii, Indiana, Iowa, Kansas, Louisiana, Maine, Nebraska, New Mexico, North Dakota, Oklahoma, South Carolina, Tennessee, Utah, Washington
Odor nuisance prohibited, but not objectively defined	Alabama, Alaska, Arkansas, California, Florida, Illinois,[a] Maryland, Massachusetts, Michigan, Mississippi, Montana, New Hampshire, New Jersey, New York, Ohio, Rhode Island, Texas
Objectionability criteria defined	Connecticut, Nevada, South Dakota, Vermont, West Virginia, Wisconsin
Scentometer limits in ambient air	Colorado, District of Columbia, Illinois,[a] Kentucky, Minnesota, Missouri, Nevada, Wyoming
Control equipment specifications for rendering[b]	Idaho, Maryland, Minnesota, Montana, Oregon, North Carolina, Pennsylvania, Vermont, Wyoming
Control equipment specifications for other sources[b]	Idaho, Montana, North Carolina, Oregon, Pennsylvania, Vermont
Odor emission standards[c]	
In odor units (o.u.)[d]	Connecticut, 120 o.u./ft³ (4238 o.u./m³); Illinois (rendering), 120 o.u./ft³ (4238 o.u./m³)
In odor concentration units (o.c.u.)[e]	Minnesota: >50 ft. (15.2 m) stack, 150 o.c.u.; <50 ft. (15.2 m) stack, 25 o.c.u.; scfm × o.c.u. <1,000,000 o.c.u./minute

[a] Illinois—rendering plant odors controlled by ambient air and source odor standards.
[b] Generally incineration at >1200°F > 0.3 second.
[c] See Table XII for list of states with emission standards for total reduced sulfur and other odorants.
[d] Mills ASTM Method *(8)*.
[e] Benforado ASTM Method *(9)*.

In Missouri, odor emission in St. Louis County is prohibited when it causes an objectionable odor:

1. On or adjacent to residential, recreational, institutional, retail sales, hotel, or educational premises.

2. On or adjacent to industrial premises when air containing such odorous matter is diluted with 20 or more volumes of odor-free air.

3. On or adjacent to premises other than those in 1 and 2 when air containing such odorous matter is diluted with 4 or more volumes of odor-free air.

The above requirements apply only to objectionable odors. An odor is deemed objectionable when 30% or more of a sample of at least 20 people exposed to it believe it to be objectionable in usual places of occupancy. If fewer than 20 people are exposed, 75% must believe it to be objectionable.

VI. Objective Standards

There are two general categories of objective emission standards, (a) those for specific chemical substances, compliance with which is established by chemical analysis of effluent air and gases; and (b) those for total solid particulate matter emissions, compliance with which is established by physically weighing the total solid particulate matter in the effluent air or gas.

A. Specific Pollutants

There are national emission standards for specific pollutants (Table IX). Table IX excludes standards for Spain, which are in Table IXA (*6a*) and organic and inorganic compounds for which there are standards in West Germany (Table X) (*10*). Japanese emission standards for nitrogen oxides from stationary sources are sufficiently extensive as to require a special table for their presentation (Table XI). There are also emission standards for specific pollutants in a number of states of the United States and provinces of Australia and Canada (Table XII) (*6*), as well as many cities of the United States. West German regulations for facilities for the production of nitric acid include two charts for emission limits for nitric oxide (Figs. 4 and 5) (*10*) and those for facilities for production of hydrofluoric acid include a chart for emission limits for hydrogen fluoride (Fig. 6) (*10*). A number of the regulations limiting emission of sulfur oxides from specific processes in the states of the

Table IX National Emission Standards for Specific Pollutants in Effluent Air or Gas from Stationary Sources[a] (6)

Substance and country	*Source*	*Standard*		*Notes[b]*
		Original units	*mg/m^3*	
Acid gases				
Australia	New H_2SO_4 manufacture	3.0 gm/m^3	3000.0	1, 2, 3
Great Britain	Superphosphate fertilizer manufacture	0.1 $grains/ft^3$	228.8	1, 4
Ireland	H_2SO_4 manufacture	4.0 $grains/ft^3$	9153.0	1
New Zealand	New H_2SO_4 manufacture	5.0 gm/m^3	5000.0	1, 3
Singapore	H_2SO_4 manufacture	6.0 gm/m^3	6000.0	1, 2, 5 •
Acrolein				
Czechoslovakia	All	3 kg/hour	—	6
Aldehydes				
West Germany	Gas-burning furnaces	mg/m^3	20.0	7
Ammonia				
Czechoslovakia	All	3 kg/hour	—	6
Antimony				
Australia	New plants	mg/m^3	10.0	2, 3, 8, 9, 10
Great Britain	Less than 5000 cfm	0.05 $grains/ft^3$	114.4	9, 11
	More than 5000 cfm	0.02 $grains/ft^3$	45.7	9, 11
Singapore	All	0.02 gm/m^3	20.0	2, 8, 9, 12
Arsenic				
Australia	All	mg/m^3	10.0	2, 3, 8, 9, 10
Czechoslovakia	All	0.03 kg/hour	—	6, 13
Great Britain	Less than 5000 cfm	0.05 $grains/ft^3$	114.4	9, 11
	More than 5000 cfm	0.02 $grains/ft^3$	45.7	9, 11
Singapore	All	0.02 gm/m^3	20.0	2, 8, 9, 12
Benzene				
Czechoslovakia	All	24.0 kg/hour	—	6

Beryllium				
Australia	All	mg/m³	0.1	2, 3, 9, 14
Cadmium				
Australia	All	mg/m³	3.0	2, 3, 8, 10, 15
Great Britain	Maximum, 30 pounds/168 hours	0.017 grains/ft³	38.9	8, 9
Japan	Cadmium pigment, cadmium carbonate, and glass manufacture; copper, lead, and cadmium refining	mg/m³	1.0	9, 16
Singapore	All	0.02 gm/m³	20.0	2, 8, 9, 12
Carbon				
West Germany	Graphite electrode manufacture	mg/m³	250.0	17
	Refuse incineration	mg/m³	50.0	49, 60
Carbon black				
Czechoslovakia	Amorphous carbon	1.5 kg/hour	—	6
Carbon dioxide				
Italy	Thermal installations	10% by volume	—	—
Switzerland	Oil burners			
	<3 kg/hour	8.0% by volume	—	3
	3–9 kg/hour	10.0% by volume	—	3
	>10.0 kg/hour	12.0% by volume	—	—
Carbon disulfide				
Czechoslovakia	All	0.3 kg/hour	—	6
Carbon monoxide				
Australia	All	0.5 gm/m³	500.0	2, 3
Czechoslovakia	All	60.0 kg/hour	—	6
France	Electric generating plants	0.05% by volume	—	—
	Incinerators	0.1% by volume	—	—
United States	Fluid catalyst regenerator	0.50% by volume	—	—
	Ferroalloy manufacture	20% by volume	—	—
West Germany	Solid fuel burning	mg/m³	250.0	61
	Gas burning	mg/m³	100.0	7

Table IX *(Continued)*

Substance and country	Source	Standard: Original units	Standard: mg/m^3	Notes[b]
	Oil burning	mg/m^3	175.0	7
	Refuse burning			
	Household	1.0 gm/m^3	1000.0	62
	Other	mg/m^3	100.0	49
West Germany (VDI 2117E)	Vaporizer oil burners	0.1 by volume	—	18
Chlorine				
Australia	All	0.2 gm/m^3	200.0	2, 3, 8, 9
Czechoslovakia	All	1.0 kg/hour	—	6
Denmark, Sweden	Mercury cell	mg/m^3	3.0	—
Great Britain	All	0.1 $grains/ft^3$	228.8	8
Japan	Ferric chloride, chlorinated ethylene, activated carbon, and other chemical manufacture	mg/m^3	30.0	19
Singapore	All	0.2 gm/m^3	200.0	2
West Germany	All sources >3 kg/hour	mg/m^3	30.0	—
	Aluminum reduction	mg/m^3	3.0	49
	Refuse burning			
	Household (<0.75 tons/hour)	kg/hour	6.0	—
	Household (>0.75 tons/hour)	mg/m^3	100.0	49
	Other	mg/m^3	100.0	63
	Chlorine manufacture	mg/m^3	3.0	—
	Chlorine manufacture	mg/m^3	6.0	20
	Chlorine manufacture—Mercury cells	1 gm/ton Chlorine	—	—
Copper				
Singapore	All	0.02 gm/m^3	20.0	2, 8, 9, 12
Formaldehyde				
Czechoslovakia	All	0.5 kg/hour	—	6

Fluorine					
Czechoslovakia	Gaseous inorganic compounds	0.3 kg/hour	—	6	
Japan	Aluminum reduction—ducts	mg/m^3	3.0		—
	Aluminum reduction—vents	mg/m^3	1.0		—
	Calcium superphosphate manufacture	mg/m^3	15.0		—
	Phosphoric acid fertilizer manufacture	mg/m^3	20.0		—
	Trisodium phosphate, phosphoric acid, and glass manufacture	mg/m^3	10.0	21	
Singapore	All	$0.1\ gm/m^3$	100.0	2, 22	
Sweden	Ferromolybdenum manufacture	1 kg/ton of product	—		—
United States	Aluminum reduction				
	Prebake plant	0.95 kg/ton Al	—		—
	Soderberg plant	1.0 kg/ton Al	—		—
	Phosphate fertilizer manufacture				
	Diammonium phosphate manufacture	30.0 gm/ton P_2O_5	—		—
	Superphosphoric acid manufacture	5.0 gm/ton P_2O_5	—		—
	Triple superphosphate				
	Manufacture	100.0 gm/ton P_2O_5	—		—
	Curing	0.25 gm/hour/ton equivalent P_2O_5	—		—
	Wet process phosphoric acid manufacture	10.0 gm/ton P_2O_5	—		—
West Germany (VDI 2286)	Aluminum reduction	$0.05\ gm/m^3$	50.0	18	
Fluorine compounds					
Australia	New aluminum reduction	$0.02\ gm/m^3$	20.0	2, 3, 22	
	All other new processes	$0.05\ gm/m^3$	50.0	2, 3, 22	
Fluorine inorganic compounds					
Singapore	All	$0.1\ gm/m^3$	100.0	2, 22	
Heavy metals (total)					
Australia	All	mg/m^3	10.0	15	
Hydrochloric acid					
Czechoslovakia	All	0.1 kg/hour	100.0	6, 23	

Table IX *(Continued)*

Substance and country	*Source*	*Standard*		*Notes*[b]
		Original units	*mg/m³*	
West Germany	Hydrogen chloride manufacture	mg/m³	10.0	—
Hydrogen chloride				
Great Britain	Alkali (salt cake) works	0.2 grains/ft³	4576.0	—
Great Britain; Ireland	Hydrogen chloride manufacture	0.2 grains/ft³	4576.0	—
Japan	Ferric chloride, chlorinated ethylene, activated carbon, and other chemical manufacture	mg/m³	80.0	19
Singapore	All	0.4 gm/m³	400.0	2
Sweden	Cable burning	mg/m³	250.0	—
West Germany	Incinerators (all)	mg/m³	100.0	24
West Germany (VDI 3451E)	Absorption	0.025 gm/m³	25.0	18, 25
	Sulfate methods	0.40 gm/m³	400.0	18, 25
	Hydrogen chloride electrolysis	0.10 gm/m³	100.0	—
	Filling and transfer	0.10 gm/m³	100.0	18, 25
	Zinc chloride manufacture	0.16 gm/m³	160.0	18, 25
	Silicon tetrachloride manufacture	0.20 gm/m³	200.0	18, 25
	Vinyl chloride manufacture	0.17 gm/m³	170.0	18, 25
	α-Chlorpropionic acid manufacture	0.18 gm/m³	180.0	18, 25
	Sintering crude phosphate	0.35 gm/m³	350.0	18, 25
	Burning organic by-products	0.30 gm/m³	300.0	18, 25
	Hydrogen chloride absorption	0.33 gm/m³	330.0	18, 25
Hydrofluoric acid				
Singapore	All	0.1 gm/m³	100.0	2, 22
Hydrogen fluoride				
Great Britain	All	0.1 grains/ft³	229.0	26

Japan	Aluminum reduction—ducts	mg/m³	3.0	—
	Aluminum reduction—vents	mg/m³	1.0	—
	Calcium superphosphate manufacture	mg/m³	15.0	—
	Phosphoric acid fertilizer manufacture, baking furnace	mg/m³	30.0	—
	Trisodium phosphate, phosphoric acid, and glass manufacture	mg/m³	1.0	21
Hydrogen fluoride (as F)				
West Germany	Incinerators, iron ore sintering	mg/m³	5.0	27
	Gases and vapors >5 kg/hour	mg/m³	5.0	59
	Refuse burning			
	Household (<0.75 tons/hour)	kg/hour	0.2	—
	Household (>0.75 tons/hour)	mg/m³	5.0	49
	Other	mg/m³	5.0	49
	Ceramic kilns	mg/m³	30.0	7, 28
	Aluminum reduction	mg/m³	2.0	—
	Closed furnaces	1 kg/ton Al	—	—
	Open furnaces	0.8 kg/ton Al	—	—
	Glass manufacture	mg/m³	15.0	—
	Sintering crude phosphate concentrates	mg/m³	10.0	—
	Slag remelting	mg/m³	1.0	—
Hydrogen sulfide				
Australia	All	mg/m³	5.0	2, 3, 29
Czechoslovakia	All	0.08 kg/hour	—	6
Great Britain; Singapore	All	5.0 ppm	7.5	—
Sweden	Kraft recovery furnace	mg/m³	10.0	30
United States	Petroleum refineries	mg/m³	230.0	31
West Germany	Refineries	mg/m³	10.0	32
	Claus sulfur plant	mg/m³	10.0	—
	Coke oven gas	1.5 gm/m³	1500.0	34

Table IX *(Continued)*

Substance and country	*Source*	*Standard*		*Notes*[b]
		Original units	*mg/m³*	
Lead				
Australia	All	mg/m³	10.0	2, 3, 8, 9, 10, 15
Canada	Secondary lead smelting	—	—	48
Czechoslovakia	Except tetraethyllead	0.007 kg/hour	—	6
Great Britain	Up to 3000 cfm of exhaust	0.05 grains/ft³	114.4	8
	3000–10,000 cfm	0.05 grains/ft³	114.4	8, 35
	10,000–140,000 cfm	0.01 grains/ft³	22.8	8, 36
	Over 140,000 cfm	0.005 grains/ft³	11.4	8, 37
Japan	Refining copper, lead, or zinc; blast and sintering furnaces	mg/m³	30.0	9
	Glass manufacture using lead oxides, baking furnace	mg/m³	20.0	9
	Pipe, sheet, wire, pigment, and storage battery manufacture and secondary refining	mg/m³	10.0	9
	Refining copper, lead, and zinc; other furnaces	mg/m³	10.0	9
New Zealand	All	mg/m³	100.0	3, 9
Singapore	All	0.02 gm/m³	200.0	2, 8, 9, 12
Manganese				
Czechoslovakia	All	0.1 kg/hour	—	6, 38
Mercury				
Australia	All	mg/m³	3.0	2, 3, 8, 9, 10, 15
Czechoslovakia	All	0.003 kg/hour	—	6, 39
Singapore	All	0.02 gm/m³	20.0	2, 8, 9, 12
Sweden	In ventilation air from new Cl manufacture	0.001 kg/ton	—	—
	In H_2 vented from new Cl manufacture	0.0001 kg/ton	—	—

Nickel				
Australia	All	mg/m^3	20.0	2, 3, 8, 9, 14, 40
Nickel carbonyl				
Australia	All	mg/m^3	0.5	2, 3, 8, 14
Nitric acid				
Australia	New HNO_3 or H_2SO_4 manufacture	1.0 gm/m^3	1000.0	2, 3, 41
	Any other process except new gas-fired power plants	0.5 gm/m^3	500.0	2, 3,41
Czechoslovakia	All	0.1 kg/hour	—	6, 23
Singapore	HNO_3 manufacture	4.0 gm/m^3	4000.0	1, 2
	Any other process	2.0 gm/m^3	2000.0	1, 2
Nitrogen oxides				
Australia	New HNO_3 and H_2SO_4 manufacture	1.0 gm/m^3	1000.0	2, 41
	New gas-fired power plants	0.35 gm/m^3	350.0	2, 3, 41
	Any other new process	0.5 gm/m^3	500.0	2, 3, 41
Czechoslovakia	All	3.0 kg/hour	—	6, 41
Great Britain	HNO_3 manufacture	1000.0 ppm	1800.0	41, 42
	All other processes	1.0 grains/ft^3	2288.3	2, 41
Japan	All	—	—	43
Singapore	HNO_3 manufacture	4.0 gm/m^3	4000.0	1, 2
	Any other process	2.0 gm/m^3	2000.0	1, 2
United States	New gas-fired power plants	0.2 pound/MMBTU	—	—
	New liquid-fuel-fired power plants	0.3 pound/MMBTU	—	—
	New solid-fuel-fired power plants	0.7 pound/MMBTU	—	—
	New HNO_3 manufacture	3.0 pounds/ton acid	—	—
West Germany	HNO_3 manufacture	—	—	44, 46
	Concentrated HNO_3 manufacture	—	—	45, 46
Organic compounds				
West Germany	Fluid incineration	mg/m^3	50.0	49
Organic gases and vapors				
West Germany	Petroleum refining	0.04% of crude	—	—

Table IX *(Continued)*

Substance and country	*Source*	*Standard*		*Notes*[b]
		Original units	*mg/m*^{3}	
Phenol				
Czechoslovakia	All	3.0 kg/hour	—	6
Silicon fluoride				
Japan	Aluminum reduction—ducts	mg/m^3	3.0	—
	Aluminum reduction—vents	mg/m^3	1.0	—
	Calcium superphosphate manufacture	mg/m^3	15.0	—
	Phosphoric acid fertilizer manufacture	mg/m^3	20.0	—
	Trisodium phosphate, phosphoric acid, and glass manufacture	mg/m^3	10.0	21
Sulfur dioxide				
Canada	Burning coke oven gas	1300 gm/metric ton of coke	—	3, 64
	Combustion of fuel oil in Arctic mining plants	1.1 gm/1000 kcal	—	—
Czechoslovakia	All	—	—	50
Denmark	Sulfite pulp mills (new)	10 kg/ton pulp	—	—
	Sulfite pulp mills (existing)	20 kg/ton pulp	—	—
	Sulfuric acid manufacture (new)	5 kg/ton acid	—	—
	Sulfuric acid manufacture (existing)	20 kg/ton acid	—	—
East Germany	All	—	—	51
Great Britain	H_2SO_4 concentration	1.5 grains/ft^3	3432.4	1
	New contact H_2SO_4 manufacture	0.5% of the sulfur burned	—	—
	Old sulfur-burning H_2SO_4 manufacture	2% of the sulfur burned	—	—
	Old H_2SO_4 manufacture other than sulfur burning	4.0 grains/ft^3	9153.2	1

Italy	Heating plants	0.20% by volume	—	52
Japan	All	—	—	53
Sweden	New H_2SO_4 manufacture	5 kg/ton acid	—	54
	Existing H_2SO_4 manufacture	20.0 kg/ton acid	—	54
	NH_3 manufacture	—	—	55
	New sulfite pulp mills	10.0 kg/ton pulp	—	—
	Existing sulfite pulp mills	20.0 kg/ton pulp	—	—
	Oil steam-electric power plants over 300 MW	20.0 kg/ton fuel	—	—
United States	New liquid-fuel-fired power plants	0.8 pound/MMBTU	—	—
	New solid-fuel-fired power plants	1.2 pounds/MMBTU	—	—
	Primary copper smelters			
	Roaster, smelting furnace, converter	0.065%	—	—
	Primary lead smelters			
	Sintering machine, electric smelting furnace, converter	0.065%	—	—
	Primary zinc smelters			
	Roasters	0.065%	—	—
	Sulfuric acid manufacture	4.0 pounds/ton	—	—
West Germany	SO_2 manufacture	mg/m^3	30.0	—
	Natural gas burning	mg/m^3	50.0	7
	Coal gas burning	mg/m^3	100.0	7
	Nonferrous smelting	3 gm/m^3	3000.0	—
West Germany (VDI 2110)	Waste coke oven gas	2.54 gm/m^3	2540.0	18
	Waste coke oven gas	0.59 gm/m^3	590.0	18, 56
	H_2SO_4 (100%) manufacture	1.5 gm/m^3	1500.0	18
Sulfuric acid				
Australia	All	0.1 gm/m^3	1000.0	3
Czechoslovakia	All	0.1 kg/hour	—	6, 23
Singapore	Other than combustion or H_2SO_4 manufacture	0.2 gm/m^3	200.0	1, 2
United States	New H_2SO_4 manufacture	0.15 pound/ton acid	—	57
West Germany (VDI 2298)	SO_3 and H_2SO_4 manufacture	mg/m^3	5.0	18
		2 kg/ton acid	—	18

Table IX *(Continued)*

Substance and country	*Source*	*Standard*		*Notes*[b]
		Original units	*mg/m*^{3}	
Sulfur trioxide				
Australia	All new plants	0.1 gm/m^3	100.0	1, 2, 3
Singapore	Other than combustion or H_2SO_4 manufacture	0.2 gm/m^3	200.0	1, 2
Sweden	New H_2SO_4 manufacture	0.5 kg/ton acid	—	54
	Existing H_2SO_4 manufacture	0.8 kg/ton acid	—	54
West Germany (VDI 2298)	SO_3 and H_2SO_4 contact manufacture	0.4 kg/ton acid	—	18, 47, 58
West Germany	SO_3 and H_2SO_4 manufacture	0.6 kg/ton acid	—	33
Sulfur trioxide and sulfuric acid mist				
Denmark, Sweden	Sulfuric acid manufacture (new)	0.5 kg/ton acid	—	—
	Sulfuric acid manufacture (existing)	0.8 kg/ton acid	—	—
Tar				
West Germany	Graphite electrode manufacture	mg/m^3	50.0	—
Vinyl chloride				
United States	Ethylene dichloride purification, vinyl chloride formation, purification, stripping, recovery, mixing, weighing, holding, venting, leakage	10 ppm	—	—
	Oxychlorination reactor	0.2 gm/kg product	—	—
	PVC (polyvinyl chloride) plants			
	Reactor opening loss	0.02 gm/kg product	—	—
	Stripping Technology			
	Dispersion PVC resins	2000 ppm	—	—
	Other PVC resins	400 ppm	—	—
	Other than stripping technology			
	Dispersion PVC resins	2 gm/kg product	—	—
	Other PVC resins	0.4 gm/kg product	—	—

[a] See also Table IXA for standards for Spain, and Table X for both organic and inorganic compounds, Federal Republic of Germany.

[b] NOTES:

1. As SO_3.
2. STP at 0°C and 1 atm (dry).
3. National guideline.
4. Or efficiency of condensation of acid gases greater than 99%.
5. Discharge free from persistent mist.
6. Emission rate above which it is necessary to submit a report to the government. Where discharge is for less than 1 hour, there is a proportionate increase in emission rate permissible without such reporting. For permissible emission, see Table XXIX.
7. As formaldehyde; at 3% O_2 by volume.
8. As the element
9. Also compounds of the element.
10. Total of antimony, arsenic, cadmium, lead, mercury, or their compounds may not exceed this limit.
11. As the trioxide.
12. Total of antimony, arsenic, cadmium, copper, lead, mercury, or their compounds may not exceed this limit.
13. Inorganic compounds except arsenic.
14. Tentative standard.
15. Addition of each metal or compound expressed as the metal in each case.
16. Glass manufacture using cadmium sulfide or carbonate as raw materials.
17. At 8% CO_2 by volume.
18. Verein Deutscher Ingenieure.
19. Includes chlorine quick cooling for chlorinated ethylene manufacture.
20. If complete liquification; also short-term peaks.
21. Glass manufacture using fluorite or sodium silicofluorate as raw materials.
22. As HF.
23. As hydrogen ion.
24. Applies to sources with 3 kg/hour hydrogen chloride or more.
25. Wet.
26. As SO_3 equivalent in original units.
27. Applies to sources with 150 gm/hour hydrogen fluoride or more.
28. If on crest or valley, 5 mg/m^3.
29. As H_2S.
30. Ninety-nine percent of the time per month for new units; 90% for existing units; also (concentration in stack gas/concentration at odor threshold) = at least 10,000.
31. Unless burned to SO_2 in a manner that prevents release of SO_2 to atmosphere.
32. 0.4% H_2S by volume must be cleaned.

33. $<6\%$ SO_2 in input and <100 tons/day.
34. Other sulfur-bearing compounds, 0.5 gm/m^3 hourly average.
35. One hundred pounds/week mass emission limit.
36. Four hundred pounds/week mass emission limit.
37. One thousand pounds/week mass emission limit.
38. As MnO_2.
39. Metallic.
40. Except nickel carbonyl.
41. As NO_2.
42. And the emission shall be colorless.
43. See Table XI.
44. See Figure 4.
45. See Figure 5.
46. Decolorization of nitric acid plant effluent $= (6100 \cdot 2.05/d)$ mg/m^3, where d = inside exit stack diameter.
47. $>6\%$ SO_2 in input.
48. Proposed national guideline $<63\%$ Pb in particulate matter. If $>63\%$, limit particulate matter to 63% of particulate matter limit for process in Table XVI.
49. 11% volume O_2 in gas.
50. See Table XXIX.
51. See Tables XXX, XXXI, and XXXII.
52. For heating installations burning liquid fuels with viscosity $>5°$ Engler and $>4\%$ S.
53. See Table XXXIII.
54. Sulfur or pyrite as raw material.
55. Equipment for releasing sulfur required.
56. By using partly desulfurized coke oven gas.
57. As H_2SO_4.
58. At least 99% SO_2 has to be recycled.
59. See Figure 6.
60. In combustible organic matter in flue gas.
61. At 13% O_2 for wood, 7% O_2 for grate-fired coal, 6% O_2 for dry-bottom pulverized coal, and 5% O_2 for wet-bottom pulverized coal.
62. At 17% O_2 for <0.75 tons/hour refuse and 11% O_2 for >0.75 tons/hour refuse.
63. At 7% CO_2.
64. 2.6 pounds/short ton (equivalent to 50 grains/standard cubic foot).

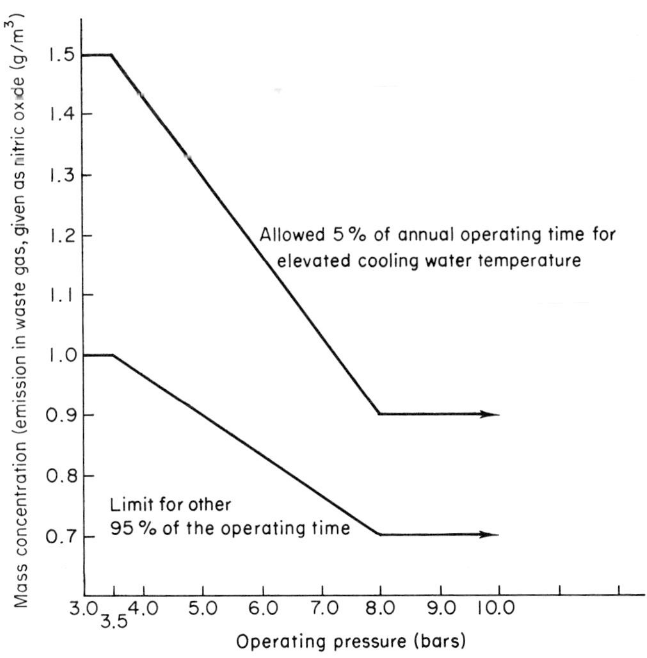

Figure 4. Emission standard for nitrogen oxides (as nitric oxide, NO) from facilities for production of nitric acid—Federal Republic of Germany (*10*).

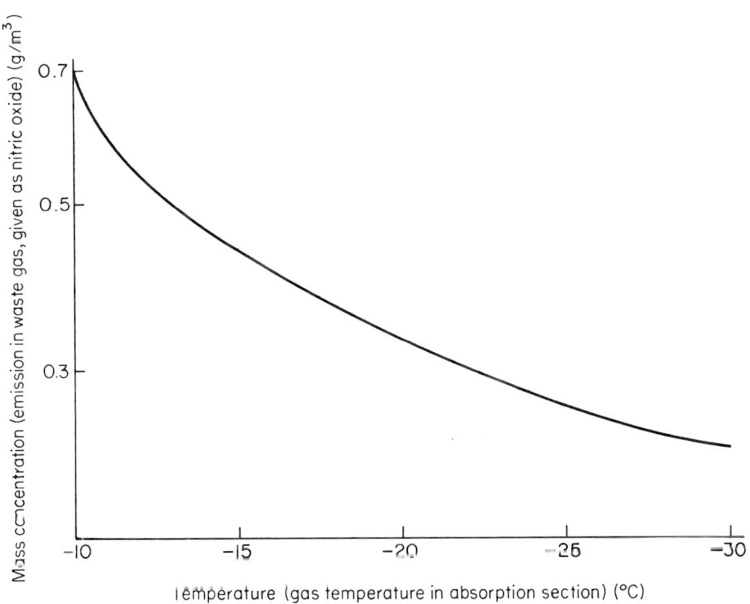

Figure 5. Emission standard for nitrogen oxides (as nitric oxide, NO) from facilities for production of highly concentrated nitric acid—Federal Republic of Germany (*10*.)

Table IXA National Emission Standards for Emission of Specific Substances in Effluent Air or Gases—Spain *(6a)*

Substance and Source	*Process*	*Installation* Existing	New	1980
			(mg/m³)	
Antimony (as Sb_2O_3)	<2500 liters/second	120	80	60
	>2500 liters/second	45	30	20
Arsenic (as As_2O_3)	<2500 liters/second	120	80	60
	>2500 liters/second	45	30	20
Cadmium	<13.6 kg/168 hours/week	40	25	17
Chlorine	Cl manufacture	230	200	150
Solvay process	—	460	300	200
Other sources	—	230	230	230
Hydrogen chloride	Copper smelting	500	300	300
Other sources	—	460	460	460
Hydrogen sulfide	Coke ovens	2500	2000	2000
Petroleum refining	—	10	7.5	5
Kraft recovery furnace	—	10[a]	10[b]	7.5
Other sources	—	10	10	7.5
Fluorides	Total	250	250	250
Lead (and its salts)	<300 m³/minute	120	100	80
	>300 m³/minute	20	15	10
Nitrogen dioxide	Chamber process[g]	3000	1000	—
Nitrogen oxides	as NO_2	3200	410	205
HNO_3 Manufacture	as NO	2000	292	146
Sulfur dioxide	Coke ovens	1000	500	500
Copper smelting	—	5700	2850	1500
Petroleum refining	Furnaces	5900	5000	4200
	Other	3400	3400	2500
Central Station	Coal	2400	2400	2400
	Lignite	9000	9000	9000
	Fuel oil	5500	4500	3000
Industrial and domestic[c]	Coal	2400	2400	2400
	Lignite	6000	6000	6000
	#2 fuel oil	6800	5000	3400
	#1 fuel oil	4200	2500	1700
	Gas oil	1700	1700	850
H_2SO_4 manufacture	Chamber process[g]	5600	4275	—
	Contact process	8550	2850	1425
Other	—	4300	4300	4300
Sulfuric acid	Chamber process[g]	615	500	—
H_2SO_4 manufacture	Contact process	500	300	150
			ppm (v/v)	
Carbon monoxide	Central stations—oil	1445	1445	1445
Petroleum refining	Catalyst regeneration	500	500	500
	Other	1500	1500	1500
Other processes	—	500	500	500
Nitrogen oxides	as NO_2	300	300	300

Table IXA (*Continued*)

Substance and Source	*Process*	*Installation* Existing	New	1980
		kg/ton of pulp		
Sulfur dioxide	Sulfite pulp	20	10	5
		kg F/ton P_2O_5		
Fluorine (HF and fluorides)	Superphosphate	0.4	0.07	0.07
	Triple superphosphate	0.3	0.05	0.05
		kg/ton of acid		
Nitrogen oxides[d]	HNO_3 manufacture	20	3	1.5
		kg/ton of alloy		
Hydrogen fluoride	Ferromolybdenum manufacture	2	1	1
		kg/ton of Al		
Fluorine (F^-)[e]	Aluminum reduction	3.6	1.2	1.0
Sulfur dioxide	Aluminum reduction	8	6	3
		tons/day[f]		
Sulfur dioxide	Petroleum refining	7*C*	5*C*	2*C*

[a] Eight minute average <10% of time/month.
[b] Eight minute average <5% of time/month.
[c] Also ferrous metallurgical operations.
[d] Two-hour average.
[e] HF and fluorides.
[f] C = capacity in 10^6 tons/year.
[g] For manufacture of H_2SO_4.

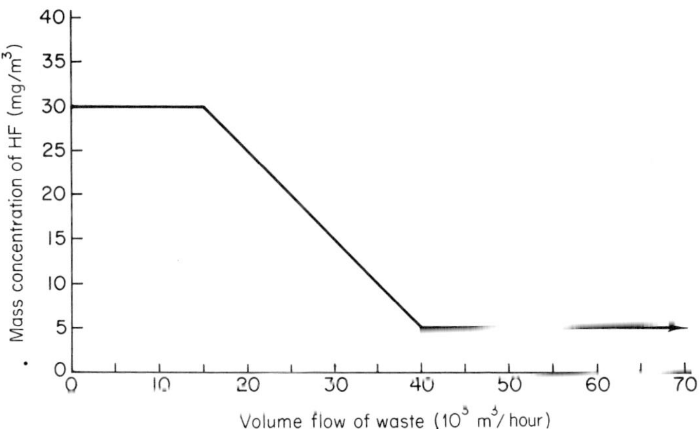

Figure 6. Emission standard for hydrogen fluoride (as HF) from facilities for production of hydrofluoric acid—Federal Republic of Germany (*10*).

Table X National Emission Standards for Organic Compounds and Inorganic Dust in Effluent Air or Gas from Stationary Sources—West Germany *(10)*

Organic Compounds[a]

CLASS I At a mass flow rate of 0.1 kg/hour or more; emission limit 20 mg/m^3

Acrolein, acrylonitrile, acrylic acids, aniline, benzene and solvents with over 1% benzene by weight, butyric acid, caproic acid, carbon tetrachloride, chloroacetic acid, chloropropionic acid, cresols, dichloroethane, dichlorophenol, diethylamine, 1,6-diisocyanate, dimethylaniline, dimethyl sulfide, dimethyl sulfate, dinitrobenzenes, epichlorohydrin, ethylacrylate, ethylenimine, ethylene oxide, formic acid, formaldehyde, furfural, ketene, mercaptans, methylacrylate, methylamine, methylisocyanate, monoethylamine, nitrobenzene, nitrocresol, nitrophenols, phenol, phosgene, polychlorinated biphenyls, pyridine, tetrachloroethane, tetraethyllead, thiocresol, thioether, thiophenol, tolylene diisocyanate, triethylamine, trimethylamine, 1,1,2-trichlorethane, trichlorophenol, valeric acid

CLASS IIa At a mass flow rate of 1 kg/hour or more; emission limit 50 mg/m^3

Organic compound dusts

CLASS IIb At a mass flow rate of 3 kg/hour or more; emission limit 150 mg/m^3

Acetaldehyde, acetic acid, *n*-amyl acetate, 1,3-butadiene, 2-chloro-1,3-butadiene, carbon disulfide, chloroform, cyclohexanone, diacetone alcohol, dibromoethane, diethanolamine, 1,1-dichloroethane, *o*-dichlorobenzene, dimethylformamide, 1,4-dioxane, ethylbenzene, ethylenediamine, 2-ethyl-1-hexanol, glycol monomethyl ether, gasoline with more than 25% C_7 and C_8, methyl acetate, methyl methacrylate, methylcyclohexanone, methylnaphthalene, morpholine, monoethanol-amine, monochlorobenzene, naphthalene, nitrotoluenes, propylene oxide, propionic acid, styrene, tetrahydrofuran, tetrahydronaphthalene, toluene, triethanolamine, trichloroethylene, trioxane, vinyl acetate, vinyl chloride, xylene

CLASS III At a mass flow rate of 6 kg/hour or more; emission limit 300 mg/m^3

Acetone, isobutanol, *n*-butanol, *n*-butyl acetate, cyclohexane, cyclohexanol, diethyl ether, 1,2-dichloroethylene, dichloromethane, dimethylsulfoxide, diisopropyl ether, ethanol, ethyl acetate, ethyl chloride, ethyleneglycol, ethylglycol, *n*-heptane, *n*-hexane, methanol, methyl ethyl ketone, methyl butyl ketone, methylcyclohexane, methyl isobutyl ketone, 2-pentanone, *n*-pentane, 1-pentanal, pinene, isopropanol, isopropyl ether, tetrachloroethane, triethyleneglycol, 1,1,1-trichloroethane

Inorganic Dusts[a]

CLASS I At a mass flow rate of 0.1 kg/hour or more; emission limit 20 mg/m^3

Arsenic,[b] asbestos, beryllium,[b] cadmium,[b] calcium arsenate, chromium compounds (hexavalent), fluorine,[b] ferrous sulfate, lead,[b] mercury[c] (except cinnabar), nickel, nickel carbonate, nickel oxide, nickel sulfide, phosphorous pentoxide, selenium,[b] tellurium,[b] thallium,[c] uranium,[c] vanadium[c]

Table X ***(Continued)***

CLASS II At a mass flow rate of 1 kg/hour or more; emission limit 50 mg/m^3

Antimony,[b] barium,[b] boron trifluoride, calcium fluoride, cobalt[b], crystabolite (<5 μm),[b] fluorspar, iodine,[c] diatomaceous earth, quartz (<5 μm); soot, silver,[d] strontium,[c] tar and pitch, tridymite (<5 μm), zinc[c]

CLASS III At a mass flow rate of 3 kg/hour or more; emission limit 75 mg/m^3

Aluminum carbide, aluminum nitride, ammonium compounds, barium sulfate, bismuth, bitumen, boron,[b] calcium cyanamid, calcium hydroxide, calcium oxide, copper,[b] ferrosilicon, magnesium hydroxide, magnesium oxide, molybdenum,[b] phosphates, silicon carbide, tungsten[c] (except tungsten carbide)

[a] If class I and II materials occur together, emission limit 50 mg/m^3; if Class II and III materials occur together, emission limit 75 mg/m^3.
[b] And its compounds soluble in the respiratory and digestive tracts, the skin, and in plants to a hazardous extent.
[c] And its compounds.
[d] Very soluble compounds only, e.g., silver nitrate.

Table XI Emission Standards for Nitrogen Oxides from Stationary Sources—Japan *(16)*

Type of source	*Percent oxygen in flue gas used as computation basis*[a]	*Emission standard (ppm)*[b]	
		New plants	*Existing plants*
Nitric acid plants	—[c]	200	200
Boiler furnaces		≥40,000 m^3/hour flue gas	≥100,000 m^3/hour flue gas
Gas-fired	5	130	170
Coal-fired >5000 kcal	6	480	600
Coal-fired ≶5000 kcal	6	480	750
Oil-fired	4	180	230
Tar-fired	4	180	280
Other furnaces		≥10,000 m^3/hour flue gas	≥40,000 m^3/hour flue gas
Metal heating	11	200	220
Petroleum heating	6	170	210

[a] Computed NO_x concentration = measured NO_x concentration × (21 − %O_2, from column/21 − actual %O_2).
[b] NO_x measured by Japanese Industrial Standard Method K 0104, time averaged where emission varies extremely with time. Nitric acid plants allowed 3 years for compliance; all others, 2 years
[c] No computation required.

Table XII States and Provinces for Which There Are State and Provincial Emission Standards for Specific Pollutants in Effluent Air or Gas from Stationary Sources *(6)*

Substance	*States of the United States*	*Provinces of Australia and Canada*
Acids (organic)	Texas	—
Acid gases	—	New South Wales, Queensland, Victoria
Acid particles	Minnesota, North Dakota	—
Alcohols	Texas	—
Aldehydes	Texas	—
Alkaline particles	Minnesota, North Dakota	—
Amines	Texas	—
Ammonia	New Mexico	—
Antimony	New Hampshire	New South Wales, Queensland, Victoria
Aromatics	Texas	—
Arsenic	New Hampshire	New South Wales, Queensland, Victoria
Asbestos	Illinois, Minnesota, New Hampshire	—
Barium	New Hampshire	—
Beryllium	New Hampshire	—
Butadiene	Texas	—
Cadmium	New Hampshire	New South Wales, Queensland, Victoria
Carbon disulfide	New Mexico	—
Carbon monoxide	All states	—
Carbon oxysulfide	New Mexico	—
Chlorine	—	New South Wales, Queensland, Victoria
Chromium	New Hampshire	—
Cobalt	New Hampshire	—
Copper	New Hampshire	Queensland
Ethylene	Alabama, Connecticut, Indiana, Kansas, Kentucky, Louisiana, Ohio, Oklahoma, Pennsylvania, Puerto Rico, Texas, Virginia	—
Ethers	Texas	—
Esters	Texas	—
Fluorine	Florida, Georgia, Mississippi, Montana, Texas	New South Wales, Queensland, Victoria
Fluorine inorganic compounds	—	New South Wales, Queensland, Victoria
Hafnium	New Hampshire	—
Hydrocarbons	Colorado, Indiana, Kansas, Oklahoma, Texas, Virginia	—
Hydrofluoric acid	—	New South Wales, Queensland, Victoria

Table XII ***(Continued)***

Substance	*States of the United States*	*Provinces of Australia and Canada*
Hydrogen chloride	New Hampshire, New Mexico, West Virginia	New South Wales
Hydrogen cyanide	New Mexico	—
Hydrogen sulfide	All states	New South Wales, Queensland, Victoria, British Columbia
Isobutylene	Texas	—
Ketones	Texas	—
Lead	New Hampshire	New South Wales, Queensland, Victoria
Lead arsenate	New Hampshire	—
Lithium hydride	New Hampshire	—
Mercury	—	New South Wales, Queensland, Victoria
α-Methylstyrene	Texas	—
Nitric acid	New Hampshire, West Virginia	New South Wales, Queensland, Victoria
Nitrogen dioxide	Massachusetts	—
Nitrogen oxides	All states	New South Wales, Queensland, Victoria
Olefins	Texas	—
Organic materials	Alabama, Colorado, Connecticut, District of Columbia, Illinois, Indiana, Kentucky, Louisiana, Maryland, North Carolina, North Dakota, Ohio, Oklahoma, Puerto Rico, Virginia, Wisconsin	—
Peroxides	Texas	—
Phosphoric acid	New Hampshire, West Virginia	Victoria
Phosphorus	New Hampshire	—
Phosphorus pentoxide	—	Victoria
Propylene	Texas	—
Selenium	New Hampshire	—
Silica	New Hampshire	—
Silver	New Hampshire	—
Styrene	Texas	—
Sulfides (organic)	Texas	—
Sulfur compounds—dioxide and oxides	All states	Victoria, British Columbia, Manitoba
Sulfur trioxide	Arkansas, Missouri, Mississippi, New Jersey, Ohio	New South Wales, Queensland, Victoria
Sulfuric acid	All states	—
Tellurium	New Hampshire	—
Thallium	New Hampshire	—

Table XII ***(Continued)***

Substance	*States of the United States*	*Provinces of Australia and Canada*
Total reduced sulfur	Alabama, Alaska, Florida, Idaho, Kentucky, Mississippi, Montana, Nevada, New Hampshire, New Mexico, Oregon, Virginia, Washington	—
Uranium	New Hampshire	—
Vanadium	New Hampshire	—
Zinc Oxide	New Hampshire	—

United States are in the form of equations relating maximum allowable emission to plant capacity, or process rate (Table XIII) (*6*) and in the form of tables (Tables XIV and XV) (*6*). Regulations that make emissions of a specific pollutant (sulfur dioxide) dependent on stack height are separately discussed in Sections VII,B and C.

B. Particulate Matter

1. All Sources

There are national emission standards for particulate matter in effluent air or gases from a large variety of stationary source processes [Tables XVI (*6*) and XVIA (*6a*)]. There are also particulate matter emission standards for such processes from a number of the states of the United States and provinces of Australia, Canada, and West Germany (Table XVII) (*6*), as well as many cities of the United States. West German regulations for the emission of particulate matter from any process include Figure 7 (*10*). The regulations of the state of Ohio includes Figure 8. Regulations that make emissions of particulate matter dependent on stack height are separately discussed in Sections VII,B and D.

2. Combustion Sources

In a number of instances, national and state particulate matter emission standards for combustion sources are stipulated in tabular, graphic, or mathematical form [Tables XVIII (*11, 11a*) and XIX (*6*); Figs. 9 (*6*) and 10 (*10*)] (see also Table XXV).

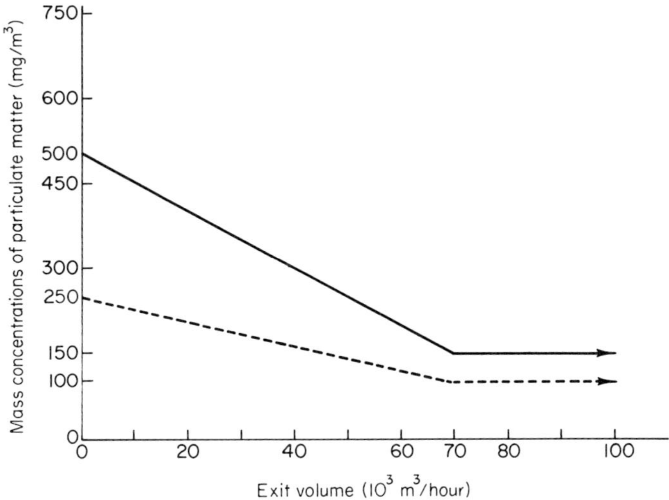

Figure 7. Emission standard for particulate matter in effluent air or gases—Federal Republic of Germany (*10*). Solid line indicates total particulate matter, broken line indicates particulates $<10\ \mu m$.

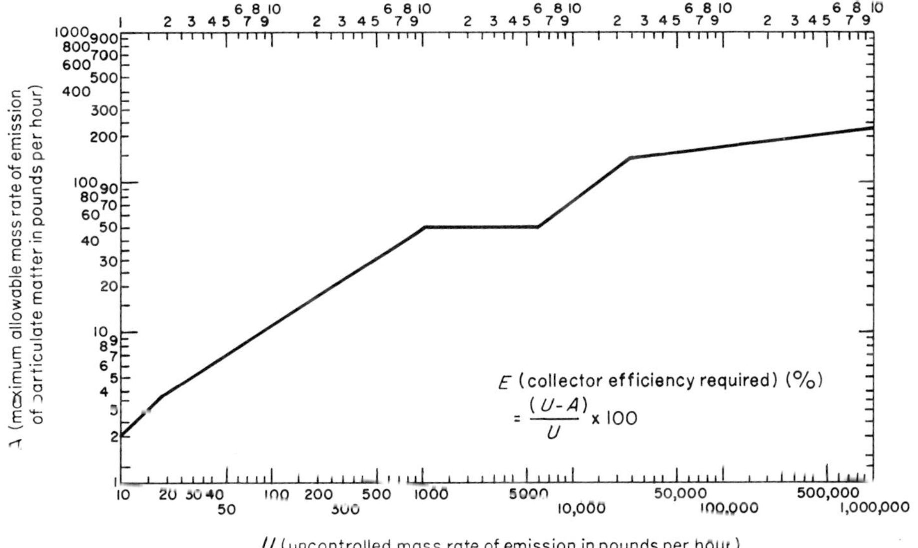

Figure 8. Emission standard for particulate matter in effluent air or gases—Ohio. E = collector efficiency required (%) = [(U − A)/U × 100]. From Regulation EP-11-11 (January 21, 1974), Ohio Environmental Protection Agency, Columbus, Ohio.

Table XIII Equations Used by States of the United States for Maximum Allowable Emission Rate and Stack Height for Sulfur Dioxide Emissions *(6)*

Process	*States*	*Maximum allowable emission rate*[a]	*Minimum stack height*[a]	*Notes*[b]
Combustion	Indiana	—	$C_{\max} = 90S_fQ_m^{0.75}n^{0.25}/ah_s$	1
Combined fuels	Illinois	—	$E = 20{,}000(H_s/300)^2$	2
Liquid fuels	Texas	—	$H_e = 0.49q^{0.5}$	3
<300 ft stack	Georgia	—	$E = 4000F(h_s/300)^3$	2
>300 ft stack	Georgia	—	$E = 4000F(h_s/300)^2$	2
≤250 MMBTU/hour	Indiana	$E_m = 17.0\ X^{-0.33}$	—	4
<250 MMBTU/hour	Nevada	$Y_s = 0.7X$	—	—
>250 MMBTU/hour	Nevada	$Y_s = 0.105X$	—	—
>50, <2000 MMBTU/hour	Pennsylvania	$E_m = 5.1X^{-0.14}$	—	—
	Special areas	$E_m = 1.7X^{-0.14}$	—	—
All rates	Virginia	$E = 2.64X$	—	5
	Virginia	$E = 1.58X$	—	5
	Virginia	$E = 1.06X$	—	5
Process operations	Indiana	$E = 19.5P^{0.67}$	$C_{\max} = 40S_pP^{0.75}n^{0.25}/ah_s$	1
Aggregate manufacture	Virginia	$E = 2.64X$	—	—
Smelters				
Nonferrous	Texas	—	$H_e = 1.8q^{0.5}$	3
	New Mexico	$Y_s = 0.1X_s$	—	—
Copper	Puerto Rico	$Y_s = 0.1X_s$	—	—
	7 states	$Y_s = 0.2X_s$	—	6, 7
Lead	7 states and Puerto Rico	$Y_s = 0.98X_s^{0.77}$	—	6, 7
New Plants	Nevada	$Y_s = 0.49X_s^{0.77}$	—	—
Zinc	7 states and Puerto Rico	$Y_s = 0.564X_s^{0.85}$	—	6, 7
New plants	Nevada	$Y_s = 0.282X_s^{0.85}$	—	—
Sulfuric acid manufacture				
Elemental sulfur feed	Texas	$E = 0.0198q$	$H_e = 0.885q^{0.5}$	3
Other feed	Texas	$E = 0.0347q$	$H_e = 1.17q^{0.5}$	3
Sulfur recovery				
$\lesseqgtr 4000\ ft^3$	Texas	$E = 123.4 + 0.091q$	$H_e = 7.4(123.4 + 0.091q)^{0.5}$	3
$>4000\ ft^3$	Texas	$E = 0.614q^{0.8042}$	$H_e = 5.8q^{0.402}$	3

[a] ABBREVIATIONS

a = plume rise factor—0.67 for <1000 MMBTU/hour, 0.8 for >1000 MMBTU/hour.
C_{max} = hourly maximum ground level concentration ($\mu g/m^3$)
D_e = stack exit inside diameter (ft).
E = maximum allowable sulfur dioxide emission (pounds/hour).
E_m = maximum allowable sulfur dioxide emission (pounds/MMBTU).
F = 0.8 when in urban area burning fuel >1% S: 2 or more sources of over 500 MMBTU/hour.
= 1.0 other sources in urban area.
= 2.0 when in rural area—fuel burning <10,000 MMBTU/hour.
= 3.0 when in rural area—fuel burning >10,000 MMBTU/hour.
F_r = reduction factor (dimensionless) = [effective stack height (h_e)/standard effective stack height (H_e)].
h_e = effective stack height (ft).
h, h_s, H, H_s = physical stack height above grade (ft).
H_e = standard effective stack height (ft).
n = number of stacks.
P = process weight (tons/hour).
P_i, P_n = percent of total emission from source.
Q_m = total equipment capacity rating, MMBTU/hour.
q = stack exit flow rate [ft^3/minute (STP)].
S_1 = maximum allowable sulfur dioxide emission (pounds/MMBTU).
S_2 = maximum allowable sulfur dioxide emission (pounds/ton process weight).
T_e = stack exit temperature (°R).
V_e = stack exit velocity (ft/second).
X = maximum heat input (MMBTU/hour).
X_s = sulfur input in feed (pounds/hour).
Y_s = maximum allowable sulfur emission (pounds/hour).

[b] NOTES:

1. C_{max} for level terrain and critical wind speed ≤ 200. Use lower values for other terrain or wind conditions.
2. Where $H_s = P_1H_1 + P_2H_2 + \cdots + P_nH_n/100$. Note: $P_1 + P_2 + \cdots + P_n = 100$; $H = h$.
3. $h_e = h - 0.083V_eD_e[1.5 + 0.82(T_e - 550/T_e)\ D_e]$. Where $h_e < H_e$, multiply E by F_r^2 to get reduced rate allowable.
4. Not to exceed 6 pounds/MMBTU input.
5. May be required by state air pollution control board to attain ambient air quality standards.
6. Also foundries in New Hampshire.
7. Connecticut, Mississippi, New Hampshire, Ohio, Oklahoma, Virginia, West Virginia.

Table XIV Emission Standards for Sulfur Recovery Plants and Sulfuric Acid Manufacture—Delaware and Virginia *(6)*

	Permissible emission (pounds per hour) (as sulfur dioxide)			
	Sulfur recovery plants		*Sulfuric acid plants—Delaware*	
Production rate (tons per day)	*Delaware*	*Virginia*	*New*	*Existing*
50	425	415	—	—
100	550	830	38	75
200	800	1660	—	—
300	1050	2490	105	210
400	1300	3320	—	—
500	1550	4150	173	345
600	1800	—	—	—
700	2050	—	240	480
800	2300	—	—	—
900	2550	—	308	615
1000	2800	—	—	—
1100	—	—	375	750
1300	—	—	443	885
1500	—	—	510	1020

Table XV Emission Standards for Nonferrous Smelters—Montana and Nevada *(6)*

	Allowable sulfur emissions (Y) (pounds/hour)		
Total feed sulfur (X) (pounds/hour)	*Copper*[a]	*Zinc*[b]	*Lead*[c]
1,000	100	100	100
5,000	500	394	348
10,000	1,000	704	593
20,000	2,000	1,270	1,000
40,000	4,000	2,310	1,000
60,000	6,000	3,210	1,000
80,000	8,000	4,120	1,000
$\geq$100,000	10,000	5,000	1,000

[a] To 10,000 pounds per hour emission: $Y = 0.1X$.
[b] To 5,000 pounds per hour emission: $Y = 0.282X^{0.85}$.
[c] To 1,000 pounds per hour emission: $Y = 0.49X^{0.77}$.

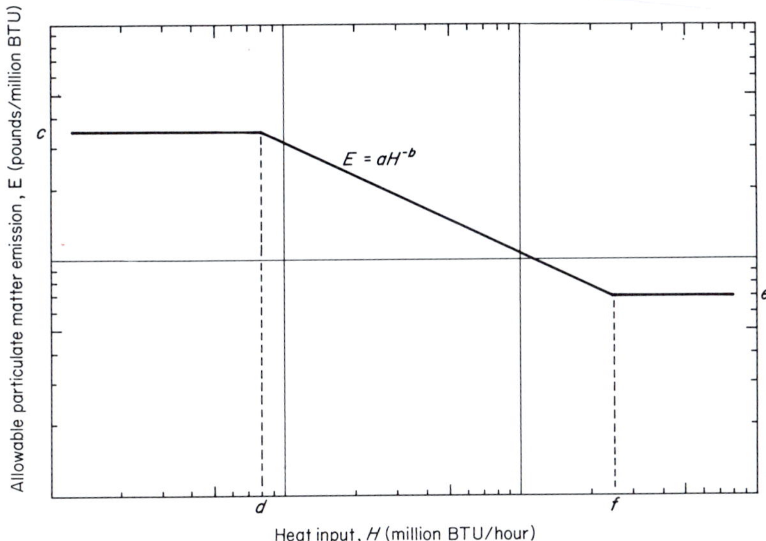

Figure 9. Figure illustrating parameters listed in Table XIX—United States (*6*).

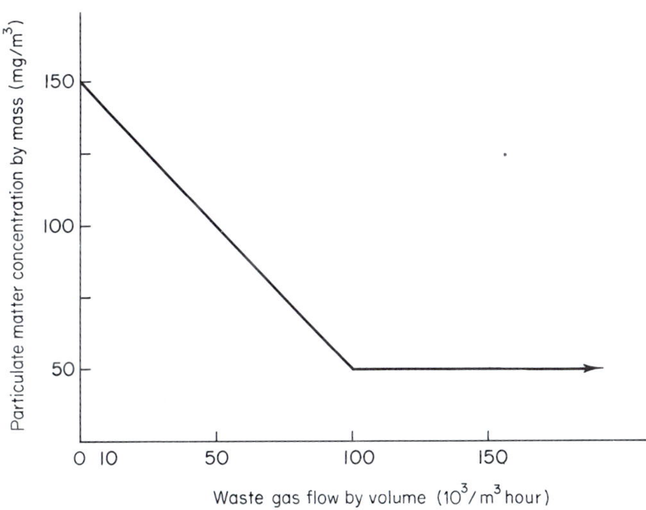

Figure 10. Emission standard for particulate matter from burning liquid fuel—Federal Republic of Germany (*10*).

3. Sources Other Than Combustion

There are particulate matter emission standards for specific processes (other than combustion) that are stipulated in tabular, graphic, or

Table XVI National Emission Standards for Particulate Matter in Effluent Air or Gas from Stationary Sources[a] (6)

Source	*Countries*	*Standard* — *Original units*	*Standard* — *mg/m³*	*Notes*
All	Australia; New Zealand	0.25 gm/m³	250.0	1, 2, 3, 4, 5
	Czechoslovakia	5.0 kg/hour	—	6
	East Germany	—	—	7
	Israel	—	—	12
	Mexico	—	—	9
	Philippines	0.40 grains/ft³	915.3	10
	Singapore	0.40 gm/m³	400.0	11
	West Germany	—	—	8
Arctic mining plants	Canada	0.04 gm/m³	40.0	—
Combustion of fuels				
All fuels	Great Britain	—	—	13
>10⁶ kcal/hour	Italy	0.25 gm/m³	250.0	14
Home heating	Spain	1.2 gm/1000 kcal	—	—
Steam power plants	United States	0.10 pound/MMBTU	—	—
Bark burning				
Kraft pulp mills	Sweden	mg/m³	250.0	15
Blast furnace gas burning	West Germany	mg/m³	50.0	—
Central stations				
>20% ash	Belgium	mg/m³	500.0	—
<20% ash	Belgium	mg/m³	350.0	—
Boiler	Japan—special	0.20 gm/m³	200.0	20
	Japan—other	0.40 gm/m³	400.0	—
>40 MMkcal/hour	Mexico	1.0 kg/MMkcal	—	—
<40 MMkcal/hour	Mexico	1.5 kg/MMkcal	—	—
<1000 kg/hour	Switzerland	mg/m³	75.0	21, 22
>1000 kg/hour	Switzerland	mg/m³	150.0	21, 22
>1000 kg/hour	Switzerland	mg/m³	100.0	23

Coal burning	Australia	0.25 gm/m^3	250.0	4, 5, 16, 17
<20% ash coal	France	0.35 gm/m^3	350.0	—
>20% ash coal	France	0.50 gm/m^3	500.0	—
Domestic heating	Belgium	mg/m^3	300.0	—
	Special protection zone	mg/m^3	150.0	—
Mixed burning	Switzerland	mg/m^3	100.0	21
Oil burning				
>2000 therms/hour	Belgium	mg/therm	250.0	—
<100 MW	Denmark	2.0 gm/kg oil	—	25
<300 MW	Denmark	1.5 gm/kg oil	—	—
>300 MW	Denmark	1.0 gm/kg oil	—	—
>63 MMkcal/hour	Mexico	45.0 gm/MMkcal	—	—
<63 MMkcal/hour	Mexico	80.0 gm/MMKcal	—	—
>50 MW	Sweden	1.5 g/kg oil	—	26, 27
>300 MW	Sweden	1.0 g/kg oil	—	28
>1000 kg/hour	Switzerland	mg/m^3	50.0	29
Programmed soot blow	Switzerland	mg/m^3	150.0	29
Hand soot blow	Switzerland	mg/m^3	200.0	29
Heavy oil burning				
<40,000 m^3/hour gas	Japan—special	0.20 gm/m^3	200.0	20
<40,000 m^3/hour gas	Japan—other	0.30 gm/m^3	300.0	—
>40,000 m^3/hour gas	Japan—special	0.05 gm/m^3	50.0	20
40,000 to 200,000	Japan—other	0.20 gm/m^3	200.0	—
>200,000 m^3/hour gas	Japan—other	0.10 gm/m^3	100.0	—
Solid fuel burning	Czechoslovakia	—	—	30
<30 MW (new)	Denmark	mg/m^3	300.0	24
<30 MW (existing)	Denmark	mg/m^3	500.0	24
>100 MW (All)	Denmark	mg/m^3	150.0	24, 89
30–100 MW (All)	Denmark	—	—	24
	West Germany	mg/m^3	300.0	31

Table XVI (Continued)

Source	Countries	Standard: Original units	Standard: mg/m³	Notes
<500,000 m³/hour	West Germany	mg/m³	150.0	31
>500,000 m³/hour				
Brown coal	West Germany	mg/m³	100.0	31
Hard coal	West Germany	mg/m³	150.0	31
Incineration of refuse				
<300 kg/hour	Australia; New Zealand	0.50 gm/m³	500.0	3, 4, 5, 16
>300 kg/hour	Australia; New Zealand	0.25 gm/m³	250.0	3, 4, 5, 16
General	Belgium	mg/m³	150.0	35
<1 ton/hour	France	1.0 gm/m³	1000.0	35, 38, 39
1–4 tons/hour	France	0.60 gm/m³	600.0	35, 38, 39
4–7 tons/hour	France	0.25 gm/m³	250.0	35, 38, 39
>7 tons/hour	France	0.15 gm/m³	150.0	35, 38, 39
Continuous furnace				
<40,000 m³/hour	Japan—other	0.7 gm/m³	700.0	—
>40,000 m³/hour	Japan—other	0.2 gm/m³	200.0	—
<40,000 m³/hour	Japan—special	0.2 gm/m³	200.0	20
>40,000 m³/hour	Japan—special	0.1 gm/m³	100.0	20
Other furnaces				
<40,000 m³/hour	Japan—other	0.70 gm/m³	700.0	—
>40,000 m³/hour	Japan—other	0.40 gm/m³	400.0	—
Batch operation	Japan—special	0.40 gm/m³	400.0	43
	Japan—other	0.70 gm/m³	700.0	—
>15 tons/hour	Sweden	mg/m³	180.0	15, 40, 41
<15 tons/hour	Sweden	mg/m³	250.0	15, 40, 41
<3 tons/hour	Sweden	mg/m³	500.0	15, 40, 42
<1 ton capacity	Switzerland	0.20 gm/m³	200.0	35

1–5 tons capacity	Switzerland	0.15 gm/m³	150.0	35
>5 tons capacity	Switzerland	0.10 gm/m³	100.0	35
Municipal incinerators	United States	0.08 grains/ft³	80.0	16
Sewage sludge incinerators	United States	mg/m³	70.0	15, 44
General	West Germany	mg/m³	100.0	19
With component recovery	West Germany	mg/m³	100.0	35
Preparation facilities	West Germany	mg/m³	150.0	—
Refuse shredding	West Germany	mg/m³	100.0	—
Asphalt plants				
Asphalt paving plants	Canada	0.1 grains/scf	230.0	5
<500 m from house				
New	Denmark	mg/m³	250.0	—
Existing	Denmark	mg/m³	500.0	—
<1 km from community and more than 2 years in one location	Denmark	mg/m³	5000.0	—
Hot-mixing plants	France	0.150 gm/m³	150.0	45
	France	0.800 gm/m³	800.0	46
All	France	2.0 gm/m³	2000.0	32
≥200 tons/hour	France	0.150 gm/m³	150.0	45, 47
<200 tons/hour	France	0.800 gm/m³	800.0	45, 47
≥150 tons/hour	France	0.150 gm/m³	150.0	45, 48
<150 tons/hour	France	0.500 gm/m³	500.0	45, 48
>500 m from built-up area				
New	Sweden	mg/m³	250.0	15, 39, 49
Existing	Sweden	mg/m³	500.0	15, 49, 87
>1 km from a single house	Sweden	mg/m³	5000.0	15, 50
All plants	United States	mg/m³	70.0	15, 50
Asphalt concrete	West Germany	mg/m³	100.0	51, 52
if feasible	West Germany	mg/m³	75.0	52
Carbon black manufacture				
Wash tower	VDI 2580	mg/m³	50.0	37

Table XVI (Continued)

Source	Countries	Standard: Original units	Standard: mg/m^3	Notes
Wet gas filter	VDI 2580	mg/m^3	60.0	37, 53
With thermal after-burner	VDI 2580	mg/m^3	40.0	37
Thermal burning in boiler	VDI 2580	mg/m^3	50.0	37
Cement production				
Clinker coolers	Canada	0.6 pound/ton clinker	—	5, 83
	United States	0.10 pound/ton feed	—	—
Crushing, grinding (new)	Denmark	mg/m^3	250.0	—
Crushing, grinding (existing)	Denmark	mg/m^3	500.0	—
	Sweden	mg/m^3	250.0	15
Crushing, grinding (new)	Great Britain	0.10 grains/ft^3	229.0	39
Crushing, grinding (existing)	Great Britain	0.20 grains/ft^3	457.6	18
Crushing, grinding	West Germany	mg/m^3	150.0	54
Dust handling	Great Britain	0.20 pound/k pound gas	457.6	—
Finish grinding	Canada	0.1 pound/ton clinker	—	5, 83
Sources other than kiln, clinker cooler, finish grinding	Canada	0.2 pound/ton clinker	—	5, 83
Coal processing				
Briquetting plants	VDI 2292	0.15 gm/m^3	150.0	37
Preparation plants	VDI 2293	0.15 gm/m^3	150.0	37
Coke manufacture (metallurgical)				
Battery stacks	Canada	0.03 grains/scf	69.0	5, 83
Charging	Canada	100.0 gm/metric ton coke	—	5, 83, 87
Crushing and Screening	Canada	0.02 grains/scf	46.0	5
Pushing	Canada	0.02 grains/scf	46.0	5, 83
Quenching	Canada	50 gm/metric ton coke	—	5, 83, 88
Electrode manufacture	West Germany	mg/m^3	150.0	59
	VDI 2100	0.15 gm/m^3	150.0	37

Fluid catalyst regeneration	United States	mg/m^3	50.0	15, 58
Furnaces				
Calcination	West Germany	mg/m^3	150.0	55
Calcium carbide manufacture	West Germany	mg/m^3	150.0	—
Catalyst regeneration	Japan—special	0.40 gm/m^3	400.0	20
	Japan—other	0.60 gm/m^3	600.0	60
Drying				
Heat treatment				
>40,000 m^3/hour	Japan—special	0.10 gm/m^3	100.0	—
<40,000 m^3/hour	Japan—special	0.20 gm/m^3	200.0	—
Heat treatment				
>40,000 m^3/hour	Japan—other	0.20 gm/m^3	200.0	—
<40,000 m^3/hour	Japan—other	0.40 gm/m^3	400.0	—
Others	Japan	Same as reaction furnace		—
Expanding clay and slate	West Germany	mg/m^3	150.0	33, 56
Gas generating	Japan	Same as catalyst regeneration furnaces		—
Gas producing	Japan—special	0.10 gm/m^3	100.0	20, 31
	Japan—other	0.20 gm/m^3	200.0	61
Glass melting	West Germany	mg/m^3	150.0	—
Tank type	Japan	Same as reaction furnace		—
Other types	Japan	0.50 gm/m^3	500.0	—
Gypsum with filters	West Germany	mg/m^3	75.0	—
Petroleum heating	Japan	Same as gas producing		—
Reaction furnace				
<40,000 m^3/hour gas	Japan—special	0.20 gm/m^3	200.0	20
<40,000 m^3/hour gas	Japan—other	0.40 gm/m^3	400.0	—
>40,000 m^3/hour gas	Japan—special	0.10 gm/m^3	100.0	20
>40,000 m^3/hour gas	Japan—other	0.20 gm/m^3	200.0	—
Refinery combustion furnace	Japan	Same as gas producing		—

Table XVI (Continued)

Source	Countries	Standard		Notes
		Original units	mg/m^3	
Kilns				
Cement (new)	Canada	0.91 pound/ton clinker	—	5
Cement (existing after 12/1/79)	Canada	1.6 pounds/ton clinker	—	5
Cement (new)	Denmark	mg/m^3	250.0	—
Cement (existing)	Denmark	mg/m^3	500.0	—
Cement	France	1.0 gm/m^3	1000.0	39, 63
<1500 tons/day	Great Britain	0.20 $grains/ft^3$	457.6	64
1500–3000	Great Britain	Sliding scale 0.1–0.2	—	—
>3000 tons/day	Great Britain	0.10 $grains/ft^3$	229.0	64
Cement (all plants)	Japan	Same as reaction furnace		—
Cement (all plants)	Spain	0.80 gm/m^3	800.0	—
Cement (all plants)	United States	0.30 pound/ton feed	—	—
High-resistivity dust	West Germany	mg/m^3	150.0	62
Low-resistivity dust	West Germany	mg/m^3	120.0	—
No electrostatic precipitator	West Germany	mg/m^3	75.0	—
Cement and lime				
<25 tons/hour product	Czechoslovakia	120.0 kg/hour	—	—
25–50 tons/hour product	Czechoslovakia	160.0 kg/hour	—	—
50–100 tons/hour product	Czechoslovakia	250.0 kg/hour	—	—
100–150 tons/hour product	Czechoslovakia	270.0 kg/hour	—	—
New plants	Denmark	mg/m^3	250.0	65
Existing plants	Denmark	mg/m^3	500.0	—
New plants	Sweden	mg/m^3	250.0	15, 65
Existing plants	Sweden	mg/m^3	500.0	15
New plants	Switzerland	mg/m^3	230.0	—
Existing plants	Switzerland	mg/m^3	155.0	—

Ceramic				
All types	West Germany	mg/m^3	150.0	33
Continuous	Japan	Same as drying, heat treatment		—
Other types	Japan—special	0.30 gm/m^3	300.0	20
	Japan—other	0.60 gm/m^3	600.0	—
Lime kilns at pulp mills	Sweden	mg/m^3	250.0	15, 65
Kraft pulp mills				
Recovery furnaces				
New plants	Sweden	mg/m^3	250.0	15, 65
Existing plants	Sweden	mg/m^3	500.0	66
Rock crushing	Denmark	mg/m^3	150.0	—
Sulfite pulp mills				
New plants	Denmark	mg/m^3	250.0	—
Existing plants	Denmark	mg/m^3	500.0	—
Trisodium phosphate manufacture	Japan	Same as reaction furnace		—
Metallurgical processes—ferrous and nonferrous				
Metal heating	Australia	0.10 gm/m^3	100.0	3, 4, 5, 17
	Japan	Same as reaction furnace		—
	New Zealand	0.10 gm/m^3	100.0	5
	Singapore	0.20 gm/m^3	200.0	4, 17
Metal melting	Japan	Same as reaction furnace		—
Metal smelting	Mexico	—	—	9, 67
Sintering plants	Japan—special	0.20 gm/m^3	200.0	20
$<40{,}000\ m^3$	Japan—other	0.40 gm/m^3	400.0	—
$>40{,}000\ m^3$	Japan—other	0.30 gm/m^3	300.0	—
Ferrous metallurgical processes				
Blast furnace gas				
Bled	West Germany	mg/m^3	20.0	—
Burned	West Germany	mg/m^3	50.0	68
Bessemer converters	Japan	Same as sintering plants		—

Table XVI (Continued)

Source	Countries	Standard: Original units	Standard: mg/m^3	Notes
Casting and shakeout				
>2,500 tons/year	Sweden	mg/m^3	150.0	15, 33
<2,500 tons/year	Sweden	mg/m^3	300.0	15, 74
Converters	Denmark, Sweden	0.3 kg/ton steel	—	90
Crucibles				
>2000 ton production/year	Denmark	—	—	92
Cupolas	West Germany	—	—	71
<2 ton capacity	France	2.0 kg/ton iron	—	39
2–10 ton capacity	France	kg/ton iron	—	39, 69
>10 ton capacity	France	kg/ton iron	—	39, 70
Existing plants	Sweden	kg/ton iron	—	73
New plants	Sweden	kg/ton iron	—	72
Upper limit	VDI 2288	2 kg/ton iron	—	37, 71
Sandblasting (foundry and nonfound' y)				
New plants	Denmark, Sweden	mg/m^3	150.0	—
Existing plants	Denmark	mg/m^3	300.0	—
Furnaces				
Blast	Great Britain	0.20 grains/ft^3	457.6	—
	Japan—special	0.05 gm/m^3	50.0	20
	Japan—other	0.10 gm/m^3	100.0	—
	Sweden	0.30 kg/ton iron	—	—
Electric arc				
New plants	Denmark	0.3 kg/ton steel	—	90
Existing plants	Denmark	1.0 kg/ton steel	—	91
New plants	Sweden	0.30 kg/ton steel	—	—
Existing plants	Sweden	0.60 kg/ton steel	—	—
All	United States	mg/m^3	12.0	—
	West Germany	mg/m^3	150.0	—

Ferroalloy					
>40% other metals	Japan—special	0.30 gm/m^3	300.0	20	
	Japan—other	0.60 gm/m^3	600.0		—
<40% other metals	Japan—special	0.20 gm/m^3	200.0	20	
	Japan—other	0.40 gm/m^3	400.0		—
Calcium silicon	United States	0.45 kg/MW-hour	—		—
Ferrosilicon					
Silicomanganese zirconium					
Silicon metal					
Calcium carbide	United States	0.23 kg/MW hour	—		—
Charge chrome					
Ferrochrome silicon					
Ferromanganese silicon					
High carbon ferrochrome					
Silicomanganese					
Silvery iron					
Standard ferromanganese					
All foundries	West Germany	mg/m^3	100.0		—
Ferrochromium affiné	Sweden	5.0 kg/ton alloy	—		—
Ferromolybdenum	Sweden	3.0 kg/ton alloy	—		—
Ferrosilicon					
New plants	Sweden	10 kg/ton alloy	—	93	—
Existing plants	Sweden	15 kg/ton alloy	—	93	—
Ferrosilicon manganese	Sweden	0.30 kg/ton alloy	—	39	
Ferrochromsilicon					
New plants	Sweden	15.0 kg/ton alloy	—		—
Existing plants	Sweden	20.0 kg/ton alloy	—		—
Furnace (electric) ferroalloy					
Ferrochrome	VDI 2576	mg/m^3	250.0	37	
Ferrosilicon					
15, 25 or 45% Silicon	VDI 2576	mg/m^3	200.0	37	
75% or 90% Silicon	VDI 2576	mg/m^3	300.0	37	

Table XVI (Continued)

Source	Countries	Standard: Original units	Standard: mg/m^3	Notes
Ferrotungsten	VDI 2576	mg/m^3	150.0	37
Silicochromium	VDI 2576	mg/m^3	250.0	37
Silicon metal, 97% Silicon	VDI 2576	mg/m^3	300.0	37
Other furnaces				
Ferrochrom "affine"	VDI 2576	mg/m^3	200.0	37
Ferrochrom "surraffine"	West Germany	mg/m^3	150.0	37
Iron smelting	France	0.15 gm/m^3	150.0	39
Open hearth				
New plants	Denmark	0.5 kg/ton steel	—	90 —
Existing plants	Denmark	1 kg/ton steel	—	93 —
All plants	Japan—special	0.20 gm/m^3	200.0	20
>40,000 m^3/hour	Japan—other	0.30 gm/m^3	300.0	—
<40,000 m^3/hour	Japan—other	0.40 gm/m^3	400.0	—
New plants	Sweden	0.5 kg/ton steel	—	—
Existing plants	Sweden	1.0 kg/ton steel	—	—
Oxygen-steel processes	West Germany	mg/m^3	150.0	51
BOF	United States	mg/m^3	50.0	15, 44
Kaldo	Sweden	0.15 kg/ton steel	—	—
LD	Sweden	0.30 kg/ton steel	—	—
Blowing cycle	France	0.12 gm/m^3	120.0	39, 75
Loading ore	France	0.15 gm/m^3	150.0	39
Loading or pouring	France	0.12 gm/m^3	120.0	39
Refining processes	Great Britain	0.05 grains/ft^3	114.4	—
Miscellaneous grit and dust	Great Britain	0.20 grains/ft^3	457.6	—
Refining processes—no oxygen	Great Britain	0.20 grains/ft^3	457.6	—
Scarfing—oxygen	Great Britain	0.05 grains/ft^3	114.4	—
Scarfing	West Germany	mg/m^3	150.0	—
Sintering plants	France	0.50 gm/m^3	500.0	39, 76

Sintering plants	Great Britain	0.05 grains/ft^3	114.4		—
All plants	Japan—special	0.2 gm/m^3	200.0	20	
>40,000 m^3/hour	Japan—other	0.3 gm/m^3	300.0		—
<40,000 m^3/hour	Japan—other	0.4 gm/m^3	400.0		—
Existing plants	Sweden	1.0 kg/ton sinter	—		—
New plants	Sweden	0.50 kg/ton sinter	—		—
Continuous operation	West Germany	mg/m^3	150.0		—
Special cases	West Germany	mg/m^3	300.0	77	
Slag remelting	West Germany	mg/m^3	150.0		—
Nonferrous metallurgical processes					
Aluminum alloy production	West Germany	mg/m^3	75.0		—
New plants	Denmark	1 kg/ton product	—		—
Existing plants	Denmark, Sweden	1.5 kg/ton product	—		—
Aluminum reduction	West Germany	—	—	57	
Alumina calcining	VDI 2286	mg/m^3	100.0	37	
Alumina grinding	West Germany	mg/m^3	150.0		—
Primary reduction	West Germany	mg/m^3	75.0		—
	VDI 2286	0.10 gm/m^3	100.0	37	
Secondary recovery					
Rotary furnaces	VDI 2441	—	—	37, 78	
Other furnaces	VDI 2241	0.3 gm/m^3	300.0	37	
Smelters	Great Britain	0.05 grains/ft^3	114.4	79	
Brass and bronze ingots					
Blast cupola furnaces					
>250 kg/hour capacity	United States	—	—	44	
Electric furnace					
>1000 kg/hour capacity	United States	—	—	44	
Reverberatory furnaces					
>1000 kg/hour capacity	United States	mg/m^3	50.0	15, 44	
Copper smelting					
New plants	Denmark, Sweden	2.0 kg/ton product	—		—
Existing plants	Denmark, Sweden	4.0 kg/ton product	—		—

Table XVI ***(Continued)***

Source	Countries	Standard: Original units	Standard: mg/m^3	Notes
Primary				
Roasting, smelting, converting	United States	mg/m^3	50.0	—
Refining furnaces	VDI 2101	0.30 gm/m^3	300.0	37
Reverberatory furnaces	VDI 2101	0.30 gm/m^3	300.0	37
Shaft furnaces	VDI 2101	0.30 gm/m^3	300.0	37
Secondary				
Blast furnaces	VDI 2102	0.30 gm/m^3	300.0	37
Converters	VDI 2102	0.30 gm/m^3	300.0	37
Refining furnaces	VDI 2102	0.30 gm/m^3	300.0	37
Copper, lead, and zinc refining				
Blast furnaces	Japan	Same as reaction furnace		—
Bessemer converters	Japan	Same as sintering plants		—
Other types	Japan	Same as gas producing		—
Drying	Japan	Same as reaction furnace		—
Foundries	West Germany	mg/m^3	100.0	—
Hydrometallurgical refining				
Cobalt calcination	VDI 2287	1.0 gm/m^3	1000.0	37, 80
Inhibition plant	VDI 2287	0.5 gm/m^3	500.0	37, 80
Roasting plant	VDI 2287	0.1 gm/m^3	100.0	37, 80
Zinc calcination	VDI 2287	—	—	37, 80
Zinc with scrubbing	VDI 2287	3.0 gm/m^3	3000.0	37, 80
With dust extraction	VDI 2287	0.5 gm/m^3	500.0	37, 80
Melting	Japan	Same as reaction furnace		—
Roasting and sintering	Japan	Same as sintering plants		—
Lead pigment manufacture				
Smelting furnaces	Japan	Same as reaction furnace		—
Lead smelting				
Blast and reverberatory furnaces, sintering machine discharge	United States	mg/m^3	50.0	—

Furnaces	West Germany	mg/m³	20.0	81
Refining furnaces	West Germany	mg/m³	400.0	81
Secondary smelting	Japan	Same as reaction furnace		—
Battery manufacture	Japan	Same as reaction furnace		—
Blast or cupola >250 kg	United States	mg/m³	50.0	15, 44
Blast furnaces, cupola, and reverberatory furnaces	Canada	0.02 grains/scf	46.0	85
Miscellaneous operations	Canada	0.01 grains/scf	23.0	85, 86
Pot furnace >250 kg	United States	mg/m³	—	44
Slag blowing	West Germany	mg/m³	100.0	81
Zinc smelting				
Distillation process	West Germany	mg/m³	200.0	82
Electrothermal process	West Germany	mg/m³	100.0	82
Rotary process	West Germany	mg/m³	500.0	82
Sintering machine	United States	mg/m³	50.0	—
Stationary retorts	West Germany	mg/m³	400.0	82
Miscellaneous processes				
Conveying, grinding, classifying, filling				
>3 kg/hour	West Germany	mg/m³	75.0	—
<3 kg/hour	West Germany	mg/m³	150.0	—
Phosphate concentrate sintering	West Germany	mg/m³	75.0	—
Wood fiber board production	West Germany	mg/m³	50.0	34, 34
	West Germany	mg/m³	150.0	34, 36

NOTES:

1. Other than incinerators burning less than 300 kg/hour and furnaces for heating of metals, except cold blast foundry cupolas.
2. At 12% CO_2 for boilers burning solid fuel and incinerators.
3. Intended for application to new plants.
4. STP at 0°C and 1 atm, dry.
5. National guideline.
6. Maximum SiO_2 content; 20% Emission rate above which it is necessary to submit a report to the government; where the discharge is for less than 1 hour, there is a proportionate reduction in emission permissible without such reporting.
7. See Tables XXX, XXXI, and XXXII and Figure 15.

8. See Figure 7.
9. See Table XXV.
10. See Table XXVI.
11. Except metal heating furnaces.
12. See Table XXV.
13. See Table XVIII.
14. See Figure 19.
15. Dry gas.
16. Proposed standards for new plants.
17. Except cold blast foundry cupolas.
18. Existing units.
19. At 11% O_2 by volume in gas.
20. See Table XXXIII.
21. As wet gas.
22. >500 kg/hour, smoke density-monitoring required.
23. Using two or more fuels, combined.
24. At 12% CO_2; linear from 30 to 100 MW.
25. For existing plants to be met by 1988.
26. Flue gas velocity greater than 8 m/second at minimum load.
27. Also a requirement for maximum soot (combustible particulate) emission.
28. Separate flues for each boiler with gas velocity greater than 25 m/second at full load.
29. Maximum average 4 hours during 24 hours.
30. See Table XXIX.
31. At 13% O_2 for wood, 7% O_2 for grate-fired coal, 6% O_2 for dry-bottom pulverized coal, and 5% O_2 for wet-bottom pulverized coal.
32. For plants >300 m from dwellings but <1000 m from agglomeration.
33. At 3% CO_2.
34. See Figure 12.
35. At 7% CO_2.
36. Chip drying.
37. VDI = Verein Deutscher Ingenieure—German Association of Engineers, Dusseldorf, West Germany.
38. Excess tolerated for a continuous 16 hours or a total of 200 hours/year.
39. New units.

40. At 10% CO_2.
41. Space should be left for scrubber.
42. Velocity of flue gases greater than 8 m/second at all operating conditions.
43. Distance to built-up area at least 1 km. Permitted only if adopted by regional planning authority.
44. 10% opacity (except for recombined water) permitted for 2 minutes per hour.
45. For plants <300 m from dwellings.
46. For plants >300 m from dwellings but <1000 m from agglomeration.
47. Effective January 1, 1977.
48. Effective January 1, 1980.
49. Applicable to dryers, conveyors, elevators, etc.
50. Accepted only for temporary location.
51. Also VDI 2112 (see Note 37).
52. At 4 volume % CO_2.
53. If diameter <30 μm, emission may be an additional 10 mg/m^3.
54. Also VDI 2094 (see Note 37).
55. Dolomite, magnesite, limestone, corundum, gypsum.
56. Additional 20 mg/m^3 allowed when organic expansion aid is used.
57. With membrane filter—pore size 3 μm—20 kg/ton aluminum.
58. 20% opacity (except for uncombined water) permitted for 3 minutes per hour; where auxiliary liquid or solid fuels are burned in an incinerator-waste heat boiler, particulate matter in excess of this standard may be permitted, except that the incremental rate of emission shall not exceed 0.18 gm/million calories (0.10 pound/million BTU) of heat input attributable to such fuel.
59. At 3 volume % CO_2.
60. Also continuous ceramic kilns.
61. Also petroleum heating furnaces; sulfur combustion furnaces in petroleum refineries; converters other than Bessemer; copper, lead, and zinc refining converters other than Bessemer.
62. Also VDI 2094 (see Note 37).
63. 0.15 gm/m^3 excess tolerated for a continuous 48 hours or for a 1.0 gm/m^3 total of 200 hours/year.
64. Older kilns, up to 0.5 grains/standard cubic foot.
65. Electrical precipitators on new kilns should have at least two independent sections, the maximum emission with one section out of operation is 500.0 mg/m^3 STP (dry gas).
66. Total, excluding wood-burning boilers and recovery stacks, monthly monitoring.
67. Smelting furnaces must be equipped with gas-cleaning equipment to remove 80% by weight of particulates from exhaust gas.

68. Also VDI 2099 (see Note 37).
69. $2.3 \times 0.15 \times$ capacity in tons (kg/ton iron).
70. $28/[(4 \times$ capacity in tons$) - 5]$ (kg/ton iron).
71. See Figure 11; >14 tons/hour iron–0.25 kg/ton iron.
72. $0.7 \leq 35/$(Annual production in thousands of tons $+ 3) \leq 7$.
73. $1 \leq 50/$(Annual production in thousands of tons $+ 3) \leq 8$.
74. Less than 2500 tons of gray iron per year production.
75. No visible color.
76. 0.15 gm/m^3 excess tolerated for a total of 200 hours/year.
77. Where, for instance, raw material is to be used in the form of fine dust and the applicant is able to show that although the present state of technical development does not permit keeping with the 150 mg/m^3 limit, no objectionable effects need be feared in the neighborhood. Also VDI 2095 (see Note 37).
78. Total dust emission not to exceed 1% of aluminum production.
79. Fume emission from use of salt as a flux.
80. Recovery after chloridizing roasting.
81. Also VDI 2285 (see Note 37).
82. Also VDI 2284 (see Note 37).
83. New plants and existing plants after December 1, 1979.
84. Pulping and conveying.
85. Proposed national guidelines.
86. Holding and kettle furnaces; lead oxide production; scrap and material handling and crushing; furnace tapping, slagging, cleaning, and casting operations.
87. 0.2 pound/short ton coke.
88. 0.1 pound/short ton coke.
89. And 99.7% dust collection efficiency.
90. Also 150 mg/m^3.
91. Also 500 mg/m^3
92. New $[35(A + 3)]$ kg/ton; existing, $[50(A + 3)]$ kg/ton where A is thousands of ton/year.
93. Also 300 mg/m^3.

Table XVIA National Emission Standards for Emission of Total Suspended Particulate Matter in Effluent Air or Gases—Spain *(6a)*

Industry and process	*Installations*		
	Existing	*New*	*1980*
	mg/m³		
Chemical industry			
Calcium carbide			
Preparation	300	150	150
Furnace	500	350	250
Carbon black manufacture	150	100	60
Fertilizer manufacture			
Phosphate	250	150	150
Nitrate	250	150	150
Organic	250	150	120
Kraft pulp—recovery furnace	500	250	150
Petroleum refining			
Furnaces	180	150	120
Catalyst regeneration	100	50	50
Sulfite pulp—liquor combustion	500	250	150
Mineral industry			
Alumina	—	150	100
Asphalt plants			
<500 m from residences	400	250	100
>2 km from construction and fixed >2 years	800	500	200
Cement			
Kilns	400[a]	250[a]	150[a]
Clinker coolers	170	100	50
Cement handling	300	250	150
Ceramic	500	250	150
Glass wool—mineral wool	300	200	150
Lime	500	250	150
Metallurgical industry-ferrous			
Blast furnaces	200	100	100
Coke ovens	200	150	150
Electric furnace			
<5 tons	500	350	250
>5 tons	200	150	120
Foundry cupolas			
3–5 tons/hour	800	600	250
>5 tons/hour	600	300	150
Martin–Siemens furnaces	200	150	120
Oxygen steel processes	250	150	120
Sintering and pelletizing ore	400	250[b]	150
Crushing coal and coke	200	150[b]	120
Metallurgical industry—nonferrous			
Aluminum—secondary	200	150	100
Copper			
Melting	400	300	150
Refining	600	500	300
Hydrometallurgical	600	500	300
Lead			
Crucibles	300	200	100
Other	200	150	50
Zinc	600	200	50
Fuel combustion			
Fuel oil			
<50 MW	250	200	175
50–200 MW	200	175	150
>200 MW	175	150	120
Coal—domestic/industrial[c]			
<500 th/hour	500	350	250
>500 th/hour	400	250	150
Coal—power plants[c]			
<50 MW	750	—	—
50–200 MW	500	—	—
>200 MW	350	—	—

Table XVIA ***(Continued)***

	Installations				Installations		
Industry and process	*Existing*	*New*	*1980*	*Industry and process*	*Existing*	*New*	*1980*
		mg/m³				*mg/m³*	
Clean zones				7–15	175	150	150
<50 MW	—	500	250	>15 tons/hour	150	150	120
50–200 MW	—	350	200	Other sources	150	150	150
>200 MW	—	200	150			*kg/ton of product*	
Polluted zones							
<50 MW	—	400	250	Metallurgical industry—			
50–200 MW	—	300	200	ferrous			
>200 MW	—	200	150	Ferrosilicon	23	15	10
Incineration of refuse				Ferrosilicon			
Clean zones[d]				chrome	30	20	15
<1 ton/hour	800	700	500	Ferrochrome—			
1–3	600	500	400	refined	8	5	5
3–7	450	400	300	Ferrosilicon—			
7–15	350	300	250	manganese	0.5	0.5	0.3
>15 tons/hour	250	250	150	Ferromolybdenum	5	3	3
Polluted zones[d]				Metallurgical industry—			
<1 ton/hour	450	350	250	nonferrous			
1–3	300	250	200	Aluminum			
3–7	225	200	200	reduction	12	9	3.5

[a] 1000 mg/m³ for 48 consecutive hours <200 hours/year.
[b] In new plants, 500 mg/m³ <200 hours/year.
[c] Special limits determined by Ministry of the Interior for plants burning coal of >1.5% S or >20% ash.
[d] At 10% CO_2 in the flue gas.
[e] th = therms.

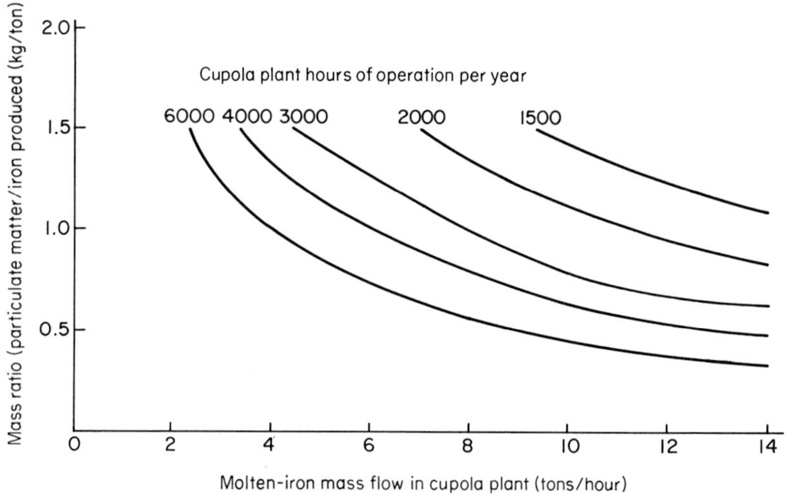

Figure 11. Emission standard for particulate matter from cupolas—Federal Republic of Germany (*10*).

Table XVII States and Provinces for Which There Are State and Provincial Standards for Particulate Matter in Effluent Air or Gas from Stationary Sources *(6)*

Process	*States of the United States*	*Provinces of Australia, Canada, and West Germany*
All sources	—	New South Wales, Queensland, Victoria, Alberta, Manitoba
Agricultural processes		
Alfalfa dehydration	Pennsylvania	—
Corn milling (wet)	Illinois, Missouri	—
Corn shelling	Delaware	—
Cotton ginning	Alabama, Georgia, South Carolina, Tennessee	—
Feed manufacture	Iowa, Virginia	—
Grain drying	Delaware	—
Grain handling	Delaware, Iowa, Pennsylvania, Wisconsin	—
Grain storage	Washington	—
Meat smoking	Iowa, Pennsylvania	—
Wood products manufacture	Virginia, Oregon	—
Chemical processes		
Ammonium nitrate manufacture	Pennsylvania	—
Carbon black manufacture	Pennsylvania	—
Catalyst regeneration	Delaware, Illinois, Missouri, Pennsylvania, Virginia, Washington	—
Charcoal manufacture	Pennsylvania, Virginia	—
Chemical kilns	Michigan	—
Chemical recovery operations	—	Alberta
Detergent drying	Pennsylvania	—
Fertilizer manufacture	North Carolina, Virginia, Georgia	—
Fluid coking operations	Delaware	—
Materials handling	Wisconsin	—
Paper manufacture	New Hampshire	—
Prill tower operations	Delaware	—
Pulp manufacture		
Kraft	Alabama, Florida, Idaho, Louisiana, Maine, Mississippi, New Hampshire, New Mexico, North Caro-	British Columbia

Table XVII (*Continued*)

Process	*States of the United States*	*Provinces of Australia, Canada, and West Germany*
	lina, Oklahoma, Oregon, South Carolina, Tennessee, Virginia, Washington	
Sulfite	New Hampshire, Oregon, Washington	—
Sewage treatment	All states	—
Metallurgical—ferrous		
Cupolas	Alabama, Connecticut, Georgia, Illinois, Indiana, Iowa, Massachusetts, Michigan, Missouri, New Hampshire, New York, North Carolina, Oklahoma, Pennsylvania, Tennessee, Wisconsin	Alberta, Ontario
Electric arc furnaces	Iowa, Michigan, Wisconsin	Alberta
Ferroalloy manufacture	Pennsylvania, Oregon	—
Iron and steel plants	All states	—
Pelletizing ore	Michigan	—
Sand handling	Connecticut, Michigan, Pennsylvania	—
Shake-out	Pennsylvania	—
Sintering	Illinois, Michigan, Pennsylvania, Wisconsin	—
Steel manufacture	Pennsylvania	—
Basic oxygen furnaces	Michigan, Wisconsin	Queensland
Blast furnaces	Michigan, Wisconsin	—
Heating furnaces	Michigan, Wisconsin	New South Wales, Queensland, Victoria
Open hearth furnaces	Michigan, Wisconsin	—
Scarfing	Pennsylvania	—
Metallurgical—Nonferrous		
Brass and bronze	Pennsylvania	—
Foundries	Massachusetts, New Hampshire	—
Primary smelters	New Mexico	—
Aluminum	Alabama, Louisiana, Oregon, Washington	—
Lead	Pennsylvania	—
Zinc	Pennsylvania	—

Table XVII (Continued)

Process	States of the United States	Provinces of Australia, Canada, and West Germany
Secondary smelters	Delaware, Virginia	—
Aluminum	Pennsylvania	—
Brass and bronze	All states	—
Lead	All states	—
Magnesium	Pennsylvania	—
Zinc	Pennsylvania	—
Mineral processes		
Aggregate manufacture	North Carolina, Virginia	—
Asphaltic roofing manufacture	Pennsylvania	—
Coal cleaning and drying	Pennsylvania, Virginia, West Virginia	—
Concrete batching	Connecticut, Iowa, Wisconsin	—
Crushing	Pennsylvania, South Carolina, Wisconsin	—
Gas Plants	New Mexico	—
Gypsum Processing	New Mexico	—
Hot-mix asphalt	All states	Alberta, Ontario, North Rhine-Westphalia
Kaolin/Fuller's earth	Georgia	—
Lime kilns	Iowa, Michigan, Pennsylvania, Wisconsin	Alberta
Mica/feldspar	North Carolina	—
Mineral kilns	Michigan	—
Paint manufacture	Pennsylvania	—
Portland cement manufacture	All states	Alberta
Pumice/mica/perlite	New Mexico	—
Combustion processes	All states	New South Wales, Manitoba
Solid fuels	—	Northrhine-Westphalia
Wood burning	—	British Columbia
Refuse incineration	All states	New South Wales, Alberta

mathematical form, e.g., cupolas [Fig. 11 (*10*) and Table XX]; fiber and particle board production [Fig. 12 (*10*)]; cotton ginning [Tables XX (*6*) and XXI (*6*)]; coal drying [Table XXII (*6*)]; hot-mix asphalt, purmice, mica, perlite, feldspar, fertilizer, and cement plants [Table XXIII (*6*)]; secondary metals operations [Table XX (*6*)]; catalytic

Table XVIII Emission Standards for Particulate Matter from Furnaces Other than Incinerators—Great Britain[a] *(11a)*

SCHEDULE 1 Furnaces rated by heat output (boiler furnaces or indirect heating appliance for gas or liquid that also falls within definition of Schedule 2 furnaces)

Maximum continuous rating [pounds of steam per hour—from and at 100°C (212°F) or in thousands of British thermal units per hour]	*Maximum permitted quantities of grit and dust (pounds/hour)*	
	Furnaces burning solid matter[b]	*Furnaces burning liquid matter*
825	1.10	0.25
1,000	1.33	0.28
2,000	2.67	0.56
3,000	4.00	0.84
4,000	5.33	1.12
5,000	6.67	1.4
7,500	8.50	2.1
10,000	10.00	2.8
15,000	13.33	4.2
20,000	16.67	5.6
25,000	20.0	7.0
30,000	23.4	8.4
40,000	30	11.2
50,000	37	12.5
100,000	66	18
150,000	94	24
200,000	122	29
250,000	149	36
300,000	172	41
350,000	195	45
400,000	217	50
450,000	239	54.5
475,000	250	57

cracking, fluid coking, and prilling operations [Table XXIV *(6)*]. Finally there are particulate matter emission standards in equation and tabular forms for a broad spectrum of sources [Tables XXV *(6)* and XXVI *(6)*].

4. Fugitive Dust Standards

Fugitive dust is defined as solid airborne particulate matter emitted from a source other than a stack, pipe, or structure designed for emission to the open atmosphere, e.g., dust emitted from unpaved roads and

Table XVIII *(Continued)*

SCHEDULE 2 Furnaces rated by heat input (indirect heating appliance or where material being heated is in contact with combustion gases but does not contribute grit or dust to them)

Heat input (millions of BTU/hour)	*Maximum permitted quantities of grit and dust (pounds/hour)*	
	Furnaces burning solid matter[b]	*Furnaces burning liquid matter*
1.25	1.1	0.28
2.5	2.1	0.55
5.0	4.3	1.1
7.5	6.8	1.7
10	7.6	2.2
15	9.7	3.3
20	11.9	4.4
25	14.1	5.5
30	16.3	6.6
35	18.4	7.7
40	20.6	8.8
45	22.8	9.8
50	25	10.9
100	45	16
200	90	26
300	132	35
400	175	44
500	218	54
575	250	57

[a] Applicable to new furnaces November 1, 1971 and to existing furnaces January 1, 1978.

[b] Not over 20% of dust >76 μm diameter, except that 33% >76 μm allowed in Schedule 1 furnaces with maximum continuous rating <16,800 lbs steam/hour or <16.8 MMBTU/hour, and in schedule 2 furnaces with heat input <25 MMBTU/hour.

streets, farms, construction sites, building demolition, quarrying, material stockpiles, and other outdoor operations. Faith (*12*) has evaluated the fugitive dust regulations of the states of the United States [Table XXVII (*12*)].

C. New Source Performance Standards—United States (*3*)

The list of new-source performance standards already promulgated by the United States will be found in reference (*3*) and the standards them-

Table XIX Emission Standards for Particulate Matter from Fuel Burning Equipment—United States *(6)*

	Paremeters for Computing Allowable Emission, E (Fig. 9)[a]					
State	*a*	*b*	*c*	*d*	*e*	*f*
Any source						
Arizona[c]	1.02	0.231	0.60	10	0.153	4,000
Arizona[c]	17.0	0.568	0.153	4,000	0.00	—
Colorado	0.5	0.26	0.50	10	0.10	500
District of Columbia	0.17455	0.23522	0.13	3.5	0.02	10,000
Georgia	1.58	0.50	0.50	10	0.10	250
Idaho	1.0261	0.2330	0.60	10	0.12	10,000
Illinois	5.18	0.715	1.00	10	0.10	250
Indiana	0.865	0.159	0.60	10	0.20	10,000
Kansas	1.026	0.233	0.60	10	0.12	10,000
Mississippi	0.896	0.174	0.60	10	0.18	10,000
Nebraska	1.026	0.233	0.60	10	0.15	3,800
Nevada	1.02	0.231	0.60	10	0.15	4,000
New Mexico[c]	0.96135	0.23471	0.96	1	0.197	1,000
New Mexico[c]	0.5431	0.14687	0.197	1,000	0.127	20,000
North Carolina	1.09	0.259	0.60	10	0.1	10,000
Oklahoma	1.09	0.259	0.60	10	0.1	10,000
Pennsylvania	3.6	0.56	0.40	50	0.10	600
Virginia	0.8425	0.2314	0.40	25	0.10	10,000
New Source						
Kentucky	1.919	0.535	0.56	10	0.10	250
Missouri	1.307	0.338	0.60	10	0.10	2,000
Montana	1.026	0.233	0.60	10	0.12[b]	10,000
New Hampshire	1.026	0.233	0.60	10	0.12[b]	10,000
North Dakota	0.811	0.131	0.60	10	—	—
Tennessee	2.1615	0.5566	0.60	10	0.10	250
Existing source						
Maryland (built before January 17, 1972)	1.026	0.233	0.60	10	0.12	10,000
Missouri	0.896	0.174	0.60	10	0.18	10,000
Montana	0.880	0.166	0.60	10	0.19	10,000
New Hampshire	0.880	0.166	0.60	10	0.19	10,000
Tennessee	1.0903	0.2594	0.60	10	0.10	10,000
Wyoming	0.896	0.174	0.60	10	0.18	10,000
Coal burning						
New York	1.02	0.219	—	10	—	10,000

[a] E is in pounds/MM BTU, H = heat input in MM BTU/hour. When $H < d$, $E = c$; When $H > f$, $E = e$; when $H > d$, and $< f$, $E = aH^{-b}$.

[b] Federal new source performance standard at 0.10 pounds/MMBTU not to be exceeded. [c] One continued function is defined.

selves in Tables IX and XVI. Over the next several years it is anticipated that the U.S. Environmental Protection Agency will have promulgated additional new-source performance standards for the following

Table XX Emission Standards for Particulate Matter from Ferrous Jobbing Foundry Cupolas (Existing), Secondary Metals Operations, and Cotton Ginning—United States *(6)*

Process weight (pounds/hour)	Maximum emission (pounds/hour)			
	Existing ferrous jobbing foundry cupolas—ten states[a]	Secondary metals operations		Cottong ginning—Alabama, Tennessee
		Delaware	Virginia	
1,000	3.05	0.75	3.05	1.6
1,500	—	—	—	2.4
2,000	4.70	1.5	4.70	3.1
2,500	—	—	—	3.9
3,000	6.35	2.25	6.35	4.7
3,500	—	—	—	5.4
4,000	8.00	3.00	8.00	6.2
5,000	9.58[c]	3.75	9.05	7.7
6,000	11.30	4.50	11.30	9.2
7,000	12.90	5.25	12.90	10.7
8,000	14.30[d]	6.00	14.30	12.2
9,000	15.50	6.75	15.50	13.7
10,000	16.65	7.50	16.65	15.2
12,000	18.70	9.00	18.70	18.2
14,000	—	—	—	21.2
16,000	21.60	12.00	21.60	24.2
18,000	23.40[e]	13.50	22.80	27.2
20,000	25.10[f]	15.00	24.00	30.1

Process weight (pounds/hour)	Maximum emission (pounds/hour)					
	Existing ferrous foundry cupolas—Indiana,[b] Missouri	Existing ferrous foundry cupolas—Georgia	Jobbing—Alabama, New York	Secondary metals operations		Cotton ginning—Alabama, Tennessee
				Delaware	Existing—Virginia	
30,000	30.0	31.30	31.3	22.5	30.0	44.9
40,000	36.0	33.76	37.0	30.0	36.0	59.7
50,000	42.0	35.40	42.4	37.5	42.0	64.0
60,000	48.0	—	—	—	42.0	67.4
70,000	49.0	—	—	—	—	—
80,000	50.5	—	—	—	—	—
90,000	51.6	—	—	—	—	—
100,000	52.6	—	—	—	—	—

[a] Alabama, Georgia, Illinois, Indiana (also existing open hearth furnaces in Indianapolis), Iowa, Missouri, New York (higher emissions acceptable if actual collection efficiency is 80%), North Carolina, Oklahoma, and Tennessee.
[b] Also existing open hearth furnaces in Indianapolis.
[c] 9.65, instead of 9.58 (Indiana and Missouri).
[d] 14.0, instead of 14.30 (Indiana and Missouri).
[e] 22.8, instead of 23.40 (Indiana and Missouri).
[f] 24.0, instead of 25.10 (Indiana and Missouri).

Table XXI Emission Standards for Particulate Matter from Cotton Ginning—Georgia and South Carolina *(6)*

Output (500 pound bales/hour)	*Permissible emission (pounds/hour)*	
	South Carolina	*Georgia*
1	—	7.0
2	—	9.9
3	—	12.12
4	12.3	14.0
5	14.4	15.65
6	16.2	17.15
7	18.0	18.52
8	19.5	19.8
9	21.2	21.0
10	22.8	22.14
11	24.2	23.22
12	25.8	24.25
13	27.1	25.24
14	28.5	26.19
15	29.9	27.11
16	31.2	28.00
17	31.2	28.86
18	31.2	29.69
19	31.2	30.51
20	31.2	31.3

Table XXII Emission Standards for Particulate Matter from Coal Drying (Thermal)—West Virginia *(6)*

Total plant volumetric flow rate (ft^3/minute)	*Maximum allowable particulate loading per dryer (grains/ft^3)*	
	Built before March 1, 1970	*Built after March 1, 1970*
≤75,000	—	0.10
111,000	—	0.09
≤120,000	0.12	—
163,000	—	0.08
172,000	0.11	—
≥240,000	—	0.07
245,000	0.10	—
351,000	0.09	—
≥500,000	0.08	—

Table XXIII Emission Standards for Particulate Matter from Hot-Mix Asphalt Plants; Pumice, Mica, and Perlite Processing; Cement Kilns; Chemical Fertilizer Plants; and Mica and Feldspar Plants—United States *(6)*

Process weight (pounds/hour)	Permissible particulate emission rate (pounds/hour)								
	Hot-mix asphalt plants[a]								
		South Carolina		*Massachusetts*					
	North Carolina	*Built before March 1, 1972*	*Built after March 1, 1972*	*New*[b]	*Existing*	*Six*[a,c] *jurisdictions*	*Cement kilns—South Carolina*	*Chemical fertilizer—North Carolina*	*Mica and feldspar plants—North Carolina*
2,000	—	—	—	—	—	—	—	—	4
10,000	10	—	—	—	—	10	—	—	—
20,000	13	—	—	—	—	15[d]	14	19	19
30,000	16	—	—	—	—	22	18	—	—
40,000	18	30	22	—	—	23	22	23	—
50,000	20	—	—	—	—	31	25	—	—
60,000	—	—	—	—	—	—	29	—	40
80,000	—	—	—	—	—	—	—	29	—
100,000	27	45	31	—	—	33	40	31	—
120,000	—	—	—	—	—	—	42	—	—
160,000	—	—	—	—	—	—	45	—	—
200,000	37	57	38	4.5	9.0	37	47	38	50
240,000	—	—	—	—	—	—	48	—	—
250,000	—	—	—	—	—	—	49	—	—
300,000	42	67	45	6.7	13.4	40	50	—	—
360,000	—	—	—	—	—	—	52	—	—
400,000	50	75	51	9.0	18.0	43	53	—	—
500,000	52	82	56	11.3	22.6	47	—	—	—
600,000	60	88	61	14.2	28.3	50[e]	—	—	—
700,000	—	94	65	15.9	31.3	—	—	—	—
800,000	—	94	65	18.1	36.2	—	—	—	—
1,000,000	—	94	65	—	—	—	—	61	—
2,000,000	—	94	65	—	—	—	—	78	80
6,000,000	—	94	65	—	—	—	—	—	90

[a] Also pumice, mica, and perlite processing (New Mexico.)
[b] Also existing plants in critical areas.
[c] Delaware, New Hampshire, New Mexico, Oklahoma City (Oklahoma), Virginia, West Virginia.
[d] New Hampshire and Oklahoma City (Oklahoma)—16, instead of 15.
[e] Not New Hampshire.

Table XXIV Emission Standards for Particulate Matter from Catalytic Cracking, Prilling, and Fluid Coking Operations—Delaware (6)

	Permissible emissions (pounds/hour)				
	Cracking[a]		*Prilling*		
Process weight (pounds/hour)	*Normal standard*	*New Castle County*[b]	*Normal standard*	*New Castle County*[b]	*Fluid coking*
5,000	—	—	25	5	15
7,000	50	5	—	—	—
10,000	—	—	50	10	30
14,000	100	10	—	—	—
15,000	—	—	75	15	50
20,000	—	—	100	20	70
21,000	150	15	—	—	—
25,000	—	—	125	25	100
28,000	200	20	—	—	—
40,000	—	—	—	—	125
42,000	300	30	—	—	—
50,000	—	—	250	50	150
56,000	400	40	—	—	—
70,000	500	50	—	—	—
75,000	—	—	375	75	—
100,000	—	—	500	100	—

[a] Coke burn-off rate.
[b] If national secondary air quality standards for particulates are exceeded in the county between July 7, 1973, and October 1, 1974.

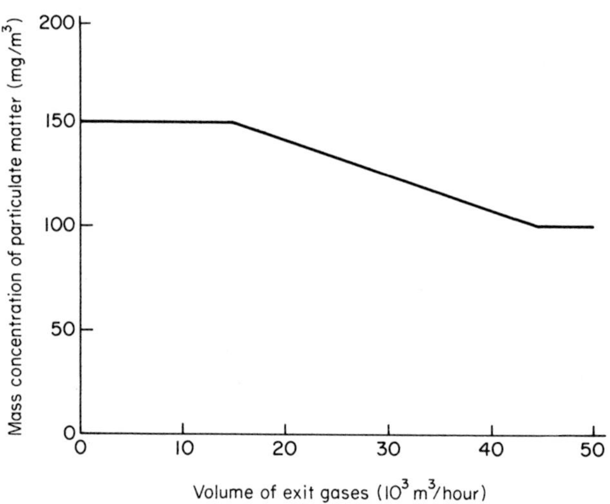

Figure 12. Emission standard for particulate matter from fiber and particle-board production units—Federal Republic of Germany (*10*).

Table XXV National and State Process Weight Emission Standards and Stack Height Standards for Particulate Matter in Effluent Air or Gases from Stationary Sources *(6)*

Process	*Process rate*	*Country or state of the United States*	*Emission standard (E)*	*Note*[a]
	P = *kg/hour*		E = *kg/hour*	
Any source	≤5000	Israel	$0.0290P^{0.60}$	—
	>5000 < 27,000	Israel	$0.0124P^{0.70}$	—
	>27,000	Israel	$3.07P^{0.16}$	—
	P = *tons/hour*		E = *kg/hour*	
New source	<28.5	Mexico	$5.805P^{0.67}$	—
	>28.5	Mexico	$75.648P^{0.11}–54.42$	—
Existing source	<28.5	Mexico	$7.740P^{0.67}$	—
	>28.5	Mexico	$100.846P^{0.11}–72.56$	—
	Q = Nm^3		E = mg/Nm^3	13
New source	100–50,000	Mexico	$3797.1Q^{-0.324}$	—
Existing source	100–50,000	Mexico	$5062.8Q^{-0.324}$	—
	P = *tons/hour*		E = *pounds/hour*	
Any source	≤20	Texas	$3.12P^{0.985}$	—
	≤30	23 states	$4.10P^{0.67}$	1
	≤30	6 states	$3.59P^{0.62}$	2
	≤3000	Mississippi	$4.10P^{0.67}$	—
	>20	Texas	$25.4P^{0.287}$	—
	>30	24 states	$55.0P^{0.11}–40$	3
	>30	6 states	$17.31P^{0.16}$	2
New source	≤30	3 states	$4.10P^{0.67}$	4
	≤30	3 states	$3.59P^{0.62}$	5
	≤450	Illinois	$2.54P^{0.534}$	—
	>30	3 states	$55.0P^{0.11}–40$	6
	>30	3 states	$17.31P^{0.16}$	7
	>30	Massachusetts	$\frac{1}{2}(55P^{0.11}–40)$	—
	>450	Illinois	$24.8P^{0.16}$	—
Existing source	≤30	4 states	$4.10P^{0.67}$	8
	≤30	New Hampshire	$5.05P^{0.67}$	—
	>30	4 states	$55.0P^{0.11}–40$	8
	>30	New Hampshire	$66.0P^{0.11}–48$	—
Aggregate	Any	Virginia	$3.51P$	—
Cement	>50	New York	$0.24P^{0.665}$	—
Cupolas—new	>45	Massachusetts	$55.0(\frac{2}{3}P)^{0.11}–40$	—
Fertilizer	≤30	Georgia	$3.59P^{0.62}$	—
	>30	Georgia	$17.31P^{0.16}$	—
Hot-mix asphalt				
New	<125	Georgia	$2.16P^{0.6}$	—
New	>125	Georgia	$14P^{0.2}$	—
Existing	<45	Georgia	P	—
Existing	>45	Georgia	$10P^{0.4}$	—
Sintering	<450	Illinois	$2.54P^{0.534}$	—

Table XXV ***(Continued)***

Process	*Process rate*	*Country or state of the United States*	*Emission standard (E)*	*Note*[a]
	P = *pounds/hour*		E = *pounds/hour*	
Refuse incineration	Any	Oklahoma	$0.01221P^{0.7577}$	—
	≤ 1000	North Dakota	$0.00515P^{0.90}$	—
	>1000	North Dakota	$0.0252P^{0.67}$	—
	P = *500 pound bales/hour*		E = *pounds/hour*	
Cotton ginning	Any	Georgia	$7(P)^{0.5}$	—
	$Q = 10^6 BTU/hour$		E = *pounds/hour*	
Combustion	<10	New Jersey	$0.6Q$	—
	$>10 <200$	New Jersey	$2.4Q^{0.398}$	—
	$>200 <10{,}000$	New Jersey	$0.1Q$	—
	$q = ft^3/minute$		E = *pounds/hour*	
Combustion	Any	Texas	$0.048q^{0.62}$	—
	Stack height (h_s) *(ft)*		E = *pounds/hour*	
Any source	<120	Georgia	$0.48h_s$	9
	>120	Georgia	$900(h_s/300)^3$	9
	>300	Georgia	$900(h_s/300)^2$	9
Any source Exit gas temperature	>500	Tennessee	$0.3h_s[Q_T \times 0.02 \times (T_s - 60)]^{0.25}$	10
$>125°F$	<500	Tennessee	$0.2h_s[Q_T \times 0.02 \times (T_s - 60)]^{0.25}$	10
$>100°$, $<124°F$	Any height	Tennessee	—	11
$<100°F$	Any height	Tennessee	$3.02 \times 10^{-4} V_s h_s^2 [d_s/h_s]^{0.71}$	12

[a] NOTES: (1) Arizona, Idaho, Indiana, Iowa, Kansas, Kentucky, Louisiana, Michigan, Minnesota, Missouri, Montana, Nebraska, Nevada, North Carolina, North Dakota, Ohio, Oklahoma, Oregon, Rhode Island, South Carolina, South Dakota, Virginia, West Virginia. (2) Arkansas, Colorado, Connecticut, Florida, Maine, Wisconsin. (3) States in Note 1 plus Maryland. (4) Alabama (priority I), Georgia, New Hampshire. (5) Alabama (priority II), Tennessee, Wyoming. (6) Alabama (priority II), Georgia, New Hampshire. (7) Alabama (priority I), Tennessee, Wyoming. (8) Illinois, Massachusetts, Tennessee, Wyoming. (9)

$$h_s = (P_1h_1 + P_2h_2 + \cdots + P_nh_n)/100$$

(Note: $P_1 + P_2 + \cdots + P_n = 100$) where P_1, P_n = percent of total emission from source. (10) Q_T = volume rate of stack gas flow (ft^3/second calculated to 60°F); T_s = temperature of stack gases at stack tip (°F). (11) Linear interpolation between results of computations for 100°F and 125°F exit gas temperatures. (12) d_s = stack exit inside diameter (ft); V_s = stack exit velocity (ft/second). (13) Nm^3 = cubic meters reduced to standard conditions of temperature and pressure.

Table XXVI Emission Standard for Particulate Matter in Effluent Air or Gases from Stationary Sources—Philippines, Vermont, District of Columbia[a] (6)

Process weight per hour (pounds)	*Maximum weight of particulate discharge per hour (pounds)*	*Process weight per hour (pounds)*	*Maximum weight of particulate discharge per hour (pounds)*
50	0.24	3400	5.44
100	0.46	3500	5.52
150	0.66	3600	5.61
200	0.85	3700	5.69
250	1.03	3800	5.77
300	1.20	3900	5.85
350	1.35	4000	5.93
400	1.50	4100	6.01
450	1.63	4200	6.08
500	1.77	4300	6.15
550	1.89	4400	6.22
600	2.01	4500	6.30
650	2.12	4600	6.37
700	2.24	4700	6.45
750	2.34	4800	6.52
800	2.43	4900	6.60
850	2.53	5000	6.67
900	2.62	5500	7.03
950	2.72	6000	7.37
1000	2.80	6500	7.71
1100	2.97	7000	8.05
1200	3.12	7500	8.39
1300	3.26	8000	8.71
1400	3.40	8500	9.03
1500	3.54	9000	9.36
1600	3.66	9500	9.67
1700	3.79	10000	10.0
1800	3.91	11000	10.63
1900	4.03	12000	11.28
2000	4.14	13000	11.89
2100	4.24	14000	12.50
2200	4.34	15000	13.13
2300	4.44	16000	13.74
2400	4.55	17000	14.36
2500	4.64	18000	14.97
2600	4.76	19000	15.58
2700	4.84	20000	16.19
2800	4.92	30000	22.22
2900	5.02	40000	28.3
3000	5.10	50000	34.3
3100	5.18	60000 or more	40.0
3200	5.27		
3300	5.36		

[a] Where the process weight per hour falls between two values in the table, the maximum weight per hour shall be determined by linear interpolation.

Table XXVII Fugitive Dust Standards—United States (Measured at Source Property Line) *(12)*

State	*Maximum particle size (μm)*	*Concentration above background (mg/m³)*	*Particles per cm³*	*Opacity (%)*	*Air quality standard not to be exceeded*
Alaska	—	—	—	—	✓
Arkansas	—	—	—	—	✓
Connecticut	—	—	—	—	✓
Illinois[a]	40	—	—	—	—
Indiana[b]	—	50	—	—	—
Kansas					
New	—	—	—	20	—
Existing	—	—	—	40	—
Maine	—	—	—	40	—
Michigan	—	—	—	40	—
Missouri	40	—	—	—	—
New York	—	—	—	20	—
North Carolina	—	—	—	—	✓
Pennsylvania	—	—	150	—	—
South Carolina	—	—	—	—	✓
Tennessee	—	—	—	—	✓
Wyoming	—	—	—	—	✓

[a] No visible emissions at source in winds <25 mph.
[b] For 1 hour.

industries: asphalt roofing, by-product coke, carbon black, crude oil production, crushed stone, detergent, dry cleaning, gas turbine, gasoline additive, gasoline marketing, grain terminal, gray iron foundry, fuel gasification, Kraft pulp, lead battery, lime, natural gas, petroleum refining, refuse incinerator, smelters, solvent degreasing, stationary internal combustion engines, surface finishing, and type melting. These standards will variously regulate carbon monoxide, hydrogen sulfide, hydrocarbons, lead, nitrogen oxides, mercury, particulate matter, sulfur dioxide, sulfur oxides, sulfides, and total reduced sulfur, but not all these pollutants in each industry.

VII. Stack Height Standards

There are several types of stack height standards:

a. Those that apply generally to all emitted substances and all processes.

b. Those that apply only to sulfur dioxide and/or particulate matter emissions but that do not require a knowledge of the specific processes involved in their emission.

c. Those that apply only to sulfur dioxide emissions from specific processes.

d. Those that apply only to specific processes but that do not require a knowledge of the specific substances emitted.

A. Generally Applicable to All Substances and Processes

Stack Height Standards for general application to all emitted substances and all processes are basically single source diffusion equations of the type discussed in Chapter 9 of Volume I in which the input parameters are specified and which are solved for the stack height by mathematical, graphical, or tabular means. One of the input parameters must be some variant of the ground-level ambient air quality standard for the specific substance or combination of substances being emitted. This can be the ambient air quality standard itself, as found in the tables of Chapter 11 of this volume, some specified percentage of the ambient air quality standard to provide a factor of safety, or a point-of-impingement-at-ground-level standard, also found in Chapter 11 of this volume.

The USSR utilizes the solution of a specified set of mathematical equations for the determination of stack height for emissions of all substances from all processes [Table XXVIII (*13*)]. In West Germany, stack height is initially determined from a nomograph [Fig. 13 (*10*)] and is corrected for the influence of buildings and structures by Figure 14 (*10*). This nomogram is used as follows:

1. Draw a horizontal line in the lower left diagram at the diameter in meters of the inside exit diameter (d) of the stack.

2. From the point of intersection of this line with the curve of appropriate stack exit gas temperature (t) in °C, draw a vertical line into the upper left diagram.

3. From the point of intersection of this line with the appropriate diagonal value of total stack effluent gas quantity (R) in cubic meters per hour STP, draw a horizontal line through the three remaining diagrams to the right.

4. From each of the three points of intersection of this line with the curves (one in each diagram) of the appropriate value of ($\bar{U}_M(Q/S)$) draw a vertical line downward to the bottom of the chart. The value of $\bar{U}_M$ is given at the top of each diagram.

Table XXVIII Stack Height Standard Computation—Soviet Union[a] *(13)* Example—Calculation of Stack Height of a Sintering Plant

No.	*Name, designation, formula, and calculation*	*Units*	*Value*
1	Background concentration of sulfur dioxide, $C_{b(SO_2)}$	mg/m^3	0
	Background concentration of dust, $C_{b(d)}$	mg/m^3	0
2	Coefficient dependent on the thermal stratification of the atmosphere, A	$sec^{2/3} \times deg^{1/3}$	160
3	$FM_{SO_2} > FM_d$ $1 \times 1960 > 2 \times 60$ The stack height is therefore determined from the emission of sulfur dioxide	gm/sec	1960 120
4	Difference between the temperature of the gas–air mixture discharged T_g, and that of the surrounding air, T_a	deg	120
	$\Delta T = T_g - T_a = 150 - 30$		
5	First approximation for minimum stack height (for $m = 1$), H	m	138
	$H = \left[\left(\frac{AM_{SO_2}Fm}{MPC}\right)\left(\frac{N}{V\,\Delta T}\right)^{1/3}\right]^{1/2} = \left[\left(\frac{160 \times 1960 \times 1 \times 1}{0.5}\right)\left(\frac{1}{300 \times 120}\right)^{1/3}\right]^{1/2}$		
6	First approximation for the parameter f	$m/sec^2\ deg$	0.32
	$f = 10^3 \frac{w_0{}^2 D}{H^2\,\Delta T} = \frac{10^3 \times 11^2 \times 6}{138^2 \times 120}$		
7	First approximation for the coefficient (m) allowing for the conditions of exit of the gas-air mixture from the stack	—	1.05
8	Preliminary value of the stack height, H	m	138
	$H = \left[\left(\frac{AM_{SO_2}\,Fm}{MPC}\right)\left(\frac{N}{V\,\Delta T}\right)^{1/3}\right]^{1/2} = \left[\left(\frac{160 \times 1960 \times 1 \times 1}{0.5}\right)\left(\frac{1}{300 \times 120}\right)^{1/3}\right]^{1/2}$		
9	Next largest size of the height of standard stacks, H	m	150
10	Parameter f	$m/sec^2\ deg$	0.27
	$f = 10^3 \frac{w_0{}^2 D}{H^2\,\Delta T} = \frac{10^3 \times 11^2 \times 6}{150^2 \times 120}$ ($f < 6$, which makes it possible to use the present instructions)		
11	Dimensionless coefficient allowing for the conditions of exit of the gas–air mixture from the stack, m	—	1.05

Table XXVIII ***(Continued)***

No.	*Name, designation, formula, and calculation*	*Units*	*Value*
12	Maximum concentration of sulfur dioxide near the underlying surface, $C_{m(SO_2)}$ $$C_{m(SO_2)} = \frac{AM_{SO_2}Fm}{H^2}\left(\frac{N}{V\,\Delta T}\right)^{1/3} = \frac{160 \times 1960 \times 1 \times 1.05}{150^2}\left(\frac{1}{300 \times 120}\right)^{1/3}$$	mg/m³	0.44
13	Maximum concentration of dust near the underlying surface, $C_{m(d)}$ $$C_{m(d)} = \frac{AM_dFm}{H^2}\left(\frac{N}{V\,\Delta T}\right)^{1/3} = \frac{160 \times 60 \times 2 \times 1.05}{150^2}\left(\frac{1}{300 \times 120}\right)^{1/3}$$	mg/m³	0.03
14	Distance at which the maximum concentration of noxious emissions is reached, x_m $$x_m = 20H = 20 \times 150$$	m	3000

[a] SYMBOLS:

A = (a) for Central Asia, Kazakhstan, the Lower Volga region, Caucasus, Siberia, and Far East, 200; (b) for the north and northwest of the European territory of the Soviet Union, Middle Volga Region, Urals, and Ukraine, 160; (c) for the central part of the European territory of the Soviet Union, 120.
D = diameter of the stack (m).
f = should satisfy the inequality $f < 6$ (f in m/sec² deg).
F = Coefficient allowing for the influence of the velocity of deposition: (a) sulfur dioxide = 1; (b) suspended particulate matter = 2.
H = stack height (physical) (m).
M = total emission: μ_{SO_2}—sulfurdioxide: μ_d—dust (gm/sec).
MPC = maximum permissible concentration for SO_2 = 0.5 (mg/m³).
N = number of chimneys situated less than two chimney heights from each other.
ΔT = temperature difference (°C) between stack gas (T_g) and ambient air (T_a).
V = total volume discharged into the atmosphere m³/sec.
w_0 = exit velocity from stack (m/sec).

Q is kilograms per hour of emission of the specific pollutant for which S is the maximum increase in ground level concentration over background concentration in milligrams per cubic meter that may be contributed by the stack in question.

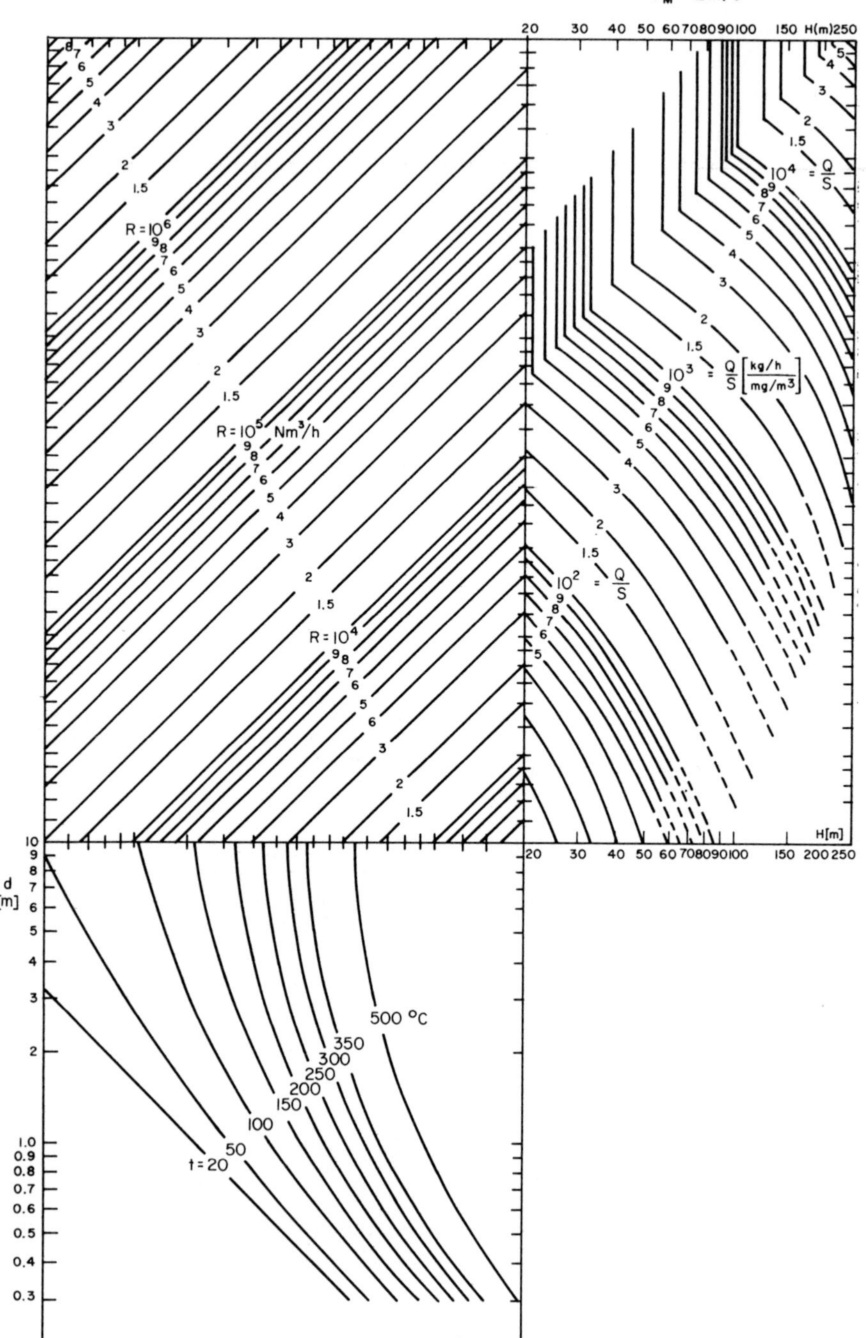

$\bar{U}_M$ = 2 m/s
20 30 40 50 60 70 80 90 100 150 H(m) 250
R = 10^6
R = 10^5 Nm^3/h
R = 10^4
10^4 = Q/S
10^3 = Q/S [kg/h / mg/m^3]
10^2 = Q/S
H[m]
20 30 40 50 60 70 80 90 100 150 200 250
d [m]
500 °C
350
300
250
200
150
100
50
t = 20

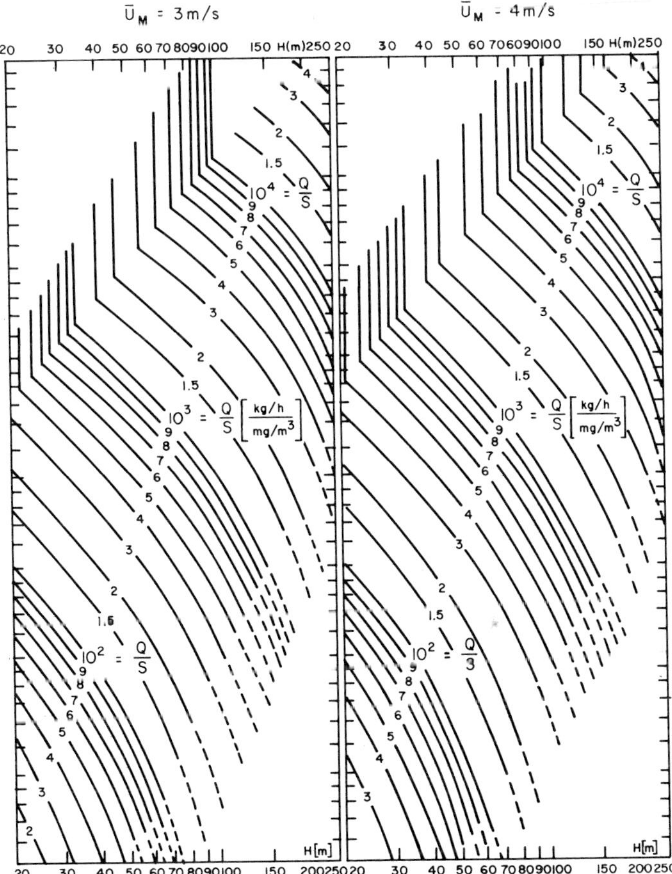

Figure 13. Nomogram for stack height calculation for inert gaseous pollutants emitted at constant quantity and temperature over flat terrain with no structural interference—Federal Republic of Germany (*10*). R = exit gas volume. *H* (m) = physical stack height (maximum, 300 m; minimum, 2 m over roof top)

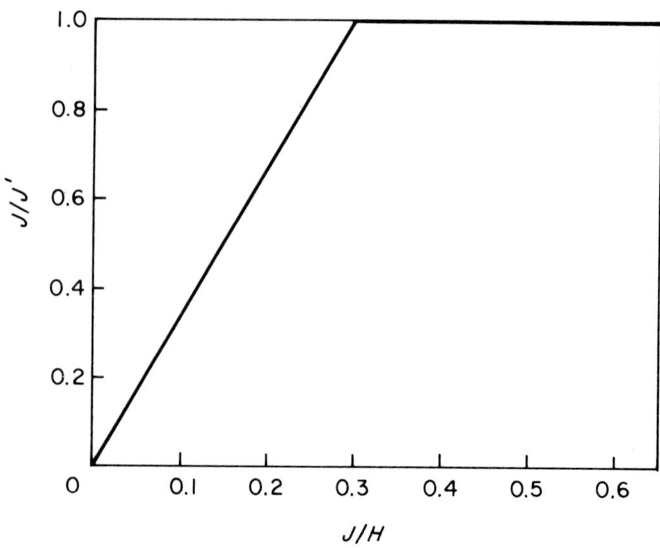

Figure 14. Graph for determining the value of J in equation $H' = H + J$, where H' is the stack height required when there are compact structures or vegetation of mean height J' on more than 5% of the area affected, and H is the stack height determined from Figure 13—Federal Republic of Germany (*10*).

5. The three points of intersection are at values of stack height (H) above ground for three different conditions of average wind velocity ($\bar{U}_M$) in meters per second about 10 m above ground, namely 2, 3, and 4 m/sec, respectively.

Czechoslovakia and East Germany determine stack height from tables [Tables XXIX (*14*) and XXX (*15*)]. The right-hand columns of these tables have general application to all pollutants, other than sulfur dioxide and fly ash, for which ambient air quality standards have been adopted in the respective countries. For these two pollutant emissions, the interior columns of the tables are used. In the case of Czechoslovakia, the use of the sulfur dioxide and fly ash columns is quite straightforward, but before the East German table can be used to determine stack height for sulfur dioxide emissions it is necessary to first determine the area classification by use of Table XXXI (*15*). For particulate matter emissions, it is necessary to determine both area classification and fine fraction–coarse fraction ratios by use of Table XXXI and to determine stack height by Figure 15 (*15*). East German stack heights determined from Table XXX (*15*) and Figure 15 (*15*) are subject to a plume rise credit (adjustment) according to Table XXXII (*15*).

Table XXIX Stack Height Standards—Czechoslovakia *(14)*

Stack height[a] *(m)*	*Permissible emission (kg/hour)*		
	From combustion of fuel		*Multiplier for K_{max} for other harmful substances*[b,c]
	Fly ash	*Sulfur dioxide*	
7	2.5	2	4
8	3	2.3	4.6
10	4	3.2	6.4
12	5	4.2	8.4
14	7	5.3	10.6
16	9	6.8	13.6
18	11.4	8.4	16.8
20	14	10	20.0
25	21	13.5	27.0
30	31	22.5	45.0
35	42	32.5	65.0
40	55	46	92.0
45	70	60	120.0
50	84	82.5	165.0
55	110	100	200
60	130	122	245.0
65	160	145	290.0
70	192	170	340.0
75	225	195	390.0
80	260	227	455
85	290	257	514
90	325	295	590
95	360	335	670
100	400	375	750
110	490	900	930
120	580	1425	1130
130	675	1950	1340
140	785	2475	1560
150	900	3000	1790
160	1010	3555	2060
170	1130	4110	2320
180	1270	4665	2600
190	1400	5220	2890
200	1550	5779	3200
220	1820	6355	3840
240	2110	6930	4500
260	2400	7510	5160
280	2700	8085	5820
300	3000	8665	6500

[a] Where the harmful substances are discharged through two or more chimneys situated within the area of a circle of 1 km in diameter, the chimneys of one and the same establishment are regarded as one chimney. For varying heights of chimneys, the method of calculation shall be established by the Ministry of Forest Administration and Water Conservation in a work instruction manual.

[b] For substances listed for Czechoslovakia in Table III of Chapter 11 of this volume, K_{max} is the concentration for 30 minute averaging time; e.g., for ammonia, $K_{max} = 0.3$ mg/m³. Therefore, the permissible emission of ammonia from a 100 m stack is $750 \times 0.3 = 225$ kg/hour.

[c] Where the discharge is for less than 1 hour, there is a proportionate reduction in allowable emission in kilograms.

Table XXX Emission Standard for Sulfur Dioxide and Multiplication Factor S for Other Gaseous and Suspended Particulate Pollutants—East Germany *(15a)*

	Permissible sulfur dioxide emission[b] (kg/hour)			
	Area class 1	*Area class 2*	*Area class 3*	
Effective stack height —H(m)[a]	*Permissible SO_2 increment (mg/m^3)* 0.4	0.3	0.2	*Multiplication factor (S)[c]*
10	4.26	3.20	2.13	10.65
15	9.59	7.19	4.79	23.96
20	17.04	12.78	8.52	42.60
25	26.63	19.97	13.31	66.56
30	38.34	28.76	19.17	95.85
35	52.19	39.14	26.09	130.46
40	68.16	51.12	34.08	170.40
45	86.27	64.70	43.13	215.66
50	106.50	79.88	53.25	266.25
60	153.36	115.02	76.68	383.40
70	208.74	156.56	104.37	521.85
80	272.64	204.13	136.32	681.60
90	345.06	258.80	172.53	862.65
100	420.00	319.50	213.00	1,065.00
120	613.44	460.08	306.72	1,533.60
140	834.96	626.22	417.48	2,087.40
160	1,090.56	817.92	545.28	2,726.40
180	1,380.24	1,035.18	690.12	3,450.60
200	1,704.00	1,278.00	852.00	4,260.00
220	2,061.84	1,546.38	1,030.92	5,154.60
240	2,453.76	1,840.32	1,226.88	6,134.40
260	2,879.76	2,159.82	1,439.88	7,199.40
280	3,339.84	2,504.88	1,669.92	8,349.60
300	3,834.00	2,875.50	1,917.00	9,585.00

[a] See Table XXXII and Figure 15.
[b] See Tables XXXI and XXXII. If the source of sulfur dioxide contributes considerably to ground-level concentration, the next higher area class restrictions may be required.
[c] See reference *15b*.

B. Applicable Only to Sulfur Dioxide and Particulate Matter, but to Any Process

1. Applicable to Both Sulfur Dioxide and Particulate Matter Emissions, and to Any Process

Japanese stack height standards apply to both sulfur dioxide and particulate matter emissions from any process. They utilize simple formulas for computation and K values from Table XXXIII (*16*).

Table XXXI Area Classification–East Germany *(15a)*

Actual air quality/ air quality standard[a]	*Pollution class*	*Class description*	*Permissible increment in SO_2 concentration*[b] *(mg/m³)*	*Q—Area class multiplying factor for settleable dust computation*[c]
≤0.5	1	Slightly polluted	0.4	0.8
>0.5–≤1.0	2	Polluted	0.3	0.6
>1.0–≤1.5	3	Overpolluted	0.2	0.4
>1.5–≤2.5	4	Considerably overpolluted	—	—
>2.5	5	Heavily overpolluted	—	—
Sulfur dioxide concentration (mg/m³)				
<0.1	—	—	0.4	—

[a] For settleable dust (i.e., >10 μm), use as air quality standard either (a) long-term arithmetic mean ≤15 gm/m²/30 days or (b) short-term maximum ≤20 gm/m²/30 days, whichever is more restrictive.

[b] See Table XXX.

[c] Multiply emission rate from Figure 15 by factor Q.

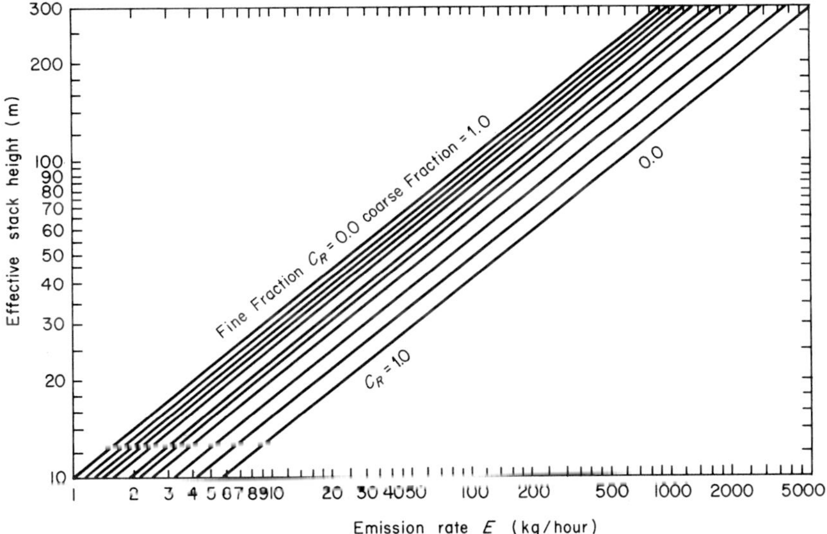

Figure 15. Stack height standard for settleable dust emission. Multiply emission rate E by area class multiplying factor Q from Table XXXI. C_R = dust >10 μm and <63 μm/dust >10 μm—Democratic Republic of Germany *15, 15a*).

Table XXXII Stack Height Adjustment—East Germany[a] *(15, 15a)*

	ΔH(m)[b]														
	Exit velocity (m/sec):														
	5					*10*					*20*				
	Exit gas temperature (°C):														
Exit volume ($10^3 nm^3$/hour)	*20*	*60*	*120*	*180*	*240*	*20*	*60*	*120*	*180*	*240*	*20*	*60*	*120*	*180*	*240*
1	—	—	—	—	—	—	—	—	—	—	—	1	1	1	1
2	—	—	—	—	—	—	—	—	1	1	1	1	1	1	1
3	—	—	—	1	1	—	1	1	1	1	1	1	1	1	1
4	—	—	1	1	1	1	1	1	1	1	1	1	1	1	2
5	—	—	1	1	1	1	1	1	1	1	1	1	2	2	2
6	—	1	1	1	1	1	1	1	1	2	1	2	2	2	2
7	—	1	1	1	1	1	1	1	2	2	1	2	2	2	2
8	—	1	1	1	2	1	1	1	2	2	1	2	2	2	3
9	—	1	1	1	2	1	1	2	2	2	2	2	2	3	3
10	—	1	1	2	2	1	1	2	2	2	2	2	2	3	3
20	1	1	2	3	3	2	2	3	3	4	2	3	4	4	5
30	1	2	3	4	5	2	3	4	5	5	3	4	5	6	7
40	1	2	4	5	6	2	3	5	6	7	4	5	6	7	8
50	1	3	4	6	7	3	4	6	7	8	4	6	7	8	10
60	1	3	5	7	8	3	5	6	8	9	5	6	8	9	11
70	2	3	6	7	9	3	5	7	9	10	5	7	9	10	12
80	2	4	6	8	10	4	6	8	10	11	5	8	10	11	13
90	2	4	7	9	11	4	6	8	10	12	6	8	11	13	14
100	2	5	7	10	12	4	7	9	11	13	6	9	11	14	15
200	3	7	12	17	21	6	10	15	19	26	9	14	18	22	25
300	5	10	17	23	29	8	14	20	25	31	11	18	25	28	34
400	6	12	21	29	37	9	17	24	32	38	13	21	27	34	41
500	7	14	25	35	44	11	19	29	37	46	15	24	32	40	43
600	8	16	29	40	50	12	22	32	42	52	16	27	36	46	55
700	8	19	31	46	57	13	24	36	43	60	18	29	40	50	62
800	9	21	36	50	65	14	27	40	50	65	19	32	44	55	67
900	10	23	40	55	70	15	29	44	58	70	21	34	48	60	74
1000	11	24	43	60	77	16	30	47	63	78	22	36	50	65	78

[a] See Table XXX and Figure 15.
[b] Effective stack height (H) = physical stack height $(h) + \Delta H$. Intermediate values to be linearly interpolated.

Danish stack height standards apply specifically to particulate matter and sulfur dioxide. They utilize a pair of nomographs (*16a*) the use of which requires consideration of the ratio of background level to the national air quality standard, and of the percentage of the air quality

Table XXXIII ***K*** **Values for Emission of Sulfur Oxides and Particulate Matter in Effluent Air or Gas from Stationary Sources—Japan**[a] *(16)*

K value	*Areas*
	General standards
3.5	Tokyo, Yokohama, Kawasaki, Nagoya, Yokkaichi, etc. (6 areas)
4.67	Chiba, Ichihara, Kurashiki, Mizushima, Kitakyushu, etc. (7 areas)
6.42	Sapporo, Muroran, Kashima, etc. (16 areas)
8.76	Tomakomai, Niigata, Shimonoseki, etc. (19 areas)
11.7	Okayama, Hiroshima, Fukuoka, etc. (16 areas)
14.6	Asahikawa, Kushiro, Shizuoka, Nagasaki, Sasebo, Kagoshima, etc. (35 areas)
17.5	Other areas
	Special standards
1.17	Tokyo A, Osaka A, Yokohoma, Kawasaki, Kobe, Amagasaki, Yokkaichi, Nagoya
1.76	Chiba, Kawaguchi, Yokosuka, Fuji, Himeji, Kurashiki, Northern Kyushu
2.34	Kyoto, Wakayama, Ube, Onoda, Omuta, Karita

[a] Formulas:

$$q = K \cdot 10^{-3} H_e^2$$
q = Stack exit flow rate (m^3/hour)
H_e = Effective stack height (m)
$$H_e = H_0 + 0.65(H_m + H_t)$$
$$H_m = \frac{0.795(QV)^{1/2}}{1 + (2.58/V)}$$
$$H_t = (2.01 \times 10^{-3})Q(T - 288)(2.30 \log J + (1/J) - 1)$$
$$J = [1/(QV)^{1/2}][1460 - 296(V/T - 288)] + 1$$
H_0 = Physical stack height (m)
Q = Rate of effluent gas at 15°C (m^3/second)
V = Emission velocity of effluent gas (m/second)
T = Temperature of effluent gas (°K)

standard an individual community is willing to have preempted by the new source (*16b*).

2. Applicable Only to Sulfur Dioxide Emissions, but to Any Process

Basic chimney heights required for sulfur dioxide emissions from any process are set forth in tabular form in Belgium (Table XXXIV) (*17*) and in both tabular (Table XXXV) (*18*) and graphic form in Great Britain. The Ministry of Housing and Local Government of Great Britain has issued a "Memorandum on Chimney Heights" (*18*). This memorandum includes nomographs for SO_2 emission covering five types of district, (A) undeveloped, (B) partially developed, (C) built-up

Table XXXIV Stack Height Standards for Sulfur Dioxide—Belgium[a] *(17)*

Degree of dilution of component sulfur gases	*Imposed minimum height (m)*	
	For gases or fumes whose temperature is over 150°C	*For gases or fumes whose temperature is under 150°C*
1/12,000	7	10
1/10,000	10	15
1/7,500	14	23
1/5,000	20	35
1/3,000	30	50
1/2,000	40	65
1/1,000	60	100

[a] These requirements apply only to factories roasting or reducing lead or zinc minerals or metals containing lead or zinc.

residential, (D) mixed industrial–residential, and (E) mixed heavy industrial–residential. There are four separate nomograms for the determination of uncorrected chimney height for the following ranges of sulfur dioxide emission in pounds per hour: 3–30, 30–100, 100–400, and 400–1800. There are a fifth and sixth nomogram for determining final chimney height from uncorrected chimney height, building height, and building length. Only two of these six nomograms are reproduced here (Fig. 16) *(18)*.

Basic chimney heights required for sulfur dioxide emission are issued in graphic form in Sweden (Fig. 17) *(19)* and as equations in the state of Georgia (Table XIII) *(6)*. The state of New Jersey has a chart relating allowable emission of sulfur dioxide, sulfuric acid mist, sulfur trioxide, and other sulfur compounds to adjusted stack height in feet (Fig. 18) *(20)*.

Table XXXV Stack Height Standards for Sulfur Dioxide (Warm Emissions)[a]—Great Britain *(18)*

Rate of emission (tons SO_2/day):	3.6	7.5	13	21	30	40
Basic chimney height (ft):	100	150	200	250	300	340

[a] For use above range of "Memorandum on Chimney Height *(18)*." To allow for interfering nearby tall buildings the correction $H = 0.625A + 0.935B$ must be applied, where H = final chimney height (ft), A = basic chimney height (ft), B = building height (ft).

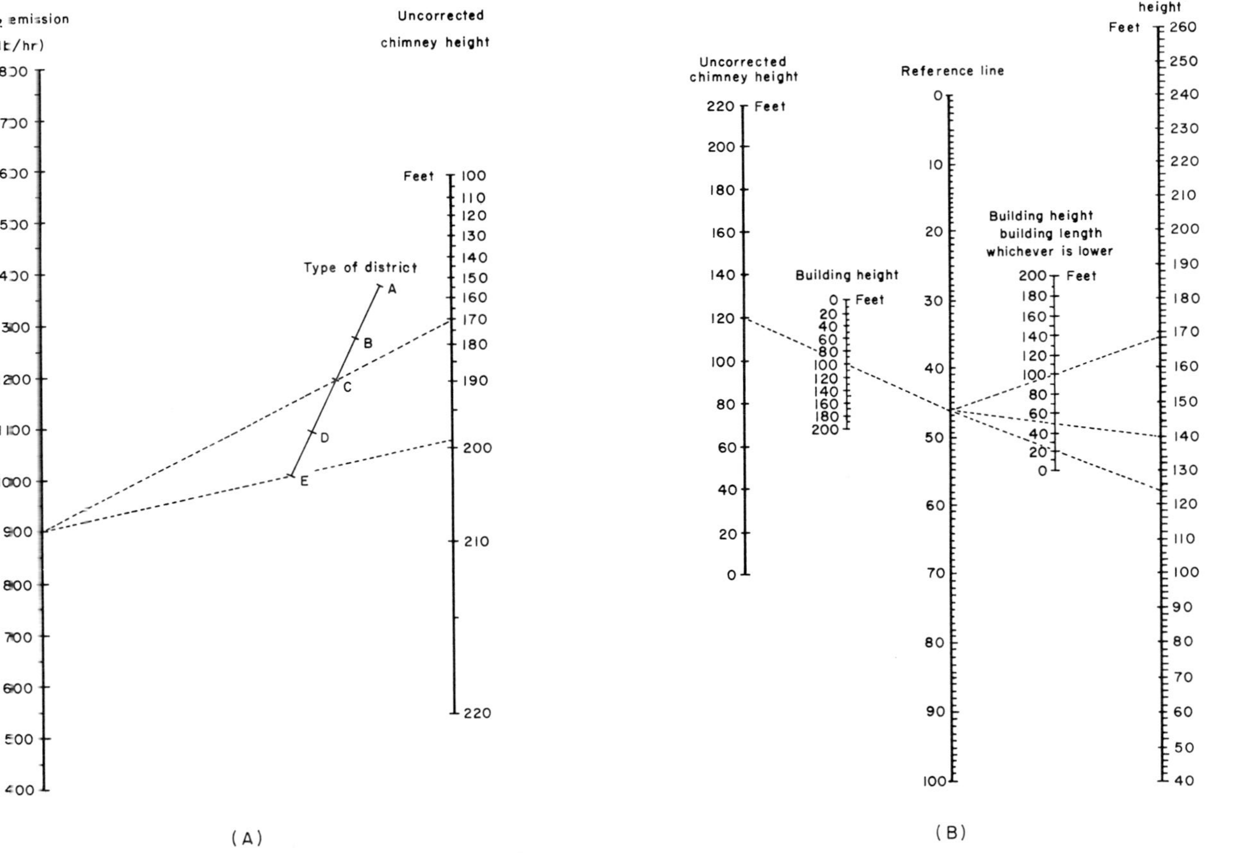

Figure 16. Stack height standard for sulfur dioxide (as SO_2) emission—Great Britain (*18a*). (A) Uncorrected height for large installations, (B) final stack height.

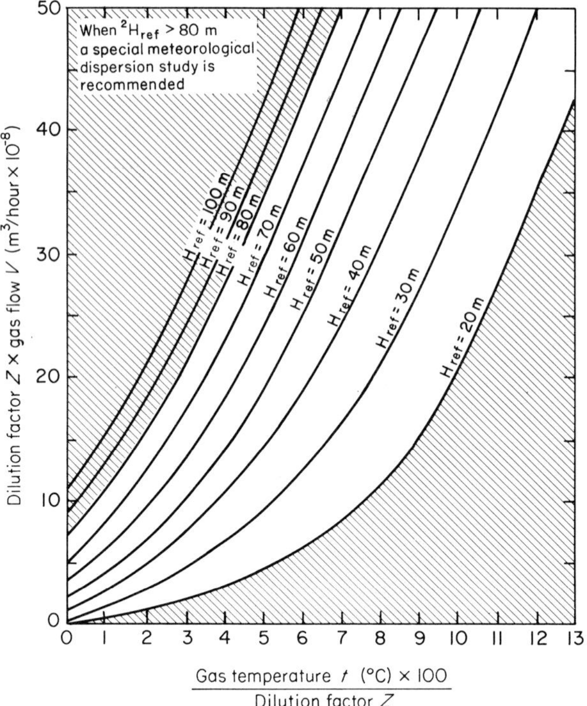

Figure 17. Stack height standard for sulfur dioxide emission—Sweden (*19, 19a*). When $H_{ref} > 80$ m, a special meteorological dispersion study is recommended. H = physical stack height = $H_{ref} + \Delta H$; Z = [emission at maximum load (SO_2/hour) × 10^6]/(V — ground-level concentration contribution); B = height of building or terrain above grade of stack base.

Distance from stack base	B/H_{ref}	ΔH (m)
$>2H_{ref}$	<0.3	0
	$>0.3 <1.0$	$(B - 0.3H_{ref})/0.7$
	>1.0	B
>2, $<20H_{ref}$	—	B

3. Applicable Only to Particulate Matter Emissions but to Any Process

Chimney heights required for particulate matter emissions from any process are in the form of equations in the states of Georgia and Tennessee [Table XXV (*6*)].

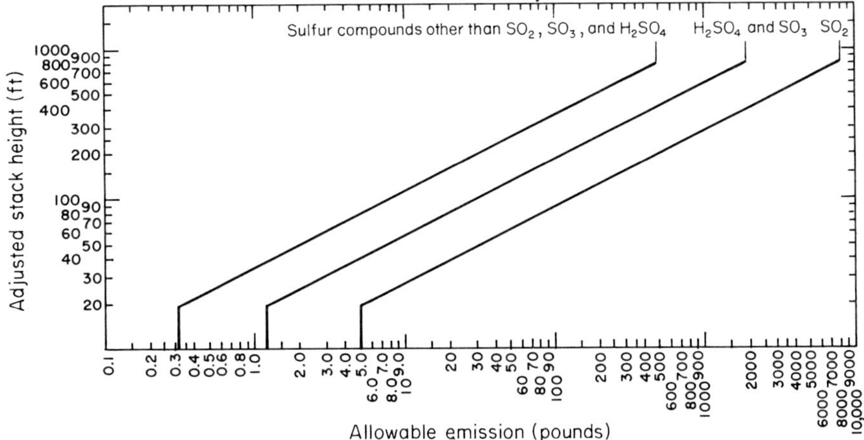

Figure 18. Stack height standard for sulfur dioxide (as SO_2), sulfur trioxide (as SO_3), sulfuric acid mist (as H_2SO_4), and sulfur compound emission—New Jersey (*20*).

C. Applicable Only to Sulfur Dioxide from Specific Processes

1. Combustion Processes

French stack height standards for sulfur dioxide emissions from combustion processes require the use of Equation (1)

$$h_p = [340q_1/C_{M_1}(1/R\ \Delta T)^{1/3}]^{1/2} \tag{1}$$

where h_p = physical stack height (m); q_1 = sulfur dioxide emission rate (kg/hour); C_{M_1} = difference between 0.25 mg/m³ and actual ambient level of sulfur dioxide (if there are no measurements of actual ambient air quality assume 0.01 mg/m³ in unpolluted rural areas, 0.11 mg/m³ in medium industrial areas and medium populated areas, 0.16 mg/m³ in urban or industrial areas); R = effluent volume rate (m³/hour); ΔT = temperature difference (°C) between stack gas and ambient air. The states of Illinois, Indiana, and Texas rely upon equations (Table XIII) (*6*).

2. Noncombustion Processes

West German stack-height standards for sulfur dioxide emissions from sinter plants are in tabular form (Table XXXVI) (*21*). Those for sulfur dioxide emissions from process operations generally in Indiana

Table XXXVI Stack Height Standard for Sulfur Dioxide from Sinter Plants—West Germany (VDI 2095) *(21)*

Sulfur dioxide emission (kg/hour)	100	500	1000	1500
Minimum stack height (meters)	45	75	90	Special agreement

and from sulfuric acid manufacture, sulfur recovery, and nonferrous smelting in Texas are in the form of equations (Table XIII) *(6)*.

D. Applicable Only to Particulate Matter from Specific Processes

Italian stack-height standards for particulate matter emissions from combustion processes require use of both an equation and a graphical aid (Fig. 19) *(22)*. French standards for this purpose require use of Equation (2):

$$h_p = [680 q_2 / C_{M_2} (n/R\ \Delta T)^{1/3}]^{1/2} \tag{2}$$

where h_p, R, and ΔT have the same meaning as the previous equation,

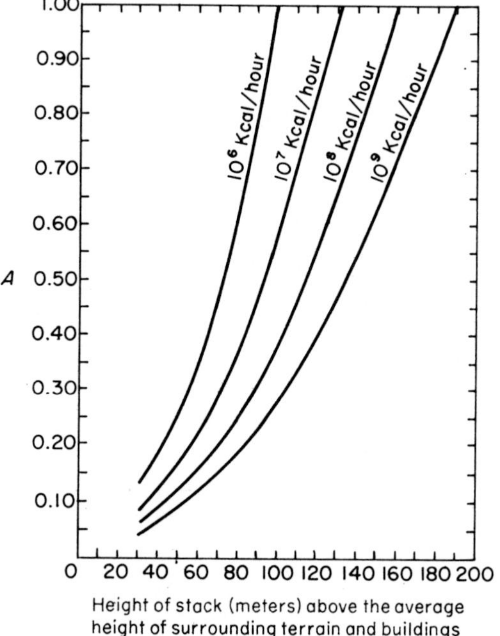

Figure 19. Stack height standard for particulate matter from combustion processes —Italy *(22)*. Allowable concentration of particulate matter in flue gas = $0.25(I + A)$ gm/m³.

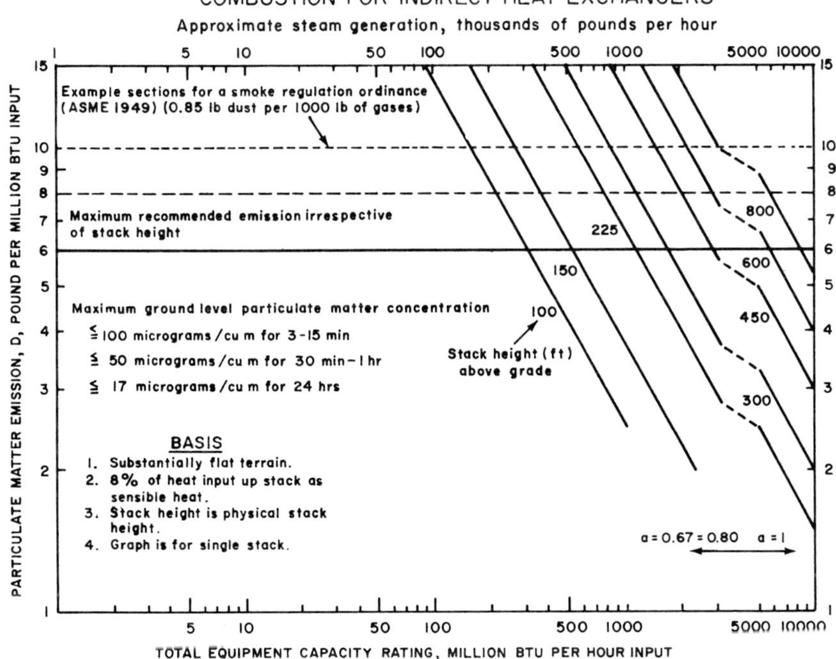

Figure 20. Stack height standard for particulate matter from combustion processes —Iowa (variants are also used in Indiana and Wisconsin) (23).

and q_2 = particulate matter emission rate (kg/hour); C_{M_2} = difference between 0.15 mg/m^3 and actual ambient level of particulate matter (if there are no measurements of actual ambient air quality assume 0.05

Table XXXVII Stack Height Standard for Cement Works—Great Britain *(24)*

Clinker throughput (tons/hour)	*Chimney height (feet)[a]*		
	Wet process	*Semidry process*	*Dry process*
30 and less	200	200	200
60	280	260	240
90	340	310	280
120	390	350	310
240	500	460	415
360	550	500	450

[a] Interpolation between 30 and 360 tons/hour on smooth curves through points in table.

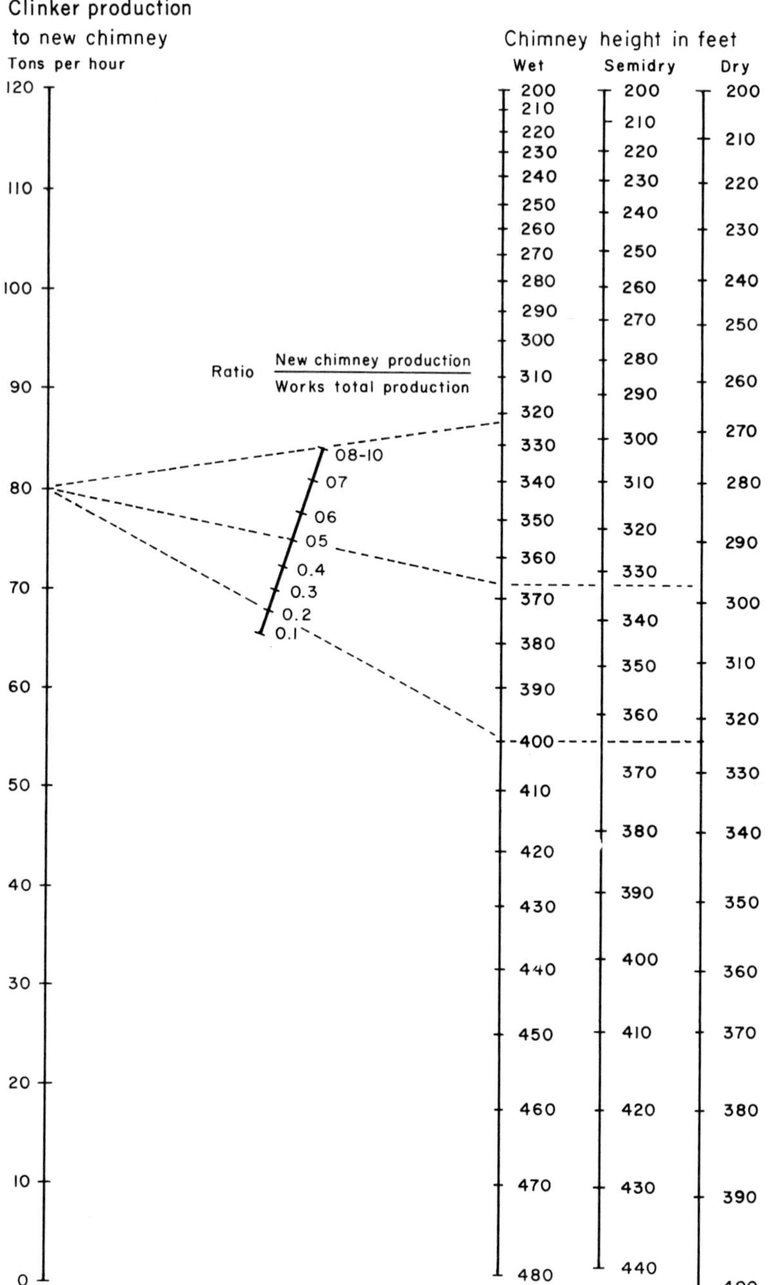

Figure 21. Stack height standard for cement works (multiple stacks)—Great Britain (*24*).

Table XXXVIII Stack Height Standard for Copper Works—Great Britain[a] *(25)*

Rate of melting[b] (tons/24 hours):	25	50	100	150	200	250	300
Basic chimney height (feet):	72	102	144	177	204	228	250

[a] The process to which these are applicable is the recovery of copper and its alloys from scrap fabricated metal, swarf, or residues. It assumes that satisfactory steps have been taken to prevent emissions of dark smoke and that the chimneys are solely to secure satisfactory dispersion of adventitious zinc oxide fume arising during melting and pouring of the copper alloys. Where there is blowing of the molten metal deliberately to remove zinc, the standards for general fume emissions are expected to be attempted.

[b] Rate of melting is the aggregate capacity of all furnaces on the works site calculated to a 24-hour day, except on a large works where groups of furnaces are so widely separated as to be able to be considered as occupying different sites.

Table XXXIX Stack Height Standard for Sulfuric Acid Contact Plants Burning Sulfur—Great Britain *(25)*

Production (tons H_2SO_4 per day)	*Basic chimney height (feet)*[a]			
	$v = 20$	*$v = 30$*	*$v = 40$*	*$v = 50$*
100[b]	104	101	99	96
200	142	138	135	132
300	175	167	163	159
400	203	197	192	188
500	226	218	214	210
600	248	241	235	230
700	267	260	253	247
800	286	278	271	266
900	304	294	287	280
1000	319	310	303	296
1100	334	325	317	310
1200	349	340	332	325
1300	363	353	344	337
1400	377	367	358	351
1500	391	381	372	364
1600	405	394	385	376
1700	417	406	397	388
1800	429	418	409	399
1900	441	429	420	409
2000	452	439	430	419

[a] These are based on a calculated 3 minute mean ground level concentration of SO_2 of 20 pphm (parts per 100 million); v = velocity of efflux of gases in feet per second.

[b] Note that the minimum height of chimney for a contact sulfuric acid plant is 120 feet.

Table XL Stack Height Standard for Nitric Acid Production Plants—Great Britain[a] *(25)*

Stack exit area (ft^2)	*Gas volume at STP* (ft^3/sec)	*Effective height* (*ft*)	*Plume rise* (*ft*)	*Basic chimney height* (*ft*)
175	14,000	205	27	180
350	28,000	287	39	250
530	42,000	353	47	300
700	56,000	412	55	350
1060	84,000	468	68	400

[a] Basis: (1) Efflux velocity 80 ft/sec. (2) Emission has no thermal buoyancy. (3) Maximum ground level concentration for a 3-minute mean is 0.16 ppm NO_2. (4) Concentration of NO_2 is 2.3 grains/ft^3 (2.0 grains SO_3). (5) No allowance made for other sources of emission interfering. (6) Wind speed taken as 20 ft/sec.

mg/m^3 for unpolluted rural areas, 0.09 mg/m^3 for medium industrialized and medium populated areas, 0.11 mg/m^3 for urban and heavily industrialized areas; n = number of stacks less than two stack heights apart. Several states of the United States use variants of the American Society of Mechanical Engineers chart APS-1 for determining stack height required for particulate matter emissions from combustion sources (Fig. 20) (*23*).

E. Applicable to Specific Processes without Regard to Substance Emitted

There are British stack height standards for cement works [Table XXXVII (*24*) and Figure 21 (*24*)], copper works (Table XXXVIII) (*25*), sulfuric acid contact plants (Table XXXIX) (*25*), and nitric acid plants (Table XL) (*25*). Iraq requires a minimum stack height of 140 feet (42 m) for a brick kiln. In computing stack height in Israel using single-source diffusion formulas, the value of ground level concentration used is 40% of the ambient air quality standard. The minimum allowable stack height is $2\frac{1}{2}$ times the height of surrounding buildings at least 2 m taller than the highest building in the neighborhood.

VIII. Fuel Standards

The several national and provincial fuel standards for sulfur and lead in liquid fuels are shown in Table XLI (*6, 26*). The states of the United

States with state fuel standards are listed in Table XLII (*6*). The cities and counties of the United States with fuel standards are too numerous to mention.

IX. Buffer Zone Standards (*6*)

A buffer zone is a required separation distance between a source of environmental nuisance or hazard and areas allowed to be lived in or frequented by the general public. There may, of course, be people who must work in or traverse the buffer zone to operate or service the source, but their exposure should be considered occupational and their employer should be expected to provide them with special protection. The usual case is that air pollution (including odor) is only one of several factors determining buffer zone width, the others being noise, vibration, glare, heat, radioactivity, and hazards from possible fire, explosion, or disease transmission. There are, however, cases where all these other factors are minimal and where buffer zone width is determined mainly or exclusively as the distance necessary to dilute air pollution emissions to an acceptable level.

When both a buffer zone and tall stacks are used to protect a community from the air pollution from a specific plant and the distance from the plant to the point where the plume comes to ground level in highest concentration is approximately the same as the width of the buffer zone separating the plant from the community, the buffer zone could serve, under certain meteorological situations, to increase, rather than decrease, pollution levels in the community.

Standards promulgated pursuant to Soviet Union labor laws require that, using the spring and summer prevailing winds as the criterion, plants expected to discharge objectionable air pollutants must be located downwind of nearby residential areas and separated therefrom by what are called "sanitary protection zones." These zones should preferably be planted with species resistant to the pollutants expected, but may not be used as public parks. Under certain conditions, nonpollution-producing industrial, commercial, and public establishments may locate in these zones. There are five categories of such zones—1000, 500, 300, 100, and 50 m minimum width, respectively, except that these minima may be doubled to protect hospitals, sanitariums, etc. Special sanitary protection zones of width 15, 20, 30, 150, 200, 400, and 700 m are added to the standard ones of 1000, 500, 300, 100, and 50 m to accomodate the requirements of sewage treatment works, veterinary surgeries, and pesticide storehouses. There are a total of 409 categories of industrial processes

Table XLI National and Provincial Liquid Fuel Standards Other than in the United States *(6, 26)*

Country	Sulfur content (%)								Lead (gm/liter)		Notes[a]
		Fuel oils					Gasoline		Gasoline		
	Gas oil	Extra light	Light	Medium	Heavy	Extra heavy	Regular	Premium	Regular	Premium	
Austria	0.8	—	1.5	2.5	3.5	—	1.0	1.0	0.4	0.4	1
Belgium	1.0	1.0	1.5	2.7	3.8	4.5	1.0	1.0	0.84	0.84	2
Voluntary	0.6	—	0.9	2.5	3.0	3.8	—	—	0.74	0.74	3
Canada	—	—	—	—	—	—	—	—	0.77	0.77	4
New Brunswick	—	0.5	0.5	1.5	2.0	3.75	—	—	—	—	—
Ontario	—	0.5	0.5	1.5	1.5	1.5	—	—	—	—	—
Denmark	0.8	0.8	0.8	2.5	2.5	2.5	0.8	0.8	0.84	0.84	5
Finland	0.8	—	—	—	—	—	—	—	0.7	0.7	3
France	0.7	0.7	2.0	2.0	4.0	—	0.7	0.7	0.45	0.45	6
Paris #1	—	0.5	2.0	2.0	2.0	2.0	—	—	—	—	7
Nord	—	2.0	2.0	2.0	2.0	2.0	—	—	—	—	8
Rhone	—	1.0	1.0	1.0	1.0	1.0	—	—	—	—	—
Greece	0.5	0.2	0.2	3.5	3.5	4.0	0.15	0.15	0.84	0.84	9
Italy	1.1	2.5	3.0	4.0	4.0	—	1.1	1.1	0.64	0.64	10
Voluntary	0.8	—	—	3.0	3.0	—	—	—	—	—	3
Zone A	—	1.1	3.0	—	—	—	—	—	—	—	11
Zone B	—	1.1	3.0	—	—	—	—	—	—	—	12
Israel	—	0.4	0.2	0.2	0.2	—	—	—	0.42	0.42	13
Japan	—	1.0	1.0	1.5	1.5	1.5	—	—	—	—	14
Netherlands	0.5	—	0.7	2.5	2.5	2.5	—	—	0.84	0.84	3
Norway	—	—	—	—	—	—	—	—	0.4	0.4	—
<700 tons/year	0.8	—	—	—	—	—	—	—	—	—	15
>700 tons/year	—	1.2	1.2	1.2	1.2	1.2	—	—	—	—	16
Portugal	0.5	—	2.0	3.0	3.5	—	—	—	0.64	0.64	—
Spain	—	—	—	—	—	—	—	—	0.72	0.72	—
Industrial	—	2.0	2.0	2.0	2.0	2.0	—	—	—	—	17
Domestic	—	1.5	1.5	1.5	1.5	1.5	—	—	—	—	17
Sweden	0.8	2.5	2.5	2.5	2.5	2.5	—	—	0.15	0.15	18
Special	—	1.0	1.0	1.0	1.0	1.0	—	—	—	—	19

Switzerland	0.5	0.5	2.0	2.0	3.5	—	—	0.5	0.4	0.4	—
United Kingdom	1.0	—	3.5	4.0	4.5	5.0	1.0	1.0	0.64	0.64	20
London	—	1.0	1.0	1.0	1.0	1.0	—	—	—	—	21
West Germany	0.5	0.8	—	—	2.8	—	0.5	0.5	0.4	0.4	22
Northrhine–Wesphalia	—	—	—	—	1.8	1.8	—	—	—	—	23

[a] NOTES:

1. Domestic heating oil in Styria and Vorarlberg provinces limited to 1% sulfur.
2. Space heating fuel in Antwerp, Brussels, Charleroi, Ghent, and Liege limited to 1% sulfur in both liquid and solid fuel.
3. Voluntary limit adopted by oil distributors.
4. Lead in gasoline limit 3.5 gm/imperial gallon.
5. Industry standard for lead in gasoline; limit of 1% sulfur in all fuels for Copenhagen and Fredericksberg temporarily suspended.
6. In January 1978, sulfur content of gasoline to decrease to 0.3%; lead content is that scheduled for January 1, 1976. The decrease from the earlier statutory 0.64 gm/liter to 0.55 gm/liter scheduled for January 1, 1974, was temporarily suspended.
7. Arrondissements 1, 2, 8, 9, and 17, and parts of 10, 16, and 18.
8. Lille, La Madeleine, Loos, Lomme, Haubourdin, Roubaix, Tourcoing, Croix, Wasquhal, and Wattrelos.
9. Sulfur limit for military gasoline, 0.25%.
10. There is a grade of gasoline with reduced tax having 0.4 gm/liter lead.
11. Zone A—central and northern Italy, towns with 70,000–300,000 inhabitants; southern and insular Italy, towns with 300,000–1,000,000 inhabitants or areas of particular importance or where adverse conditions exist.
12. Zone B—central and northern Italy, towns >300,000 inhabitants; southern and insular Italy, towns >1,000,000 inhabitants or where adverse conditions exist. In Zone B, fuel oils heavier than medium with over 4% sulfur can be burned only with municipal agreement.
13. Maximum ash content: diesel fuel, 0.01%; light fuel oil, 0.1%; heavy fuel oil, 0.2%.
14. Lead content of gasoline, 0.3 cm^3/liter.
15. Oslo and Drammen.
16. Oslo and Drammen (in winter); Drammen (in summer), 2.5% sulfur.
17. For Madrid and Barcelona only; also, a limit of 4% on sulfur and 15% on volatile matter in coal.
18. Lead limit suspended; presently at 0.7 gm/liter interim value.
19. Stockholm, Goteberg, Malmo only.
20. Lead content of gasoline by voluntary limit adopted by oil distributors. Proposed reduction to 0.45 gm/liter by January 1, 1976, temporarily suspended.
21. City of London new installations; existing installations to comply by January 1, 1987.
22. Lead content was to be decreased to 0.15 gm/liter on January 1, 1976. Sulfur content of gas oils and gasoline was to be decreased to 0.3% on January 1, 1977.
23. >800,000 kcal/hour in special areas of Northrhine-Wesphalia.

Table XLII States of the United States for Which There Are State Standards for the Sulfur Content of Fuels *(6)*

Type of fuel	*States with sulfur limit on fuel*
Any	Connecticut, Georgia, Michigan, Puerto Rico, Rhode Island, Vermont
Liquid fuel	Idaho, Maryland, New Hampshire, New Jersey, Oklahoma, Oregon, Utah, West Virginia
Solid fuel	District of Columbia, Idaho, Maryland, New Jersey, Oregon, Utah, West Virginia

covered by the Soviet Union list, some classified into more than one zone category, depending upon production or flow rate, storage quantity, or raw material composition, e.g., percent of sulfur.

Poland classifies 280 processes and activities into these same five standard protection zone widths. Israel classifies 575 processes and activities into groups with buffer zone widths of 2000, 1000, 500, 150, and 50 m, respectively. The Landes of Northrhine-Westphalia in the Federal Republic of Germany classifies 37 processes and activities into seven buffer zones—2000, 1500, 1200, 1000, 800, 500, and 300 m wide, respectively.

In Delaware, open burning is prohibited in residential, commercial, industrial, and rural areas, within 1 mile (1.6 km) of the corporate limits of any municipality; state boundaries; military, commercial, county, municipal, or private airports or landing strips; primary highways; areas within $\frac{1}{2}$ mile (0.8 km) of secondary highways; national reservations; state parks; state forests; wildlife areas; or sites of three or more residences. In addition, such burning is prohibited between 1 hour after sunset and 1 hour before sunrise and during meteorological conditions declared adverse by the commission. Other states have similar restrictions.

REFERENCES

1. "National Emission Standards Study," Report of the Secretary of Health, Education, and Welfare to the United States Congress in compliance with Public Law 90-148, The Air Quality Act of 1967. Senate, 91st Congress, 2nd Session, Document No. 91-63. U.S. Government Printing Office, Washington, D.C., 1970.
2. A. C. Stern, *J. Air Pollut. Contr. Ass.* **20**, 524 (1970).
3. "Standards of Performance for New Stationary Sources," U.S. Environmental Protection Agency *Code Fed. Regul.* **40**, 60 (1976), *Fed. Regist.* **36**, 24876 (1971), amended by *ibid* **37**, 14877 (1972); **38**, 13562 and 28564 (1973); **39**, 9308, 13776, 15396, 20790, 37987 and 39874 (1974); **40**, 2803, 18169, 33152, 43850 and 46250 (1975); **41** 2232, 2331, 3825, 8346, 18497 and 20659 (1976), Washington, D.C.

3a. Regulations on National Emission Standards for Hazardous Air Pollutants, U.S. Environmental Protection Agency, *Code Fed. Regul.* **40**, 61 (1976), *Fed. Regist.*

38, 8820 (1973), amended by *ibid* **39,** 15396 and 37987 (1974); **40,** 18169 (1975), **41,** 46560 (1976), Washington, D.C.

3b. National Emission Guidelines, Department of the Environment. The Canada Gazette, 3868 and 4563 (1974); 1284 and 2219 (1975). Queen's Printer for Canada, Ottawa, Ontario.

4. Code of Federal Regulations, Title 40, Chapter 1, Subchapter C, Part 52, Subpart A, Section 52.21; *Fed. Regist.* **39,** 42510 (1974).

4a. A. C. Stern, *J. Air Pollut. Contr. Ass.* **27,** 440–453 (1977).

5. T. L. Montgomery, J. W. Frey, and W. B. Norris, *Environ. Sci. Technol.* **9,** 528–533 (1975).

6. W. Martin and A. C. Stern, "The World's Air Quality Management Standards," Vol. 1 (EPA-650/9-75-001-a). and Vol. 2 (EPA-650/9-75-001-b). United States Environmental Protection Agency, Washington, D.C., 1974.

6a. "Decree of Application of the Ambient Air Protection Law," No. 96, pp. 8391–8416. Boletin Oficial del Estado, Madrid, Spain, 1975 (in Spanish).

7. G. Leonardos, *J. Air Pollut. Contr. Ass.* **24,** 456 (1974).

8. "Standard Method for Measurement of Odor in Atmospheres (Dilution Method)," ASTM D-1391-57 (reapproved 1967); American Society for Testing and Materials, Philadelphia, Pennsylvania.

8a. J. L. Mills, A. T. Walsh, K. D. Luedtke, and L. K. Smith, *J. Air Pollut. Contr. Ass.* **13,** 467 (1963).

9. ASTM D-1391-57 (*8*), plus panel test procedure of D. M. Benforado, W. J. Rotella, and D. L. Horton, *J. Air Pollut. Contr. Ass.* **19,** 101 (1969).

10. "Erste Allgemeine Verwaltungsvorschrift zum Bundes—Immissions—schutzgesets vom 28, August 1974," Gemeinsames Ministerial blatt, pp. 426 and 452. Technische Anleitung zur Reinhaltung der Luft Der Bundesminister des Innern, Bonn, Federal Republic of Germany, 1974 (in German).

11. "Standard Levels of Emissions (75-187), The Measurement of Emissions from Boiler and Furnace Chimneys", Her Majesty's Stationery Office, London, England (1967).

11a. "Statutory Instruments, (162), The Clean Air (Emission of Grit and Dust from Furnaces) Regulations", Her Majesty's Stationery Office, London, England (1971).

12. W. L. Faith, "An Evaluation of Fugitive Dust Regulations," Paper No. 74–80 (presented at 67th Annual Meeting, Denver, Colorado) Air Pollution Control Ass., Pittsburgh, Pennsylvania, 1974.

13. "Regulations for Calculating Harmful Substance Concentrations in the Air (Dust and Sulfur Dioxide) Contained in Industrial Enterprise Emissions," SN 369–67. Gidrometeorizdat, Leningrad, USSR 1967 (in Russian).

14. o opoatřeních proti znečišťování ovzduši, Sbirka zákonu Československé socialistické republicky, 35. Zakon zedne 7, pp. 118–124. Dubna 1967 (in Czech).

15. Fünfte Durchführungsverordnung zum Landeskulturgesetz—Reinhaltung der Luft (Jan. 17, 1973), Gesetzblatt der Deutchen Demokratischen Republik, Part I, No. 18, p. 157, Berlin, Democratic Republic of Germany, 1973 (in German).

15a. Erst Durch Führungsbestimmung zur fünfte Durchführungsverordnung zum Landeskulturgesetz—Reinhaltung der Luft—Begrenzung und Uberwachung der Immissionen und Emissionen (Luftverun—reinigungen), Gesetzblatt der Deutchen Demokratischen Republik, Part I, No. 18, p. 162, Berlin, Democratic Republic of Germany, 1973 (in German).

15b. Allowable emission in kg/hour = S (Ambient air quality standard (MIK_k), in mg/m^3 listed in Table III in Chapter II of this volume for "East Germany" for 30-minute averaging time); e.g. for acetaldehyde—MIK_k — 0.03 mg/m^3. Therefore the permissible emission of acetaldehyde from a 100-m stack is 1,065.0 × 0.03 = 31.95 kg/hour. When nitrogen dioxide, phenol, hydrogen fluoride, or SiF_4 coexist with sulfur dioxide, their emission shall be based on the sulfur dioxide area class.
16. "Enforcement Ordinances of Air Pollution Control Law." Air Quality Bureau, Environment Agency, Tokyo, Japan, 1972 (in English).
16a. Vejledning fra miljøstyrelsen, "Begraensning af luftforurening fra virksomheder," Nr. 7. København, Denmark, 56 pp., 1974.
16b. W. L. Lee and A. C. Stern, *J. Air Pollut. Contr. Ass.* **23**, 505 (1973).
17. "Règlement général pour la protection du travail," Chapter II, Sect. I, Art. 364–373, Brussels, Belgium.
18. "Chimney Heights, 2nd Ed." (75–115), Clean Air Act (1956 Memoranda) Ministry of Housing and Local Government, Her Majesty's Printing Office, London, England, 1967.
18a. "Method of Calculating Chimney Heights—Appendix to Memoranda on Chimney Heights" (75-115-1), Ministry of Housing and Local Government, Her Majesty's Printing Office, London, England, 1967.
19. "Ordinance concerning the restriction of sulfur content of fuel oil (551)". The National Swedish Environment Protection Board, Solna, Sweden, 1968.
19a. "Royal ordinance amending the 1968 ordinance (551) concerning the restriction of sulfur content of fuel oil" (621). The National Swedish Environment Protection Board, Solna, Sweden, 1970.
20. "Control and Prohibition of Air Pollution from Sulfur Compounds", New Jersey Administrative Code 7:27–7.2. New Jersey Department of Environmental Protection, Trenton, New Jersey, 1966.
21. "Kommission Reinhaltung der Luft," Richtlinien. Verein Deutscher Ingenieure, Dusseldorf, Federal Republic of Germany.
22. "Recante provvedimenti contro l'inquinamento atmosferico, limitatamente al settore degli impianti termici, Regolamento perlèsecuzione della legge 13 luglio 1966, N. 615, Decreto del Presidente della Repubblica, October 24, 1967, No. 1288, Supplemento ordinario No. 6 alla Gazzetta Ufficiale della Repubblica Italiana, Part Prima Gennaio 9, Rome, Italy, 1968 (in Italian).
23. "Control of Dust Emission—Combustion for Indirect Heat Exchangers," ASME Std. APS-1. American Society of Mechanical Engineers, New York, New York, 1966.
24. "Notes on Best Practicable Means for Cement Works Emissions," Appendix V. Alkali etc. Works Annual Report by the Chief Inspector, Great Britain, London, 1967.
25. E. A. J. Mahler, "Standards of Emission under the Alkali Act," Appendix V. Alkali etc. Works Annual Report by the Chief Inspector, Great Britain, London, 1966.
26. "Existing Environmental Regulations of Concern to the Oil Industry in Western Europe," Report 2/74. Stichting CONCAWE, The Hague, Netherlands, 1974.

Subject Index

A

B

C

D

E

H

I

Q

R

S

ENVIRONMENTAL SCIENCES

An Interdisciplinary Monograph Series

EDITORS

Arthur C. Stern, editor, AIR POLLUTION, Second Edition, Volumes I–III, 1968; Third Edition, Volumes I, III, 1976; Volumes II, IV, V, 1977

L. Fishbein, W. G. Flamm, and H. L. Falk, CHEMICAL MUTAGENS: Environmental Effects on Biological Systems, 1970

Douglas H. K. Lee and David Minard, editors, PHYSIOLOGY, ENVIRONMENT, AND MAN, 1970

Karl D. Kryter, THE EFFECTS OF NOISE ON MAN, 1970

R. E. Munn, BIOMETEOROLOGICAL METHODS, 1970

M. M. Key, L. E. Kerr, and M. Bundy, PULMONARY REACTIONS TO COAL DUST: "A Review of U. S. Experience," 1971

Douglas H. K. Lee, editor, METALLIC CONTAMINANTS AND HUMAN HEALTH, 1972

Douglas H. K. Lee, editor, ENVIRONMENTAL FACTORS IN RESPIRATORY DISEASE, 1972

H. Eldon Sutton and Maureen I. Harris, editors, MUTAGENIC EFFECTS OF ENVIRONMENTAL CONTAMINANTS, 1972

Ray T. Oglesby, Clarence A. Carlson, and James A. McCann, editors, RIVER ECOLOGY AND MAN, 1972

Lester V. Cralley, Lewis T. Cralley, George D. Clayton, and John A. Jurgiel, editors, INDUSTRIAL ENVIRONMENTAL HEALTH: The Worker and the Community, 1972

Mohammed K. Yousef, Steven M. Horvath, and Robert W. Bullard, PHYSIOLOGICAL ADAPTATIONS: Desert and Mountain, 1972

Douglas H. K. Lee and Paul Kotin, editors, MULTIPLE FACTORS IN THE CAUSATION OF ENVIRONMENTALLY INDUCED DISEASE, 1972

Merril Eisenbud, ENVIRONMENTAL RADIOACTIVITY, Second Edition, 1973

James G. Wilson, ENVIRONMENT AND BIRTH DEFECTS, 1973

Raymond C. Loehr, AGRICULTURAL WASTE MANAGEMENT: Problems, Processes, and Approaches, 1974

Lester V. Cralley, Patrick R. Atkins, Lewis J. Cralley, and George D. Clayton, editors, INDUSTRIAL ENVIRONMENTAL HEALTH: The Worker and the Community, Second Edition, 1975

A
B 7
C 8
D 9
E 0
F 1
G 2
H 3
I 4
J 5